D1523045

Broadband Hybrid Fiber/Coax Access System Technologies

Series in Telecommunications

Series Editor
T. Russel Hsing
Bell Communications Research
Morristown, NJ USA

Broadband Hybrid Fiber/Coax Access System Technologies
Winston I. Way
Department of Communications Engineering
National Chiao-Tung University
Hsinchu, Taiwan, ROC

Other Books in the Series

Ali N. Akansu and Richard A. Haddad, *Multiresolution Signal Decomposition: Transforms, Subbands, Wavelets*. 1992.

Hseuh-Ming Hang and John W. Woods, *Handbook of Visual Communications*. 1995.

John J. Metzner, *Reliable Data Communications*. 1997.

Tsong-Ho Wu and Noriaki Yoshikai, *ATM Transport and Network Integrity*. 1997.

Ali N. Akansu and Richard A. Haddad, *Multiresolution Signal Decomposition: Transforms, Subbands, Wavelets, 2nd Edition*. In preparation.

Broadband Hybrid Fiber/Coax Access System Technologies

Winston I. Way
Department of Communications Engineering
National Chiao-Tung University
Hsinchu, Taiwan, ROC

ACADEMIC PRESS
San Diego London Boston
New York Sydney Tokyo Toronto

This book is printed on acid-free paper. ♾

ACADEMIC PRESS
525 B Street, Suite 1900, San Diego, CA 92101-4495
http://www.apnet.com

ACADEMIC PRESS
24-28 Oval Road, London NW1 7DX

Library of Congress Cataloging-in-Publication Data
Way, Winston I.
Broadband hybrid fiber/coax access system technologies / Winston I. Way.
p. cm. — (Telecommunications)
ISBN 0-12-738755-2
1. Broadband communication systems. 2. Fiber optic cables. 3. Coaxial cables. 4. Multiplexing. I. Title. II. Series: Telecommunications (Boston, Mass.)
TK5103.4.W39 1998
004.6'4—dc21 98-18156
CIP

Printed in the United States of America
98 99 00 01 02 IP 9 8 7 6 5 4 3 2 1

ERRATA

Broadband Hybrid Fiber/Coax Access System Technologies

Page	For	Read
p.37, Eq.(2.3)	$CNR=P_{in}/(kTB)$	$CNR=P_{in}/(kTBF)$
p.38, Eq.(2.8)	$NLD=-20\log(10^{-NLD_1/20} + 10^{-NLD_2/20} + ...)$	$NLD=20\log(10^{NLD_1/20} + 10^{NLD_2/20} + ...)$
p.94, Eqs.(4.12)&(4.13)	α_m	α_{mir}
p.96, Fig.4.8(a)	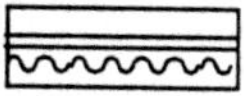	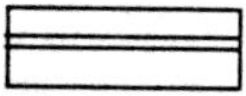
p.99, Eq.(4.18)	m_n^*	m^*
p.109, Eq.(4.35)	$ej^{\{...\}} = ...$	$e^{j\{...\}} = ...$
p.132, Table P4.1	units of τ_p and τ_n: none	units of τ_p and τ_n: second
p.188, line 7	$< i_{sig-sp}^2 >$	$< i_{sp-sp}^2 >$
p.217, Eq.(7.2)	$\Delta n \approx -n^3 r_{33}\Gamma(V/(2G_p),$	$\Delta n \approx -n^3 r_{33}\Gamma(V/(2G_p)),$
p.231, Eq.(7.22)	$... \approx \frac{1}{PC \cdot \left(\frac{\chi}{2}\right)^4}$	$... \approx \frac{1}{N_{CTB} \cdot \left(\frac{\chi}{2}\right)^4}$
p.240, Eq.(7.46)	$...(P_{in,p} - kP_{in,s})...$	$...(P_{in,p} - cP_{in,s})...$
p.240, Eq.(7.46)	$... \frac{(\pi V/V_\pi)_3}{6} ...$	$... \frac{(\pi V/V_\pi)^3}{6} ...$
p.344, Eq.(9.47)	$...A \cdot \exp(i\Omega_o t + \phi(t)),$	$...A \cdot \exp[i(\Omega_o t + \phi(t))],$
p.359, line 1	the dotted βs	the β_i' s
p.359, line 3	$\dot{\beta}(\Omega_o)z$	$\beta_1(\Omega_o)z$
p.367, line 1	η	η_s
p.367, Eq.(9.103)	CSO=....	CTB=...
p.372, Eq.(9.113)	$CSO = N_{CSO} \cdot \{...f_d^2\}$	$CSO = N_{CSO} \cdot \{...(2\pi f_d)^2\}$
p.393, Eq.(10.1)	$... = ... - \Delta G$	$... = ... - \Delta G + \frac{2}{3}\log(BW)$
p.396, Fig.10.9, horizontal title	Detected photocurrent I_D	Detected photocurrent I_P

To my parents
and
to my wife, Jenny, and our children,
Serena, Joanna, and Alan

Contents

Preface

"Broadband access" has been a fancy technology in the telecommunications industry ever since I joined Bellcore in 1984. I have seen many world-class engineers and scientists, both inside and outside Bellcore, who were dedicated to designing opto-electronic devices, systems, and networks, hoping that broadband access can one day become a real part of our life. Through their continuous efforts, a fundamental understanding of various opto-electronic device physics and feasibility studies of numerous lightwave access system technologies were pretty much established by the early 1990s. However, their invention and development pace may have been so fast that those achievements were not appreciated. One of the reasons was the constantly asked, yet unanswerable, question, "Who needs broadband in the access networks?" This question later even became a nightmare for anyone who tried to obtain opto-electronic research and development funding. At that time, *broadcasting* multiple channels of CATV video signals seemed to be the only justifiable and realistic *access* application for lightwave systems. The other reason, which related more to economic reality, was that the "pure" electronic empire struck back, claiming they could use electronic digital signal processing techniques to reuse the existing twisted-pair copper wires and provide evolutionary *switched* broadband access. Later their efforts indeed produced broadband access technologies such as asymmetrical digital subscriber line (ADSL).

Today, with the killer applications Internet and Intranet so popular, "Who needs broadband access?" is no longer a meaningful question. However, economic consideration is still an important factor in selecting various broadband access technologies such as ADSL, fiber-to-the-home/fiber-to-the-customer, LMDS (local multipoint distribution service) at 28 GHz, and hybrid fiber/coax (HFC)—the subject matter of this book. On the other hand, we all know that once a technology becomes more widely accepted or understood by engineers and scientists, more innovations can be created through intensive research and development momentum, which can eventually make the technology economically viable. This is actually one of the main purposes of this book because there are essentially no books covering in-depth HFC technologies today.

HFC is a relatively new technology which is formed after the majority of North American and European CATV systems have replaced their coaxial-cable trunk lines with optical fibers. In mainland China, which has a population of more than 1.2 billion, newly installed HFC systems are also gaining popularity due to their largely community-based living environment. Whether rebuilt or newly installed, HFC systems are now evolving from a simple multichannel video broadcasting system to a two-way interactive multimedia network and are becoming an important part of the telecommunication infrastructure worldwide.

An HFC system is "hybrid" in the following senses: (1) it is based on both optical fiber and radio-frequency (RF) technologies; (2) it uses star- and ring-based optical fiber architecture and tree-and-branch-based coaxial cable architecture; and (3) it is used to transport both existing analog video signals and evolving digital/interactive multimedia signals. Because of its "hybrid" nature, HFC access system technologies involve multidisciplinary engineering fields, including subcarrier multiplexed (SCM) lightwave components and systems, RF modem designs, and medium-access-control (MAC) protocols.

The purpose of this book is to provide a systematic in-depth understanding of the above multidisciplinary fields to graduate students, engineers, and scientists who can potentially contribute more to this broadband access technology. The materials organized in this book have been taught in my graduate school for the past 5 years, and have been updated many times due to rapid changes in optical fiber, RF modem, and MAC protocol technologies.

The volume is divided into three parts: *Outlook and Overview*, *Components and Modules*, *Systems and Protocols*. Part I gives an outlook of various broadband services that an HFC system can provide, and overview of the conventional coaxial cable system technologies, and how the optical fiber system architecture may evolve. Part II provides a basic physical understanding of all necessary lightwave components and modules, and RF modem design principles. Unlike most of the available textbooks on optical fiber communications, this part covers the lightwave components and modules from the standpoint of SCM lightwave systems, that is, analyses and explanations are concentrated on device linearity and intensity noise (e.g., in laser diodes, external modulators, optical amplifiers, and optical receivers), and on available output optical power (e.g., in high-power erbium-doped fiber amplifiers). In addition, because of my own hands-on experience in this area for the past 14 years, I tend to cover only those devices that could be of practical use in SCM systems; although some potentially promising, but not yet practical, optical components may also be covered. In the discussions of RF modem designs, we have kept the technical contents aligned to those of *standardized* RF modems built on digital circuits, so that readers will not be overwhelmed by the numerous old modem technologies which have been developed since the 1970s. However, the coverage of equalizers, timing, and carrier recovery circuits is by no means exhaustive, even though examples were cited from those state-of-the-art modems as much as possible. Part III is the key part of the book from the viewpoint of real system applications, because it is important to know how to use those components and modules in Part II successfully in real-world HFC systems. In discussing the MAC protocols, we first give a historical overview of those protocols that were used in early CATV systems, and then summarize the operation principles of today's standardized cable modem MAC protocols.

The book can be used either as a graduate-school textbook or as a reference book. When it is used as the former, a course of "Subcarrier Multiplexed Lightwave Systems and Components for Video Distribution" can be taught for a semester, with Chapters 8, 10, and 11 used on a reference basis. A set of selected problems at the end of each chapter can be used by the students to grasp the main concepts presented in each chapter. Conversely, those students and engineers who do not want to study optical fiber technologies in great detail can use Chapters 1, 2, 8, and 11 as an entrance to a basic understanding of cable modem technologies.

I appreciated very much the research environment which Bellcore provided me in the period from 1984 to 1992, and the opportunities to work with many world-class optical fiber experts in numerous pioneering lightwave system researches. Many of my colleagues who were associated with Bellcore are now core members in the field of fiber optic research and developments worldwide. I also appreciated the opportunities to be involved in down-to-the-earth development experience on cable modems after I came back to Taiwan. With significant funding sponsored by National Science Council (NSC), Microelectronics Technology Inc., Microtek Inc., and UFOC Inc., I was able to expose myself to RF modem and MAC protocol designs. I would also like to thank NSC and New Elite Technology, Inc., for their funding support since 1997, so that we can carry out long-distance SCM system research projects. I believe that not many people have the opportunity to learn both fiber optics and cable modem technologies in the path of their career, which motivates me to write this book with the hope of making some contributions to this growing field.

I am grateful to my parents, Yung-Peng and Lan-Ying, who constantly encouraged me when I felt that I could never finish the book. I would like to thank my Ph.D. students C. Tai, B. W. Wang, P. Y. Chiang, M.Wu and C. C. Hsiao who proofread the manuscript, my secretary H. Chang for helping prepare the manuscript, and the Office of Research and Development and Center for Telecommunication Research at NCTU for supporting my computer facilities while writing the book. Having finished the book, I feel great relief on one hand, but on the other hand feel that I am in deep debt to my wife Jenny and kids, Serena, Joanna, and Alan, for their whole-hearted support and love while I was preoccupied with book preparations the last two years.

Winston I. Way
August 1998

Broadband Hybrid Fiber/Coax Access System Technologies

PART I

Outlook and Overview

Chapter 1

An Overview of Hybrid Fiber Coax (HFC) Networks

Today's telecommunication networks can generally be classified into three main categories: the public switched telephone network (PSTN), the local-area computer network, and the emerging community antenna television (CATV) network, as shown in Fig.1.1. These are three truly disparate networks: today's telephone networks with centralized switches provide mainly voice-oriented communication services; today's local-area computer network, although it is becoming multimedia-oriented with the ever-increasing computing power available at the desktop, still delivers mainly data; and today's CATV network delivers mainly one-way broadcast entertainment video signals. However, realizing the stagnant revenue growth rate, both the telcos and the CATV industry are pushing ahead to provide new services: Telephone companies want to provide video services, and CATV companies want to provide voice- and data-oriented services. Needless to say, every party is eager to play a significant role in providing Internet-related services, especially since the U.S. 1996 Telecommunications Act was passed [1]. The Act deregulated all segments of the communication industry: Telephone companies, cable operators, wireless operators, and broadcasters are all free to enter each other's market. The European Commission has also proposed moving forward the date for lifting restrictions on the provision of services over CATV networks. In addition, worldwide computer manufacturers, utility companies, and even railway operators are eager to get their hands on communication networks. Mergers in the telecommunications industry are taking place almost every day now, and it seems that all three networks shown in Fig. 1.1 may eventually merge in terms of the involved technologies and the services (e.g., Internet) they provide. However, the most difficult part of bringing broadband to residential areas is the so called "last-mile access." In building the last-mile networks, economic factors are more important than technology considerations. Therefore, it is the purpose of this book

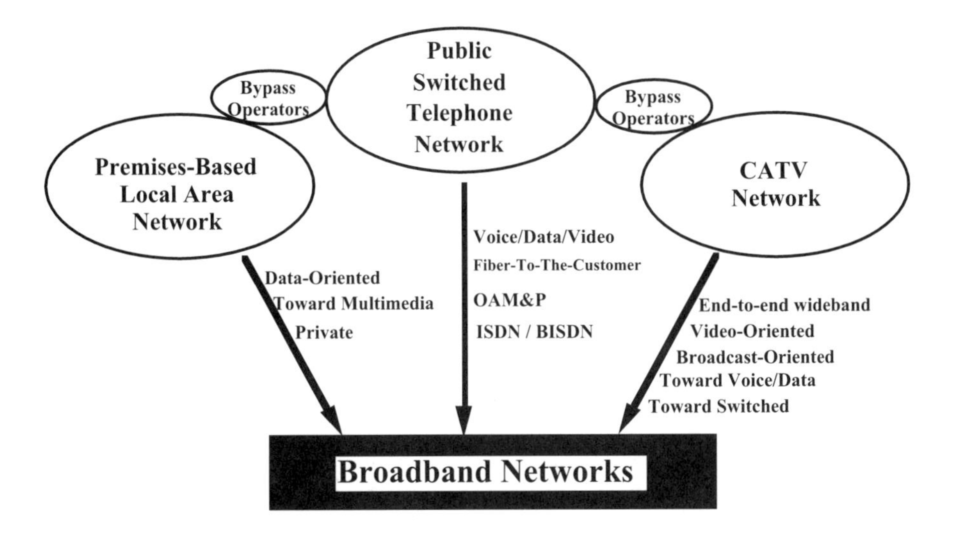

Figure 1.1: The merging of three networks to become tomorrow's broadband network.

to introduce to readers all aspects of the hybrid fiber coax (HFC) access system technology, which could offer the lowest initial installation cost of the various access technologies, for communication services via optical fiber and coaxial lines (see Fig. 1.2, specifically Fig. 1.2a).

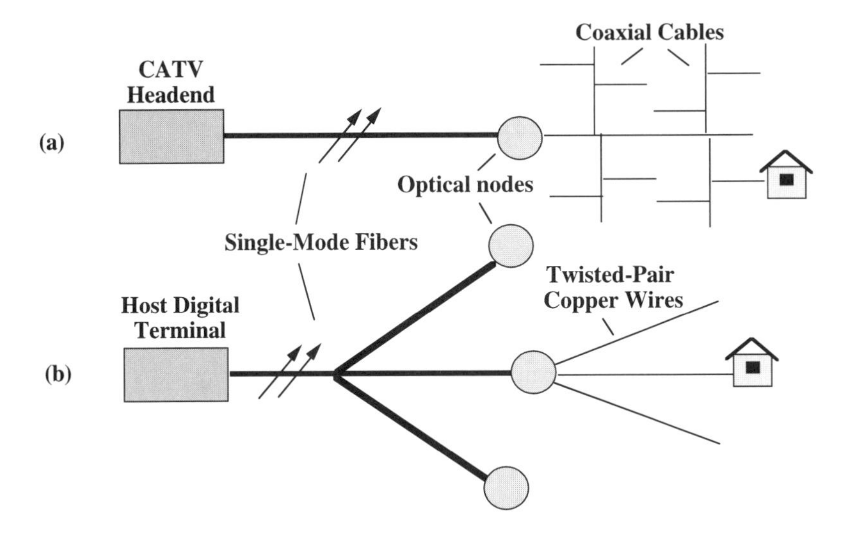

Figure 1.2: Basic architecture of (a) an HFC network, and (b) a fiber-in-the-loop (FITL) network.

1.1 Introduction

HFC networks are naturally based on the next generation of CATV networks, which are capable of providing a variety of telecommunication services as their infrastructure evolves into an all-digital highway. Telephone industries in the U.S. and Europe have various strategies to include CATV networks in their infrastructure, and computer industries also look forward to being a part of CATV networks (as evidenced by the collaborations between CATV equipment vendors and Microsoft, Apple, Silicon Graphics Inc., etc.). Hollywood studios such as MCA, Paramount, and Disney, as well as broadcast television networks, have all begun spending vast sums on developing new services built around multimedia technology, anticipating that the cable network will provide the mass market reach they need to achieve return on such huge investments.

The increasing promise of using CATV networks for broadband communications stems from the fact that more and more optical fiber systems have been deployed in CATV networks over the past few years. In addition, the dramatic demand for Internet access has pushed the development of cable modems and digital set-top boxes, which include the technologies of RF modem design and medium-access-control (MAC) protocol design. This leads to our definition of HFC access system technologies: HFC access technologies are based on four research and development areas, i.e., subcarrier multiplexed (SCM) lightwave system technology, radio frequency (RF) modulator and demodulator (modem) technology, coaxial cable transmission technology, and MAC protocol technology, as shown in Fig. 1.3. SCM lightwave system technology [2–4] uses frequency-division-multiplexed (FDMed) RF channels to modulate a laser diode (Chapter 4) or an external modulator (Chapter 7). These RF carriers are termed "subcarriers" in relation to an optical "main"

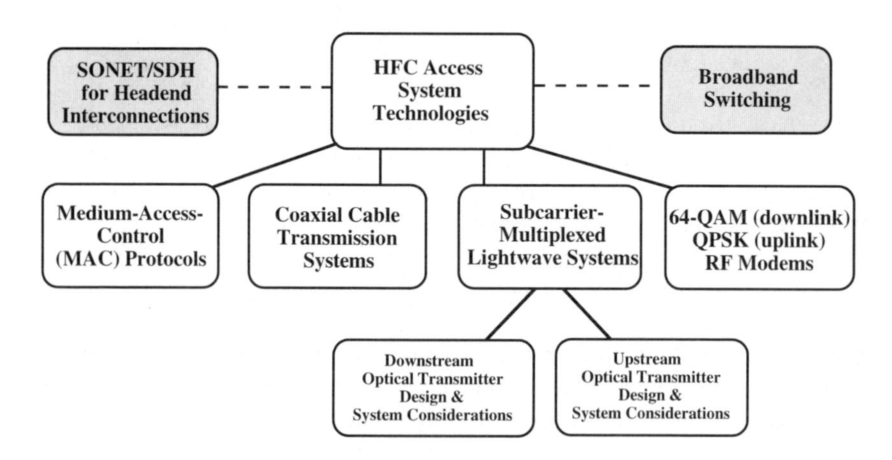

Figure 1.3: Transmission, switching, and access technologies associated with HFC systems.

carrier. Optical devices and components that have been or will be used in SCM lightwave systems are the subject of the second part of this book. The SCM system design issues to achieving high linearity and signal-to-noise ratio, HFC MAC protocols, and wireless access via HFC networks are discussed in the third part of this book.

The RF channels can transport analog and digital signals by using analog and digital modems, respectively. The latter have been of great interest because of their important applications in delivering a large number of digital video and audio signals to homes. Therefore, we will give a brief overview of RF modem design principles in Chapter 8.

To achieve (asymmetrical) bidirectional broadband communications, such as Internet access via TCP/IP, in a tree-and-branch architecture such as HFC networks (Fig. 1.2a), a well-designed MAC protocol is very important to reach high network throughput and efficiency. Therefore, in Chapter 11, MAC protocol design principles are reviewed.

1.2 Competing Access Technologies

HFC is not the only access technology to bring broadband to homes. Here "broadband" essentially means Internet and video services. Several access technologies are currently competing with HFC systems, as shown in Fig. 1.4:

- Fiber-in-the-loop (FITL) [5] which uses star-architecture-based optical fibers (see Fig.1.2b): passive optical networks (PONs) [6]; fiber-to-the-home (FTTH), fiber-to-the-curb (FTTC), fiber-to-the-zone (FTTZ), etc., all belong to this category of access technology. Most of the FITL access techniques employ baseband digital signals as opposed to HFC's RF-carried signals. Another main difference between FITL and HFC is "switched" versus "broadcast and selected." In FITL systems, a switched broadband digital signal, through a central-office-based broadband switch, is directed to a subscriber's home upon request [7]. In HFC systems, by way of contrast, all broadband signal channels are broadcast to all subscribers. Subscribers then use their CATV tuners on premises to select the particular channel(s) that they are interested in. Therefore, encryption or scrambling techniques must be applied to block unauthorized users, and security issues may have to be handled more carefully in HFC systems than in FITL systems.
- Asymmetric digital subscriber line (ADSL) [8,9] which uses twisted-pair copper wires: Transmission technologies such as quadrature-amplitude-modulation (QAM), carrierless AM/PM (CAP), and discrete multitone can be considered for ADSL to transmit high-speed data below about 600 KHz. State-of-the-art ADSL technology can send 1.5–9 Mb/s to the home and 16–640 kb/s from the home to the service provider over a distance of 12,000 feet.
- Multipoint multichannel distribution services (MMDS) which uses 2.5-GHz microwave links [10] and local multipoint distribution services (LMDS) [11], which uses 28-GHz millimeter-wave links: MMDS can provide about 200 MHz to

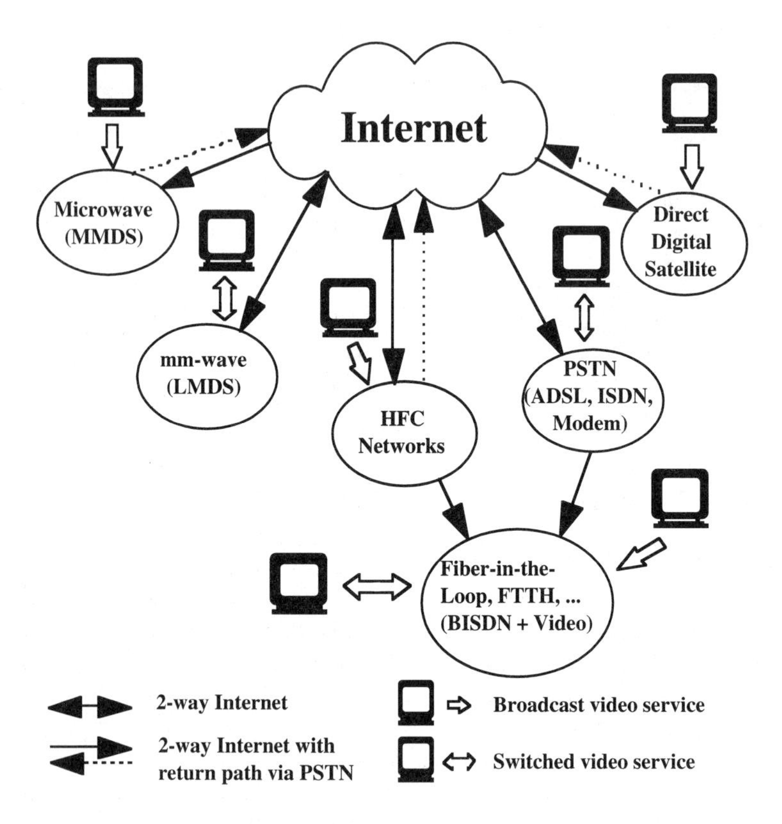

Figure 1.4: Competing technologies which can be used to provide both internet and video services.

deliver 100 or more MPEG-II channels using 64-QAM modulation (see Chapter 8), and MMDS provides mainly broadcast video services including pay-per-view (using phone lines for upstream control). LMDS can provide asymmetric two-way services, with 850 MHz for downstream (hub to subscriber) services and 150 or 125 MHz for upstream (subscriber to hub) services. A summary of MMDS/LMDS system specifications is shown in Table 1.1.

- Digital broadcast satellites (DBS): Digital broadcast satellite represent another powerful downstream technology which poses serious competition to the cable TV industry as far as one-way services are concerned. Digital satellite systems operate by relaying digital programs from a geostationary satellite directly to the home. The home units have an antenna and a special subscription number so the transmitted program can be descrambled and viewed. Satellite systems can transmit as many as 300 digital television channels to users and can also use telephone lines for upstream control.

Table 1.1: MMDS and LMDS System Characteristics.

	MMDS	**LMDS**
Frequency	2.5 GHz	27.5-28.35 GHz & 29.10-29.25 GHz
Bandwidth	200 MHz	1000 MHz
Path Loss	130 dB at 30 Km	121 dB at 1 Km
Hub Radius	20-50 km	1-2 km
Transmitter height	60-70 m	20-30 m
House passed	1 to 2 million	4,000 to 6,000
Line of sight required	Yes	Yes
Downstream capacity	64 QAM, 1Gbps	QPSK, 1.5 Gbps
Upstream capacity	Use phone lines	QPSK, 200 MHz

1.3 One-Way Broadcast Services

Let us first briefly review the essence of conventional one-way broadcast services. The existing services can be categorized into (1) network stations' programs, (2) movies and entertainment programs, (3) pay channels, and (4) special-interest group programs (e.g., Congress, stock market, home shopping, horse racing). Besides these existing services, near-video-on-demand (NVOD) for broadcasting newly released movies is a near-term feasible new service. This new service, however, consumes a large number of available channels, and therefore can only be used in large-bandwidth CATV systems (at least 550 MHz), and preferably using compressed digital video signals such as MPEG*-2 [12]. For example, a typical movie that lasts 90 minutes will have to use 6 channels if it is to be shown every 15 minutes.

In terms of transmission signal formats, both analog and digital signals will be available on broadcast channels. Analog signals are the dominant signals today, mainly because of their compatibility with several million existing television sets. Although analog TV sets will continue to be with us for many more years, we are already seeing the benefits that digital video/audio brings:

- Many more channels available (see discussions in Chapter 8 on digital modems) and better signal qualities.
- More flexible system operations: for example, any number of different audio languages can be sent with a given video service; auxiliary data services can be

*MPEG stands for "Motion Picture Expert Group" which was formed in 1988 to establish standards for coding of moving pictures and associated audio for various applications. MPEG-1 is for compact disc storage, low-bit-rate environment, MPEG-2 is for broadcast TV quality video, and MPEG-4 is for very low bit-rate applications such as video phone, electronic newspaper, etc.

associated with a video channel, providing interactive on-screen display; users can select only the portion of videos, pictures and text that are of interest to him, without wasting time on the portion in which they are not interested.

In the next few years, digital signal formats will be playing a more and more important role, and Table 1.2 [13] lists the bandwidth requirements for various real-time streaming applications which could be transported by HFC networks.

1.4 Two-Way Communications Services (Symmetric and Asymmetric)

Some cable operators in the United States deployed two-way-ready cable plants during early 1980s. Two-way communications in CATV networks were started with applications such as impulse buying service and plant status monitoring. In these applications, two-way communications were carried out by the combination of an upstream signaling channel which is located between 5 and 30 MHz and a downstream video channel. However, it was found that only a very small portion of the 25 MHz of upstream bandwidth originally assigned was required by these simple applications. Therefore, almost all operators found the two-way communications channels went unused or drastically underutilized. In the late 1990s, however, there are several new driving forces that make CATV networks important resources for two-way communications: Internet access, IP telephony, two- or three-party teleconferencing/telecommuting, file transfer, as shown in Table 1.2.

Table 1.2: Bandwidth requirements of real-time streaming applications [13] (G.x and H.x represents the assigned standard document number in ITU).

	Bandwidth requirements (after compressions)	
Real-time streaming applications	**Downstream**	**Upstream**
Audio		
CD-quality stereo: 10 Hz–20 kHz	256 kb/s	
Broadcast quality (G.722): 50Hz–7 kHz	64/56/48 kb/s	
POTS (PCM, G.711): 0.2–3.4 kHz	64 kb/s	64 kb/s
Low-bit-rate POTS (G.723.1)	6.4/5.3 kb/s	6.4/5.3 kb/s
Video		
HDTV	~ 20 Mb/s	
Video on demand, MPEG-2	~ 4–6 Mb/s	
Video on demand, MPEG-1	~ 1–2 Mb/s	
ISDN p× 64 video conferencing (H.261)	64 kb/s–2 Mb/s	64 kb/s–2 Mb/s
Low-rate videoconferencing (H.263)	Optimized for < 28.8 kb/s	< 28.8 kb/s

In the following, we will examine what has been or will be implemented on two-way HFC communications:

1.4.1 Internet Access and Data Communications via Cable Modems

Internet access is probably the most promising two-way communication service on HFC networks. The reasons are (1) Internet access and communications have shown exponential growth in the past few years and have become a killer application in the so called "global information infrastructure," and (2) cable modem can provide a downstream data rate of >30 Mb/s, which is > 200 times faster than the basic ISDN rate (128 kb/s) and >1000 times faster than that of a typical telephone modem (28.8 kb/s), and (3) superior economic performance compared to ISDN [14]. There are two types of cable modems: symmetric and asymmetric. Symmetric modems with equal upstream and downstream data rate are mainly used for CATV systems which serve as a LAN or MAN backbone to support baseband subnetworks, bridges, and routers (i.e., LAN interconnection) and can replace T1 circuits, dedicated, and dial-up phone lines. Cable data networks for 10 Mb/s Ethernet service up to 70 cable miles are available. Symmetric cable modems can be shared by multiple networked personal computers and can be a very cost-effective way to penetrate the data over cable market in business corporations and school campuses. Asymmetric cable modems (e.g., downstream 30 Mb/s, upstream 128 kb/s) allows centrally controlled accessing and will dominate the cable modem market in residential areas [15].

Note that data over cable services can be provided via two-way HFC networks or a one-way HFC network in combination with a return-path public switched telephone network (PSTN), as shown in Fig. 1.5 (see Sec. 11.3 for more descriptions). The latter is a quick way to provide Internet services by using lower-cost cable modems (since no upstream RF modem and medium access protocol is needed) without upgrading the present HFC networks to two-way, while the former can bring in more revenue for long-term operations (with higher upstream capacity and no need to lease T1/E1 lines).

1.4.2 Telephony over CATV Networks

1.4.2.1 IP Telephony IP telephony has the potential to open a quick path into a highly featured second-line business for cable operators that avoids the headaches and costs of implementing circuit-switched voice over HFC networks [16]. Instead of using centralized, large circuit switches, telco/MSO operators and ISPs (Internet Service Providers) can use multiple distributed, small router switches or gigabit routers to perform the same voice-switching functions [17]. This is a very important telecommunication concept for the next century, especially because the cost of hardware equipment and software OAM&P (operation, administrative, maintenance and provisioning) of circuit switches is currently a major financial burden to telecommunication network operators. However, to make such services viable on a wide-scale basis, there must be points of entry based on commonly used protocols and

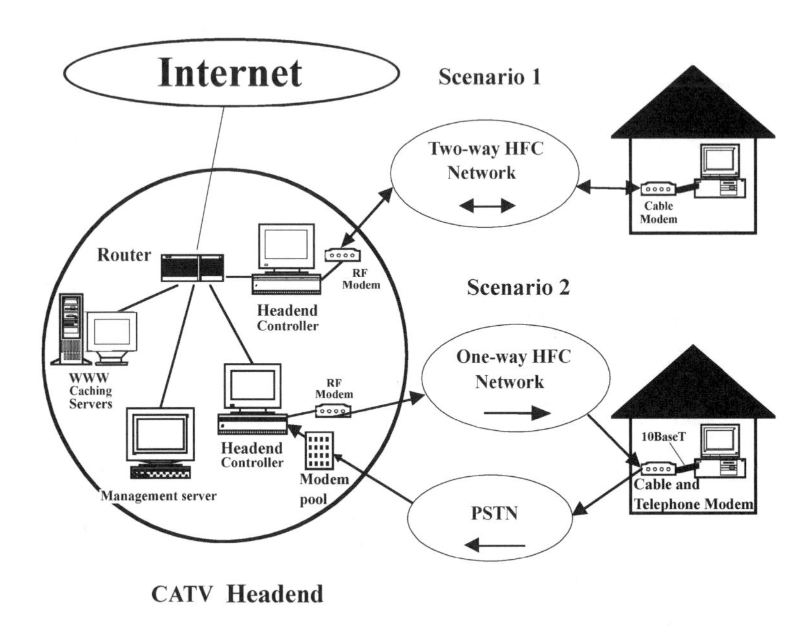

Figure 1.5: System architecture for two- and one-way HFC data services.

standardized customer-premises equipment at all the regional levels, allowing servers to hand off calls destined for circuit service customers, while sending along calls in data formats to customers using cable's data-based voice services. End-to-end connectivity can be carried out using gigabit Ethernet switches or ATM switches (IP over ATM).

Note that the packet length used in IP telephony significantly affects the delay incurred in filling up the packet with voice samples. The larger the packet, the longer the delay to accumulate sufficient voice samples. For example, a 1500-byte packet will incur a packetization delay of 1.875 s (3.75 s round-trip) for a 6.4 kb/s voice codec. This is definitely not acceptable because the end-to-end round-trip delay for telephone conversation should be limited to about 500 ms [18]. This problem has been overcome in today's commercial products by keeping the packets short.

1.4.2.2 Conventional Telephony As will be shown in the overall architecture in Chapter 2 (see Fig. 2.22), CATV headends can be easily connected via SONET/SDH rings, and therefore CATV networks can become a part of the telephone network quite naturally. The presently available return channel bandwidth of about 35 MHz (i.e., from 5 to 40 MHz) can be used sufficiently for delivering upstream telephony signals (if not via IP data services). For example, let us first assume an optical fiber node serving 500 subscribers, with each user allocated 64 kb/s for voice plus 2 kb/s for remote control. Then if the penetration of CATV telephony service can reach

25%, with a one in five active user ratio and using QPSK modulation that has a practical spectral efficiency of 1.5 bit/sec/Hz (see Chapter 8), the required spectrum is only about 1.1 MHz (66kb/s × (500×0.25×0.2)/1.5 = 1.1 MHz). However, as mentioned in the previous section, the possibility of wide-scale IP telephony can significantly change the design concept of conventional cable telephony network.

1.4.2.3 Personal Communications Services The biggest challenge in providing personal communications services (PCS) via HFC systems is that the wireless signals will have to travel through a return-channel path full of ingress and impusive noise (see Sec. 2.4 and [19]), in addition to multipath fading in a typical microcellular environment. Simulcast and delay spread problems will also have to be carefully resolved [20]. System design considerations for PCS signal transmission via HFC networks will be discussed in Chapter 10.

Note that both conventional and PCS telephony currently require circuit switching functionality. Two approaches to switching are being considered: (a) traditional class 5 central office switching with centrex capabilities and (b) distributed switching. The distributed switching approach offers the cable operator a "pay as you go" deployment methodology that adds switching fabric as customers join the network incrementally. However, because switching is the second-largest cost factor (next to transmission cost) in providing telephony services via CATV networks, MSO operators have recently become more interested in router switching [16,17].

1.4.3 Video-on-Demand and Set-Top Boxes

Video-on-demand (VOD) service is similar to the currently popular video-tape rental services. A customer can signal the network to select a movie, and the network will start transmission within a reasonable time. The customer can stop and continue the transmission. The customer can also signal the network to rapidly move forward or backward some number of basic time units. A set-top box decodes the information (e.g., video and audio) at the subscriber premises and provides subscribers with interactive capabilities. (See [21–23].)

Generally, a customer will use a set-top box to send his or her control signals in services such as VOD. Set-top boxes are considered to be a crucial piece of hardware for the coming age of interactive television [22]. A simplified block diagram of a digital set-top box is shown in Fig. 1.6. Key components such as tuner and 64-QAM demodulator will be covered in Chapter 8.

To give a few examples of VOD, we see that encyclopedias and other references may be kept on-line for ready access. Pictures and text can be mixed into a common interactive data stream. This allows reference materials to be kept current, whereas CD-ROMs and books run the risk of becoming obsolete. Services can charge on a pay-as-you-go basis, making the service much more affordable than the purchase of an entire encyclopedia.

The feasibility of providing VOD service from a centralized location over CATV networks depends on several factors, including database access, switching,

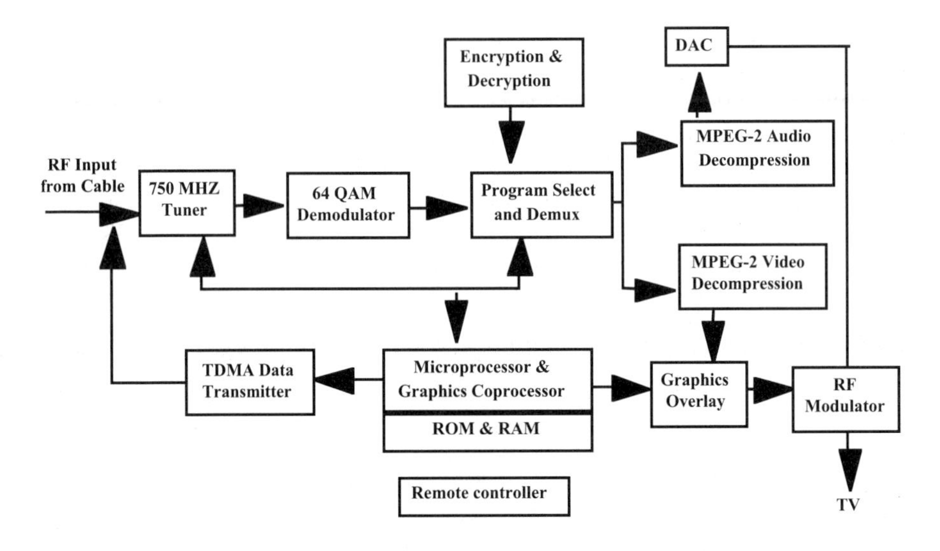

Figure 1.6: Digital set-top box. After Ref. [21] (© 1995 IEEE).

and transmission. Three-level memory hierarchy may be required for storing the video database: (a) a very large, long-access, low-transfer-rate *library;* (b) a medium-sized fast-quantized-access *copier memory;* and (c) a per-user *stop–start buffer*. The library subsystem holds several thousand reels of tape or video disks, each containing a single movie. Loading a tape to a player can be done by operators. The copier memory, used as a cache, could be implemented as a group of digital read/write disk memories, and a disk should be large enough to store a significant portion of a 100-minute movie (which requires about 1.125 Gbytes). One-gigabyte disks with a transfer rate of 30 Mb/s are commercially available today. The stop–start buffer is a small FIFO buffer. Increasing the number of heads per movie on the copier memory would reduce in inverse proportion the size of the stop–start buffer. At some point, as the number of heads in the copier memory becomes large and the amount of video material under each head diminishes, the stop–start buffer may no longer be needed.

For a single central office switch to serve a large number of customers, say 10,000, with each customer watching a 1.5-Mbps encoded video, the switch capacity must be around 15 Gbps, which is within today's technological reach. However, VOD has not yet taken off because of the high cost of large-capacity servers.

1.4.4 Interoperability and Standardization Activities

Complete system interoperability in a CATV communication network should be defined with respect to end-to-end interface compatibility and functional interoperability. Interface compatibility includes the specification of signal levels and waveforms, bit rates, bit framing and packet formats, protocols, and so on. Key dimensions of functional interoperability are compression, encryption, multiplexing, modulation, error correction, return channel characteristics, access control, and other transmission signal processing.

A number of standards already exist for digital video and audio formats for end-to-end digital CATV systems such as IEEE 802.14, Multimedia Cable Network System (MCNS), Data-Over-Cable Service Interface Specifications (DOCSIS), and Digital Audio and Video Council (DAVIC) [24].

1.4.5 Multimedia Communications and ATM Technology

CATV networks can employ asynchronous transfer mode (ATM) switching technology and bring multimedia communications to the mass market. ATM can certainly be used to maximize the functionality of a high-bandwidth medium in a star/bus architecture such as in a CATV network. The question before us is whether we can find subsets of ATM switches that are cable specific. In its field trial, Time Warner installed an ATM switch to handle routing of DS3 channels over their network. ATM functionality is built into the end-to-end connection of the multimedia bit stream, from the server through the headend to the subscriber terminal.

Generally speaking, ATM may be a good transport mechanism in the SONET/SDH ring [25] and on the fiber-trunk portion of the cable distribution plant, but its use all the way to the home requires dramatic reduction of ATM customer premises equipment (CPE) cost.

References

[1] "The coming telescramble," Special Report, Business Week, pp.38–53, April 8, 1996.

[2] R. Olshansky, V. A. Lanzisera, and P. M. Hill, "Subcarrier multiplexed lightwave systems for broadband distribution," IEEE J. Lightwave Technol., vol.7, pp.1329–1342, September 1989.

[3] W. I. Way, "Subcarrier multiplexed lightwave system design considerations for subscriber loop applications," IEEE J. Lightwave Technol., vol.7, pp.1806–1818, November 1989.

[4] T. E. Darcie, "Subcarrier multiplexing for lightwave networks and video distribution systems," IEEE J. Selected Areas Commun., vol.8, pp.1240–1248, September 1990.

[5] Bellcore Technical Report, TR-NWT-000909, *Generic Requirements and Objectives for Fiber in the Loop*, Issue 1, December 1991.

[6] D. Minoli, *Video Dialtone Technology*, McGraw-Hill, New York, 1995.

[7] K. Rath, D. H. Wanigasekara-Mohotti, R. G. Wendorf, and D. C. Verma, "Interactive digital video networks: Lessons from a commercial deployment," IEEE Commun. Magazine, pp.70–74, June 1997.

[8] W. Y. Chen and D. L. Waring, "Applications of ADSL to support video dial tone in the copper loop," IEEE Commun. Magazine, pp.102, May 1994.

[9] G. Young, K. T. Foster, and J. W. Cook, "Broadband multimedia delivery over copper," B T Tech. J., vol.13, no.4, pp.78–96, October 1995.

[10] D. Cervanka, "MMDS standing tall on digital technology, RBOC $$," Commun. Eng. and Design, pp.58–60, July 1996.

[11] D. A. Gray, "Broadband wireless access systems at 28 GHz," Commun. Eng. and Design, pp.46–56, July 1996.

[12] L. Chariglione, "MPEG: A technological basis for multimedia applications," IEEE Multimedia, vol. 2, pp. 85–89, Spring 1995; J. L. Mitchell, W. B. Pennebaker, C. E. Fogg, and D. J. Legall, *MPEG Video Compression Standard*, Chapman & Hall, New York, 1997; B. G. Haskell, A. Puri, and A. N. Netravili, *Digital Video: An Introduction to MPEG-2*, Chapman & Hall, New York, 1996; also, see http://www.cselt.stet.itlufvAeonardo/mpeg/index.htm.

[13] T. C. Kwok, "Residential broadband Internet services and application requirements," IEEE Commun. Magazine, vol.35, pp.76–83, June 1997.

[14] S. E. Gillet, "Connecting homes to the Internet: an engineering cost model of cable vs. ISDN," MIT, 1995, available at ftp://ftp.tns.lcs.mit.edu/pub/papers/MIT-LCS-TR-654.word5.1.

[15] E. Moura, "Demystifying asymmetric data networks," Commun. Eng. and Design, pp.50–57, August 1996.

[16] F. Dawson, "National data net could be key to new services," Commun. Eng. and Design, pp.116–121, June 1997.

[17] P. Newmann, G. Minshall, T. Lyon, and L. Huston, "IP switching and gigabit routers," IEEE Commun. Magazine, pp.64–69, January 1997. Also see "A framework for multiprotocol label switching," http://search.ietf.org/internet-drafts/draft-ietf-mpls-framework-0.2.txt.

[18] N. Kitawaki and K. Itoh, "Pure delay effects on speech quality telecommunications," IEEE J. Selected Areas Commun., vol.9, pp.586–593, May 1991.

[19] C. A. Eldering, N. Himayat, and F. M. Gardner, "CATV return path characterization for reliable communications," IEEE Commun. Magazine, vol.33, no.8, pp.62–69, August 1995.

[20] R. W. Donalson and A. S. Beasley, "Wireless CATV network access for personal communications using simulcasting," IEEE Trans. Vehicular Technol., vol.43, pp.666–671, August 1994.

[21] W. S. Ciciora, "Inside the set-top box," IEEE Spectrum, pp.70–75, April 1995.

[22] B. Furht, D. Kalra, F. L. Kitson, A. A. Rodriguez, and W. E. Wall, "Design issues for interactive television systems," IEEE Computer, pp.25–38, May 1995.

[23] T. S. Perry, "The trials and travails of interactive TV," IEEE Spectrum, pp.22–28, April 1996.

[24] MCNS (http://www.cablemodem.com), DAVIC 1.0 Specification (http://www.davic.org/DOWNI.htm), IEEE 802.14 (http://walkingdog.com/catv/index.html).

[25] T. H. Wu and N. Yoshikai, ATM Transport and Network Integrity, Academic Press, Chestnut Hill, Massachusetts, 1997.

Chapter 2

General Technical Background: *Modulation Signal Formats, Coaxial Cable Systems, and Network Architecture Evolutions*

Information transported by HFC systems follows four evolutionary stages: (1) broadcast multichannel analog signals, (2) a mixture of broadcast RF analog and digital signals, (3) a mixture of two-way digital signals on RF carriers and broadcast analog/digital signals, and (4) a mixture of broadcast and two-way digital signals (i.e., all analog signals are replaced). From a network architecture perspective, HFC systems begin with fiber backbone or fiber overlay; then their fibers penetrate deeper to the feeder or bridger amplifier section of the network, and finally to the taps. To provide more reliable communication services, ring architecture may have to be considered, and overall network architecture, including headend interconnections, should be planned. This chapter introduces the necessary technical background for various modulation signal formats and network architecture evolutions.

2.1 Analog and Digital Video Signal Formats and Standards

2.1.1 Analog NTSC and AM-VSB Video Signals

The most commonly used color television standards are the National Television Standard Committee (NTSC), the Sequentiel Couleur Avec Memoire (SECAM) and Phase Alternation Line (PAL). The NTSC television signal standard is in use in North America, Japan, and a total of 30 countries for terrestrial broadcasting and cable system delivery of television signals. The SECAM standard is in use in France and Russia, and PAL standard is used in Germany, Italy, and many other European countries. In this section, we will concentrate on NTSC television signal specifications and standards. The two major components of the NTSC color television signal are the picture carrier and the audio subcarrier. The luminance and the 3.58-MHz chrominance subcarrier amplitude modulates an IF carrier (e.g., 45.75 MHz) that is then asymmetrically filtered such that the upper band edge is 1.25 MHz above the IF carrier and the lower band edge is 4.75 MHz below the IF carrier. The format is referred to as amplitude modulation with vestigial sideband (AM-VSB) [1].

The audio signal frequency modulates a 4.5-MHz subcarrier. The deviation of this subcarrier by the audio signal is limited to ±25 kHz. This audio signal is then added to the IF signal. Note that the aural carrier level must be maintained between 10 and 17 dB below the picture carrier [2]. Frequency domain waveform of an AM-VSB video signal at RF (>54 MHz) is shown in Fig. 2.1. Note that it is an image version of the NTSC signal at IF because of the up-conversion process. The conventional television receiver uses the full upper sideband and a small portion of the filtered lower sideband (the vestigial sideband) to reconstruct the television video.

The time-domain waveform of a baseband NTSC color television signal [3] is shown in Fig. 2.2. Note that the time for each vertical scanning cycle (i.e., the time for the electron beam to complete its cycles of 525/2 = 262.5 horizontal lines from top to bottom and back to top again) is 1/60 second; therefore, the time for each horizontal line scanning time is 1/60/262.5 = 63.5 μs, as shown in Fig. 2.2. The baseband video signal is 1V peak-to-peak (M = 1 V in Fig. 2.2). For measurement, this range is divided into 140 units, called IRE units (1/100 of the luminance, or 1/100 of the

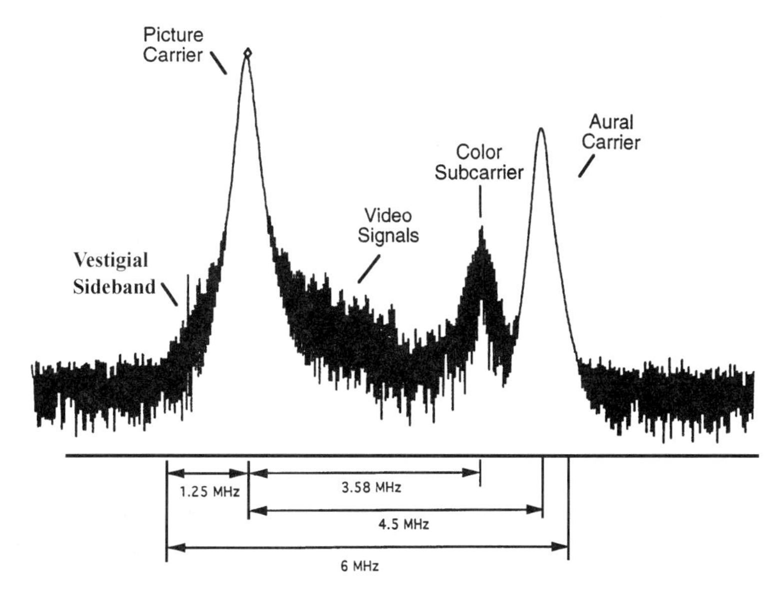

Figure 2.1: Frequency components contained in an NTSC AM-VSB video signal at RF. After Ref. [8].

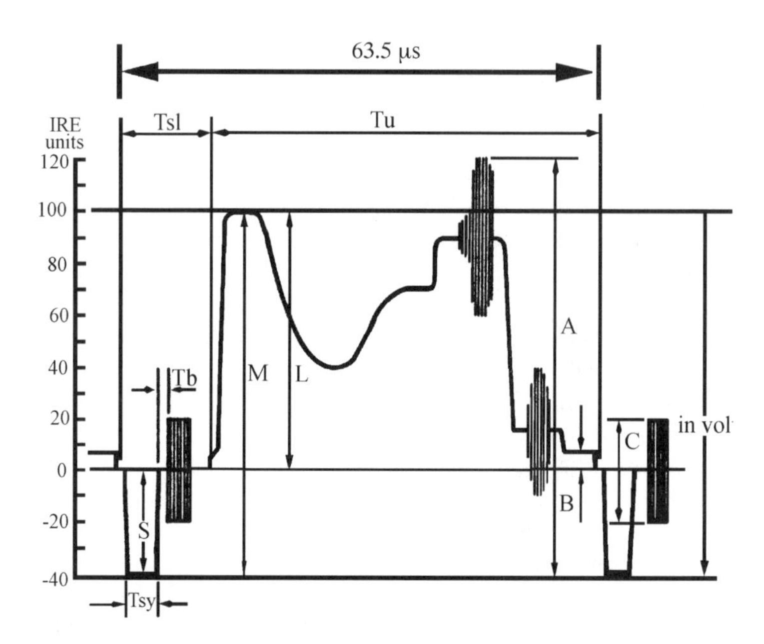

Figure 2.2: Time-domain waveform of an NTSC video signal. A, the peak-to-peak amplitude of the composite color video signal; B, the difference between black level and blanking level (setup); C, the peak-to-peak amplitude of the color burst; L, luminance signal-nominal value; M, monochrome video signal peak-to-peak amplitude (M = L + S); S, synchronizing signal amplitude; Tb, duration of breezeway; Tsl, duration of line blanking period; Tsy, duration of line synchronizing pulse; Tu, duration of active line period. (After Ref. [3])

blanking to reference white range). The seven components contained in the television waveform are (1) horizontal line synchronization pulses, (2) color sync (bursts), (3) setup (black) level, (4) luminance (gray) levels, (5) color hue (tint), (6) color saturation (vividness), and (7) field synchronization pulses (not shown).

The NTSC waveform is a complicated, periodic signal. The luminance and chrominance signals are picture dependent and, therefore, time dependent. All these make quantitative measurement of video performance difficult. Therefore, video performance measurements are often made with a certain test signal which has a periodic luminance signal substituted for the video signal [3]. With these test signals, many parameters can be used to characterize the quality of an NTSC signal. Several parameters that are frequently quoted in the literature or product specifications to measure the qualities of an NTSC video signal after transmission are briefly explained next.

Signal-to-(Weighted) Noise Ratio The signal-to-noise ratio (SNR) is the ratio between the normal picture signal level, 714 mV or 100 IRE, and the rms amplitude of the "weighted" noise component contained in the signal. Because of the response characteristics of the human eye, viewers perceive certain noise frequencies to cause greater degradation in the viewed picture than other frequencies. To compensate for this visual response characteristic, a weighting filter correlates electrical noise to visually perceived picture degradation. The noise referred to is predominantly thermal noise in the 10-kHz to 4-MHz range (the 4-MHz range can be roughly observed in Fig. 2.1 for the video signal portion). SNR is approximately equal to CNR in AM-VSB video signals. According to U.S. Federal Commission Rules and Regulations [2], a video signal received at a subscriber TV terminal through CATV systems should meet the requirement that its CNR shall not be less than 43 dB, that is, the signal-to-weighted noise ratio (CNR, dB) = 20 log (peak–peak signal amplitude/rms weighted noise) ≥ 43 dB. For a point-to-point radio link, the NCTA [3] requires that SNR should be no less than 53 dB.

Differential Gain Differential gain (DG), more precisely described as differential chroma gain, is a difference in amplitude response at the color subcarrier frequency (3.58 MHz) as the luminance signal level changes. The test signal used for DG measurements is the staircase signal with chroma added to the risers, as shown in Fig. 2.3. The distance from zero voltage to the mean of each equal-amplitude chroma-burst packet is the luminance (level L in Fig 2.2). Any difference in the chroma-burst amplitudes of the chrominance component after transmission (the

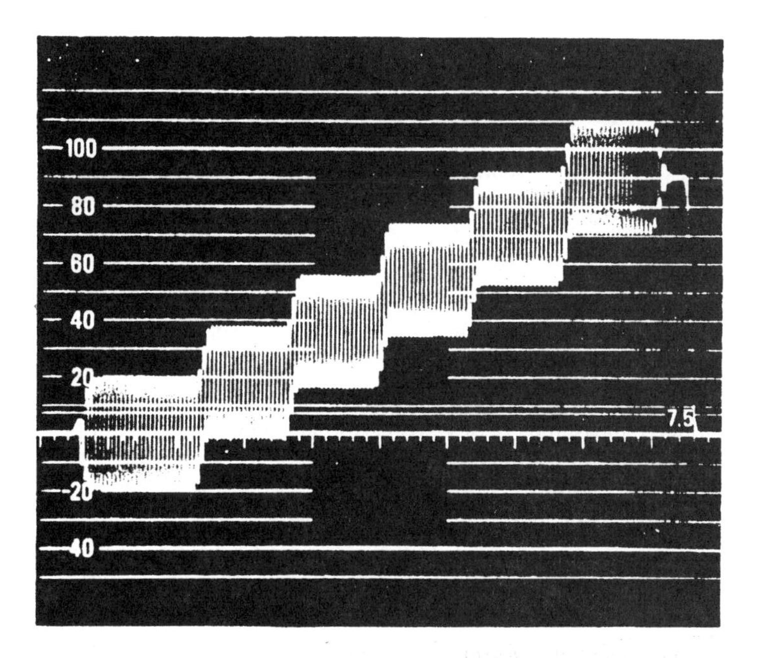

Figure 2.3: Staircase signal (with chroma added to the risers) used for differential-gain (DG) and differential-phase (DP) testing. (After Ref. [3])

difference is due to the different levels of luminance signal and nonlinear gain in the system) is the DG. According to U.S. Federal Commission Rules and Regulations [2], the DG shall not exceed 20% for the received signal at a subscriber's TV set. For a point-to-point radio link, the NCTA [3] requires that the DG be less than 15%. Typical optical fiber links for single-channel video transmission can achieve a DG of about 2%.

Differential Phase A change in the phase of the color subcarrier as a result of changes in the amplitude of the luminance signal is known as differential phase (DP). The same test signal shown in Fig. 2.3 is used to measure DP. With the subcarrier signal viewed on a vector scope, any phase difference between the chrominance on the steps indicates differential phase. The difference, in degrees, between the two steps farthest apart is a measure of the differential phase. According to U.S. Federal Commission Rules and Regulations [2], the differential phase shall not exceed 10° for the received video signal at a subscriber's TV set. For a point-to-point radio link, the NCTA [3] requires that DP shall not exceed 5°. Typical optical fiber links for single-channel video transmission can achieve a DP of about 3°.

From the preceding descriptions, it is clear that both DG and DP are used to quantize the nonlinear distortions in a single video channel. Therefore, changes in color intensity and color hue as brightness changes are due to poor DG and DP, respectively. When an optical fiber system is used to transport even just a single analog video signal, the DG and DP of the received video signal must both be within the specifications.

2.1.2 FM Video Signals

In discussing analog video signals, so far we have only talked about baseband NTSC and AM-VSB video signals. However, frequency-modulated (FM) video signals have been extensively used in long-distance CATV and satellite broadcast systems. From the viewpoint of a multichannel transmission system, FM signals are much easier to transport than AM-VSB signals, mainly because of their low carrier-to-noise ratio (CNR) requirement. The CNR requirement of an AM-VSB signal is approximately equal to its SNR and can be as high as 43–53 dB, whereas the CNR requirement of an FM video signal can be as low as 9–17 dB, depending on the applied frequency deviation. The relation of CNR and SNR of an FM video signal is [4]

$$\mathrm{SNR} = \mathrm{CNR} + 10\log(3m_f^2(1+m_f)) + W + P \tag{2.1}$$

where m_f is the FM modulation index given by $\Delta f_{\mathrm{peak}}/f_v$, Δf_{peak} is the peak deviation (typically 10.75 MHz), f_v is the maximum video modulating frequency (4.2 MHz for NTSC), W is the weighting factor used to account for the nonuniform response of the eye to white noise in the video bandwidth, and P is the preemphasis factor ($W+P$ values ranging from 18 to 26 dB are quoted in the literature). The overall improvements in SNR over CNR range from 36.5 to 44.5 dB in a typical satellite network TV [4]. Therefore, if a 53-dB SNR is required at the receiver end, the required

CNR can be as low as 8.5 to 16.5 dB, which is much lower than that of an AM-VSB video signal. The improvement may be lower for terrestrial FM optical fiber trunks depending on the applied frequency deviation, *W*, and *P* factors. Note that the bandwidth (B) required for each FM video is given by Carson's rule, $2(f_m + \Delta f_{\text{peak}})$, where f_m is the top audio subcarrier frequency, and the resulting *B* is typically 30 to 36 MHz. This bandwidth is much wider than the 6-MHz bandwidth of an AM-VSB signal.

Despite the fact that multichannel FM video signals are much easier to transport than multichannel AM video signals, there are currently very few newly installed multichannel FM video optical links. This is due to the incompatibility of FM video signals with a typical NTSC TV set (because an FM demodulator is required at a residence), and to the fast development of cost-effective multichannel AM-VSB optical fiber systems.

2.1.3 Digital Video Signals

Digital video signals have the following advantages over their analog counterparts: (1) They can provide better picture and audio quality, (2) they are much less susceptible to transmission-system-induced noise and nonlinear distortions, (3) they are much easier to integrate with other services (e.g., they can be made compatible with SONET/SDH signal for transmission), and (4) they can be readily encrypted to provide security and control. To facilitate transmission and switching of digital video signals in an HFC network, a significant amount of compression must be done. To get a feeling for the amount of compression required, let us look at an example: The digital coding of the NTSC composite signal normally involves sampling at 3 or 4 times the color subcarrier frequency ($3.58 \times 3 = 10.74$ or $3.58 \times 4 = 14.32$ MHz). Assuming A/D conversion in the 8- to 9-bit range, the resulting raw bit rate is 86–129 Mbps. This is obviously too many bits per second to be accommodated in a 6-MHz CATV channel, and that is why a compression technique must be applied. The first result of standardization efforts with respect to compression was ITU-T (formerly CCITT) Recommendation H.261, supporting video coding at $p \times 64$ kbps ($p = 1,2,...,30$) (see also Table 1.2). The second was a standard originated by the MPEG (Moving Picture Experts Group) Committee (see Table 1.2). The first standard, MPEG-1, provides video coding for digital storage media with a rate of 2 Mbps or less. H.261 and MPEG-1 standards are not sufficiently flexible to support high-action video signals such as sporting events. As a result, MPEG-2 standards [5] with compression of bit rates in the range of 4 to 6 Mbps for NTSC, PAL, and SECAM material and with "network" quality became available in 1994. MPEG-2 has video quality higher than that of the current NTSC, PAL, and SECAM broadcast systems and is widely accepted by CATV industry as the targeted digital video format.

To transmit digital video signals such as MPEG-2 through CATV systems by RF carriers, modulation formats such as 64-QAM have been standardized by ITU J.83, MCNS, IEEE 802.14, and DAVIC [6]. The theoretical bit-error-rate (BER) versus CNR ($\approx$SNR) requirement for QAM modulation is given in Fig. 2.4 (see

Chapter 8 for more detailed discussions). We see that a theoretical CNR of about 28 dB (30 dB in practice) is required for a 64-QAM signal to achieve a bit-error rate of 10^{-9} when no error-correction coding is used. This CNR requirement is much lower than that of an AM-VSB signal. Even when the modulation format is upgraded to 256-QAM, the CNR requirement is only about 35 dB without error correction coding.

Although the quality of analog NTSC signals degrades gradually as the CNR goes below a certain level, the quality of digital NTSC signals degrades drastically as the CNR goes below a "threshold" level. Threshold levels are defined for the onset of digital picture block errors, the unusable point, and the loss of digital signal, or for the SNR level below which the analog picture quality is unacceptable. Qualitative results comparing analog satellite reception (FM NTSC) to digital satellite reception (QPSK with MPEG-2) were obtained [7]. The results are shown in Table 2.1.

It can be seen that above the threshold, the digital baseband signal has a perfect 63-dB baseband video SNR, whereas the analog signal has a degraded baseband SNR of 48 dB. In addition, the at-threshold video SNRs and CNRs are 15 and 3.5 dB different—all in favor of digital. It can also be observed from Table 2.1 that quality of digital video signals on QPSK changes abruptly, because at a CNR of 6 dB the SNR is still as good as 63 dB, but at a CNR of 3.5 dB, the digital signal is completely lost.

Among all signal formats mentioned thus far, AM-VSB modulation requires a much higher CNR than those of FM and digital QPSK/QAM modulations. In addition, AM-VSB modulation requires very stringent second- and third-order nonlinear distortions (to be discussed in Section 2.3). This is why the high linearity and low

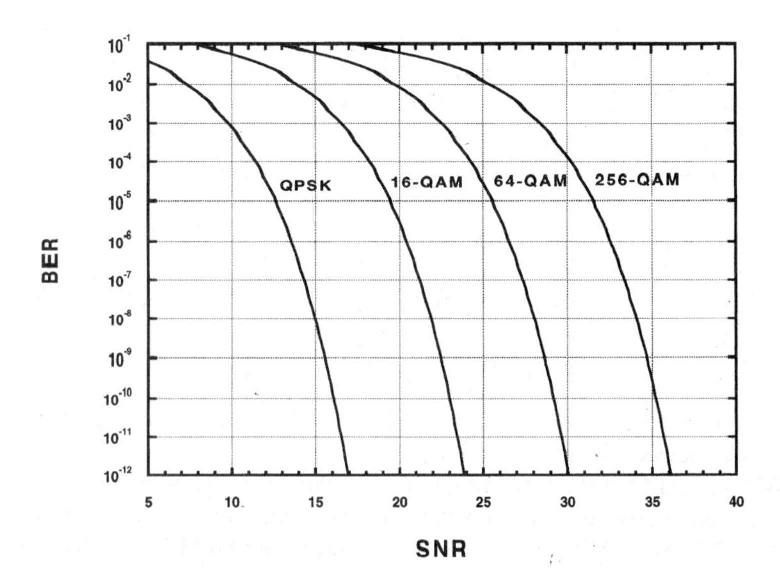

Figure 2.4: Bit-error-rate (BER) versus signal-to-noise ratio (SNR) for (uncoded) digital modulations of QPSK, 16-, 64-, and 256-QAM, respectively.

Table 2.1: Comparison of analog FM and digital QPSK (with MPEG-2) CNR/SNR performances. After Ref. [7].

	CNR at first appearance of impulse noise (analog) or error blocks (digital)	Baseband Video SNR at threshold	Loss of signal CNR
Analog FM	10 dB	48 dB	7 dB
Digital QPSK	6 dB	63 dB	3.5 dB

noise requirements imposed on all optical fiber components (laser diode, photodiode, connectors, couplers, etc.) in the optical fiber trunk systems for multichannel AM-VSB signal transmission can guarantee future upgrades to RF-carried digital video signals.

Digital channels, however, are not transmitted without price. They are sensitive to RF carrier phase noise and channel frequency/phase responses. The former can cause smearing of the constellation diagram in QAM transmissions. The latter sets stringent requirements on the cables, connectors, amplifiers, etc., in a cable system [8]. This will be discussed in more detail in Sec. 2.2.4 and Chapter 8.

2.1.4 CATV Frequency Plans

Broadcast television stations in the United States are assigned channels 2 through 83 by the Federal Communication Commission (FCC). Table 2.2 shows the carrier frequencies of each channel. We can see that the neighboring channels are separated by 6 MHz, except for an added 4-MHz gap between channels 4 and 5.

In standard CATV carrier frequency allocation, which is the most common and similar to the broadcast allocation, all picture carriers except channels 5 and 6 are placed 1.25 MHz above 6-MHz multiples. Channels 5 and 6 are placed 0.75 MHz below 6-MHz multiples. Video carriers in the bands 108–137 and 225–400 MHz must operate at frequencies offset from $(6N + 1.25)$ MHz, which may be used by aeronautical radio services (FCC 76.612 Cable Television Frequency Separation Standards). Therefore, we can see from Table 2.2 that carriers in 108–137 MHz are offset by 25 kHz, and carriers in 225–400 MHz are offset by 12.5 kHz.

In Table 2.2, IRC stands for incremental related carriers, where all channels except channels 98 and 99 are placed 1.2625 MHz above 6-MHz multiples. HRC stands for harmonically related carriers, where the picture carriers are placed very close to the 6-MHz multiples. Both IRC and HRC can reduce composite triple-beats in nonlinear devices, as will be discussed in Section 2.3.

With optical-fiber trunk and feeder lines in place, various vintages of equipment from competing cable equipment manufacturers provides upper frequency limits of 450, 550, 750, 860, and even 1000 MHz. Broadcast digital MPEG-2 channels and two-way communications services have been proposed to be provided via various spectral ranges such as 450–750, 550–750, 550–860, and 550–1000 MHz [9]. An example of such a frequency plan is shown in Fig. 2.5.

Table 2.2: CATV frequency plans for standard, IRC, and HRC video carriers.

	Conventional Channel Number	EIA Channel Number	Standard Video Carrier	New IRC Video Carrier	New HRC Video Carrier
Subband	T7		7.0000		
	T8		13.0000		
	T9		19.0000		
	T10		25.0000		
	T11		31.0000		
	T12		37.0000		
	T13		43.0000		
	T14		49.0000		
Low band	2	02	55.2500	55.2625	54.0027
	3	03	61.2500	61.2625	60.0030
	4	04	67.2500	67.2625	66.0033
	A8	01		73.2625	72.0036
	5	05	77.2500	79.2625	78.0039
	6	06	83.2500	85.2625	84.0042
	A5	95	91.2500	91.2625	90.0045
	A4	96	97.2500	97.2625	96.0048
	A3	97	103.2500	103.2625	102.0051
	A2	98	109.2750	109.2750	108.0054
Midband	A1	99	115.2750	115.2750	114.0057
	A	14	121.2625	121.2625	120.0060
	B	15	127.2625	127.2625	126.0063
	C	16	133.2625	133.2625	132.0066
	D	17	139.2500	139.2625	138.0069
	E	18	145.2500	145.2625	144.0072
	F	19	151.2500	151.2625	150.0075
	G	20	157.2500	157.2625	156.0078
	H	21	163.2500	163.2625	162.0081
	I	22	169.2500	169.2625	168.0084
High band	7	07	175.2500	175.2625	174.0087
	8	08	181.2500	181.2625	180.0090
	9	09	187.2500	187.2625	186.0093
	10	10	193.2500	193.2625	192.0096
	11	11	199.2500	199.2625	198.0099
	12	12	205.2500	205.2625	204.0102
	13	13	211.2500	211.2625	210.0105
Superband	J	23	217.2500	217.2625	216.0108
	K	24	223.2500	223.2625	222.0111
	L	25	229.2625	229.2625	228.0114
	M	26	235.2625	235.2625	234.0017
	N	27	141.2625	241.2625	240.0120
	O	28	247.2625	247.2625	246.0123
	P	29	253.2625	253.2625	252.0126
	Q	30	259.2625	259.2625	258.0129

	R	31	265.2625	265.2625	264.0132
	S	32	271.2625	271.2625	270.0135
	T	33	277.2625	277.2625	276.0138
	U	34	283.2625	283.2625	282.0141
	V	35	289.2625	289.2625	288.0144
	W	36	295.2625	295.2625	294.0147
Hyperband	AA	37	301.2625	301.2625	300.0150
	BB	38	307.2625	307.2625	306.0153
	CC	39	313.2625	313.2625	312.0156
	DD	40	319.2625	319.2625	318.0159
	EE	41	325.2625	325.2625	324.0162
	FF	42	331.2750	331.2750	330.0165
	GG	43	337.2625	337.2625	336.0168
	HH	44	343.2625	343.2625	342.0171
	II	45	349.2625	349.2625	348.0174
	JJ	46	355.2625	355.2625	354.0177
	KK	47	361.2625	361.2625	360.0180
	LL	48	367.2625	367.2625	366.0183
	MM	49	373.2625	373.2625	372.0186
	NN	50	379.2625	379.2625	378.0189
	OO	51	385.2625	385.2625	384.0192
	PP	52	391.2625	391.2625	390.0195
	QQ	53	397.2625	397.2625	396.0198
	RR	54	403.2500	403.2625	402.0201
	SS	55	409.2500	409.2625	408.0204
	TT	56	415.2500	415.2625	414.0207
	UU	57	421.2500	421.2625	420.0210
	VV	58	427.2500	427.2625	426.0213
	WW	59	433.2500	433.2625	432.0216
	XX	60	439.2500	439.2625	438.0219
	YY	61	445.2500	445.2625	444.0222
	ZZ	62	451.2500	451.2625	450.0225
Upper band	AAA	63	457.2500	457.2625	456.0228
	BBB	64	463.2500	463.2625	462.0231
	CCC	65	469.2500	469.2625	468.0234
	DDD	66	475.2500	475.2625	474.0237
	EEE	67	481.2500	481.2625	480.0240
	FFF	68	487.2500	487.2625	486.0243
	GGG	69	493.2500	493.2625	492.0246
	HHH	70	499.2500	499.2625	498.0249
	III	71	505.2500	505.2625	504.0252
	JJJ	72	511.2500	511.2625	510.0255
	KKK	73	517.2500	517.2625	516.0258
	LLL	74	523.2500	523.2625	522.0261
	MMM	75	529.2500	529.2625	528.0264
	NNN	76	535.2500	535.2625	534.0267
	OOO	77	541.2500	541.2625	540.0270
	PPP	78	547.2500	547.2625	546.0273

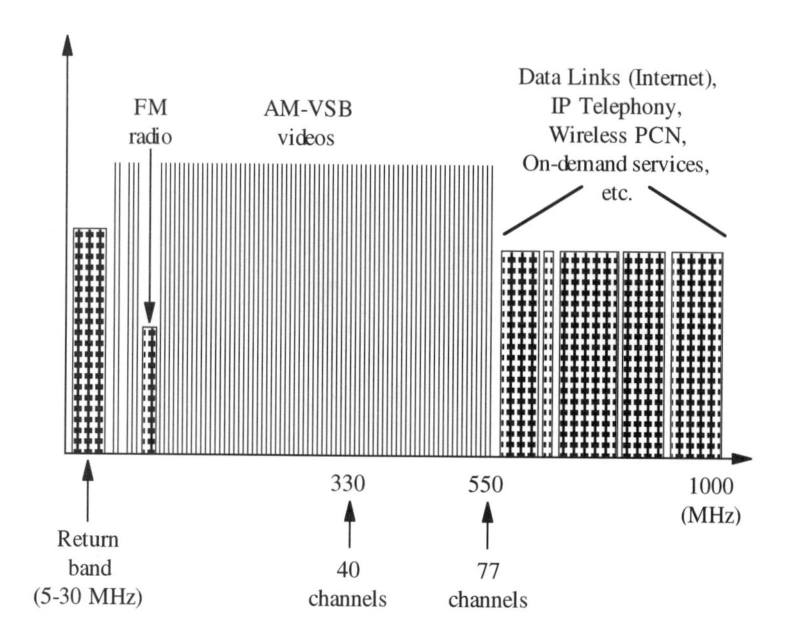

Figure 2.5: An example of a frequency plan which includes various two-way communication services.

It is unfortunate that there is no activity aimed at standardizing spectrum allocation in CATV systems, even though frequency-agile RF modulators can be built to fit any frequency plan changes. Aside from economic considerations, the difficulty stems from the lack of sufficient field test data for advanced digital transmission equipment to support any frequency allocation plan.

2.2 CATV Coaxial Cables, Components, and Systems

In designing a conventional CATV system for delivering multichannel analog video signals, the major factors that must be considered include transmission loss, frequency response, amplifier gain, noise, and nonlinear distortions. When digital RF channels are added, linear distortions (group delay and frequency response) in each 6-MHz channel must also be considered. A brief overview of those design factors is given in this section.

2.2.1 Coaxial Cable

Three types of cable are used in the distribution system: trunk, feeder, and drop. Typical trunk cable is approximately ½- to ¾-inch in diameter. Typical ¾-inch trunk cable loss is around 1.15 dB per 100 m at 50 MHz and 4.10 dB per 100 m at 550

MHz. The corresponding attenuations for 1/2-inch cable are 1.71 and 5.97 dB, respectively. Feeder cable is usually somewhat smaller than trunk cable. The drop cable (the cable run from the tap to the subscriber terminal is referred to as the drop), such as RG-59 (4C), is rather lossy: 5.84 dB and 18.8 dB per 100 m at 50 and 550 MHz, respectively. Note that the cable attenuation loss is proportional to the square root of the signal frequency; for instance, the cable loss at 450 MHz is about three times that at 50 MHz. The cable loss is also a function of temperature. A general rule of thumb is a 1% increase in attenuation at all frequencies for an increase in temperature of 10°F. A typical amplitude loss factor for trunk cables is shown in Table 2.3.

2.2.2 Amplifiers

Depending on where an amplifier is located, CATV RF amplifiers can be categorized into trunk, bridging, and line-extender amplifiers. Their respective locations in a typical coaxial CATV system are shown in Fig. 2.6. Typical trunk amplifier spacing is 20–22 dB at the high end of the frequency range in order to obtain a balance between high SNR and low nonlinear distortions [10,11]. The output power of each trunk amplifier is typically 30–32 dBmV.[1]

Knowing the cable loss at the maximum operation frequency, we can estimate the inter-trunk-amplifier distance. For example, for a 550-MHz system, the inter-trunk-amplifier distance is (22 dB/4.10 dB) × 100 m, or about 540 m of ¾-inch

Table 2.3: Trunk cable loss (dB/100 meters) as a function of frequency for four different cable diameters (maximum values).

Frequency (MHz)	0.5 in.	0.75 in.	0.875 in.	1.0 in.
5	0.52 dB	0.36 dB	0.30 dB	0.26 dB
30	1.31	0.89	0.75	0.69
50	1.71	1.15	0.98	0.89
108	2.53	1.71	1.48	1.35
300	4.30	2.92	2.56	2.36
400	5.02	3.44	2.99	2.76
450	5.35	3.67	3.18	2.95
550	5.97	4.10	3.58	3.31
600	6.27	4.30	3.74	3.48
700	6.82	4.69	4.07	3.81
750	7.09	4.86	4.23	3.97
800	7.35	5.05	4.40	4.13
862	7.64	5.25	4.59	4.33

[1]dBmV is a term commonly used in CATV systems. The definitions of dBmV and dBm are $20\cdot\log(V_{rms}/10^{-3})$ and $10\cdot\log(V_{rms}^2/(10^{-3}\cdot R))$, respectively, where R is the characteristic resistance. Therefore, 0 dBmV is equivalent to −48.75 dBm in a 75-Ω system.

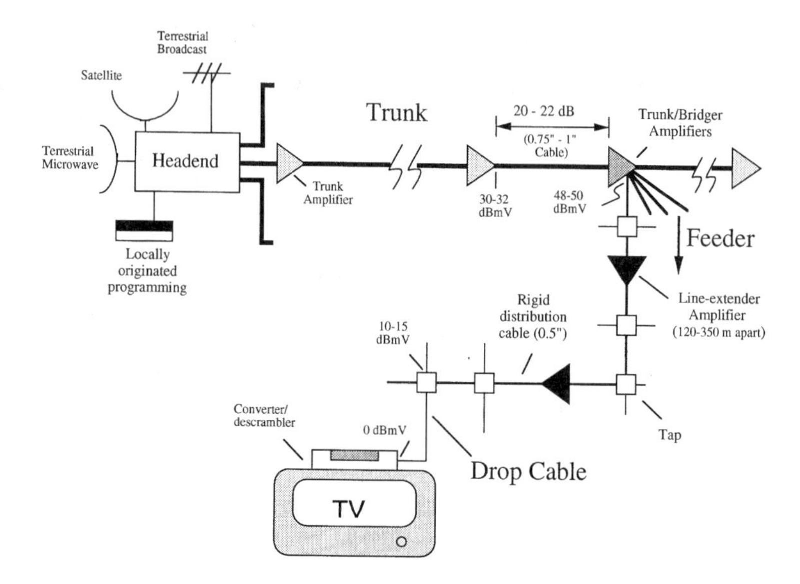

Figure 2.6: Tree-and-branch architecture of a coaxial-cable-based CATV system. Typical output power levels of different types of amplifiers are shown in the figure.

cables. In the feeder portion, bridger amplifiers are used not only to forward deliver signals, but also to split the signals to up to four "feeders" (see Fig. 2.6). Therefore, the output power of a bridger amplifier is typically in the range of 48–50 dBmV which is about 18 dB higher than that of trunk amplifiers. Because of this high output power, nonlinear distortions in the bridger amplifiers and in the distribution line extender amplifiers are higher than those in the trunk amplifiers. To avoid nonlinear distribution accumulation and to maintain the flatness of frequency response over the entire band, a maximum of two or three line-extenders are used (typically spaced between 120 and 350 m, depending on how many taps are between line extenders) in the distribution portion.

For two-way cable systems, a typical two-way trunk/bridger amplifier is shown in Fig. 2.7. The forward signal inputs to the high-frequency port (marked by "H" in Fig. 2.7) of the input diplexer, through an equalizer and an amplifier module, and then through the high-frequency port of the output diplexer to the trunk output port. A portion of the forward signal is coupled to a bridger module, which drives the feeders. The diplexer is a three-port device, with a common port and a high and low port. The common-to-low port is essentially a lowpass filter and the common-to-high port is essentially a highpass filter.

In some cable systems, the status of AC and DC supplies, forward and return RF signal levels, station temperature, etc., are monitored. Status monitoring transceivers are installed in some or all trunk amplifiers. The transceivers (each with a unique address) are periodically polled from a controller at the headend, and the

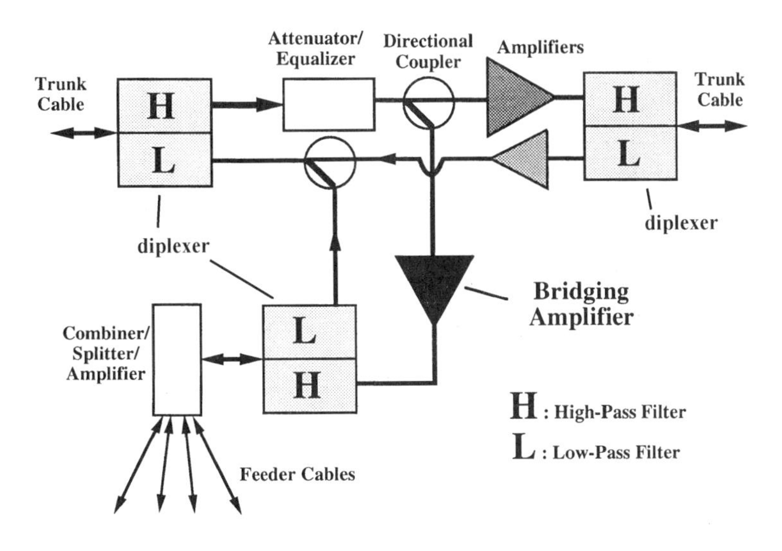

Figure 2.7: A two-way trunk/bridger amplifier.

transceivers report back the status of certain station conditions. Queries are typically sent to the transceivers in the forward cable system FM frequency band (88–108 MHz) on an unoccupied FM channel. Responses from the transceiver are generally in the sub-split frequency band (5–~42 MHz).

AGC and ASC A cascade of 20 amplifiers in a given trunk span has $22 \times 20 = 440$ dB of cascaded loss and gain. Because the cable loss is frequency dependent, the loss must be amplitude equalized to avoid significant nonflatness across the frequency band. This is accomplished by installing in each amplifier plug-in equalizers (see Fig. 2.7) to introduce additional attenuation at lower frequencies. An equalizer may be an automatic gain control (AGC) and/or automatic slope control (ASC) amplifier. Typical gain control and slope control ranges are both ±4 dB at the upper frequency. Certain predetermined program channels carried on the cable system are used as pilot channels. A module in the trunk amplifier detects the signal level of the selected AGC or ASC pilot channel sampled at the amplifier output and feeds a control voltage to the appropriate gain or slope network in the forward trunk module.

Several different types of amplifiers have been used to improve the amplifier dynamic range and the maximum achievable CNR. For example, the push–pull amplifier shown in Fig. 2.8a is used to minimize second-order distortions. In Fig. 2.8a, V_1 and V_2 are 180° out of phase, so that $V_{o1} = a_1 V_1 + a_2 V_1^2$, $V_{o2} = a_1(-V_1) + a_2 V_1^2$. At the second transformer, V_{o1} and V_{o2} are then recombined, again 180° out of phase, to obtain $2a_1V_1$ (a result of "push" and "pull") and zero second-order terms. Depending on the balance of the two amplifiers in Fig. 2.8a, typical push–pull amplifiers can reduce second-order distortions by about 20 dB. However,

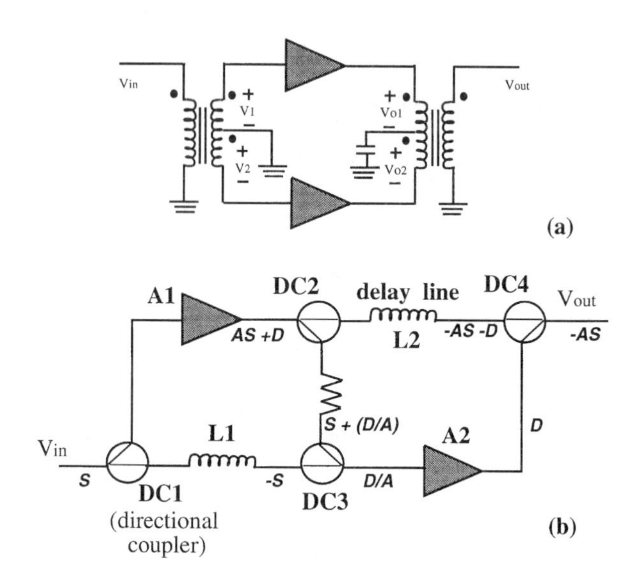

Figure 2.8: (a) Push–pull amplifier. (b) Feedforward amplifier. (S: Signal, D: Nonlinear distortions, A: Amplifier gain, DC: directional coupler)

push–pull amplifiers cannot cope with third-order nonlinear distortions and cross-modulations (defined in Section 2.3). Therefore, to operate in a wide band, the feedforward amplifier shown in Fig. 2.8b is preferred. In Fig. 2.8b, A1 and A2 are two push–pull amplifiers. Input signals are applied to A1 and L1 (delay line #1) through an directional coupler. L1 and L2 are adjusted such that they cause S and S + D to be delayed 180°. Following the signal flows in the figure, we can see that, theoretically, the output signal contains no nonlinear distortions. Feedforward amplifiers are often used in super trunks, and the composite-triple-beat reduction can generally be about 20 dB. Second-order distortions can also be significantly reduced. A typical feedforward amplifier can boost the output signal level of a trunk amplifier to 32–38 dBmV (originally 30–32 dBmV in Fig. 2.6).

2.2.3 Taps

Taps in Fig. 2.6 are used to split the signals to drop cables, with a tapping power level of about 10–15 dBmV. About a 0-dBmV signal level is expected to be received at the set-top converter box. Typically, taps are available with two, four, or eight subscriber ports. The power loss from an input port to an output port is called insertion loss, whereas the power loss from an output port to another output port is called isolation loss. High isolation loss is important for two-way communications to keep the upstream signals sent from customer A from leaking into the receiver of customer B. Typical port-to-port isolation in a tap is about 20 dB.

2.2.4 Overall Coaxial Cable System Noise, NLDs, and Frequency Responses

Figure 2.6 shows the overall coaxial cable system configuration, which consists of trunk amplifiers, bridging amplifiers, line-extender amplifiers, taps, and set-top boxes at subscriber premises. The typical output power levels for various types of amplifiers and interstage power loss are also marked in the figure. Note that this conventional coaxial cable system is based on a tree-and-branch architecture, and as many as 80% of the amplifiers may be line extenders because a substantial portion of the total cable system mileage is feeder. Major noise contributions to the entire coaxial cable system are from the trunk amplifiers and the set-top box [12], and most nonlinear distortions are contributed by bridging amplifiers and line extenders. Therefore, by replacing the coaxial trunk and bridging amplifiers with optical fibers, both the noise and nonlinear distortion accumulations can be minimized, provided we use as few line extenders in the drop cable as possible.

For the transmission of QAM-modulated digital signals, end-to-end frequency response (including amplitude and phase responses) has significant impact on the QAM modem equalizer design and the corresponding bit-error-rate performance. The ability of a QAM modem to cope with the poor frequency response of the coaxial cables, especially the drop portion of the network, has been used as one of the major figures of merit in choosing different brands of QAM modems. Therefore, it is important to establish a practical channel model (in both computer and hardware implementation) to simulate an actual HFC transmission system. Examples of such efforts are shown in Fig. 2.9, Table 2.4, and Table 2.5 [13]. From these CableLab channel models, we can see that typical microreflections in the coaxial part of an

Table 2.4: Parameters for four coaxial channel models. After Ref. [8].

	Delay (ns)	Power (dB)
Model#1 (CableLabs 1)	300	–18
Model#2 (CableLabs 2)	2500	–20
Model#3 (CableLabs 3)	600	–20
Strong reflection model	600	–13

Table 2.5: Parameters for an ensemble channel model. After Ref. [8].

Delay (ns)	Power (dB)
0	0
–200	–19
80	–22
150	–17
300	–22
600	–19

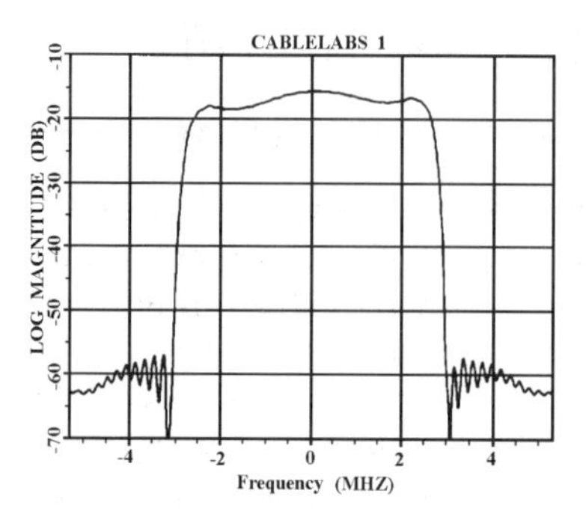

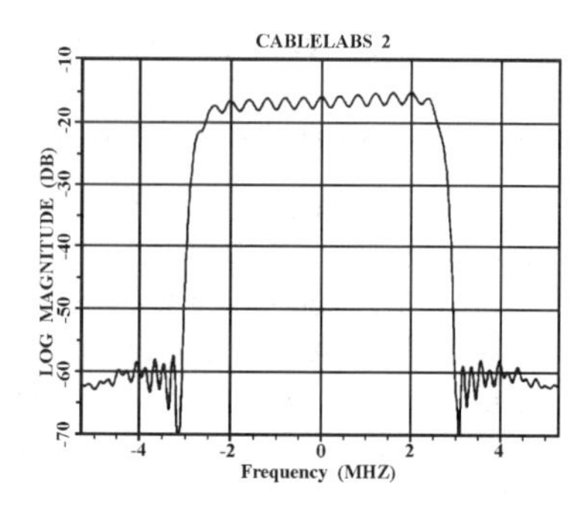

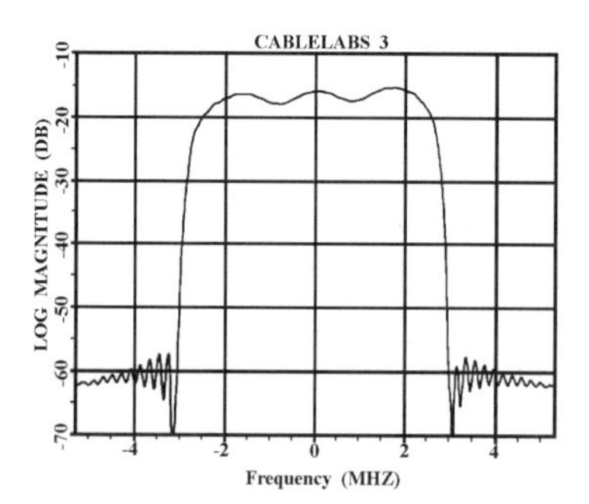

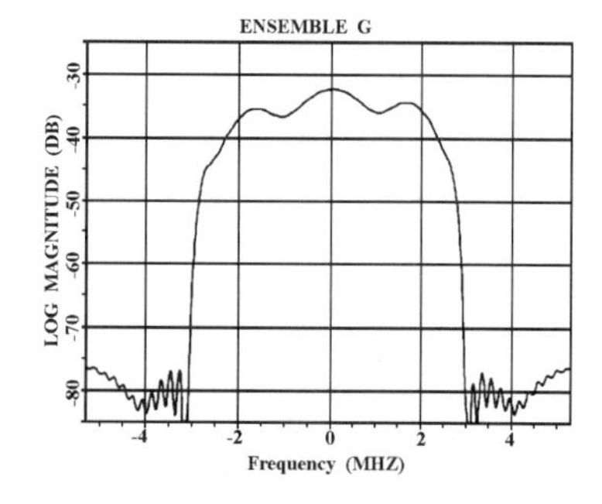

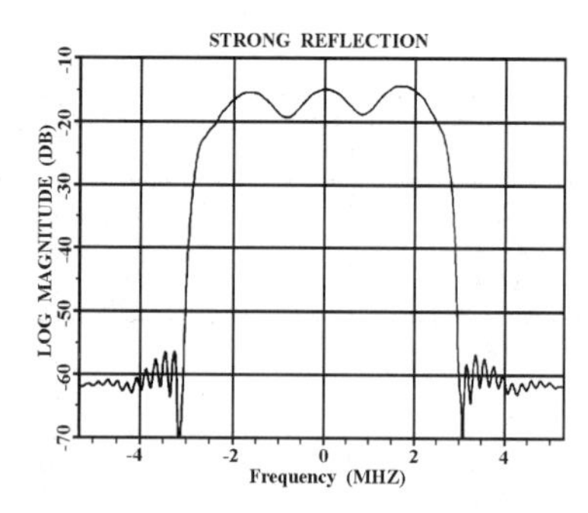

Figure 2.9: Channel frequency response of five models: CableLab1, CableLab2, CableLab3, microreflections ensemble model, and a strong-reflection model. After Ref. [8].

HFC system have a delay range of 80 to 600 ns, and a reflectivity range from –22 to –13 dB. The equalizer of a well-designed QAM modem (see Chapter 8) should be able to take care of all these environments.

2.3 Multichannel System with Cascaded Amplifiers

2.3.1 Nonlinear Distortions in an Amplifier: CTB, CSO, and XM

When multichannel signals are transmitted through a nonlinear device such as a trunk amplifier, various nonlinear distortions (NLDs) (illustrated in Fig. 2.10a) will be generated. The second-order intermodulation distortions (IMDs) with A − B or A + B type, where A and B stand for two arbitrary RF frequencies, respectively, are the dominant second-order NLDs. The triple beat products of the A + B − C type which occurs within signal bands most frequently are the dominant third-order NLDs in multichannel systems. The relative power levels of different types of nonlinear distortions are shown in Fig. 2.10b. We can see that the A ± B type second-order NLDs

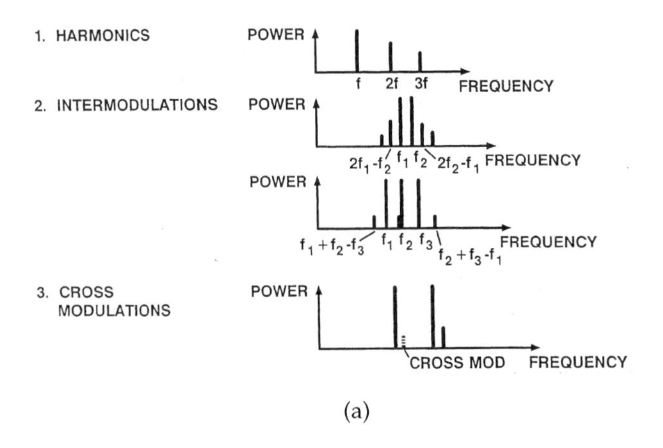

(a)

	Nonlinear Distortion Type Term	No. Of Components Per Term	dB Of Terms Above Harmonic
2nd order	2A	1	0
	A+B	2	6
	A - B	2	6
3rd order	3A	1	0
	2A+B	3	9.54
	2A - B	3	9.54
	A-2B	3	9.54
	A+B - C*	6	15.56
	A - B - C	6	15.56
	A+B+C	6	15.56

* Dominant form of the 3rd order intermodulation products

(b)

Figure 2.10: (a) Illustrations of various nonlinear distortions. (b) Relative power levels of second- and third-order nonlinear distortions.

are always 6 dB higher than second-order harmonics, and the A ± B ± C type third-order intermodulation products are always 6 dB higher than 2A ± B type NLDs. Note that the number of two-tone third-order IMDs of the 2A – B type increases with $N(N-1)$, whereas the number of triple beats of the form A + B – C increases with $N(N-1)(N-2)/2$. Therefore, triple-beat IMDs are the dominant third-order NLDs for a large number of channels. In a CATV system, the summation of all possible triple-beat products and second-order IMDs falling into a particular channel is termed composite triple beat (CTB) and composite second order (CSO), respectively. The locations where CSOs and CTBs occur in a standard CATV channel are illustrated in Fig. 2.11a. We can see that CSO beats are above and below the visual carrier 0.75 and 1.25 MHz, while CTB beats fall right on the visual carrier frequencies. The CSO beats at ±0.75 MHz are the result of standard allocation channels 5 and 6, which are offset 2 MHz from 6 MHz multiples (see Table 2.2), mixing with other channels. As for the CSO beats at ±1.25 MHz and the CTB beats, their frequency locations can be easily verified by mixing two or three standard channels given in Table 2.2

In systems using an IRC frequency plan (Table 2.2), CSOs fall exactly ±1.25 MHz around the carrier, with no ±0.75 MHz beats. In systems using an HRC frequency plan, both CSO and CTB fall exactly on the carrier frequency.

CSO usually appears as swimming diagonal stripes in the TV picture, while CTB often appears as horizontal streaks covering one or more lines of video [14]. The distribution of CTBs and CSOs in a 78-channel CATV system is shown in Fig. 2.11. Note that the third-order IMDs are the highest in the center channels, and A – B and A + B type CSOs are the highest at low- and high-frequency edges, respectively. Product count in Fig. 2.11 is defined as the total number of intermodulation beats that fall into a particular channel. Product count can be conveniently used to deduce multichannel CSOs and CTBs due to static (frequency-independent) nonlinearities according to the measured results of single-tone second-order harmonics (HD_2) and the two-tone third-order intermodulation products (IM_3), respectively, as follows:

$$\begin{aligned} \mathrm{CSO} &= \mathrm{HD}_2 + 10 \log N_{\mathrm{CSO}} + 6\ (\mathrm{dB}) \\ \mathrm{CTB} &= \mathrm{IM}_3 + 10 \log N_{\mathrm{CTB}} + 6\ (\mathrm{dB}) \end{aligned} \qquad (2.2)$$

where N_{CSO} and N_{CTB} are the product counts of second- and third-order intermodulations in a particular channel, respectively as illustrated in Fig. 2.11 (b). Six decibels were added in both cases, because HD_2 is lower than second-order intermodulation by 6 dB, and IM_3 (i.e., 2A – B or 2A + B) is lower than triple beats by 6 dB, as was shown in Fig. 2.10b. For example, if IM_3 is measured after an optical receiver to be –100 dBc when a laser transmitter is modulated by two tones with 4% modulation index each, then when there are 78 channels of carriers (also with 4% modulation index each), the worse-case CTB in the middle channel is $-100 + 10 \cdot \log(2140) + 6 = -61$ dBc, where we have assumed the product count in the middle channel to be about 2140, as shown in Fig. 2.11. This method, although analytically convenient, may not be very practical because it is very difficult to measure IM_3 on a spectrum analyzer that is –100 dB below the fundamental carriers. On the other hand, it could be slightly easier to use this method to derive second-order intermodulation products in a direct modulation optical fiber system, because the

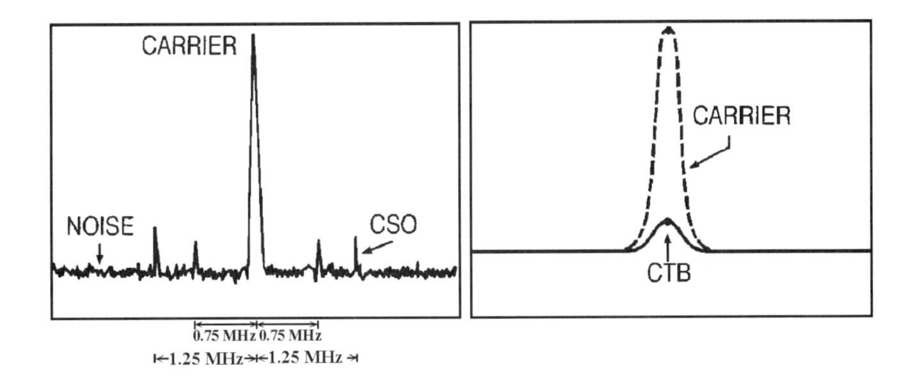

Figure 2.11 (a): Product counts versus channel frequencies for CSO and CTB, respectively. The short markers above the distribution curves indicate the frequency of each video carrier.

second-order distortions are higher than third-order distortions, and a typical laser diode could generate HD_2 in the range of –85 to –75 dBc.

Recall that, in some channels, there is a slight frequency offset (e.g., 12.5 or 25 kHz) from the nominal carrier frequency to prevent interference with aeronautical navigation signals, and so not all intermodulation products will fall exactly on the same frequency. Then how can Eq. (2.2), which uses the concept of product count, be valid? The reason is that CTB and CSO are generally measured with a spectrum analyzer whose resolution bandwidth is set to 100 kHz (for unmodulated carriers [14]), and 100 kHz is wide enough to include all beat products around the nominal frequency under consideration. Thus, the beat products can be added up in power even they are offset by a few tens of kilohertz.

Note that CTB distortion is perceptible because the beats among channels fall near and not exactly on the interfered carrier. This is owing to the fact that the carriers of the various channels have small frequency variations from the precise channel assignment, that is, $(6L + 1.25 + \Delta f_1) + (6M + 1.25 + \Delta f_2) - (6N + 1.25 + \Delta f_3) = (6K + 1.25 + \Delta f_4)$, where L, M, N, K are integers, and Δf_i's ($i = 1,2,3,4$) represent phase noise terms of each carrier. The subjective effects of CTB distortions can be reduced by locking all channel carriers to harmonics of a single reference [15–17]. This technique is referred to as harmonically related carriers (HRCs). HRCs can help increase about 5 dB additional trunk output signal level for a subjectively equivalent CTB distortion. Note that HRC systems lock all carriers to a harmonic of 6 MHz. This requires placing the channel 2 carrier on 54 MHz instead of the normally assigned frequency of 55.25 MHz. This problem can be avoided by using an incrementally related carrier (IRC) system, where all carriers are locked to $6N + 1.2625$ MHz. Note, however, that only 10% to 20% of the North American CATV operators use HRC or IRC systems because of the cost issues.

The visual effects due to A + B and A – B type CSOs are quite different. The reason is that A + B type CSO falls directly into the actual video portion of a signal.

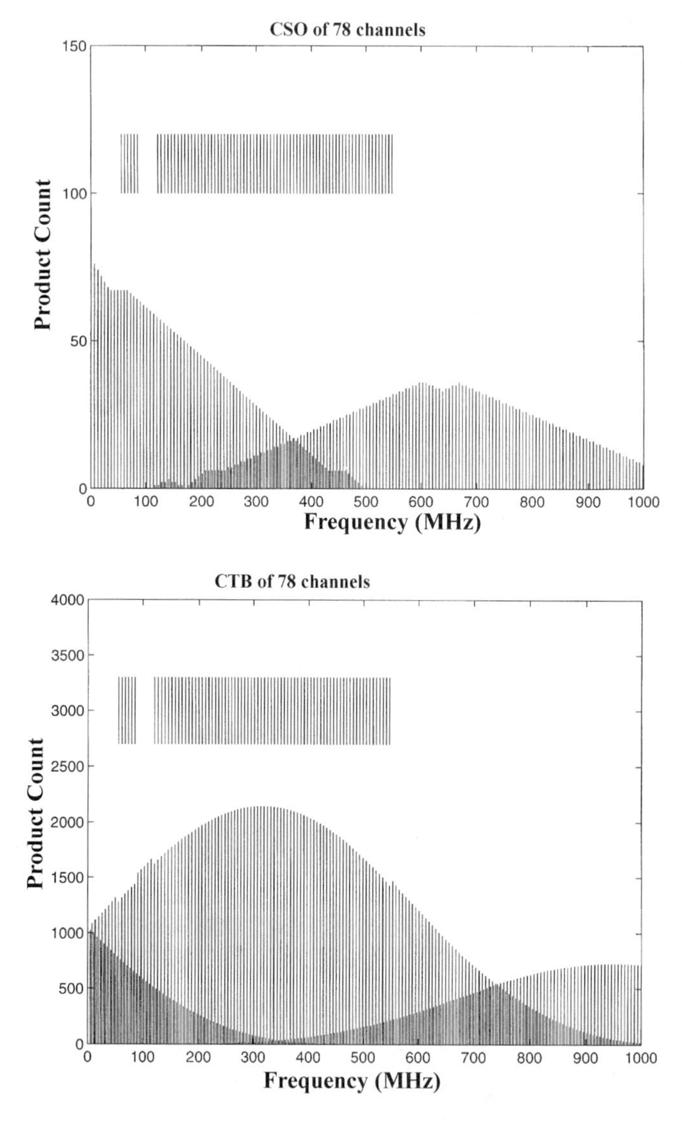

Figure 2.11 (b): Frequency locations of CSOs and CTBs in a standard CATV channel.

This can be seen from the fact that $(6L + 1.25) + (6M + 1.25) = 6N + 2.5$, which shows that the CSO occurs 2.5 MHz from the edge of an AM-VSB signal. On the other hand, A – B type CSOs occur right at the edge of an AM-VSB signal because $(6L + 1.25) - (6M + 1.25) = 6N$, which is less visible and therefore not as damaging.

Cross modulation (XM) is the transfer of modulation from one or more carriers to another in a multichannel system; it occurs as a result of the triple beat due to, for instance, a signal carrier, an interfering carrier, and its sideband. For example, the triple beat among $(6L + 1.25)$, $(6M + 1.25)$, and $(6M + 1.25 + \Delta)$ (where Δ ranges from 0 to ~4 MHz) can occur in $(6M + 1.25 + \Delta) - (6M + 1.25) + (6L + 1.25) = 6L + 1.25 + \Delta$ and can degrade the video quality in the frequency band of $6L + 1.25$ to $6L + 5.25$ MHz.

Generally speaking, CTB, CSO, and XM should all be below about –61 to –65 dBc for trunk systems, and below about –54 dBc for distribution to customers' premises.

2.3.2 CNR and Nonlinear Distortions in a Single Amplifier

The CNR for a given single amplifier is

$$\mathrm{CNR} = P_{\mathrm{in}}/(kTB) = 59 - F + V_{\mathrm{in}}\ (\mathrm{dBmV}), \tag{2.3}$$

where P_{in} and V_{in} are the power and rms amplitude of an input signal, respectively, k is Boltzmann's constant (1.38×10^{-23} joule/K), T is the absolute temperature, B is the signal noise bandwidth (4 MHz), and F is the amplifier noise figure. Note that the number 59 was obtained from $-10 \log(kTB) - 48.75$ (where 48.75 was used to convert dBm to dBmV). Typical noise figures of trunk amplifiers range from 7 to 9 dB; typical input level of signal is +10 dBmV. Therefore, for an amplifier with a noise figure of 7 dB, its achievable CNR is 62 dB.

Typical CTB and CSO – 85 dBc for single trunk amplifier, and 60 dB for bridging and line-extender amplifiers.

2.3.3 Cascaded Amplifiers

Consider a trunk system with n identical cascaded amplifiers, each with a noise figure of F and a gain G and followed by a length of cable with loss L, as shown in Fig. 2.12. Let $G = L^{-1}$ and $F >> 1$; then, from conventional transmission system theory [18], the total noise figure for n cascaded amplifier stages is given by

$$\begin{aligned} F_n &= F_1 + \frac{G-1}{G} + \frac{F_2 - 1}{GL} + \frac{G-1}{GLG} + \ldots \\ &\approx nF + n - (n-1) \\ &\approx nF \end{aligned} \tag{2.4}$$

Therefore, we have

$$NF_n = NF_1 + 10 \log n \tag{2.5a}$$

and

$$\mathrm{CNR}_n = \mathrm{CNR}_1 - 10 \log n. \tag{2.5b}$$

Note that $NF_i = 10 \log(F_i)$. For example, if the CNR_1 given by a single amplifier is 61 dB, then cascading 10 of these amplifiers (gain = interstage loss) will give a total CNR_{10} of $61 - 10 \log 10 = 51$ dB.

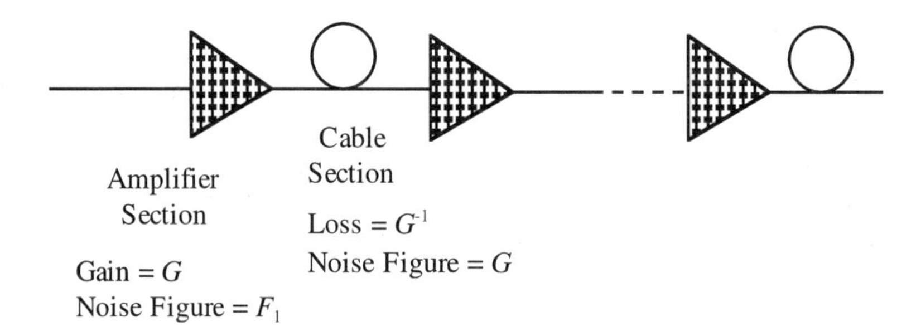

Figure 2.12: Configuration for calculating effective noise figure of cascaded amplifiers.

To estimate the CNR of a system which consists of unlike amplifiers, one can simply follow the calculation

$$\mathrm{CNR} = -10 \log \left(10^{-\mathrm{CNR}_1/10} + 10^{-\mathrm{CNR}_2/10} + \cdots\right). \tag{2.6}$$

On the other hand, the nonlinear distortions build up in a cascaded amplifier chain with respect to the voltage. Assuming identical nonlinear distortions in all n amplifiers, then

$$\mathrm{NLD}_n = \mathrm{NLD}_1 + 20 \log n. \tag{2.7}$$

For example, assume that an individual amplifier produces –85 dBc CTB. Then, a cascade of 10 amplifiers will produce CTB of –65 dBc. To combine amplifiers with unequal NLDs, the resultant NLDs can be obtained from

$$\mathrm{NLD} = -20 \log \left(10^{-\mathrm{NLD}_1/20} + 10^{-\mathrm{NLD}_2/20} + \cdots\right). \tag{2.8}$$

For example, when an amplifier with XM of –59.5 dBc and a second amplifier with XM of –54.5 dBc are cascaded, the resultant XM is –50.6 dBc.

2.4 Characteristics of Current CATV Return Path

The biggest problem in providing reliable two-way communication services with today's cable system is that the 5–30 or 5–42 MHz return band (5~65 MHz in Europe) is not only too narrow, but also full of all kinds of uncorrelated and correlated noise. Because the currently available return-path spectrum is narrow, the number of subscribers that each fiber node serves should not exceed an upper limit, which depends upon how many different services (data, telephony, etc.) will be provided and how busy the traffic may be. As far as noise and ingress is concerned, a return path in 5–42 MHz generally has funneling noise arriving at the headend due to all noise terms which were generated from many subscribers' houses and/or picked up by antenna-like properties of the distribution plant (mainly due to poorly connectorized and poorly shielded in-house cables). About 70% of the noise comes from the subscribers' homes, 25% from the tap to the ground block, and 5% from the hard coaxial plant [19]. This noise and ingress can be classified into the following categories:

- Ingress noise (duration > 1 ms) due to narrowband shortwave signals, AM radio, ham radio, land mobile radio, paging signals, etc. The intensity of these ingress noise depends on the varying conditions of ionosphere from time to time, and is expected to increase significantly at the next peak of the 11 year-sunspot cycle [20] which will be around early 2000s.
- Impulse noise (duration << 1 ms) [21] due to corona discharges from power lines (often located on the same poles or conduits as the CATV cable); discharges across corroded connector contacts; automobile ignition; and the turn-on and turn-off of electric facilities at homes. Shown in Fig. 2.13 is a measured short burst lasting for a few microseconds due to a vacuum cleaner at a subscriber's house. Measurement across the return-path spectrum of a cable plant shows that the majority of impulse noise has a burst duration less than about 5 μs [22].

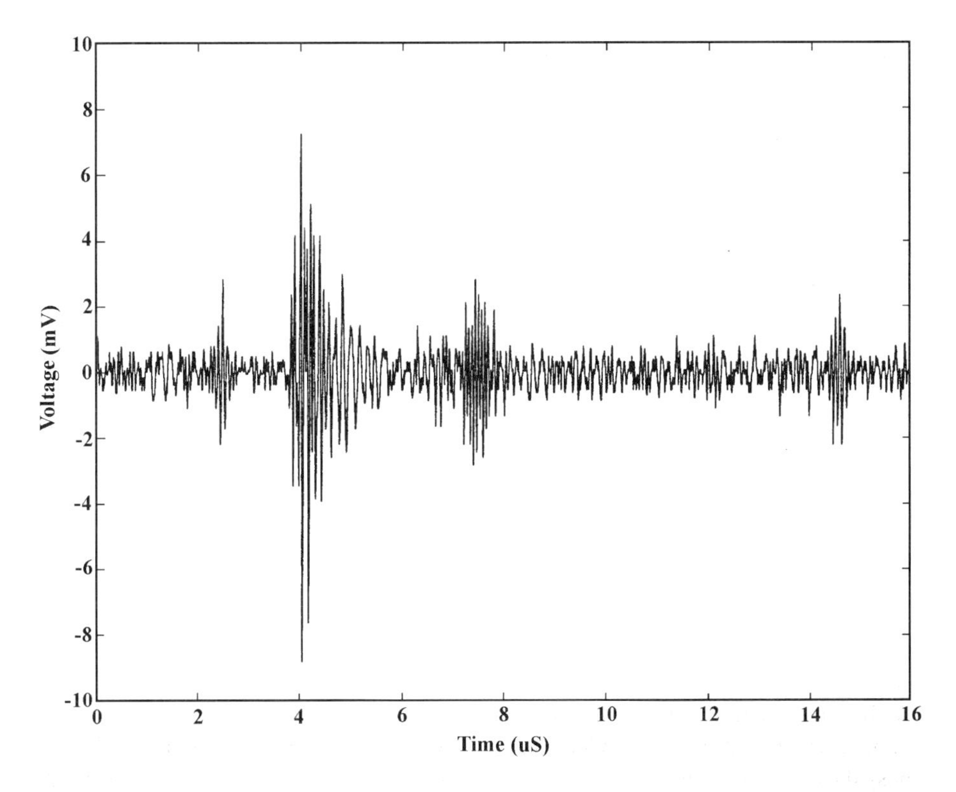

Figure 2.13: Measured short burst caused by turning on a vacuum cleaner in a subscriber's house. The burst signal was measured at a CATV headend via a return path consisting of eight amplifiers.

- Common-mode distortion due to nonlinearities in the plant (e.g., oxidized connectors) [23].

A typical long-term swept return-channel spectrum (measured at headend) is shown in Fig. 2.14 [23]. Observing the high levels of ingress noise, we can see that not all 5–42 MHz spectrum is available for two-way communication services (the frequency region above 20 MHz is cleaner). According to Fig. 2.14, the frequency range above about 15 MHz is more suitable for adding new two-way services.

It is important to carry out spectrum sweeping and monitoring for certain length of time before deciding where the active channels should be located. A statistical ingress frequency distribution measured over 168 hours is shown in Fig. 2.15. It can be seen that although it is still possible to have ingress occur above ~15 MHz, its probability is very low.

Currently, there are several techniques and proposals to overcome the funneling impulse and ingress noise problem. For impulse noise, field-proven techniques include adding interleavers and/or forward-error-correction (FEC) codes (see Chapter 8) to return-data channels, or the active channel can automatically hop to another frequency where noise level is low (frequency agility). For ingress noise, more com-

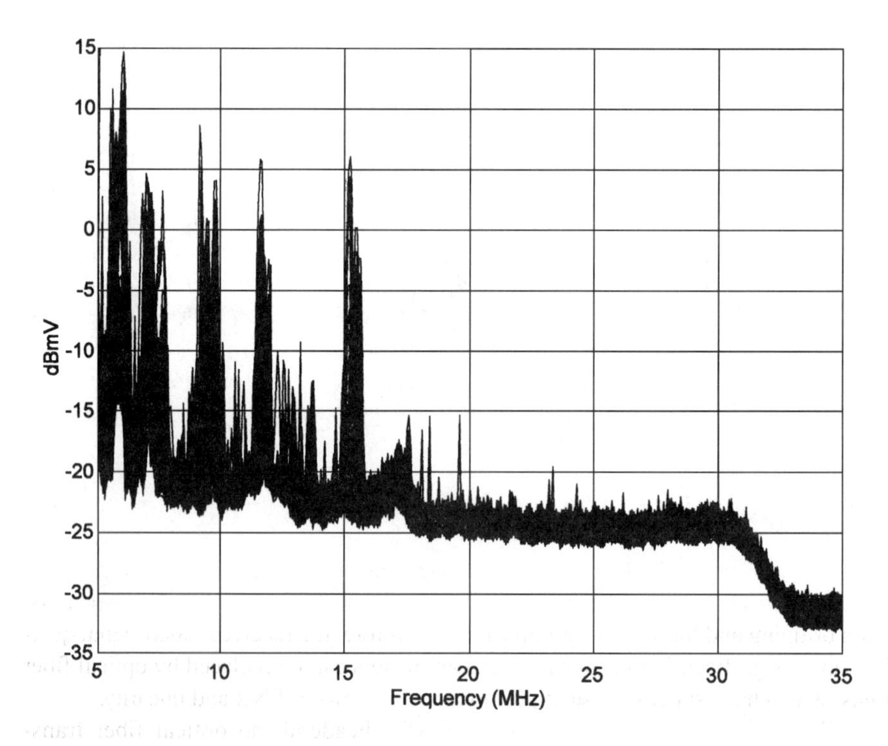

Figure 2.14: Measured ingress noise at a CATV headend via a return path consisting of eight amplifiers. The measured period was 24 hours.

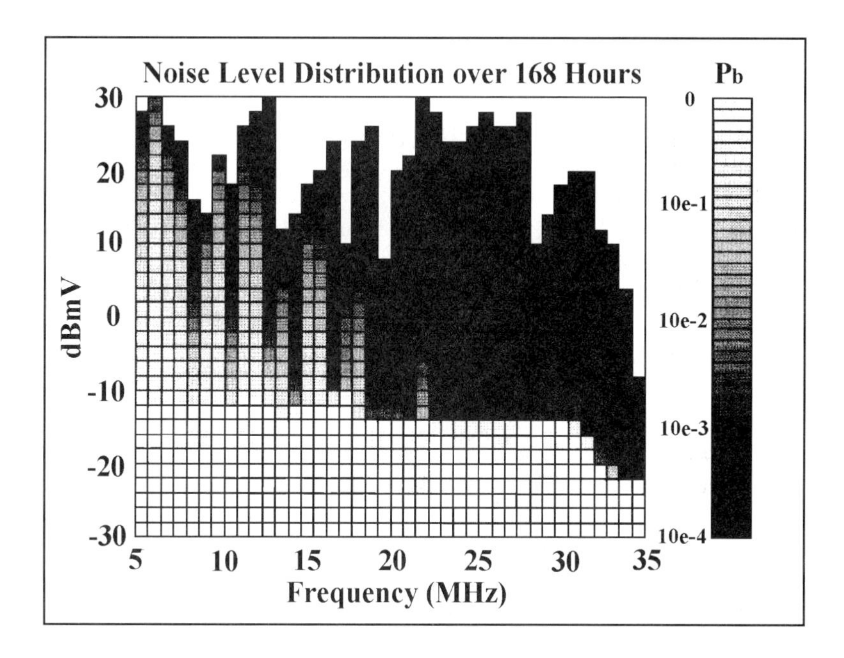

Figure 2.15: Probabilistic frequency distribution of ingress noise measured over 168 hours.

plicated proposals have also been set forth, such as using synchronous code-division-multiple-access, synchronous discrete-multi-tone [24], or transmission sampling gates (located near the return-path amplifiers) which would deter upstream flow of noise impairments by only opening when a subscriber modem is transmitting upstream bursts [25]. Longer-term proposals include using the bandwidth beyond 550 or 750 MHz for return-channel path, because there is very little radio interferences in that frequency range.

An upstream channel model, which includes all the factors described in Sec. 2.2.4, 2.3 and 2.4, is proposed in IEEE 802.14 [21] and is shown in Fig. 2.16.

2.5 System Upgrade by Optical Fibers

As mentioned in Section 2.2, most of the noise components in a CATV distribution system are from trunk amplifiers, and most of the nonlinear distortions are generated from bridging and line-extender amplifiers. Therefore, the received video signal quality can be significantly improved if these amplifiers can be replaced by optical fiber links, which have superior system performance in terms of SNR and linearity.

Obviously, for customers near the CATV headend, no optical fiber transmissions are necessary. The question then is, what is the threshold distance to use optical fibers instead of coaxial cables? Today's 80-channel optical-fiber link

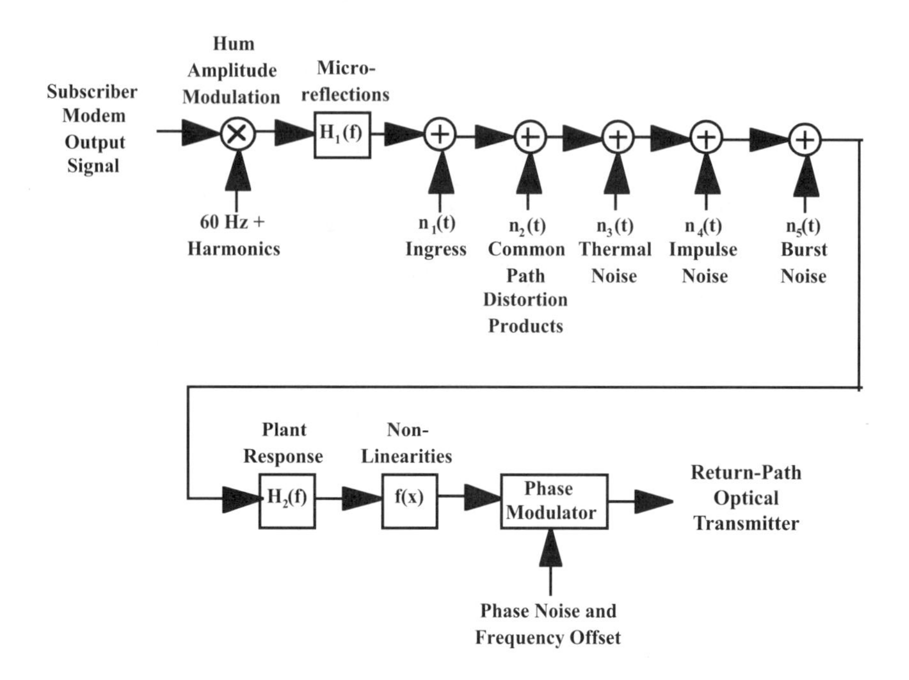

Figure 2.16: Upstream channel model for the coaxial portion of an HFC system. (After Ref. [21])

performance has typical parameters such as (after optical receiver) CNR > 51 dB, CTB/CSO < –63 dB; today's typical trunk amplifiers exhibit CNR > 61 dB and CTB/CSO < –85 dBc. Therefore, according to Eqs. (2.5) and (2.7), 10 cascaded trunk amplifiers could achieve about the same performance as a typical CATV optical-fiber link. That is, when a system requires more than 10 trunk amplifiers to reach customers, it is about time to use optical fiber links.

However, the deployment of deeply penetrated optical fibers cannot be achieved in a short period of time, for economic reasons; it must be done stage by stage. In this section, we will discuss the evolutionary processes of optical fiber deployment: from existing coaxial cable systems to future brand-new systems. For existing systems, we can reuse most of the existing amplifiers in the following two different architectures [26]:

Fiber-backbone: The purpose of this architecture was to divide the network into smaller service zones and in the same time reuse the original trunk amplifiers. Each service zone was fed by an optical link from the headend. As shown in Fig. 2.17, some of the original trunk amplifiers were reversed. The number of cascaded trunk amplifiers after the optical receiver is limited to 4–8. Therefore, because of the decreased number of cascaded number of amplifiers and the high-quality AM optical

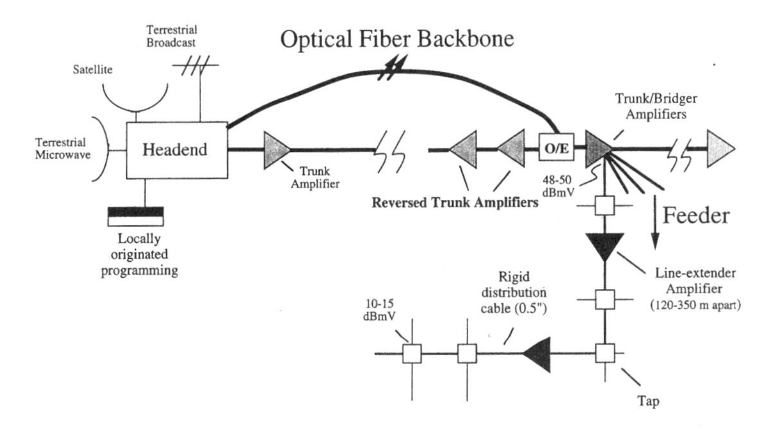

Figure 2.17: Optical-fiber backbone.

fiber system, the received video quality can be improved. In the mean time, system upgrade by using fiber-backbone can be much more economical than completely rebuilding a new optical-fiber system.

Fiber-overlay: Networks with optical fiber overlay use the original trunk amplifiers as a backup in case the optical fiber system fails. Its architecture is shown in Fig. 2.18, from which we can see that all trunk amplifiers remain in the forward direction, and not reversed as in the fiber-backbone architecture. Each optical node serves more than 1000 customers. There is a sensing switch in each optical receiver. If the switch senses that the optical signal is lost, it will switch back to the original coaxial cable plant. By selecting a route for the fiber cable which does not duplicate

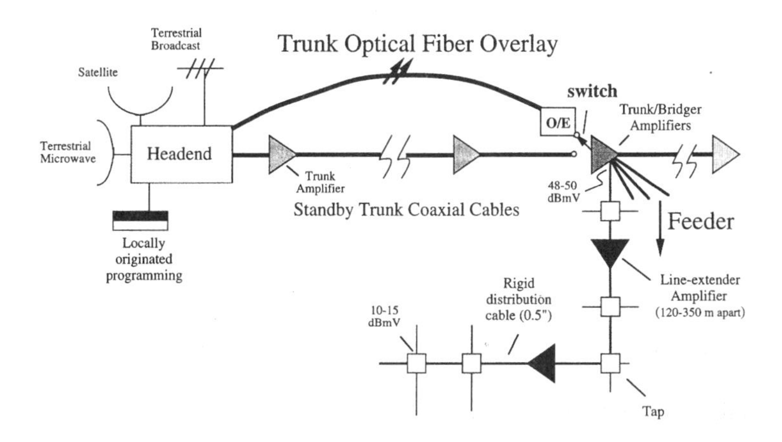

Figure 2.18: Optical-fiber overlay.

the path of the coaxial trunk system, the chance of both cables being cut at the same time is dramatically reduced.

For cable systems that are degraded to a point that the cables and the amplifiers cannot be reused, we may want to try one of the following advanced architectures:

Fiber-to-the-feeder: The architecture is shown in Fig. 2.19. We have a fiber link to each service area where 300 to 500 customers are served. The output of the optical receiver provides four outputs with the same voltage levels as a bridger amplifier. Conventional feeder line amplifiers or distribution amplifiers are used to increase the area which can be fed by a given node (no large trunk cables or trunk amplifiers are used in this architecture). Note that the number of cascaded feeder amplifiers is limited to three or four; therefore, very good picture quality and system reliability can be achieved. Another important factor to consider is that lower-quality AM laser transmitters and receivers can be used because the number of cascaded amplifiers after the optical node is limited to fewer than four. This gives great potential for cost savings.

Passive fiber coax (fiber-to-the-line extender and fiber-to-the-tap): Both the fiber-to-the-line extender architecture and the fiber-to-the-tap architecture (Fig. 2.20) do not use any active components after the optical node, hence the name "passive fiber coax." The number of customers who can be served is around 50–100 for the former, and around 4–8 for the latter. It is obvious that these architectures should be considered only for completely new construction. For fiber penetration to such a deep level, both signal quality and reliability will be excellent, but the cost will surely be increased even though we can use a even lower grade of AM laser transmitters and receivers.

Ring architecture: The reliability of service provided to subscribers is a significant problem in cable systems. Not only the tree-and-branch architecture, but also the star architecture is unreliable when optical-fiber cables are cut accidentally. Al-

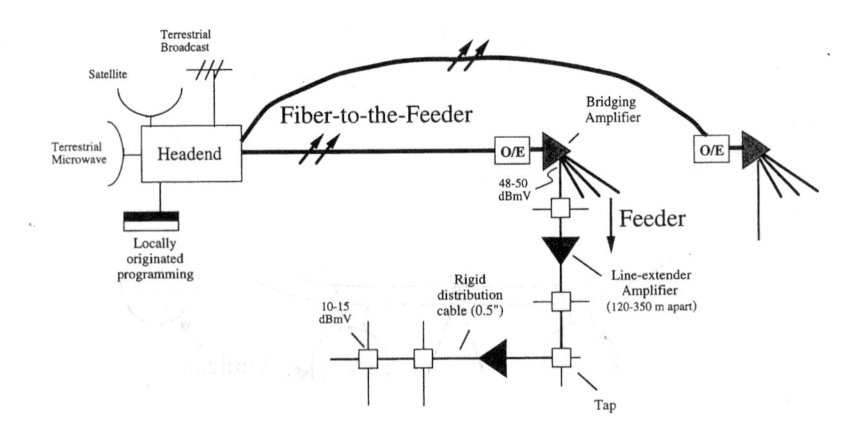

Figure 2.19: Fiber to the feeder.

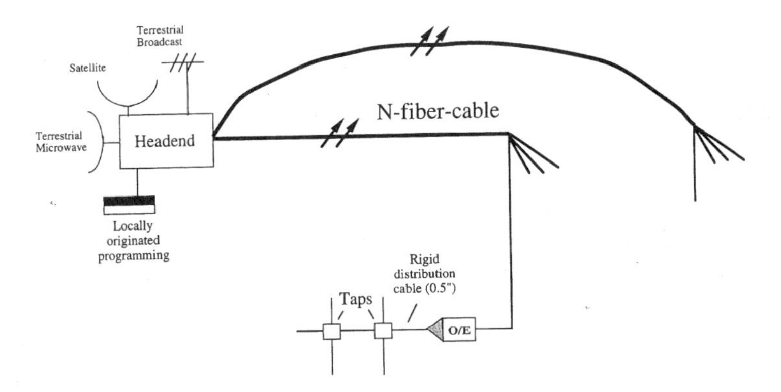

Figure 2.20: Passive fiber coax architecture (fiber to the line-extender or fiber to the tap).

though a star-architecture-based optical fiber trunk and feeder systems can minimize the number of subscribers affected, the capability to detect and repair in an extremely short time (e.g., 50 ms by SONET/SDH standards) any failures in the last few hundred feet of coaxial cables is critical, in order to compete with today's highly reliable telephone networks. It is for this reason that ring architecture in the subscriber loop for HFC systems can be a more reliable and economical alternative to the passive fiber coax networks shown in Fig. 2.20. This ring architecture is illustrated in Fig. 2.21. Each optical node has an optical receiver with a 1×2 optical switch at the input. Once a "loss of signal" is detected in the active fiber ring, the

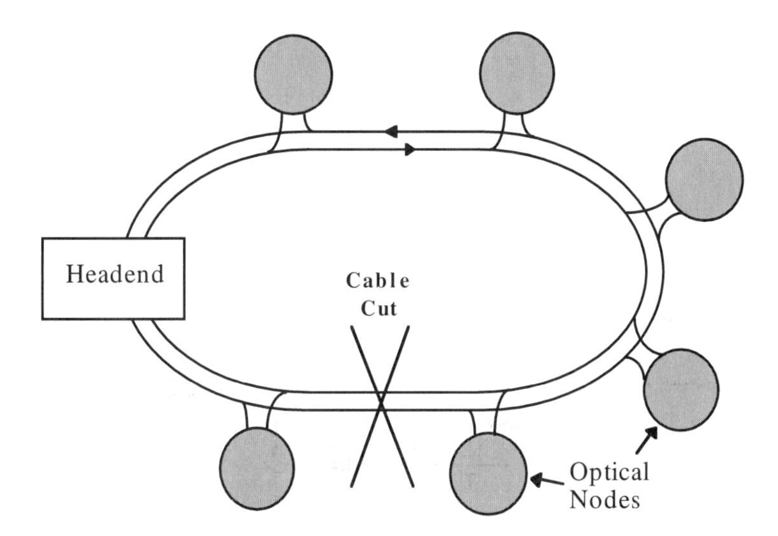

Figure 2.21: Ring architecture which is self-healing under a cable-cut condition.

switch automatically turns to the standby fiber ring. We can see that no matter where the fiber cut may occur, all optical nodes will continue to receive signals from the CATV headend. This kind of architecture, of course, will require a transmitter with much higher-output optical power, because of the higher power budget needed for longer distance transmission and for power splitting along the ring.

So far we have not specifically address how return-path transmitters can be used in the various architectures. A basic fact to remember, however, is that due to the advance of WDM technologies (see Chapter 3) it is likely that for downstream optical transmitters in the 1.55 μm region, the corresponding return-path transmitters should be operating in the 1.3 μm region, and vice versa. By using WDM techniques, the number of required single-mode fibers in HFC networks can be significantly saved.

2.6 Next-Generation Cable Network Architecture

The current CATV architecture is just a simple tree-and-bus configuration, and there is basically no direct communications between any cable operators. In order to establish a cable communication network, this "isolated" operation mode must be changed. Today, we can see that there are many islands of information in the form of corporate, educational, and health-care campuses which have extensive information-processing, storage, and networking capabilities. The opportunity for the CATV industry is to extend, expand, and connect these islands into a more contiguous and ubiquitous infrastructure providing services to more people "off-campus" by utilizing the CATV network as an information highway.

Currently, a CATV operator provides services without any dependence on other operators, even though the cable operators may obtain some or most of their source material from the same programming providers. Hence, a duplication of satellite feeds, off-the-air equipment, microwave facilities, and tens to hundreds of VCRs/modulators is required. The result is slow implementation of new service opportunities because of the amount of money that has to be invested. To change this condition, centralization of capital-intensive investments for a range of advanced functionalities at a "super hub" or "regional hub," in combination with a metropolitan-area-networking (MAN) or wide-area-networking (WAN) interconnection, allows the cable operators to spread the investments across a wider base. It also provides a platform for offering a common set of functionalities to large and small cable operators. It allows multiple cable operators to share the investments and risks associated with providing advanced applications such as IP telephony, video-on-demand, and PCS switching and cross-connecting.

In addition, with the centralization of advanced functionality in the regional hub, it is economical to have duplicate copies of systems, such as mass storage, video servers, and switching. This system redundancy is important to increase the operation reliability in the new architecture.

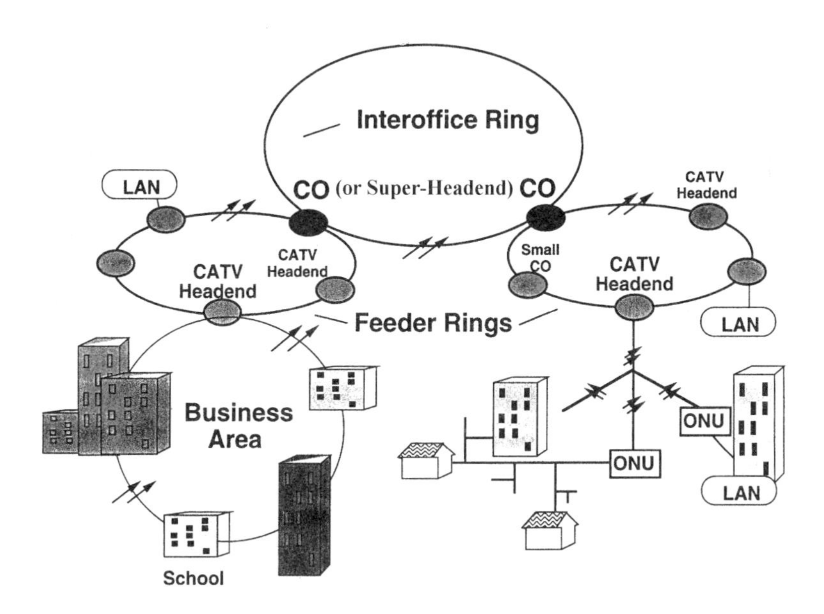

Figure 2.22: Next-generation HFC architecture with headends interconnected by high-capacity rings such as SONET/SDH self-healing rings. (ONU: optical network unit or optical node)

Since the regional hub or super hub serves a gateway to other networks, that is, the Internet, the inter-exchange carriers, the local exchange carriers, etc., the gateway must be interoperable with other network types. The choice of transport protocols, therefore, is important to ensure interoperability. SONET or SDH may be used for the survivable, self-healing rings (see Fig. 2.22). The choice of modulation formats, forward-error-correction codings, bit rates, etc., has been standardized by various standard bodies [6], but would need further convergence on many details.

Problems

1. Design a computer program to plot a diagram similar to Fig. 2.11b for 78 channels. Channel number is assigned as shown in Table 2.2 except that channels between 88 and 120 MHz should be left unused. Note that we assume all channels have picture carriers at 6X + 1.25 (MHz), where X is a positive integer. Show that the dominant CTBs are from the type $f_a + f_b - f_c$.

2. Besides push-pull and feedforward amplifiers, parallel hybrid and quad amplifiers are often used in CATV systems. Survey the literature and describe the operation principle of these two types of amplifiers.

3. Using Eqs. (2.6) and (2.8), calculate how many line-extenders can be used after an optical fiber trunk system and a bridger amplifier. Assume the following

conditions: After the optical receiver, CNR = 51 dB, CTB/CSO = –63 dBc. A parallel hybrid bridger amplifier with a noise figure of 7.5 dB, CTB = –66 dBc, CSO = –63 dBc, is used immediately after the optical receiver (see Fig. 2.19). Each line extender that will be used after the bridger amplifier exhibits a noise figure of 9 dB, both CSO and CTB = –67 dBc. We further assume that the required signal quality at a subscriber's premises are CNR = 46 dB and CTB/CSO < –54 dBc. The power of the input signals to the bridger amplifier and line extenders are 16 and 14 dBmV, respectively.

4. Survey the literature and related textbooks, and briefly describe the operation principles of synchronous code-division-multiple-access (S-CDMA) and discrete-multitone (DMT) techniques, respectively, which could potentially find important applications in upstream HFC systems to combat strong ingress noise. Do you have any new ideas about coping with the ingress and impulsive noise originated from subscribers' premises?

5. To characterize laser transmitter or CATV amplifier linearities, there are two types of testing signal sources: multichannel modulated video signals or multichannel unmodulated CW carriers. It was experimentally found that, for the same CNR performance, CTB and CSO can be lowered by about 10 and 6 dB, respectively, when using modulated video signals. Therefore, a fiber system measured using CW carriers for CNR = 51 dB, CTB = –65 dBc, and CSO = –62 dBc, would yield performance specifications of CNR = 51 dB, CTB = –75 dBc, and CSO = –68 dBc when using modulated carriers. In this case (with modulated carriers), suppose we want to increase the laser RF drive level to improve CNR performance while in the same time maintain CTB ≤ –65 dBc and CSO ≤ –62 dBc, what is the best CNR we can achieve?

References

[1] A. S. Taylor, "The vestigial sideband and other tribulations," NCTA 37th Annual Convention and Exposition Technical Papers, Washington, D.C., April/May 1988.

[2] *Code of Federal Regulations, Title 47, Telecommunications, Part 76, Cable Television Service*. Federal Communications Commission Rules and Regulations, 1990.

[3] *NCTA Recommended Practices for Measurements on Cable Television Systems*, 2nd Edition, National Cable Television Association, Washington, D.C., October 1993.

[4] T. Pratt and C. W. Bostian, *Satellite Communications,* Wiley, New York, 1986.

[5] D. Minoli, *Video Dialtone Technology*, Chapter 6, McGraw-Hill, New York, 1995; L. Chariglione, "MPEG: A technological basis for multimedia

applications," IEEE Multimedia, vol.2, pp. 85–89, Spring 1995; also, see http://www.cselt.stet.itlufvAeonardo/mpeg/index.htm.

[6] ITU-T Recommendation J.83 (Digital multi-programme systems for television sound and data services for cable distribution); MCNS (http://www.cablemodem.com); DAVIC 1.0 Specification (http://www.davic.org/DOWNI.htm); IEEE 802.14 (http://walkingdog.com/catv/index.html).

[7] C. Cuttner, V. Conanan, and J. Vartanian, "Real world results: Digital video compression field tests—analog versus digital," NCTA 42nd Annual Convention and Exposition Technical Papers, San Francisco, California, pp. 402–410, June 1993.

[8] K. Laudel, E. Tsui, J. Harp, A. Chun, and J. Robinson, "Performance of a 256-QAM demodulator/equalizer in a cable environment," Technical Papers, pp. 283–304, 43rd NCTA Annual Convention and Exposition, New Orleans, Louisiana, 1994.

[9] R. Olshansky, "Options for video delivery," OFC'94 Tutorial.

[10] G. Luettgenau, A. Morawski, and W. Inouye, "400 MHz: A challenge for the hybrid amplifier," Technical Papers, 29th NCTA Annual Convention and Exposition, Washington, D.C., 1980.

[11] K. A. Simons, "The optimum gain for a CATV line amplifier," Proc. IEEE, vol.58, pp.1050–1056, July 1976.

[12] L. Thompson, R. Pidgeon, and F. Little, "Supertrunking alternatives in CATV," IEEE LCS, vol.1, pp.26–31, Feb. 1990.

[13] CableLabs, Inc., Test Procedures for the FCC Advisory Committee for Advanced Television Service (ACATS), Louisville, Colorado.

[14] J. L. Thomas, *Cable Television Proof-of-Performance*, Hewlett-Packard Professional Books, Prentice Hall PTR, Englewood, New Jersey, 1995.

[15] W. Krick, "Improvement of CATV transmission using an optimum coherent carrier system," 11th International TV Symposium, Montreux, Switzerland, May 1979.

[16] I. Switzer, "Phase phiddling," NCTA 23rd Annual Convention Official Transcript, Washington, D.C., 1974.

[17] G. D. Alley and Y. L. Kuo, "Optimal control of intermodulation distortion in hybrid fiber–coaxial CATV systems," Tech. Digest, 49th ARFTG Conference Digest, pp.46–55, June 1997.

[18] Members of the Technical Staff, *Transmission Systems for Communications*, 5th Edition, Bell Telephone Laboratories, Holmdel, New Jersey, 1982.

[19] T. J. Staniec, "Return Systems 102: What goes around . . . ," Commun. Eng. and Design, pp.62–75, December 1996.

[20] National Geophysical Data Center (http://www.ngdc.noaa.gov).

[21] B. Currivan and B. Xenakis, "Considerations for upstream channel communications in CATV systems," Electron. Design, pp.67–74, February 1997.

[22] K. Haelvoet, D. De Bal, K. De Kesel, B. Vanlandschoot, and L. Marrtens, "Procedure for measurement and characterization of upstream channel noise in CATV networks," Tech. Digest, 49th ARFTG Conference, pp.19–23, Denver, Colorado, June 1997.

[23] C. A. Eldering, N. Himayat, and F. M. Gardner, "CATV return path characterization for reliable communications," IEEE Commun. Magazine, pp.62–69, August 1995.

[24] R. Gross, "The HI_PHY," *Commun. Eng. and Design Magazine*, pp.40–50, March 1998.

[25] R. Wells, "Return path noise: Testing tool aids diagnosis," *Commun. Eng. and Design*, pp.24–37, January 1997.

[26] J. A. Chiddix, J. A. Vaughan, and R. W. Wolfe, "The use of fiber optics in cable communications networks," J. Lightwave Technol., vol.11, pp.154–166, January 1993.

PART II

Components and Modules

Chapter 3

Principles of Passive Optical Fiber Components and WDM Filters

Passive optical fiber components include single-mode optical fibers, couplers, wavelength-division multiplexers (WDMs), fiber-Bragg-grating (FBG) based filters, optical fiber connectors and splices, optical attenuators, optical fiber isolators, and circulators. This chapter will cover the basics of these components. In addition, planar lightwave circuits (PLCs) based on silica on silicon technologies [1–3] will be constantly referred to because of their increasingly important role in the local-access passive optical networks (PONs).

3.1 Single-Mode Optical Fibers

Multimode fibers cannot be used to transport multichannel AM-VSB and *M*-ary QAM channels with targeted CNR, because the associated modal noise [4] and inter-modal dispersion can severely degrade the system performance. Therefore, HFC systems must use single-mode (SM) fibers. Millions of miles of "conventional" (which means "dispersion-unshifted" or "1310-nm-zero-dispersion") SM fibers have been installed worldwide, and typical specifications such as loss, wavelength-dependent loss variations, chromatic dispersion, and macrobending loss can be found in Table 3.1 [5] and Fig. 3.1 [6], respectively. Optical loss in state-of-the-art SM fibers is primarily caused by the intrinsic Rayleigh scattering of the silica glass, which decreases with the inverse fourth power of wavelength, as can be seen in Fig. 3.1. Beyond 1600 nm, a rapidly increasing absorption is due to the intrinsic infrared tail of the Si–O and/or Ge–O vibrations. Figure 3.1 also shows that the average losses obtained in a manufacturing environment are closer to 0.35 and 0.21 dB/km at 1310 and 1550 nm, respectively. Wavelength-dependent loss variations should be as small as 0.1 dB/km for a range of 45 nm near 1310 nm, and smaller than 0.05 dB/km for a range of 50 nm near 1550 nm. The chromatic dispersion

Table 3.1: Single-mode Fiber Specifications[a]

- The maximum attenuation in the wavelength region from 1285 to 1330 nm is no more than 0.1 dB/km greater than the attenuation at 1310 nm. The maximum attenuation from 1525 to 1575 nm is no more than 0.05 dB/km greater than the attenuation at 1550 nm.
- The maximum fiber attenuation at 780 nm is 3 dB/km.
- The zero-dispersion wavelength (λ_0) is 1310 ± 10 nm. The maximum dispersion is 2.8 ps/nm/km in 1285–1330 nm. The maximum dispersion at 1550 nm is 18 ps/nm/km. The maximum value of the dispersion slope at λ_0 is 0.092 ps/nm^2/km.
- Macrobending loss: The attenuation of 100 turns of fiber 75 mm in diameter does not exceed 0.5 dB at 1550 nm and 0.1 dB at 1310 nm.

[a]After Ref. [5].

(caused by the dependence of index of refraction on wavelength in glass fibers) of SM fibers consists mainly of material and waveguide dispersion and is defined as

$$D = (\frac{1}{l})(\frac{dt_g}{d\lambda})\ (\text{ps/nm/km}), \tag{3.1}$$

where l is the length of the fiber, and dt_g is the incremental group delay due to a wavelength difference $d\lambda$. As shown in Fig. 3.2 [7], the zero material dispersion of a fused silica SM fiber occurs at 1273 nm. Near this wavelength region, waveguide dispersion which results from the modal velocity being a function of a/λ (a is the radius of the fiber) becomes important [8]. When a negative waveguide dispersion is added to the material dispersion, we can obtain a zero dispersion either around 1310

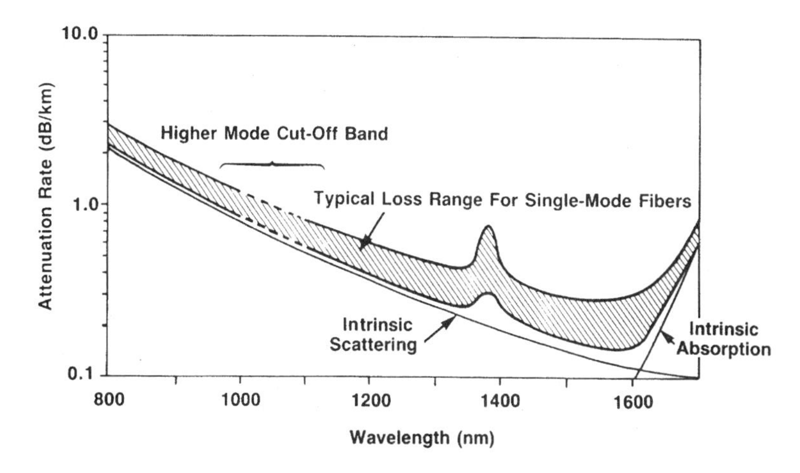

Figure 3.1: Optical attenuation (dB/km) versus wavelength. After Ref. [6].

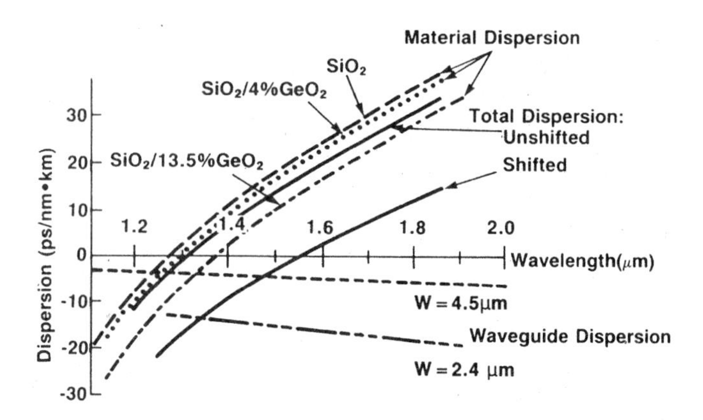

Figure 3.2: Optical fiber material and waveguide dispersion as a function of optical wavelength. The two solid lines represent the total dispersions for conventional (unshifted) and dispersion-shifted optical fibers, respectively. After Ref. [6].

nm for dispersion-unshifted fiber, or around 1550 nm for dispersion-shifted fiber. Waveguide dispersion becomes more pronounced for smaller core diameters and/or index differences. Since the mode-field radius (the radius at which the optical intensity decreases to e^{-2} of its peak value) for a dispersion-shifted fiber ($W \sim 2.4\ \mu$m) is much smaller than that of a dispersion-unshifted fiber ($W \sim 4.5\ \mu$m), the former has more negative waveguide dispersion values. Maximum dispersion values specified near 1285–1330 nm must be less than 2.8 ps/nm/km [5]. Note that when one uses a dispersion-unshifted fiber and operates at 1550 nm, not only much higher dispersion (up to 18 ps/nm/km) but also more bending loss may be experienced, as can be seen in Fig. 3.3 [7]. Bending resistance can be improved at 1550 nm if one uses a dispersion-

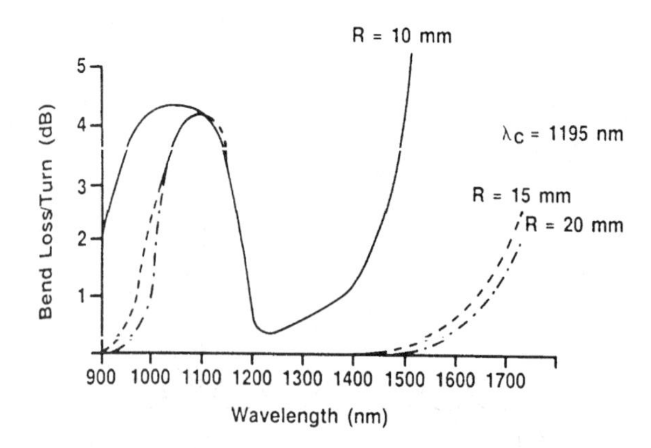

Figure 3.3: Bending loss as a function of wavelength for an unshifted fiber with different bending radius of curvature. (After Ref. [7], with kind permission from Kluwer Academic Publishers.)

shifted fiber instead. Note that the cutoff wavelength(λ_c) marked in Fig. 3.3 is defined as the wavelength above which a given fiber becomes single-mode.

A typical trunk optical cable can have 6 to 240 optical fibers, depending on the penetration depth of optical fibers in an HFC system. A distribution breakout cable can have 4 to 36 optical fibers. The expected service life of an optical cable in the operational environment (buried or aerial cable) is 20 years. The end-of-service-life condition is defined as an 0.2 dB/km attenuation increase at the wavelengths of interest [5].

3.2 Optical Fiber Couplers

A fiber coupler fuses two optical fibers with a fixed coupling length L, as shown in Fig. 3.4, so that the power coupling between the two fibers can be estimated using basic coupled-mode analysis [9] to obtain

$$\begin{aligned} P_1 &= P_0 \cos^2(\kappa \cdot z) \\ P_2 &= P_0 \sin^2(\kappa \cdot z), \end{aligned} \tag{3.2}$$

where P_0 is the power launched into fiber 1 at $z = 0$, and κ is the coupling coefficient between the two fibers. It can be seen that at $z = 0$, $P_1 = P_0$ and P_2 is zero, at $z = \pi/(2\kappa)$, $P_2 = P_0$ and P_1 is zero, etc.; this shows that the two fibers exchange power after a distance L_c given by $\pi/(2\kappa)$, where L_c is the "spatial period" with which the guided energy is periodically transferred back and forth between the two fibers. By varying the coupling length L_c, one can obtain couplers with different coupling power ratios, such as 50/50, or 90/10, in the two output fiber ports.

For a $1 \times N$ fiber coupler, we need $N - 1$ $(= (N/2) + (N/4) + \ldots + 1)$ 1×2 fiber couplers. To build an $N \times N$ fiber coupler, the brute-force way is to cascade two $1 \times N$ couplers back to back. In this case, $2(N-1)$ couplers are required and a huge loss of 3 dB $\times$ 2 $\log_2 N$ is incurred. To reduce the loss through an $N \times N$ fiber coupler, other alternative architectures can be used [10]. An example is that if we let $N = m \times n$ and we use m $n \times n$ fiber couplers cascaded with n $m \times m$ fiber couplers, then a signal going from any input port to any output port passes through $\log_2 N$ stages and the total loss is reduced to 3 dB $\times \log_2 N$, and the total number of 2×2 couplers is $(N/2) \log_2 N$. An example of 8×8 is shown in Fig. 3.5.

The preceding method apparently will make the physical size of an $N \times N$ coupler too cumbersome when N is large. In addition, because of the different excess

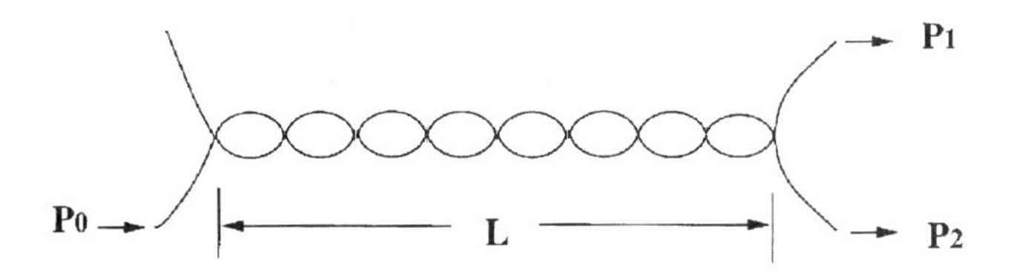

Figure 3.4: A 2×2 fused fiber coupler.

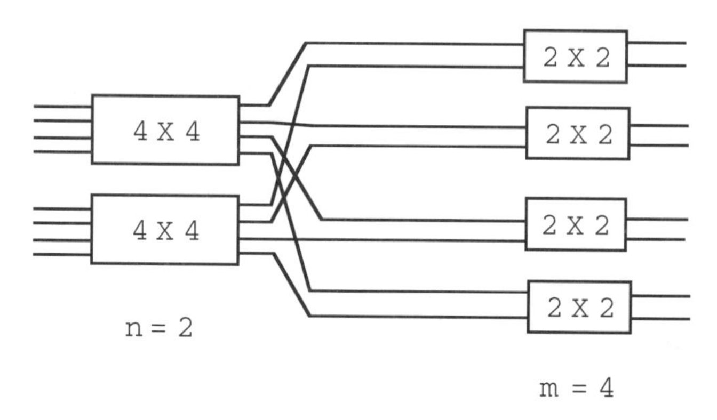

Figure 3.5: An 8×8 fiber coupler made of three stages of 2×2 fiber couplers.

loss of each 2×2 coupler, very nonuniform output power may be presented at the N output ports. This problem can be avoided by using couplers made of silica glass waveguide, which are commercially available, as shown in Fig. 3.6. Its physical size is much more compact than fused couplers and the power levels at each output port are fairly uniform (the power variation is within 1 dB for a 1×8 coupler, 2.5 dB for a 1×32 coupler). However, in this type of coupler, additional optical reflections due to the fiber–waveguide interface must be carefully minimized (commercial products with return loss <–55 dB are available today).

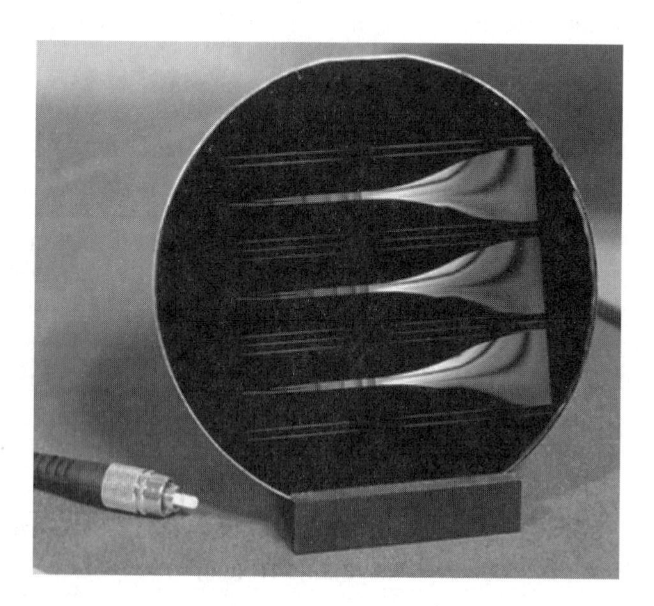

Figure 3.6: A silicon wafer with three 1×32 splitters. (Courtesy of PIRI Inc.)

3.3 Wavelength-Division Multiplexers

Figure 3.7 shows that there are two low-loss windows, around 1.3 and 1.55 μm, in a single-mode optical fiber for telecommunications applications. The total available frequency range is about 60 and 110 nm, respectively. The corresponding electrical bandwidth available can be calculated according to

$$\Delta f = -(f^2 / c)\, \Delta\lambda, \tag{3.3}$$

which gives a Δf of 175 and 125 GHz around the 1.3- and 1.55-μm regions, respectively, for $\Delta\lambda = 1$ nm. Therefore, a single-mode fiber can provide a tremendous bandwidth of $175 \times 60 + 125 \times 110 = 24250$ GHz or ~24 terahertz. It is impossible, at least in the near future, for a single laser transmitter or a photoreceiver to have such a huge bandwidth. Therefore, in order to fully utilize this almost infinite capacity, WDM technology, which uses multiple single-wavelength laser diodes (or fiber lasers) and the corresponding WDM multi/demultiplexers (mux/demuxes) or optical filters, is deemed the most promising approach that can be practically used (other capacity-exhaustive techniques include >100 Gb/s TDM and optical code-division multiplexing).

In the next generation HFC systems, WDM technologies can be used for super-headend and/or headend interconnections (see Fig. 2.22) that require a very large transmission capacity for video programs and Internet traffic.

Mux/demuxes are naturally the essential components of WDM systems. Depending on how fine a wavelength spacing ($\Delta\lambda$) must be resolved, there are various WDM techniques available, as will be introduced next. It should also be noted that for access networks, generally speaking, passive WDM filters are preferred. Nevertheless, we will also briefly review some of those active filters.

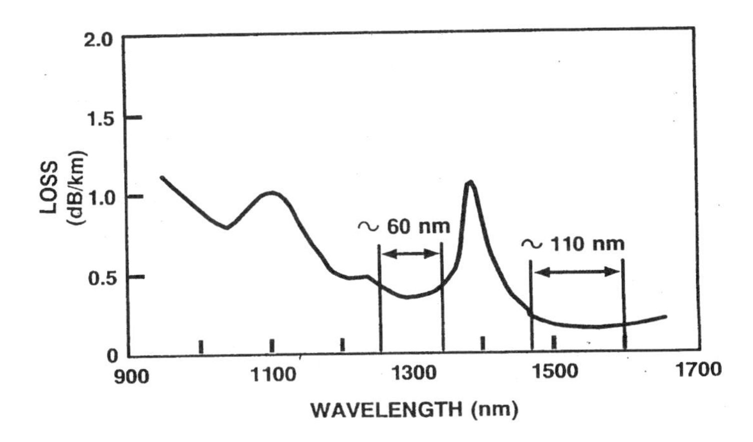

Figure 3.7: Two SMF transmission "windows" near 1310 and 1550 nm.

3.3.1 Coarse WDM Mux/Demuxes or Filters

These devices are used for 1.3/1.55 μm WDM or WDM channels with a channel spacing $\Delta\lambda \approx 5$–100 nm.

Today's most common commercially available coarse WDMs are based on fused-fiber or planar lightwave circuit directional couplers. The same coupled-mode equations given in Eq. (3.2) can be used to describe the characteristics of this type of WDM devices, except that the coupling coefficient κ is wavelength dependent (κ is nearly constant for a certain wavelength range in a 3-dB coupler). If κ and the length L_c of the fused fiber is designed in a way (e.g., changing the distance between the two cores, the shape of the core, and the refractive index profile) such that $\kappa L_c = 2n\pi + \pi/2$ for wavelength λ_1 and $\kappa L_c = 2m\pi + \pi$ for wavelength λ_2 (where m and n are integers), then we have $P_2(L_c) = P_0$ for λ_1, and $P_2(L_c) = 0$ for λ_2, that is, only one wavelength is passed through at output port 2.

Another simple structure built on PLC is the Mach–Zehnder interferometer (MZI), as shown in Fig. 3.8 [11,12], where we can see that there are two input ports, two output ports, two 3-dB couplers, and two waveguide arms with a *fixed* length difference of ΔL. Since the phase difference of the two arms is $\Delta\phi = (2\pi/\lambda)n\,\Delta L$ (n is the waveguide refractive index), the output power is related to the input power by

$$P_{out,1'} = P_{in,1} \cos^2 (\Delta\phi/2) \tag{3.4a}$$

$$P_{out,2'} = P_{in,1} \sin^2 (\Delta\phi/2). \tag{3.4b}$$

When $\Delta\phi/2$ is designed such that $(\pi/\lambda_1)n\Delta L = (m + ½)\pi$ and $(\pi/\lambda_2)n\Delta L = m\pi$ (where m is an integer) for lights launched into port 1, then λ_1 comes out of port 2' and λ_2 comes out of port 1', as shown in Fig. 3.8. Therefore, the wavelength spacing $\Delta\lambda$ is given by

$$\Delta\lambda = \lambda_2 - \lambda_1 = \lambda_1 \lambda_2/(2n\Delta L) \tag{3.5}$$

and the corresponding frequency spacing is given by

$$\Delta f = \frac{c}{2n \cdot \Delta L}. \tag{3.6}$$

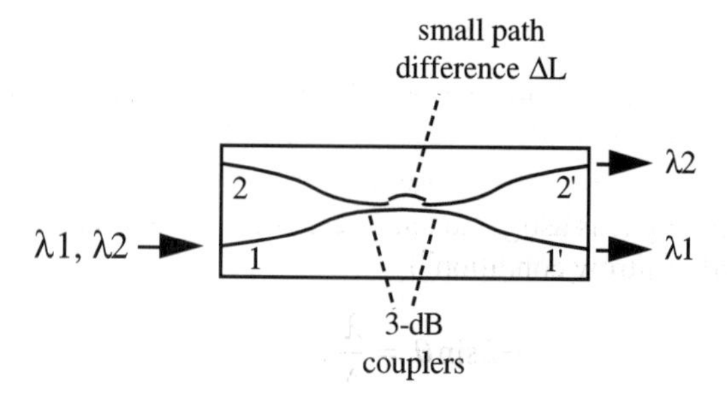

Figure 3.8: A basic Mach–Zehnder interferometer filter with a small ΔL and wide wavelength spacing.

For example, for ΔL = 2.7 μm, λ_1= 1.3 μm, and λ_2= 1.55 μm; then $\Delta\lambda \approx$ 0.25 μm = (1.55–1.3) μm. For ΔL = 15.5 μm, λ_1= 1.50 μm, and λ_2= 1.55 μm, then $\Delta\lambda \approx$ 0.05 μm = (1.55–1.50) μm.

Besides directional couplers and MZI, many of the commercially available WDM filters are multilayer dielectric stacks placed between a pair of GRIN rods [13]. These filters can be either reflective or transmission types [13,14]. The multiple layers of dielectric thin film are assumed to have a thickness of a quarter wavelength. Depending on the relative phase of the multiply reflected (or transmitted) light waves from different layers of the dielectric stack, at a certain wavelength these waves can be constructively added, which consequently exhibits bandpass filtering characteristic. Therefore, this type of WDM filters are often called interference filters. By cascading several interference filters, an 8-channel 200 GHz-spacing commercial demux can achieve a performance of 5.5 dB insertion loss, 0.3 dB polarization sensitivity and >0.5nm bandwidth [13].

Both the 1.3/1.55 μm thin-film filters and the directional-coupler type filters have been very useful in planar lightwave circuits for access PONs. Examples are shown in Figs. 3.9 a and b where we can see that a thin-film filter is inserted into a trench formed on a silica PLC platform [15], or a directional coupler [16] is directly formed on the PLC, respectively.

3.3.2 Dense WDM Filters

These filters are for WDM channels with a channel spacing of 0.5–5 nm. A bulk-type approach for dense WDM (DWDM) filters is typically a diffraction grating-based angular-dispersive filter as shown in Fig. 3.10. A grating is a reflecting element with a series of close parallel lines engraved or etched into the reflection surface.

This type of filter reflects light in a particular direction according to the optical wavelength and the angle of incidence. For incident light of wavelength λ that arrives at an angle θ_i from the plane of the grating and leaves at a wavelength-dependent angle θ_d, from conservation of momentum we have [17]

$$\frac{2\pi}{\lambda}\sin\theta_d + \frac{2\pi}{\lambda}\sin(-\theta_i) = \frac{2\pi}{\Lambda}, \tag{3.7}$$

where the spatial wavelength Λ, shown in Fig. 3.10b, is somewhere between 0.2 and 10 μm. The Littrow configuration shown in Fig. 3.10a is such that the incident and reflected light beams follow virtually the same path, thereby maximizing the grating efficiency and minimizing lens astigmatism. Therefore, Eq. (3.7) can be modified to become (the so-called "Littrow condition")

$$-2\sin\theta_i = \frac{\lambda}{\Lambda}. \tag{3.8}$$

Figure 3.10a shows that a single input fiber and multiple output fibers are arranged on the same focal plane of the lens, and the demultiplexed λ's can be obtained from

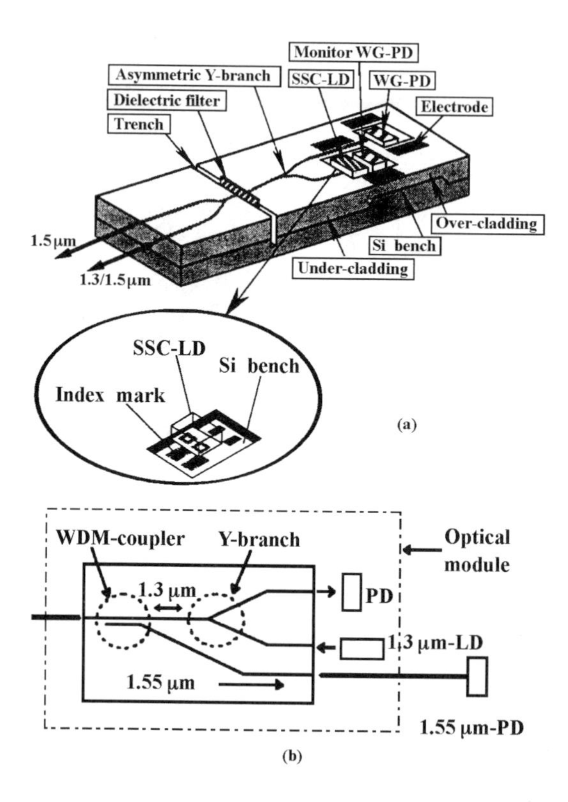

Figure 3.9: A PLC transceiver module for access networks using (a) a 1.3/1.55 μm dielectric filter, and (b) a directional-coupler WDM. In (a), SSC-LD = spot-size converted laser, WG-PD = waveguide photodiode. The index mark on Si bench is used for passive alignment to save packaging cost. After Refs. [15] and [16], respectively (© 1996 IEEE).

different output fibers. Using this kind of approach, a 32-channel DWDM filter with 1-nm spacing and 0.7-nm bandwidth have been demonstrated [18].

The second approach of DWDM filter is again using the MZI principle described in Fig. 3.8. However, to obtain a much larger ΔL, and in the same time a finer wavelength resolution, a configuration shown in Fig. 3.11 is used. In this configuration, a large path difference, say, $\Delta L = 20$ mm, can provide an optical frequency spacing Δf, according to Eq. (3.6), as narrow as 5 GHz, which corresponds to a wavelength spacing of 0.04 nm at 1.55 μm. As a matter a fact, the silica-based integrated optic MZI WDM can provide channel spacing from 0.01 to 250 nm for different ΔL [12]. The large path difference in Fig. 3.11 has been realized by using high-Δ silica waveguides which allow a minimum bending radius of 5 mm [1]. Note also in Fig. 3.11 that a thin-film heater is placed on one of the arms. It acts as a phase shifter for precise frequency tuning because the light-path length of the heated waveguide arm changes with the refractive index change (1×10^{-5}/°C) [2]. The other

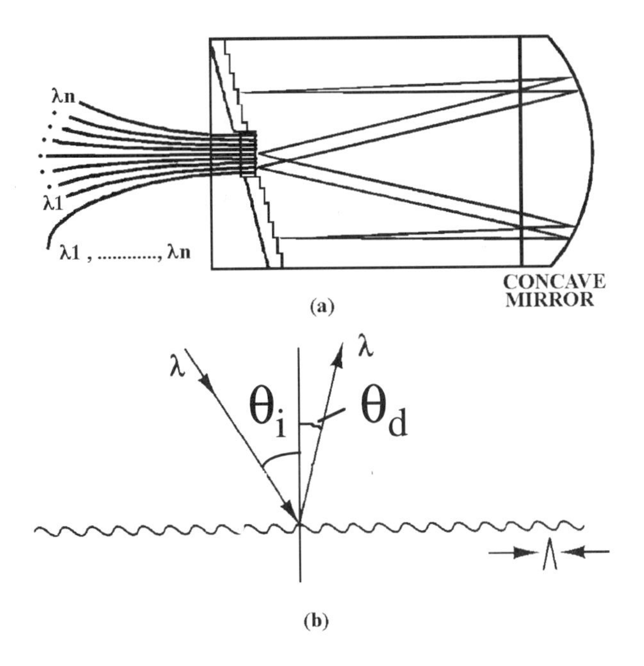

Figure 3.10: (a) Diffraction-grating based WDM multiplexer with Littrow configuration. (b) Incident and diffraction angles. After Ref. [17].

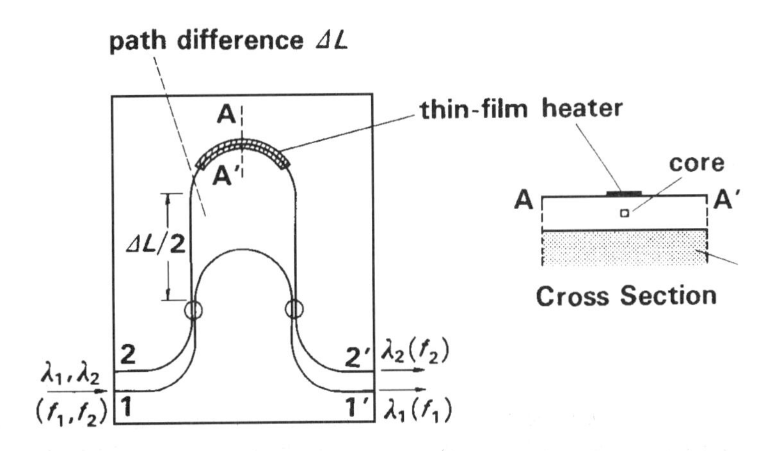

Figure 3.11: A DWDM multi/demultiplexer with an MZI configuration and a large ΔL. The thin-film heater is used to adjust the phase between the two arms so that the output can be tuned to one of the multiple wavelengths. After Ref. [1].

MZI arm is sometimes loaded with a stress-applying film to balance the waveguide birefringence between the long and short arms [1].

The Fabry–Perot interferometer is yet another approach to achieve DWDM multiplexing and demultiplexing. Input light bounces back and forth in the Fabry–Perot cavity formed by etalon mirrors [17], and the multiply reflected interference beams can constructively add at a particular wavelength region and form a passband. Examples of such filters include fiber Fabry-Perot (FFP) filters (for $\Delta\lambda \approx$ 0.5–10 nm) [19,20] (see Fig. 3.12), and liquid-crystal filters [21–25], both of which are actively tunable and are commercially available today.

Research and development on integrated-optic DWDM multi/demultiplexers has increasingly been focused on grating- and phase-arrayed-based arrayed waveguide gratings (AWGs) [1,26,27]. In the bulk diffraction-grating based devices mentioned previously, an etched reflection grating provides the focusing and dispersive properties required for demultiplexing. In AWG devices, as shown in Fig. 3.13, these properties are provided by an array of waveguides, the length of which is designed such the optical path length difference ΔL between adjacent waveguides equals an integer multiple of the central wavelength (λ_c) of the demultiplexer, that is,

$$\frac{2\pi}{\lambda_c} n_g \cdot \Delta L + \frac{2\pi}{\lambda_c} n_{FPR} \cdot d_a(\theta_i + \theta_o) = 2m\pi, \quad (3.9)$$

where the first and second terms on the left represent the neighboring path phase differences in the waveguide and free-propagation-region (FPR), respectively. In Eq. (3.9), n_g and n_{FPR} are the effective indices of waveguide and FPR modes, respectively; d_a is the spatial separation of array waveguides; and θ_i and θ_o are the incident and diffracted angles at the input and output aperture of the AWG (θ_o can be seen in fig. 3.13). For λ_c, the fields in the individual waveguides will arrive at the output aperture with equal phase (apart from an integer multiple of 2π), and the

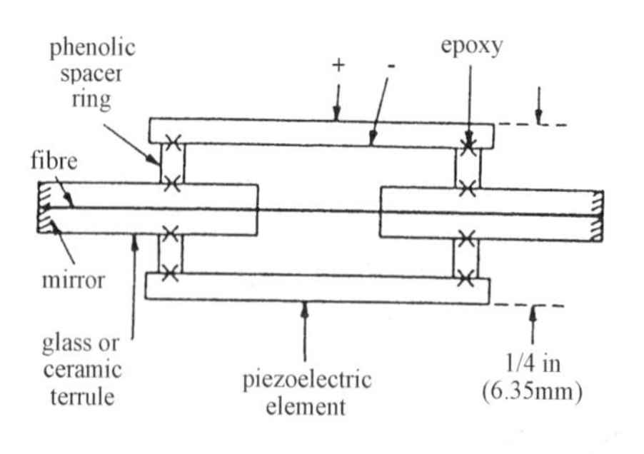

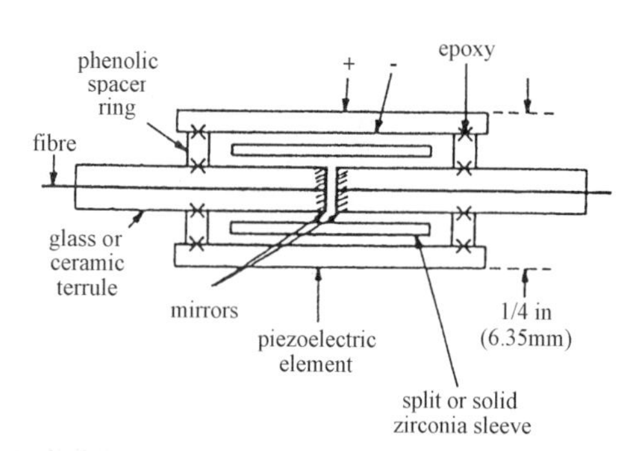

Figure 3.12: Two types of fiber Fabry–Perot filters (with different cavity lengths). After Ref. [19].

field distribution at the input aperture will be reproduced at the output aperture. The angular dispersion for different input wavelengths is achieved by linearly increasing the length of array waveguides. As a consequence, the outgoing beam for different wavelengths will be tilted and the focal point will shift along the image plane at the output. By placing receiver waveguides at proper positions along the image plane, spatial separation of the different wavelength channels is obtained. The dispersion of the array can be described as the angular displacement of the focal spot along the image plane per unit frequency change at the array center ($\theta_i = 0$, $\theta_o = 0$) and can be obtained from Eq. (3.9) as [26,27]

$$\frac{d\theta_o}{df}\bigg|_{\theta_o,\theta_i=0} = \frac{1}{f_c}\cdot\frac{n_g}{n_{FPR}}\cdot\frac{\Delta L}{d_a}, \tag{3.10}$$

from which we can easily obtain the channel spacing Δf as a function of the lateral displacement of the receiving fiber ($= R_a\Delta\theta_o$, where R_a is shown in Fig. 3.13 b),

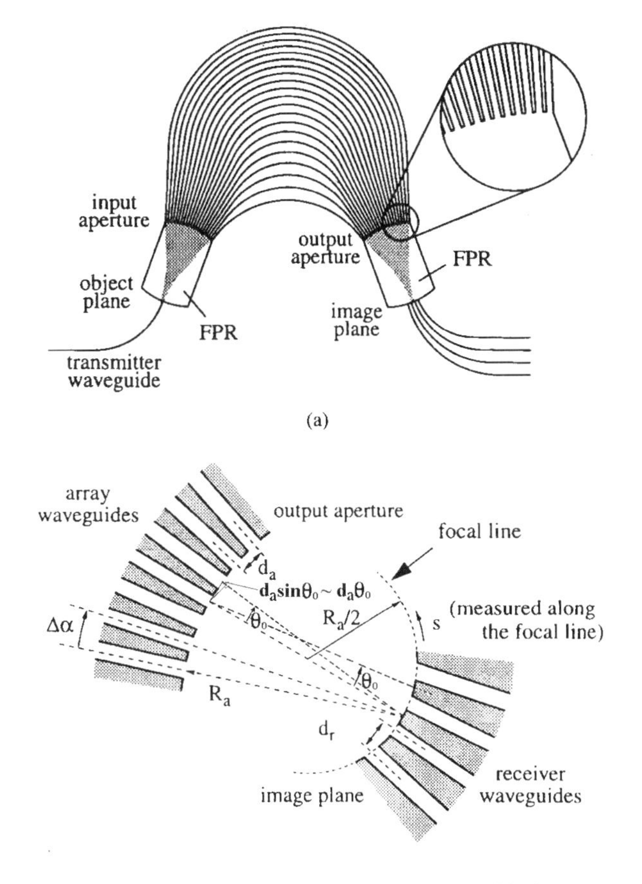

Figure 3.13: (a) Schematic layout of AWG demultiplexer. (b) Geometry of the receiver site. FPR = free propagation region. After Ref. [26] (© 1996 IEEE).

lateral spacing of the waveguides d_a, and the differential length difference ΔL. The free spectral range of the AWG can also be obtained from Eq. (3.9) as ($\theta_i = \theta_o = 0$) [27]

$$FSR = f_1 - f_2 = \frac{c}{n_g \cdot \Delta L}. \tag{3.11}$$

where f_1 and f_2 correspond to the frequencies at orders m and m + 1, respectively. AWG mux/demuxes are now commercially available from 4 to 32 λ's, with channel spacing as narrow as 0.8 or 1.6 nm, in conformance with the ITU standard [28]. The $1 \times N$ device was also extended to $N \times N$ devices, which are the so-called wavelength routers [29,30].

3.3.3 Fiber Bragg Gratings

Optical filters based on Bragg gratings written in photosensitive optical filters [31–33] (by using UV lights or excimer laser in combination with phase mask) have become very important for commercial applications because of their compactness and potential low cost. Their cost can be low because gratings can be photoimprinted in the fiber core using ultraviolet irradiation through a phase mask [31–33]. They could be used to replace bulk optical filters, which we mentioned previously. They can also be used to compensate fiber dispersions in a transmission link [34–36], or to stabilize single-frequency and pump laser sources [37,38]. Therefore, in the following, we will go into details regarding the operation principles of photosensitve (germanium) fiber-based grating filters.

A photosensitve fiber with a grating written in the middle section with a length of L is shown in Fig. 3.14. When the gratings are formed by an external light source, a small perturbation of the fiber refractive index $n(z)$ also occurs, that is; $n(z) = n + \Delta n \cos kz$, where k is the grating wave number and $\Delta n/n << 1$. This fiber grating will reflect all lights in a certain wavelength region when a "Bragg" condition is met. As shown in Fig. 3.15a, when waves are launched into a grating medium, the reflected waves recombine constructively only if the following condition is satisfied:

$$2\, \Lambda \sin\theta = \lambda / n. \tag{3.12}$$

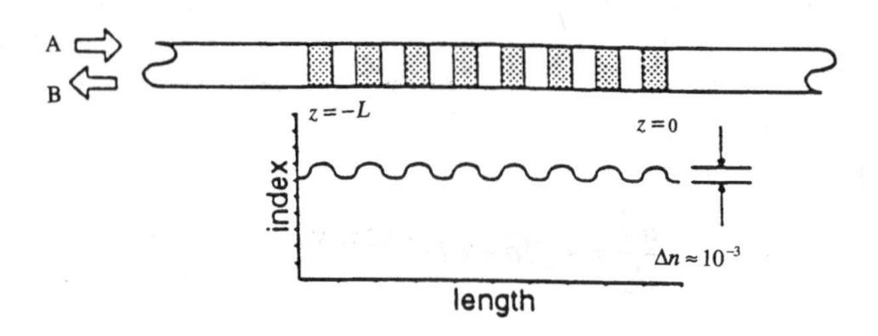

Figure 3.14: A photosensitive fiber grating with resultant refractive-index gratings.

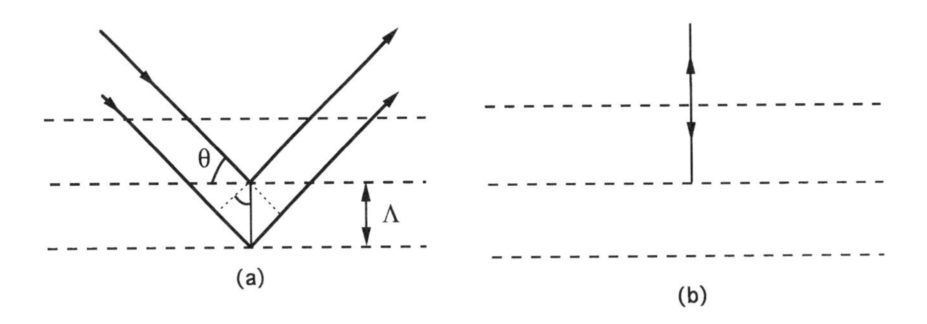

Figure 3.15: Illustrations of incident and diffracted light beams on fiber grating planes for (a) arbitrary incident angle θ, and (b) $\theta = \pi/2$. Λ is the spatial period of the gratings.

Here, Λ is the spatial period of the grating, and λ/n is the wavelength in the propagation medium. This equation is the so-called Bragg condition. Now consider the case when θ approaches $\pi/2$, as shown in Fig. 3.15 b, we have

$$2\,\Lambda = \lambda\,/\,n. \tag{3.13}$$

This is the Bragg condition for coupling oppositely traveling waves with propagation constant $\pm\beta$ at an angular frequency ω. Now let $\lambda_B = \lambda/n$, and define the Bragg wavelength as

$$\lambda_B = 2\,\Lambda \tag{3.14}$$

The Bragg propagation constant is

$$\beta_B = \beta\,(\omega_B) = 2\,\pi\,/\lambda_B = \pi\,/\,\Lambda, \tag{3.15}$$

and the Bragg frequency is

$$\omega_B = c\;\beta_B\,/n. \tag{3.16}$$

To reflect lights at 1.55 μm in the opposite direction, for example, we can let $\Lambda \approx 0.53$ μm, considering $n \approx 1.46$ in a typical single-mode fiber (from Eq. (3.13)).

Let us denote the wave with positive group velocity, that is, the "forward" wave in the guiding structure, by $\overline{a}$, and the wave with negative group velocity, that is, the "backward" wave, by $\overline{b}$. Assume $\overline{a}$ has a spatial dependence of $\exp(-j\beta z)$, $\overline{b}$ has a spatial dependence of $\exp(+j\beta z)$, and the coupling between $\overline{a}$ and $\overline{b}$ is introduced by the coupling coefficient κ (proportional to Δn). We can obtain the coupled wave equations [39]

$$\begin{aligned} \frac{d\overline{a}}{dz} &= -j\beta\overline{a} + \kappa\overline{b}e^{-j(2\pi/\Lambda)z} \\ \frac{d\overline{b}}{dz} &= j\beta\overline{b} + \kappa^{*}\overline{a}e^{+j(2\pi/\Lambda)z} \end{aligned} \tag{3.17}$$

where κ^* is the complex conjugate of κ. Letting $\bar{a} = A(z)e^{-j(\pi/\Lambda)z}$, $\bar{b} = B(z)e^{j(\pi/\Lambda)z}$ we have

$$\begin{aligned}\frac{dA}{dz} &= -j(\beta - \frac{\pi}{\Lambda})A + \kappa \cdot B \\ \frac{dB}{dz} &= j(\beta - \frac{\pi}{\Lambda})B + \kappa^* \cdot A \ .\end{aligned} \tag{3.18}$$

Note that if we expand β in the neighborhood of the angular frequency ω_B at which $\beta(\omega_B) = \pi / \Lambda$, then

$$\beta = \beta(\omega_B) + \frac{d\beta}{d\omega}\Big|_{\omega_B} (\omega - \omega_B) = \beta(\omega_B) + \frac{\omega - \omega_B}{v_g}, \tag{3.19}$$

where v_g is the group velocity of the unperturbed medium at ω_B. Now if we let $\delta = \beta - (\pi / \Lambda)$ be the detuning parameter, from Eq. (3.19) we have

$$\delta = \frac{\omega - \omega_B}{v_g}. \tag{3.20}$$

Now Eqs (3.18) can be simplified to become

$$\begin{aligned}\frac{dA}{dz} &= -j \cdot \delta \cdot A + \kappa \cdot B \\ \frac{dB}{dz} &= j \cdot \delta \cdot B + \kappa^* \cdot A \ ,\end{aligned} \tag{3.21}$$

which in turn gives

$$\frac{d^2 A}{dz^2} = (|\kappa|^2 - \delta^2)A \ . \tag{3.22}$$

If $|\delta| < \kappa$, we have a solution $A(z) \sim \exp(-\gamma z)$ where $\gamma = \sqrt{|\kappa|^2 - \delta^2}$; if $|\delta| > \kappa$, $A(z) \sim \sin(\gamma z)$. Therefore, we see that the coupling produces a transmission stop-band, as shown in Fig. 3.16 [39], centered at ω_B, within which the wave decays exponentially according to $\exp(-\gamma z)$. In Fig. 3.16, the reflection coefficient $|\Gamma| = |B(-L)/A(-L)|$ was calculated at $z = -L$ when the envelope of the reflected wave B is assumed to be zero at $z = 0$ (solve Problem 3.1). Therefore, the stop band is within a range of 2κ. In other words, the reflection coefficient in this frequency range is high and the fiber grating acts as a reflection mirror. The peak reflectivity of a grating at $z = -L$ is given by $\tanh^2(\kappa L)$ when $\delta = 0$ (solve Problem 3.1), and is shown in Fig. 3.17. Therefore, we can control the coupling coefficient κ and the length of the grating L to achieve the desired reflectivity. However, from Figs. 3.16 and 3.17, we can see that there is a trade-off in choosing the coupling coefficient κ ($\propto \Delta n$) when the fiber grating length L is fixed. The larger κ is, the higher the peak reflectivity, but the bandwidth of the filter (δ) becomes wider. The bandwidth of the filter may become so wide that it is not suitable for adding/dropping a wavelength in DWDM systems.

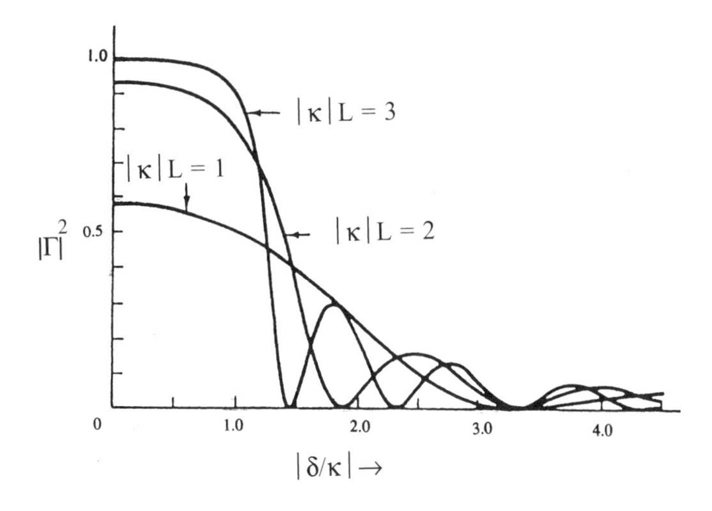

Figure 3.16: Reflection as a function of frequency detuning parameter for uniform distributed feedback structure of length L. After Ref. [39].

In order to build a band pass filter based on fiber gratings, arrangements shown in Fig. 3.18 a or b can be used. The advantage of using an optical circulator (see Section 3.5.2) is its lower insertion loss (typically <0.7–0.8 dB). We can see that the reflected light is the filtered passband signal.

Besides a simple bandpass filter, optical fiber Bragg grating (FBG) can be used as a fiber dispersion-compensation device in digital [34] and subcarrier-multiplexed lightwave systems [35,36]. In these applications, it is necessary to maintain a linear group delay (within a given bandwidth) which is the opposite of that caused by transmission fibers. To achieve this purpose, the FBG must be "linearly chirped"

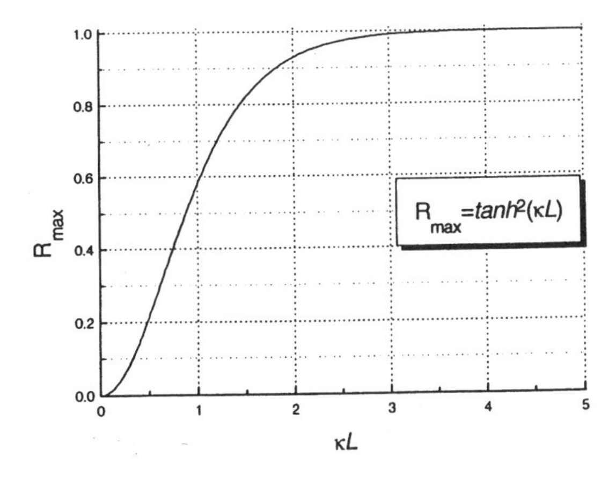

Figure 3.17: Peak reflectivity of a fiber grating versus κL.

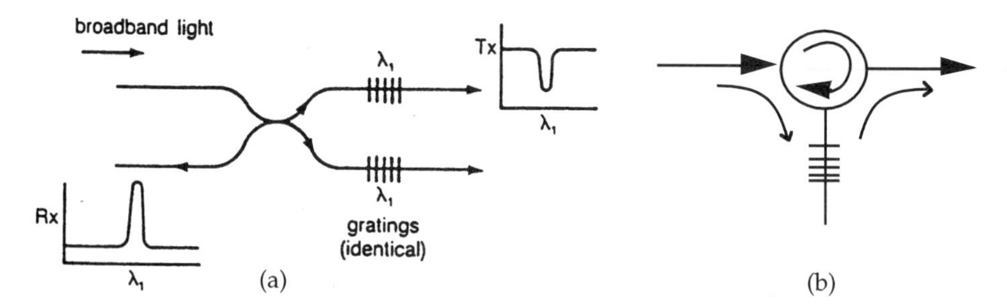

Figure 3.18: Optical bandpass filters based on fiber gratings in combination with (a) an optical fiber coupler, or (b) an optical circulator (see Section 3.5.2).

[40], which means that when a modulated light beam with a finite bandwidth is launched into the FBG, lower frequency components will experience lower delay than higher frequency components. This is illustrated in Fig. 3.19.

Furthermore, to reduce the amplitude and phase ripple of the dispersion-compensation FBG filter, the amplitude function must be "DC-apodized [41]." The DC index apodization eliminates the sidelobes by keeping a constant average index in the grating so the local Bragg wavelength is unchanged through the length of the device. The reflectance of apodized (tapered) linearly chirped gratings (TLCGs) can then be calculated by modifying Eq. (3.21) as

$$\frac{dA}{dz} = -j \cdot \delta \cdot A - j\kappa(z) \cdot B \cdot e^{-j\phi(z)}$$
$$\frac{dB}{dz} = j \cdot \delta \cdot B + j\kappa^{*} \cdot A \cdot e^{j\phi(z)}, \tag{3.23}$$

where $\phi(z)$ represents the newly added phase function given by

$$\phi(z) = F \cdot (z/L)^2, \tag{3.24}$$

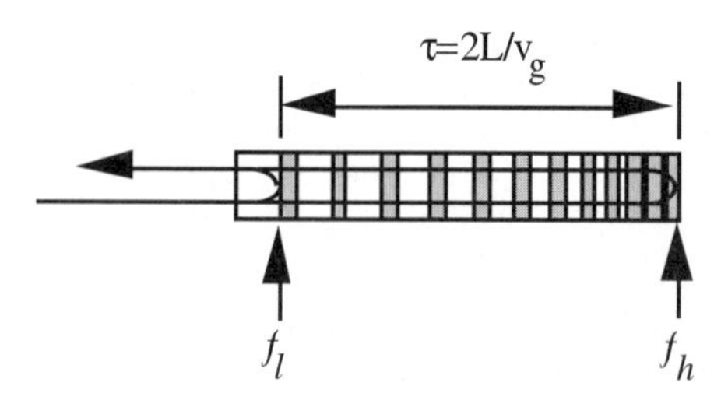

Figure 3.19: Illustration of a linearly chirped fiber grating. The lower frequency f_l is reflected near the input (and output) end, while the higher frequency f_h is reflected near the tail of the fiber grating. The relative delay between the two frequency extremes is given by $2L/v_g$.

where F is the chirp coefficient and L is the fiber grating length. As for the coupling coefficient $\kappa(z)$, we have several different options [35]:

Gauss function:

$$\kappa(z) = \kappa_o \cdot \exp\left(-X \cdot \left(\frac{z}{L}\right)^2\right) \tag{3.25}$$

Hyperbolic tangent (tanh):

$$\kappa(z) = \frac{\kappa_o}{2} \cdot \left\{1 + \tanh[Y(1 - 2|2z/L|^X)]\right\} \tag{3.26}$$

Sinc function:

$$\kappa(z) = \kappa_o \cdot \mathrm{sinc}^X\left[\frac{1}{2}\left(\frac{2z}{L}\right)^Y\right], \tag{3.27}$$

where X and Y are the taper's adjustable parameters, and κ_0 is the value for the coupling coefficient at the middle of the fiber grating. These symmetric functions (with respect to z) are just a few examples; other mathematical functions can also be designed to meet a specific system requirement. For a given optical fiber link distance l and a targeted 3–dB bandwidth B of the TLCG, the minimum length of the fiber grating and the optimum chirp coefficient F_{opt} which can provide sufficient delay compensation can be obtained from [40]

$$\boldsymbol{L}_{\min} = \boldsymbol{lB}\left(\frac{\lambda^2 \boldsymbol{D}}{2\boldsymbol{n}_o}\right) \tag{3.28}$$

$$\boldsymbol{F}_{opt} = \frac{8\boldsymbol{L}_{\min}^2 \boldsymbol{n}_o^2 \pi}{\lambda^2 \boldsymbol{l}\ \boldsymbol{cD}}, \tag{3.29}$$

where n_o is the refractive index and λ is the center wavelength. For those values of F lower than F_{opt}, there is not a full fiber delay compensation, whereas for those values of F higher than F_{opt}, the delay introduced by the equalizer exceeds that of the fiber. Also, as noted in [35], the targeted bandwidth B of the TLCG should cover the FM sidebands of the N-channel of SCM signals in a direct modulation system.

Figure 3.20 compares the relative delay versus passband frequency of four types of FBG filters. We see that without apodization (Fig. 3.20 (a)), both the amplitude and phase exhibit significant ripples. The Gauss (Fig. 3.20 (b)), and sinc (Fig. 3.20 (c)) functions can offer an almost linear delay within the filter bandwidth and should provide better dispersion compensation. However, it is known that all of today's TLCGs still have slight nonlinear group-delay characteristics (ripples) which are introduced by the fabrication process [42]. Therefore, experimental verifications of the feasibility of using TLCGs to reduce the dispersion-induced nonlinear distortions in subcarrier-multiplexed lightwave systems are still to be demonstrated.

It should be noted that optical fiber gratings are very sensitive to both temperature and strain. Temperature affects the Bragg response through the thermo-optic effect and expansion or contraction of the grating periodicity. Strain affects the Bragg

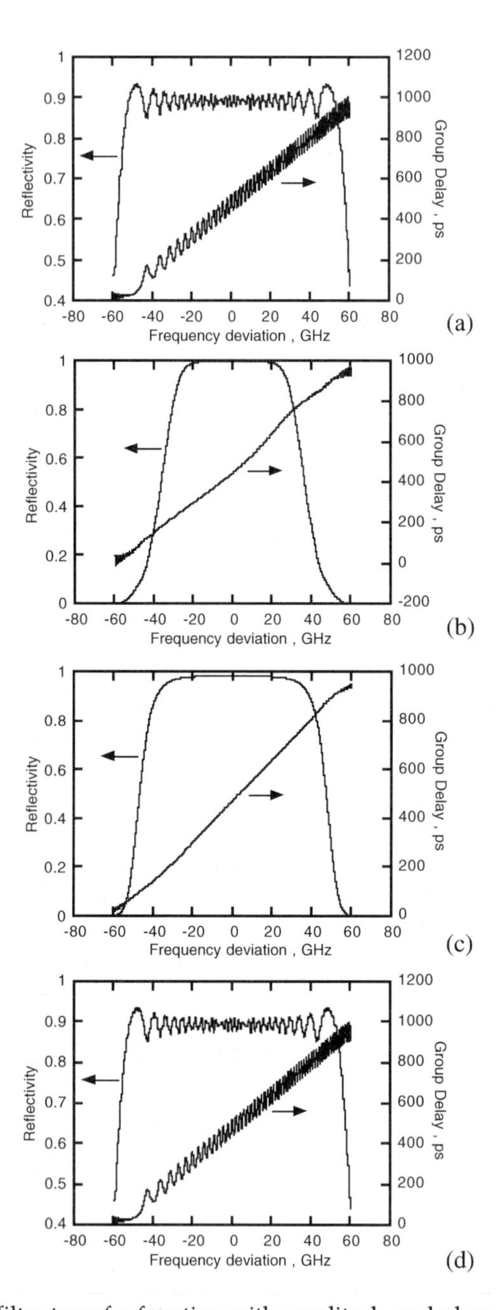

Figure 3.20: FBG filter transfer function with amplitude and phase responses for different apodization functions (L_{min}= 10 cm, $F = 718.7718$, $\int_{-L}^{0} \kappa(z)dz = 16$): (a) no apodization, (b) Gauss function ($X = 15$), (c) sinc function ($X = 9$, $Y = 3$), and (d) hyperbolic tangent function ($X = 3$, $Y = 4$).

response through expansion or contraction of the grating periodicity and through the strain-optic effect. Both sensitivities have been measured to be approximately linear over a wide temperature range. A few packaging techniques have been proposed to cope with these problems [43,44].

3.4 Optical Fiber Connectors and Splices

Connectors are convenient for disconnection and reconnection (1) between a transmitter output fiber (pigtail) and a transmission fiber (or a coupler), (2) between a transmission fiber (or coupler) and a receiver input fiber (pigtail), and (3) between transmission fibers. The design goals of single-mode optical fiber connectors regarding insertion losses and reflectances are shown in Table 3.2 [45]. We can see that the objective mean insertion loss is 0.2 dB while the maximum insertion loss must be less than 0.6 dB per connection. The mean reflectance should be less than –43 dB (or return loss > 43 dB) while the maximum reflectance must be less than –40 dB (or return loss > 40 dB). The reason why the reflectance requirement is set at –40 dB or below will become clear in Chapter 9. A connector is usually composed of an alignment sleeve and two interconnecting plugs, as shown in Fig. 3.21. The plug is basically a (ceramic/metal) ferrule which has a precision-drilled hole in a cylinder where a stripped-single-mode fiber is inserted and end-polished. Lateral offset, angular tilt, end separation, and end quality all contribute to extrinsic loss and extra reflections at a connector joint (see Fig. 3.22). Note that when there is an air gap between end faces, as in Fig. 3.22c, the reflectance is $10 \cdot \log\{[(n_2 - n_1)/(n_2 + n_1)]^2\}$ or about –15 dB, where $n_2 = 1.46$ and $n_1 = 1$ are the refractive indices of a single-mode fiber and air, respectively. The reflectance can become even worse when the fiber end faces have rough surfaces, as shown in Fig. 3.22d. The important concept which we should obtain from Fig. 3.22 is that *physical contact* of the 8- to 10-μm diameter fibers is crucial to a good fiber connector.

A thorough survey of commercially available optical fiber connectors is given in [46] and is shown in Fig. 3.23. Out of those connectors in Fig. 3.23, the most popular connector types used in today's HFC systems include SC, ST, FC/PC, and

Table 3.2: Specifications of single mode fiber connectors (R: required, O: objective) (After Ref. [45])

	Performance(dB)	
Measurement	**(R)**	**(O)**
Maximum loss	0.60	0.30
Mean loss	<0.50	<0.20
Maximum reflectance	–40	–50
Mean reflectance	<–43	—

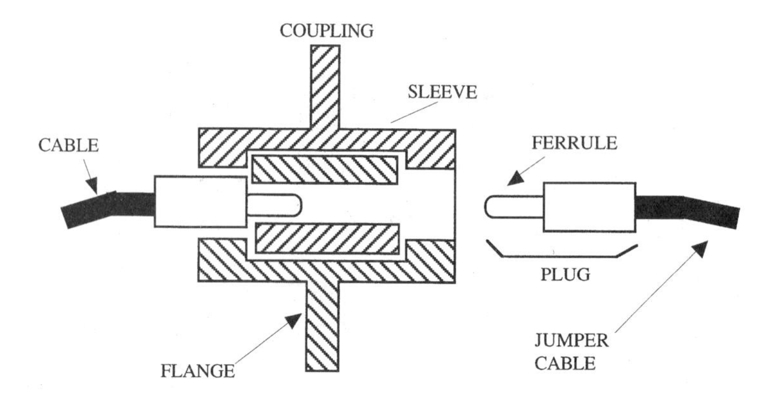

Figure 3.21: A generic optical fiber connector. (After Ref. [45])

angled-FC/PC (not shown in Fig. 3.23) connectors, whose performances all meet those in Table 3.2 and are now described in more details.

The type SC (square/subscriber) connector was developed by NTT around 1986. It employs a rectangular cross-section of molded plastic and has a push-to-insert and pull-to-remove locking mechanism which prevents rotational misalignment. An audible click lets a user know that the connector is fully engaged. The SC has a good packing density and is useful in multicable installations.

The ST connector was introduced in early 1985 by AT&T Bell Laboratories. It uses a spring-loaded *twist-and-lock* bayonet coupling, similar to BNC connectors used with coax cables. This feature inhibits the fiber/ferrule from rotating during multiple connections, thereby ensuring a more consistent insertion loss. It has a cylindrical ferrule 2.5 mm in diameter, which may be made of ceramic, stainless steel, glass, or plastic.

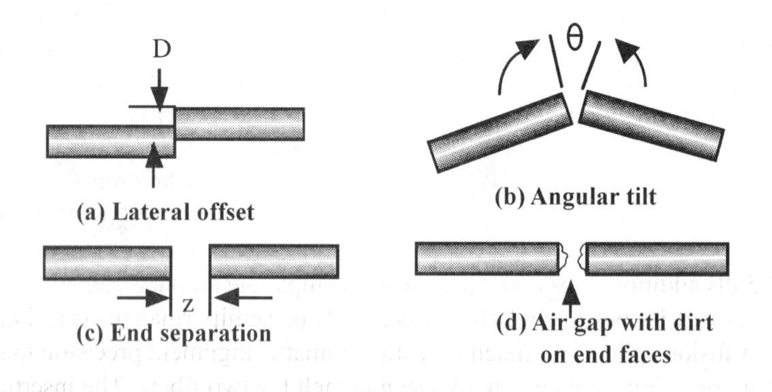

Figure 3.22: Illustrations of imperfect alignment at an optical fiber connector joint. These are the basic mechanisms which cause insertion loss and reflections.

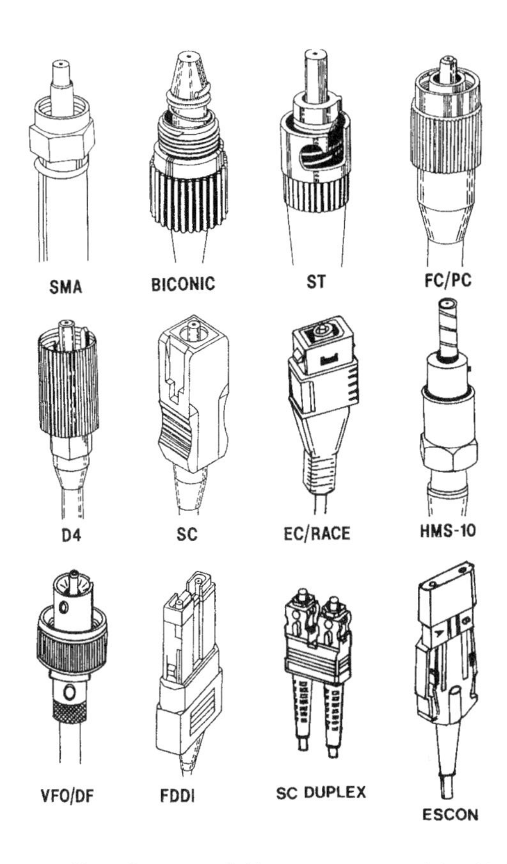

Figure 3.23: Popular optical-fiber connectors. After Ref. [46].

The FC/PC (face contact/physical contact) connector was a modified version of the FC connector first developed by NTT. FC means that a flat end face on the ferrule can provide "face contact" between joining connectors. FC/PC connectors further use "physical contact" between fiber ends to reduce insertion loss and to increase return loss. To obtain a superior performance on return loss (> 60 to 65 dB), which is critical to analog AM-VSB CATV and multi-gigabit transmissions, FC/APC connectors were developed. APC means "angled polished/physical contact" and uses a ferrule that combines an 8° to 10° angled geometric design with PC convex polishing at the end face. However, APC connectors have higher insertion loss (about 0.5 dB additional loss) and are often incompatible among vendors.

Splices can be permanent (fusion-spliced) or semipermanent (mechanically spliced). A fusion splicer is a machine with automatic alignment precision to submicrons, that uses a high-voltage arc to heat and melt the two fibers. The insertion loss due to a fusion splice is typically less than 0.1 dB. The corresponding reflectance is essentially zero. A mechanical splice uses a fixture such as a V-groove to bring two fibers to contact and then an epoxy or a cover plate to fasten the two aligned fibers. Typical insertion loss and reflectance due to a mechanical splice is around 0.2 and –40 dB, respectively.

3.5 Optical Isolators and Circulators

3.5.1 Optical Isolators

Optical isolators can be classified into two types: polarization-sensitive and polarization-insensitive. The operation principles of both types are illustrated in Figs. 3.24a and b, respectively [47]. The optical isolator in Fig. 3.24a is composed of a front polarizer, a Faraday rotator, and a back polarizer. When a light beam enters the optical isolator, its vertically (or horizontally) polarized component passes through the front linear polarizer. The Faraday rotator with an appropriate magnetic field intensity and length will then rotate the direction of polarization by 45°. The back polarizer is aligned with this 45° polarization, so that the light beam passes with little loss. However, when an undesired light beam is reflected back to the output port of the isolator, the selected 45° polarized light beam will be rotated another 45° by the Faraday rotator. Therefore, the polarization direction of the reflected light becomes orthogonal to the front linear polarizer and cannot pass to the input port. Optical isolation is thus achieved. The disadvantage of this technique is that because of the existence of the first linear polarizer, an average 3-dB loss may exist because the input light is generally randomly polarized.

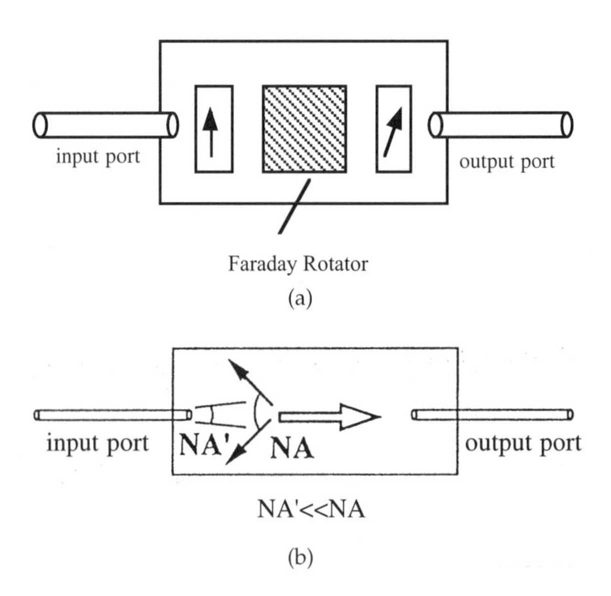

Figure 3.24: Optical isolators: (a) type A and (b) type B. After Ref. [47] (© 1994 IEEE).

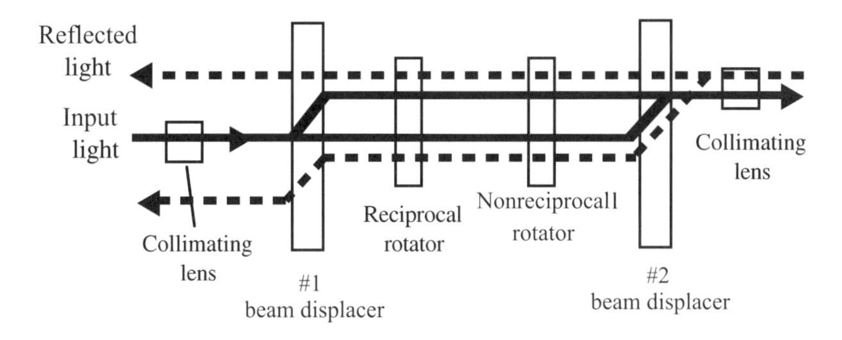

Figure 3.25: A method of implementing Type B optical isolator. (After Ref. [48])

An improved polarization-insensitive optical isolator is shown in Fig. 3.24b. The arrangement of the lenses in the isolator must ensure that the effective numerical aperture[1] of the returning light is much higher than that of the input light. This improved version typically can reduce the insertion loss of the optical isolator to about 0.5 dB. A specific design of this category is shown in Fig. 3.25. This structure consists of a pair of input and output coupling lenses, a pair of input and output beam displacers (these are birefringent crystals which displace one of the polarizations by a small distance, and let the other polarization passes through without displacement), and reciprocal (RR) and nonreciprocal rotators (NRR) (RR rotates light in both direction by 45°, while NRR rotates light by 45° and −45° in the forward and backward directions, respectively). As shown in Fig. 3.25, the two polarizations of the input light (solid line) are directed into two separate paths by the #1 beam displacer. The vertical and horizontal polarizations travel through the upper and lower paths, respectively, each rotated 90° after the RR and NRR, and finally recombine at the output of the collimating lens. The reflected light (dashed line) has its horizontal and vertical polarizations pass through the upper and lower paths of the #2 beam displacer, respectively. The upper path vertical beam is then rotated −45° by the NRR and +45° by the RR and arrives at the #1 beam splitter still with horizontal polarization. Therefore, it leaves the displacer without offset, as shown in the figure. Similarly, the lower path arrives the #1 beam displacer with vertical polarization, and therefore it leaves the displacer with an offset. Consequently, the reflected light cannot come back to the input fiber. Today's commercial optical isolators can achieve a typical isolation of 40 (single-stage) to 50 (double-stage) dB with an insertion loss of ~0.5 dB (polarization dependent loss <0.1 dB).

[1] The numerical aperture (NA) of a fiber is the sine of the maximum allowable incident angle ϑ_{max} of incident light rays, that is, $NA = \sin(\vartheta_{max})$.

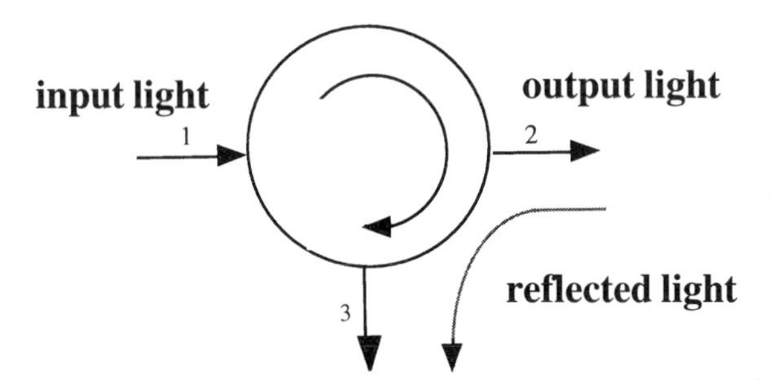

Figure 3.26: Basic configuration of an optical circulator.

3.5.2 Optical Circulators

The basic function of an optical circulator is shown in Fig. 3.26. A light beam launched into port 1 is directed to port 2, as desired, but the reflected light beam which enters port 2 is directed to port 3 instead of going back to port 1. Many system applications such as optical amplifier supervision [49] and optical time reflectometry [50] have used optical circulators to prevent the reflected light coming back to the input port and at the same time direct the reflected light to a different path for other uses. Several types of optical circulators have been proposed [51,52]. An example is shown in Fig. 3.27. The proposed structure is composed of two polarization beam splitters (PBSs), two pentagonal prisms, two nonreciprocal rotators (NRRs), and two walkoff PBSs. The function of each NRR is to rotate the polarization plane of light propagating in the $+z$ and $-z$ directions by $-45°$ and $+45°$, respectively. The walkoff PBSs, which are birefringent crystal blocks, are used to let the reflected beam in the $-z$ direction "offset" or "walk off" relative to the input beam path so that the reflected light can be redirected to fiber 3 in Fig. 3.27a. Walkoff PBS1 is set such that $-45°$ azimuth linearly polarized light propagating in the $-z$ direction walks off simultaneously along the x and y axes by $-d$ and $+d$, respectively. Similarly, walkoff PBS2 is set such that $+45°$ azimuth light in the $-z$ direction walks off simultaneously along the x and y axes, both by $-d$.

The operation principle of the circulator can now be explained as follows: An unpolarized light beam from fiber 1 is first divided into two linearly polarized light beams by PBS1, as shown by broken lines in Fig. 3.27a. The upper beam passes directly through NRR1, and the lower beam is reflected by the pentagonal prism 1 and then passes through NRR1. Both beams are rotated $-45°$ by NRR1 and pass through walkoff PBSs without walkoff (because they are in the $+z$ direction). The following NRR2 again rotates both beams by $-45°$, which results in two emerging beams with $-90°$ and $0°$ polarizations, respectively (see solid arrows in Fig. 3.27b for illustration). These two beams are then recombined at PBS2 and output at fiber 2. For an unpolarized beam reflected back into fiber 2, the beam is divided into two linearly polarized beams by PBS2. The paths of these beams are shown by dotted lines in

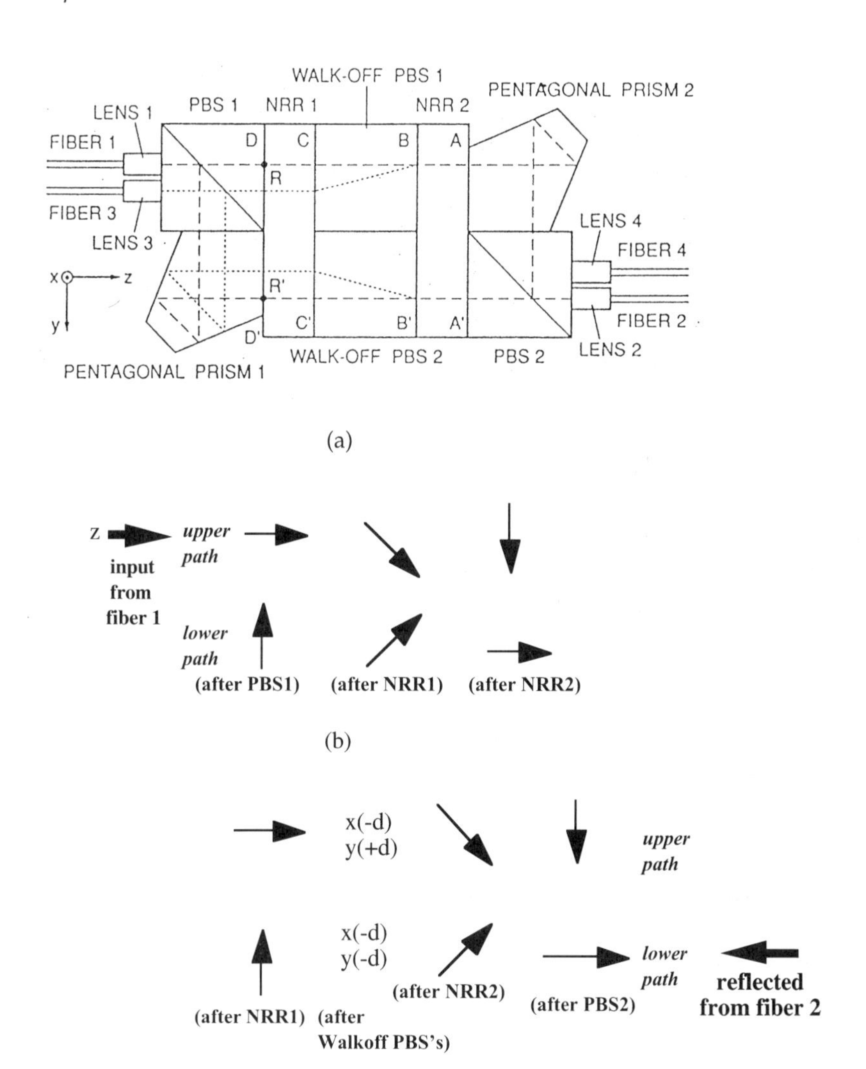

Figure 3.27: An example of optical circulator. (a) Structure, and polarization evolutions in forward (©1992 IEEE) (b) and backward (c) directions. (After Ref. [51]).

Fig. 3.27a. The lower beam passes through the NRR2 directly, while the upper beam is reflected by pentagonal prism 2 before going into NRR2. After NRR2, the polarizations of the upper and lower beams change to –45° and +45°, respectively, as shown in Fig. 3.27c. Subsequently, the upper beam walks off $-d$ in the x direction and $+d$ in the y-direction in walkoff PBS1, and the lower beam walks off $-d$ in both

x and y directions in walkoff PBS2. Following the PBSs is the NRR1 which further rotates each beam by +45°, and results in two beams with 0° and 90°, respectively. The two orthogonal beams are then recombined at PBS1 and exited at fiber 3, as planned. The insertion loss and isolation of such a device have been measured to be ≤ 1.7 and ≥ 48.6 dB, respectively, at 1.3-μm wavelength range. Today's commercial circulators can achieve a typical insertion loss < 0.7 dB and a typical isolation > 50 dB.

3.6 Optical Attenuators and Mechanical Switches

Optical attenuators use one of the following techniques to decrease light intensity: (1) use light-absorption materials [53], (2) use angular displacement of single-mode fibers, or (3) use two in-line fiber tapers with fixed or adjustable axial separation [54]. Since optical reflections must be minimized in optical fiber links delivering analog video signals, a multilayered dielectric thin-film coating is usually applied to the surfaces of the absorption materials. Obliquely polished surfaces can also be used. There are fixed and tunable optical attenuators. The former is frequently used in practical systems to avoid the saturation of an optical receiver, whereas the latter (which is more expensive) is mostly used in laboratory environment to measure the receiver sensitivity or dynamic range.

Optical mechanical switches can be used in optical fiber links of an HFC system to provide redundancy protection. For example, a 1×2 optical mechanical switch can be used before an optical receiver. Under normal operation conditions, the optical signal is received from path A. When path A is broken accidentally, the optical switch will be directed to receive optical signals from path B. The switching time usually is several milliseconds. Discussions of the construction of optical mechanical switches can be found elsewhere [53].

Problems

1. Continue the derivations of Eqs. (3.21) and (3.22) to obtain Fig. 3.16 and Fig. 3.17.
2. Using the apodization functions given in Eqs. (3.25–3.27), see if you can plot the amplitude and group delay transfer functions given in Fig. 3.20.
3. Use the components introduced in this chapter to design WDM wavelength add-drop multiplexers (ADMs). ADMs can be used in an interoffice ring (Fig. 2.22) to drop one or more wavelengths to an office (headend), and insert one or more wavelengths from an office (headend). Explain the advantages and disadvantages of various designs of WDM-ADM that you surveyed or designed.
4. (a) Use two-beam interfering structure and silica waveguides to design Mach–Zehnder filters with frequency spacing $\Delta f_1=10$ GHzand $\Delta f_2 = 20$ GHz, assuming optical wavelength is 1.55 μm.

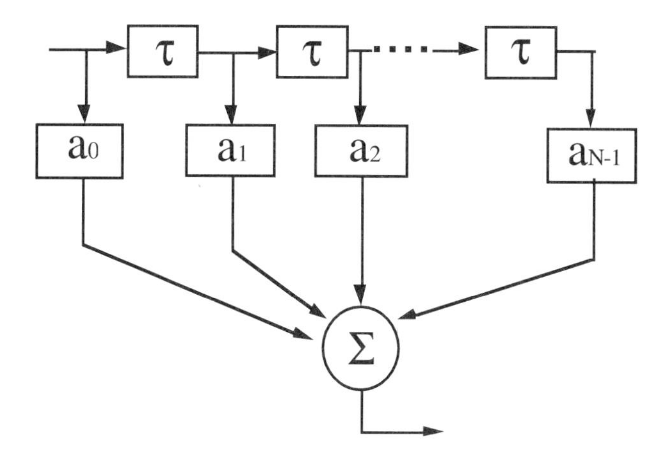

Figure 3.P.1: Find the frequency characteristics and transmittance of the filter. Discuss how this transversal filter can be used to equalize the optical gain spectrum in a transmission system.

(b) Using the results in (a), design a two-stage Mach–Zehnder filter that is capable of resolving four FDM signals which are equally spaced with $\Delta f = 10$ GHz.

5. An optical transversal filter consists of delay lines and variable power dividers. The filter configuration is shown in Fig. 3.P.1. Find the frequency characteristics and transmittance of the filter.

References

[1] M. Kawachi, "Silica waveguides on silicon and their application to integrated-optic components," Optical Quantum Electron., vol.22, pp.391–416, 1990.

[2] M. Kawachi, "Recent progress in silica-based planar lightwave circuits on silicon," IEE Proc.-Optoelectron., vol.143, pp.257–262, October 1996.

[3] Y. P. Li and C. H. Henry, "Silica-based optical integrated circuits," IEE Proc.-Optoelectron., vol.143, pp.263–280, October 1996.

[4] R. E. Epworth, "Modal noise–causes and cures," Laser Focus, pp.109–115, 1981.

[5] Bellcore Technical Advisory, "Generic requirements for optical fiber and optical fiber cable," TA-NWT-000020, Issue 8, December 1991.

[6] P. Kaiser and D. B. Keck, "Fiber types and their status," in *Optical Fiber Telecommunications II*, Eds., S. E. Miller and I. P. Kaminow, Academic Press, Orlando, Florida, 1988.

[7] F. P. Kapron, "Transmission properties of optical fibers," in *Optoelectronic Technology and Lightwave Communication Systems*, Ed., Chinlon Lin, Van Nostrand-Reinhold, Princeton, New Jersey, 1989.

[8] G. Keiser, *Optical Fiber Communications*, 2nd Edition, McGraw-Hill, New York, 1991.

[9] D. Marcuse, "The coupling of degenerate modes in two parallel dielectric waveguides," Bell System Tech. J. vol.50, No.6, p.1791, July–August 1971.

[10] I. P. Kaminow, "Photonic local networks," in *Optical Fiber Telecommunications II*, Eds., S. E. Miller and I. P. Kaminow, Academic Press, Orlando, Florida, 1988.

[11] B. H. Verbeek, C. H. Henry, N. A. Olsson, K. J. Orlowsky, R. F. Kazarinov, and B. H. Johnson, "Integrated four-channel Mach–Zehnder multi/demultiplexer fabricated with phosphorous doped SiO2 waveguides on Si," J. Lightwave Technol., vol.6, pp.1011–1015, June 1988.

[12] N. Takato, T. Kominato, A. Sugita, K. Jinguji, H. Toba, and M. Kawachi, "Silica-based integrated optic Mach-Zehnder multi/demultiplexer family with channel spacing of 0.01–250 nm," IEEE J. Selected Areas in Commun., vol.8, pp.1120–1127, August 1990.

[13] B. Nyman, "Passive Components," OFC '98 Tutorial.

[14] M. Born and E. Wolf, *Principles of Optics*, Chapter 7, Pergamon, 1980.

[15] H. Okano, M. Okawa, H. Uetsuka, T. Teraoka, S. Aoki, and S. Tsuji, "Passive aligned hybrid integrated optical module using planar lightwave circuit platform," IEEE LEOS, paper MH3, 1996.

[16] J. Yoshida, "Low-cost optical modules for fiber-to-the-home," IEEE LEOS, paper MH1, 1996.

[17] J-P Laude and J. M. Lerner, "Wavelength division multiplexing/demultiplexing (WDM) using diffraction gratings," SPIE-Application, Theory and Fabrication of Periodic Structures, vol.503, pp.22–28, 1984.

[18] D. R. Wisely, "32 Channel WDM multiplexer with 1 nm channel spacing and 0.7 nm bandwidth," Electron. Lett., vol.27, pp.520–521, 1991.

[19] J. Stone and L. W. Stulz, "Pigtailed high-finesse tunable fibre Fabry-Perot interferometers with large, medium, and small free spectral ranges," Electron. Lett., vol.23, pp.781–783, July 1987.

[20] C. M. Miller and J. W. Miller, "Wavelength-locked, two-stage fiber Fabry-Perot filter for dense wavelength division demultiplexing in erbium-doped fibre amplifier spectrum," Electron. Lett., vol.28, pp.216–217, January 1992.

[21] S. R. Mallinson, "Wavelength-selective filters for single-mode fiber WDM systems using Fabry-Perot interferometers," Appl. Opt., vol.26, pp.430–436, February 1987.

[22] K. Hirabayashi, H. Tuda, and T. Kurokawa, "Narrow-band tunable wavelength-selective filters of Fabry-Perot interferometers with a liquid crystal intracavity," IEEE Photon. Technol. Lett., vol.3, pp.213–215, 1991.

[23] K. Hirabayashi, H. Tuda, and T. Kurokawa, "New structure of tunable wavelength-selective filters with a liquid crystal for FDM systems," IEEE Photon. Technol. Lett., vol.3, pp.741–743, 1991.

[24] Y. Bao, A. Sneh, K. Hsu, K. M. Johnson, J. Y. Liu, C. A. Miller, and Y. Morita, "High-speed liquid crystal fiber Fabry–Perot tunable filter," Tech. Digest, Optical Fiber Commun. Conf., paper WA1, 1996.

[25] K. Y. Wu and J. Y. Liu, "Liquid crystal and wavelength routing switches," Conf. Proc., IEEE LEOS 10th Annual Meeting, paper WC4, 1997.

[26] M. K. Smit and C. V. Dam, "Phasar-based WDM-Devices: Principles, design and applications," IEEE J. Selected Topics Quantum Electron., vol.2, pp.236–250, June 1996.

[27] Y. P. Li and C. H. Henry, "Silica-based optical integrated circuits," IEE Proc.-Optoelectron., vol.143, pp.263–280, October 1996.

[28] "Optical interfaces for multichannel systems with optical amplifiers," Draft G.MCS, International Telecommunication Union (ITU), COM 15-24-E, February 1997.

[29] C. Dragone, "An N × N optical multiplexer using a planar arrangement of two star couplers," IEEE Photon. Technol. Lett., vol.3, pp.812–815, September 1991.

[30] C. Dragone, C. A. Edwards, and R. C. Kistler, "Integrated optics N × N multiplexer on silicon," IEEE Photon. Technol. Lett., vol.3, pp.896–899, October 1991.

[31] R. J. Campbell and R. Kashyap, "The properties and applications of photosensitive germanosilicate fibre," Int. J. Optoelectron. vol.9, pp.33–57, 1994.

[32] K. O. Hill, B. Malo, F. Bilodeau, D. C. Johnson, and J. Albert, "Bragg gratings fabricated in monomode photosensitive optical fiber by UV exposure through a phase mask," Appl. Phys. Lett., vol.62, no.10, pp.1035–1037, March 1993.

[33] K. O. Hill, "Optical fiber gratings," Tech. Digest, 1st Optoelectron. Commun. Conf., paper 18C1, Chiba, Japan, July 1996.

[34] W. H. Loh, R. I. Laming, N. Robinson, A. Cavaciuti, F. Vaninett, C. J. Anderson, M. N. Zervas, and M. J. Cole, "Dispersion compensation over distances in excess of 500 km for 10-Gb/s systems using chirped fiber gratings," IEEE Photon. Technol. Lett., vol.8, pp.944–946, July 1996.

[35] J. Marti, D. Pastor, M. Tortola, J. Capmany, and A. Montero, "On the use of tapered linearly chirped gratings as dispersion-induced distortion equalizers in SCM systems," J. Lightwave Technol., vol.15, pp.179–187, February 1997.

[36] D. Pastor, J. Capmany, and J. Marti, "Reduction of dispersion induced composite triple beat and second-order intermodulation in subcarrier multiplexed systems using fiber grating equalizers," IEEE Photon. Technol. Lett., vol.9, pp.1280–1282, September 1997.

[37] C. R. Giles, T. Erdogan, and V. Mizrahi, "Simultaneous wavelength-stabilization of 980-nm pump lasers," IEEE Photon. Technol. Lett., vol.6, pp.907–909, August 1994.

[38] A. Hamakawa, T. Kato, and G. Sasaki, "1480 nm pump fiber-grating external-cavity-laser with two fiber gratings," Tech. Digest, 2nd Optoelectron. Commun. Conf. (OECC), paper 9D2–5, July 1997.

[39] H. A. Haus, *Waves and Fields in Optoelectronics*, Chapter 8, Prentice-Hall, Englewood Cliffs, New Jersey, 1984.

[40] F. Quellette, "Dispersion cancellation using linearly chirped Bragg grating filters in optical waveguide," Optics Lett., vol.12, no.10, pp.847–849, 1987.

[41] P. S. Cross and H. Kogelink, "Sidelobe suppression in corrugated-waveguide filters," Optics Lett., vol. 1, no. 1, pp.43–45, 1977.

[42] K. Ennser, R. I. Laming, M. N. Zervas, M. Ibsen, and M. Durkin, "Effects of non-ideal group delay and reflection characteristics of chirped fibre grating dispersion compensators," Conference Proceedings, European Conference on Optical Communications (ECOC), paper WeP.06, September 1997.

[43] T. E. Hammon, J. Bulman, F. Ouellette, and S. B. Poole, "A temperature compensated optical fiber Bragg grating band rejection filter and wavelength reference," Tech. Digest, 1st Optoelectron. Commun. Conf. (OECC), paper 18C1-2, July 1997.

[44] T. Iwashima, A. Inoue, M. Shigematru, M. Nishimura, and Y. Hattori, "A novel temperature compensation technique for fiber Bragg gratings using liquid crystalline polymer tubes," Tech. Digest, 2nd Optoelectron. Commun. Conf. (OECC), paper 9D2-4, July 1997.

[45] Bellcore Technical Reference, "Generic requirements for optical fiber connectors," TR-NWT-000326, Issue 3, June 1992.

[46] R. G. Ajemian, "A selection guide for fiber optic connectors," Optics Photonoics News, pp.32–36, June 1995.

[47] K. Kikushima, K. Suto, H. Yoshinaga, and E. Yoneda, "Polarization dependent distortion in AM-SCM video transmission systems," J. Lightwave Technol., vol. 12, pp.650–657, April 1994.

[48] G. E. Lano and C. Pinyan, "Optical isolators direct light the right way," Laser focus World, pp.125–127, July 1995.

[49] W. I. Way, Y. W. Lai, and Y. K. Chen, "The effect of transient gain compression in saturated EDFA on optical time domain reflectometry testing," IEEE Photon. Technol. Lett., vol.6, pp.1200–1202, 1994.

[50] Y. Sato and K. Aoyama, "Optical time domain reflectometry in optical transmission lines containing in-line erbium-doped fiber amplifiers," J. Lightwave Technol., vol.10, pp.78–83, 1992.

[51] Y. Fujii, "High-isolation polarization-independent quasi-optical circulator," J. Lightwave Technol., vol.10, pp.1226–1229, September 1992.

[52] M. Koga and T. Matsumoto, "High-isolation polarization insensitive optical circulator for advanced optical communication systems," J. Lightwave Technol., vol.10, No.9, pp.1210–1217, September 1992.

[53] N. Kashima, *Passive Optical Components for Optical Fiber Transmission,* Artech House, 1995.

[54] A. Benner, H. M. Presby, and N. Amitay, "Low-reflectivity in-line variable attenuator utilizing optical fiber tapers," *J. Lightwave Technol.* **8**, 7–10, January 1990.

Chapter 4

Fundamentals of Semiconductor Laser Diodes, Their Modulation, Noise, and Linearity Characteristics

Semiconductor laser diodes are the key light sources for HFC systems, both for downstream and upstream transmissions. HFC systems have gained their current momentum mainly because highly linear and low noise distributed feedback (DFB) lasers were developed during the late 1980s and early 1990s. In the near future, when two-way HFC systems become prevalent, low-cost, temperature-insensitive return-path laser diodes may become another type of key optical device for HFC systems. Therefore, it is important to understand the basic physics and modulation characteristics of semiconductor laser diodes.

In this chapter, we will first review the basic physics of a semiconductor laser diode, including Fabry–Perot (FP), DFB, and quantum-well (QW) laser diodes, in Sections 4.1–4.7. The modulation, noise, and linearity characteristics of laser diodes are described in Sections 4.8, 4.9, and 4.10, respectively. A transmitter design example for HFC systems is illustrated in Section 4.11.

4.1 Basic Physics

A semiconductor laser diode, as its name shows, is basically formed by bringing p-type and n-type semiconductors into contact. When the p–n junction is forward biased, the built-in electric field is reduced, which results in electric current flow due to the diffusion of electrons and holes across the junction. In this case, electrons and holes are present simultaneously in the depletion region and can recombine through spontaneous and stimulated emission and generate photons in an appropriate semiconductor material. In order to achieve lasing, two conditions must be met:

(1) optical gain equals optical loss, and (2) optical feedback is provided (by cleaved facets in Fabry–Perot or FP lasers, and by internal distributed gratings in DFB lasers).

Let us consider the two atomic energy states in a semiconductor *p–n* junction—the excited state (the conduction band for electrons) and the ground state (the valence band for electrons), as shown in Fig. 4.1. Figure 4.1 is the simplest band structure consisting of parabolic conduction and valence bands in the energy-wavevector space (E–**k** diagram). The reason why the curves are parabolic is because $E = P^2/(2m)$, where m is the atom mass and P is the momentum given by $h\mathbf{k}/(2\pi)$. Here h is Planck's constant, ~6.626×10^{-23} Joule-sec. Therefore, $E = (h^2\mathbf{k}^2)/(8\pi^2 m)$ and we see that E is proportional to $\mathbf{k}^2$. Electrons can be injected into the excited state or the conduction band (and holes into the valence band) by gaining energy from optical pumping or from forward-biasing the *p–n* junction. The excited electrons eventually return to their normal ground state and emit light through two fundamental processes mentioned earlier: spontaneous emission and stimulated emission. The spontaneous emission is an *incoherent* process because photons are emitted in random directions with no phase relationship among them. The stimulated emission, initiated by existing photons, is a *coherent* process because the emitted photon matches the original photon not only in energy (or equivalently, in wavelength) but also in propagation direction. Both spontaneous emission and stimulated emission are the results of *radiative* electron–hole

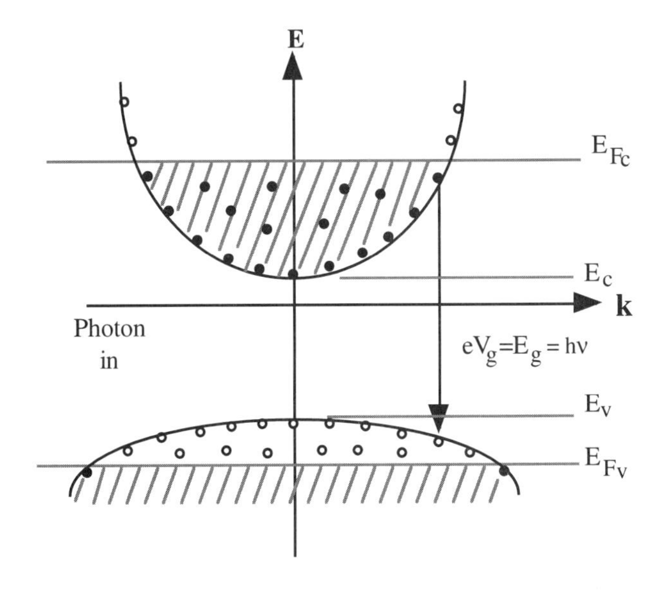

Figure 4.1: The valence and conduction band states, at electron energies E and momenta **k**. The quasi-Fermi levels are E_{F_c} and F_{F_v} Bandgap energy E_g is equal to the emitted photon energy $h\nu$.

recombination. Stimulated emission forms the basis for laser action. However, stimulated emission must overcome material absorption so that an significant amount of light can be emitted. This is achieved only when more population is in the excited state than in the ground state, which is called *population inversion* and is the prerequisite for laser operation. Population inversion can be efficiently realized by current injection across the *p–n* junction. Once the population inversion is achieved, the quasi-Fermi-level (the energy level at which the probability of occupancy is 50%) separation between the conduction band and the valence band exceeds the bandgap under forward biasing of the *p–n* junction; that is, $E_{Fc} - E_{Fv} > E_g$ in Fig. 4.1, and the active region is said to exhibit *optical gain*. It is commonly stated in the literature that when the optical gain is high enough, the gain peak g_p is linearly related to the injected carrier density N as

$$g_p = \overline{g_o}\,(N - N_0) \tag{4.1}$$

where N_0 is the transparent carrier density (typical value is in the range of $1–1.5 \times 10^{18}$ cm^{-3} for InGaAsP lasers) and $\overline{g_o}$ is the spatial gain constant in cm^2. Figure 4.2 [1] shows the calculated gain for a 1.3-μm InGaAsP active layer at different values of the injected carrier density N. We can see that when the injected carrier density N

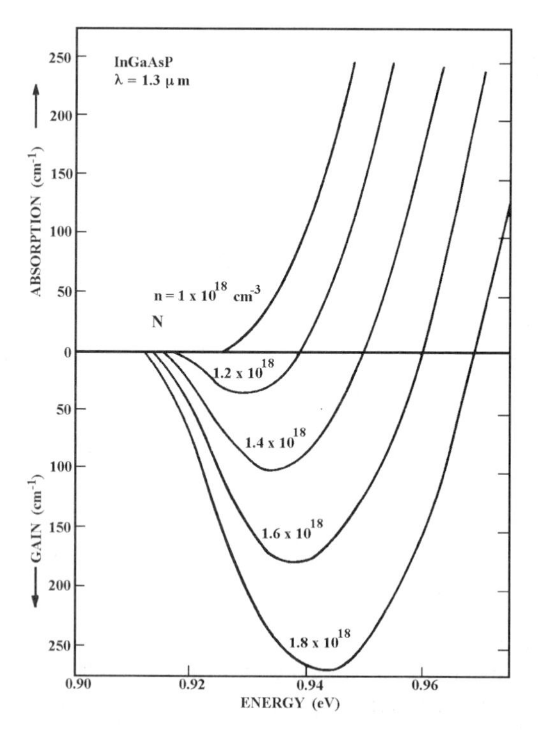

Figure 4.2: Calculated gain or absorption per unit length (g or α) versus λ of the emitted radiation for various values of carrier density N. Reprinted with permission from Ref. [1]. Copyright 1980 American Institute of Physics.

exceeds 1×10^{18} cm^{-3}, a positive optical gain g can be obtained. Note that as N increases, not only does g increase, but also the gain peak shifts toward shorter wavelength (or higher photon energy).

One may wonder why semiconductor materials such as Si and Ge do not emit light even when the population inversion condition is reached. This has to do with the predominant *nonradiative* recombination in these materials. Nonradiative recombination of electron–hole pairs will not generate photons. Nonradiative recombination mechanisms include recombination at defects, surface recombination, and the Auger recombination. Auger recombination is particularly important for long-wavelength (1.3–1.6 μm) semiconductor optical sources [2,3]. In the Auger recombination process, the energy released during electron–hole recombination is given to another electron or hole as kinetic energy, rather than producing light [3]. Because of the existence of nonradiative recombination, the efficiency in producing light is reduced. This effect can be described by the *internal quantum efficiency* (η_{int}), which indicates what fraction of the injected carriers is converted into photons. It is given by

$$\eta_{int} = \frac{R_{rr}}{R_{rr} + R_{nr}} , \tag{4.2}$$

where R_{rr} is the radiative recombination rate ($= R_{spon} + R_{stim}$), and R_{nr} is the nonradiative recombination rate. For Si and Ge, η_{int} is about 10^{-5}, mainly because of their *indirect* bandgap (note that a semiconductor is said to have a direct bandgap if the conduction-band minimum and the valence-band maximum occur at the same value of the electron wavevector, as shown in Fig. 4.1). This is why the materials such as Si and Ge used for electronic devices cannot produce light. For direct-bandgap materials such as GaAs or InP, η_i approaches 100% when stimulated emission dominates. The probability of radiative recombination is large in such a semiconductor because it is easy to conserve both energy and momentum during electron–hole recombination. By contrast, indirect-bandgap semiconductors release the energy from the electron–hole recombination to lattice phonons through a lattice vibration in a crystal which results in much lower radiative recombination rate.

4.2 Gain-Guided versus Index-Guided

Until the 1960s, semiconductor lasers used homostructure injection lasers, as shown in Fig. 4.3a. They exhibit high threshold current densities (J_{th} ~50 – 100 kA/cm^2 for GaAs homostructure lasers at room temperature) and cannot operate continuously at room temperature. Two major factors contribute to these high threshold currents: (1) the lack of carrier (electrons and holes) confinement, and (2) the poor optical confinement. In 1970's, these problems were solved by using *heterostructure* devices, as shown in Fig. 4.3b.

This kind of device is basically a layer of semiconductor material sandwiched between two cladding layers of another semiconductor material that has a relatively wider bandgap. The bandgap difference helps to confine electrons and holes to the

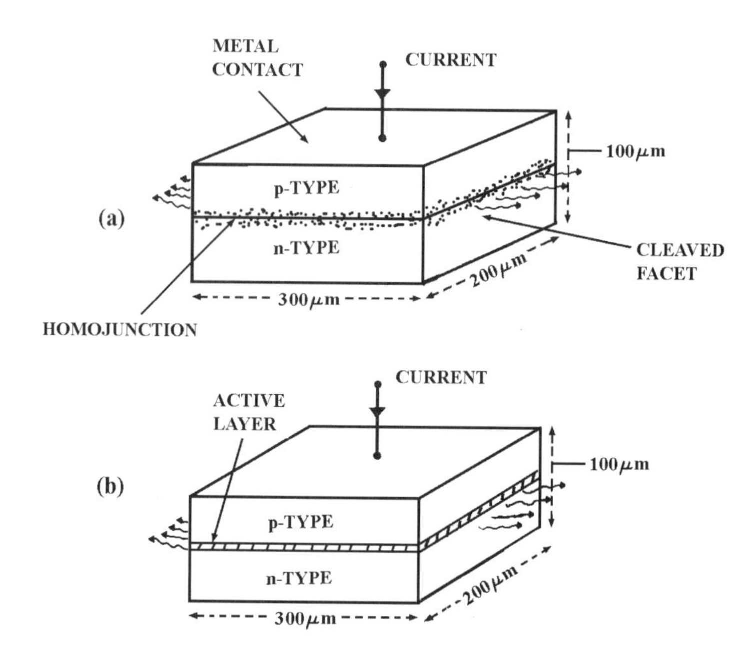

Figure 4.3: Schematic illustration of (a) homostructure and (b) double-heterostructure semiconductor lasers with their typical physical dimensions. The dotted area represents the depletion region in the vicinity of the homojunction. The hatched area shows the (~0.2 μm) active layer of a semiconductor material whose bandgap is slightly lower than that of the surrounding cladding layers. (After Ref. [3], with kind permission from Kluwer Academic Publishers.)

active layer, where they recombine to produce optical gain. This process is schematically illustrated in Fig. 4.4. Note that owing to the fact that cladding layers have lower refractive indices compared to that of the active layer, the optical mode is confined well in the active layer.

A double-heterostructure (DH) semiconductor laser such as the one shown in Fig. 4.3b does not incorporate any mechanism for the lateral confinement of the injected current or the optical mode. The threshold current of such a laser is directly proportional to the area of the current-pumped region. For example, for a typical DH GaAs laser at room temperature with J_{th} of 1 kA/cm^2, cavity length L of 250 μm, and contact width W of 100 μm, the threshold current is as high as about 250 mA $(= J_{th} \bullet L \bullet W)$.

To further reduce the threshold current, the stripe geometry which limits the injected carriers and the emitted photons to a narrow stripe (≤10 μm) was adopted. In gain-guided lasers, the injected carriers are laterally confined to a narrow stripe by means of a narrow contact. In index-guided structure, the photons and the carriers are laterally confined by using a two-dimensional heterostructure configuration which incorporates a built-in index distribution in the lateral direction.

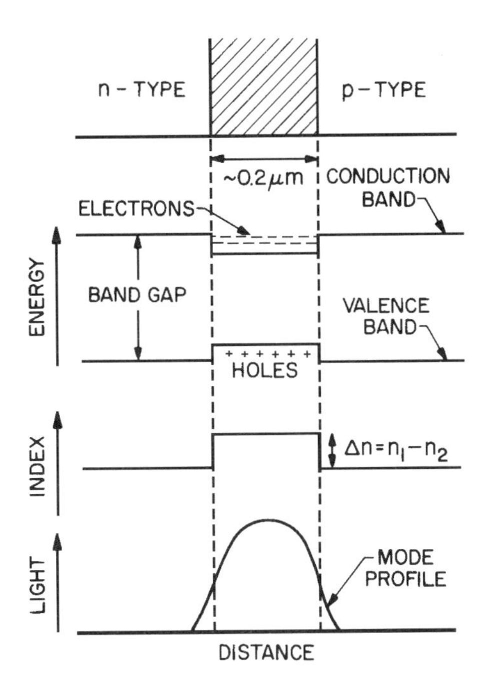

Figure 4.4: Schematic illustration of the simultaneous confinement of the charge carriers and the optical mode to the active region occurring in a double-heterostructure semiconductor laser. The active layer has a lower bandgap and a higher refractive index that those of the cladding layers. (After Ref. [3], with kind permission from Kluwer Academic Publishers.)

The lateral distribution of the injected carriers, the carrier density, the gain, and the optical power in a gain-guided laser are shown in Fig. 4.5. The width of the injected carrier distribution can clearly be much greater than the actual contact stripe width, because of the current spreading in the upper cladding layer, and because of carrier diffusion and drift in the active layer.

The difference in spectral characteristics of index-guided and gain-guided lasers arises from different nature of their lateral waveguiding. As will be seen in Section 4.4, the ratio of the intensities of the longitudinal modes is determined by the spontaneous emission rate into each mode and the excess loss at the oscillation frequency of each mode. It has been shown that the effective spontaneous emission rate into the modes in gain-guided lasers can be substantially larger than that of index-guided lasers [3]. Therefore, a much wider spectral envelope is observed for gain-guided lasers.

An example of index-guided laser is the buried-heterostructure (BH) laser shown in Fig. 4.6. In this heterostructure configuration, lateral and transverse optical confinements can both be achieved. Note that the active region is vertically sandwiched by two reversely biased *p–n* junctions, so that the injection current is

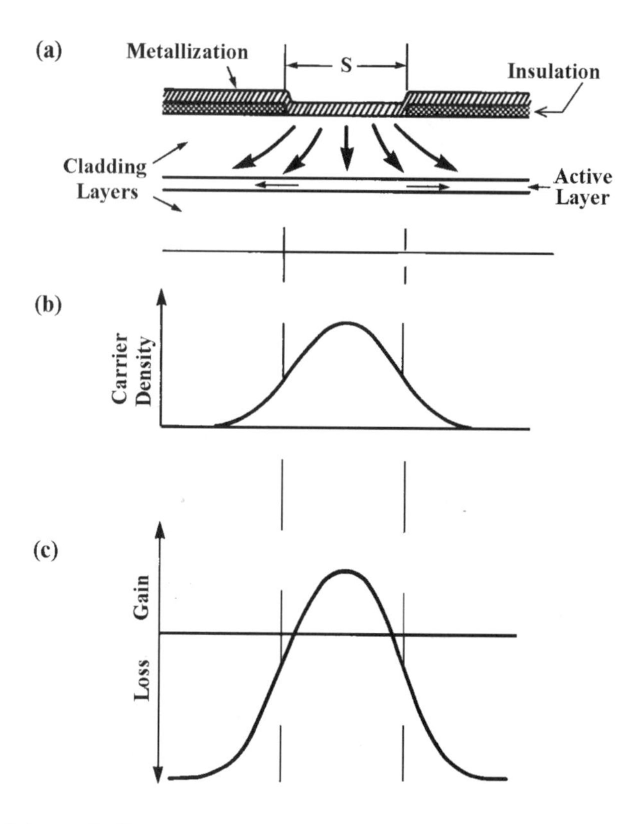

Figure 4.5: Schematic illustration of lateral distributions in a gain-guided stripe geometry laser: (a) laser cross-section with arrows showing current spreading in cladding layer and carrier diffusion and drift in active layer; (b) injected carrier distribution; (c) gain distribution; (d) optical power distribution. After Ref [4].

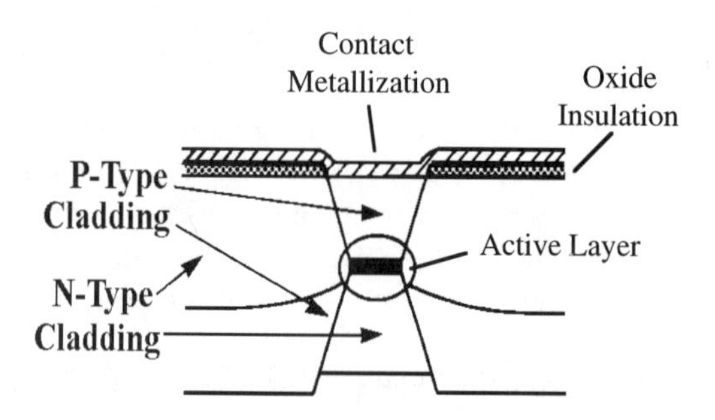

Figure 4.6: Schematic cross section of an index-guided buried-heterostructure laser. After Ref. [4].

confined and directed to the active region without wide spreading. BH lasers have a very linear light intensity versus bias current (*L*–I) curve and a low threshold current of 10 to 15 mA, can be operated at high output power (≥10 mW) in the fundamental transverse mode, and have a modulation bandwidth in excess of multigigahertz [2].

4.3 Semiconductor Materials

Not all semiconductors with direct bandgaps can be used in a heterostructure laser diode because possible lattice defects can be formed between two materials with lattice constants that differ by more than 0.1% [2]. To achieve this goal, ternary (e.g., AlGaAs) and quaternary compounds (e.g., InGaAsP), in which a fraction of the lattice sites in a binary semiconductor (e.g., GaAs or InP) is replaced by another element, can be used as confining materials. In the case of GaAs, a ternary compound $Al_xGa_{1-x}As$ can be made by replacing a fraction x of Ga atoms by Al atoms. For example, GaAs can be sandwiched between $Al_xGa_{1-x}As$ and form an optical source with a wavelength corresponding to 0.87 μm. GaAs has a lower bandgap energy than the cladding layer $Al_xGa_{1-x}As$, as is necessary for confining the carriers within the active region. Note that the emitted wavelength can be easily calculated according to $\lambda = hc/E_g$ or simply $1.24/E_g$ (μm), where E_g is in electron volts of the sandwiched layer. The same thing is true for a long-wavelength laser, in which case an $In_{1-x}Ga_xAs_yP_{1-y}$ alloy can be sandwiched between InP layers and emit light in the 0.93- to 1.65-μm region.

4.4 Lasing Threshold and Fabry–Perot Modes

In Section 4.1 we have seen that high injected carrier density can help achieve optical gain in a direct-bandgap semiconductor. This is the basic condition to obtain a semiconductor optical amplifier. However, a laser diode is an optical *oscillator* instead of an optical amplifier. Therefore, some sort of optical feedback mechanism is necessary to help achieve the oscillation condition. This was usually done by cleaving both end facets of a gain medium, or by applying high-reflection coatings to one or both of the facets. In the former case, the facet reflectivity is around 0.32 for InGaAsP lasers. The facet reflectivity loss must be overcome by the material optical gain in order to achieve lasing condition. The minimum optical gain that is required gives us the so called "lasing threshold." This threshold gain can be obtained as follows.

The electrical fields inside a laser cavity with a cavity length of L can be expressed as

$$E^{(+)}(z) = E^{(+)}(0) \bullet \exp\{\left[\frac{1}{2}\left(\Gamma g - \alpha_{\text{int}}\right) - j\beta\right] z\} \tag{4.3}$$

$$E^{(-)}(z) = E^{(-)}(L) \bullet \exp\{-\left[\frac{1}{2}\left(\Gamma g - \alpha_{\text{int}}\right) - j\beta\right](z - L)\}, \tag{4.4}$$

where g is the optical gain, Γ is the optical confinement factor, α_{int} is the sum of internal losses in the active and cladding layers, and β is the propagation constant. Note that the one-pass optical gain obtained from a *traveling-wave* optical amplifier (which has anti-reflection coated facets) is given by $\exp\left(\frac{1}{2}(\Gamma g - \alpha_{int})\,L\right)$. The threshold condition is achieved when the optical field travels a roundtrip in the optical cavity; it restores itself both in amplitude and in phase. This yields

$$\sqrt{R_1 R_2}\,\exp\left\{\left[\frac{1}{2}(\Gamma g - \alpha_{int}) - j\beta\right]2L\right\} = 1, \tag{4.5}$$

where R_1 and R_2 are the reflectivities of the front and back facets of the laser diode, respectively. Equating the real and imaginary parts of both sides, we obtain

$$\begin{cases} \sqrt{R_1 R_2}\,\exp\left\{\left[\Gamma g_{th} - \alpha_{int}\right]L\right\} = 1 \\ 2\,\beta_m\,L = 2\,m\,\pi \end{cases} \tag{4.6a}$$

or

$$\begin{cases} \Gamma g_{th} = \alpha_{int} + \dfrac{1}{2L}\ln\dfrac{1}{R_1 R_2} = \alpha_{int} + \alpha_{mir} \equiv \alpha \\ \lambda_m = \dfrac{2nL}{m}\quad. \end{cases} \tag{4.6b}$$

The first condition in Eq. (4.6b) states that the gain at threshold, Γg_{th}, should be equal to the total loss α, which includes the internal loss α_{int} and the average mirror loss per unit length $\alpha_{mir} = (1/2L)\ln(1/R_1R_2)$. For a typical reflectivity of the cleaved mirrors in an InGaAsP semiconductor laser, $R_1 = R_2 = 0.32$, and if $L = 250\ \mu$m, the mirror loss per unit length is 45 cm^{-1}. Typical $\Gamma = 0.5$, $\alpha_{int} = 30$ cm^{-1}; hence, $g_{th} \cong$ 150 cm^{-1}.

The second condition in Eq. (4.6b) shows that lasing may occur only in a set of discrete wavelengths λ_m, which are referred to as the longitudinal, or Fabry–Perot (FP), laser modes. The wavelength separation of the FP modes is obtained from

$$\frac{dm}{d\lambda} = \frac{d}{d\lambda}\left(\frac{2nL}{\lambda}\right) = \frac{2L}{\lambda^2}\left(\lambda\frac{dn}{d\lambda} - n\right) = -\frac{2\mathcal{N}L}{\lambda^2}, \tag{4.7}$$

where n is the refractive index of the active layer. $\mathcal{N}$ is the group index and is equal to $n - \lambda\frac{dn}{d\lambda}$. Thus, Eq. (4.7) can be written as

$$\Delta\lambda = \frac{\lambda^2}{2\mathcal{N}L}\Delta m, \tag{4.8}$$

which means

$$\text{FP mode separation} = \frac{\lambda^2}{2\mathcal{N}L} \cong \frac{\lambda^2}{2nL} \tag{4.9}$$

For an InGaAsP laser diode, $\mathcal{N} \cong 4$; hence, if $L = 250\ \mu$m and λ is in the wavelength range 1.3–1.6 μm, then $\Delta\lambda \cong 1$ nm.

If a laser diode has m longitudinal modes, the rate of change of photon density S_m in the mth mode oscillating at a frequency ω_m is given by a simplified rate equation [2]

$$dS_m / dt = (G_m - \gamma_m)\, S_m + R_{sp}(\omega_m), \tag{4.10}$$

where $G_m = \Gamma\, g\, v_g$ (gain due to stimulated emission), $\gamma_m = \alpha\, v_g$ (loss due to internal and mirror loss), v_g is the group velocity, and $R_{sp}(\omega_m)$ is the rate of spontaneously emitted photons into this mth mode. Recall that for a bulk laser diode, the gain curve (g) as a function of wavelength (shown in Fig. 4.2) reveals that the FP mode spacing in a bulk laser diode is much smaller than the spectral width of the gain profile (i.e., ~1 nm versus a few tens of nanometers), and it seems that the gain difference between main mode and its side modes can be very small (~0.1 cm^{-1}). However, the actual power difference among longitudinal modes can be 5–10 dB because under steady state, Eq. (4.10) gives

$$S_m = R_{sp}(\omega_m) / (\gamma_m - G_m), \tag{4.11}$$

and we see that the actual power difference among modes is determined by the balance between the gain and loss in the laser cavity (see Fig. 4.8a). Therefore, even though the difference among G_m may be small, the difference among S_m, which depends on $(\gamma_m - G_m)^{-1}$, can be large. For the central mode, the gain G_m is almost equal to the loss γ_m, and its power is the largest. Since G_m is smaller for the side modes, the amplitudes of the side modes are less than the central mode. In addition, because of the index-guiding effect which limits the spontaneous emission rate into side modes, the power difference between main mode and side modes is much more pronounced in index-guided lasers than in gain-guided lasers.

The multiple FP modes of an index-guided laser may spread over a few nanometers, and this spread in wavelength can cause higher-intensity noise (see Section 4.9.2) and higher fiber dispersion-induced nonlinear distortions (because each frequency component of the input wave arrives at the photodetector at different instant of time, and as a result, the recombined waveform is distorted; see Section 9.4. for detailed discussions). Therefore, it is desirable to have a laser diode operate only in a single longitudinal mode for systems with high linearity and low noise requirements. Although we can shorten the laser cavity to increase the mode spacing and enhance the gain difference between central and side modes, the side-mode suppression is usually not adequate. In order to obtain single longitudinal mode operation, many techniques have been proposed, such as external cavity laser, cleaved-coupled-cavity (C^3) laser, distributed Bragg reflector (DBR) laser, and DFB laser [5]. In recent years, almost all downstream HFC links have used multi-quantum-well (Section 4.7) DFB lasers as the directly modulated light source, and some systems have used CW high-power multi-quantum-well DFB lasers as the light sources for external modulation systems (see Section 7.5). Indeed, DFB lasers, which will be discussed further in Section 4.6, have proved to be a very reliable and high-performance light source for HFC systems.

4.5 Quantum Efficiency and Characteristic Temperature

Usually, the laser optical output power versus the injection current (*L–I*) characteristic is the very first property of a laser diode that a system designer must pay attention to. The slope and threshold current of an *L–I* curve are of practical importance, as both have to do with how efficiently the injected current can be converted to emitted optical power, especially under high-temperature operation conditions. An example of typical 1.3-μm laser diode *L–I* curves under different operation temperatures is shown in Fig. 4.7. We note that as temperature increases, the slope of the *L–I* curve decreases and the threshold current (I_{th}) increases (exponentially). To understand these phenomena, we first note that the slope efficiency of the *L–I* curve for $I > I_{th}$ is [3]

$$\frac{dP_o}{dI} = \eta_{facet} \frac{h\nu}{q} \cdot \frac{\eta_{\text{int}} \alpha_m}{\alpha_m + \alpha_{\text{int}}}, \tag{4.12}$$

where η_{facet} is the fraction of optical power emitted from the laser front facet, $h\nu$ is the photon energy, q is the electron charge, and all other terms have been defined previously. When the operation temperature increases, η_{int} [defined in Eq. (4.2)] decreases, because Auger recombination rate and heterobarrier leakage increase exponentially with temperature in InGaAsP lasers. Consequently, the slope efficiency of the *L–I* curves is decreased accordingly.

Note that it is quite common in the literature that the slope efficiency is specified in terms of the *differential quantum efficiency,* defined by

$$\eta_d = \frac{\eta_{\text{int}} \alpha_m}{\alpha_m + \alpha_{\text{int}}}. \tag{4.13}$$

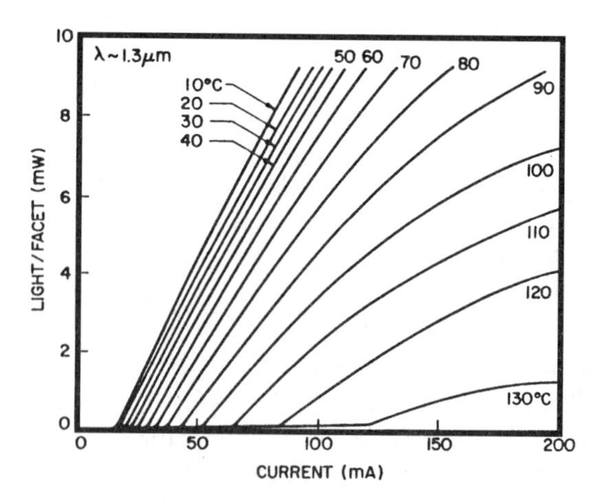

Figure 4.7: *L–I* curves at several operation temperatures for a 1.3-μm BH laser. After Ref. [2]

Note that the difference between dP_o/dI and η_d is $\eta_{facet} h\nu / q$.

Having discussed the temperature-dependent slope efficiency and differential quantum efficiency, we now turn our attention to the strong dependence of threshold current on temperature in long-wavelength lasers. The threshold current is found experimentally to increase exponentially with temperature as

$$I_{th} = I_o \exp (T/T_o), \tag{4.14}$$

where T_o is a constant and is a *characteristic temperature* often used to express the temperature sensitivity at threshold current. For InGaAsP lasers, T_o is typically in the range of 50–100 K. By contrast, T_o is in the range of 100–150 K for GaAs lasers [6]. The exponential increase of threshold current with temperature can be attributed to optical gain decrease ($g \propto 1/T$), internal loss increase ($\alpha_{int} \propto T$), overflow of injected carriers into confinement layers (in quantum-well lasers) due to Auger recombination increase ($\propto \exp(\gamma_c T)$), and carrier leakage rate increase ($\propto \exp(\gamma_l T)$) [6]. Among these factors, the temperature dependence of the optical gain is considered a dominant effect in the long-wavelength semiconductor [7,8].

InGaAsP lasers have a typical temperature-dependent wavelength shift of about 0.1 nm/°C. Because of the temperature sensitivity of InGaAsP lasers, it is often necessary to control their temperature through a built-in thermal-electric (TE) cooler (see Section 4.11). "High-temperature lasers" which exhibit small temperature dependence of *L–I* slope efficiency and threshold current have been built using multi-quantum-well lasers, as will be discussed in Section 4.7.

4.6 DFB Laser

A Fabry–Perot laser uses end reflectors (at the end facets) to provide optical feedback. On the other hand, if a periodic medium is provided with sufficient gain at frequencies near the Bragg wavelength λ_B (see Section 3.3), which satisfies the condition

$$m \lambda_B = 2 n \Lambda, \tag{4.15}$$

oscillation can result without the benefit of end reflectors. In Eq. (4.15), the integer m represents the order of Bragg diffraction (usually $m = 1$ is used, but sometimes $m = 2$ is also used), n is the index of the mode in the laser, and Λ is the grating period. This phenomenon has been discussed in Section 3.3 (fiber gratings) and the operation principle of a fiber grating can be applied to a DFB laser. A grating is now etched into a cladding layer on either side of the active layer, as shown in Fig. 4.8b. Although the grating is physically removed from the active region, the feedback can be provided by the continuous coherent backscattering from the periodic perturbation (we can deem the refractive index being modified as $n(z) = n + \Delta n \bullet \cos(Kz)$, where K is the grating wave number and $\Delta n/n \ll 1$). The periodic DFB lasers can also be made with cleaved facets (in addition to distributed gratings) to provide necessary optical feedback [9], but we will just point out the operation principle of a DFB laser based only on distributed-grating feedback.

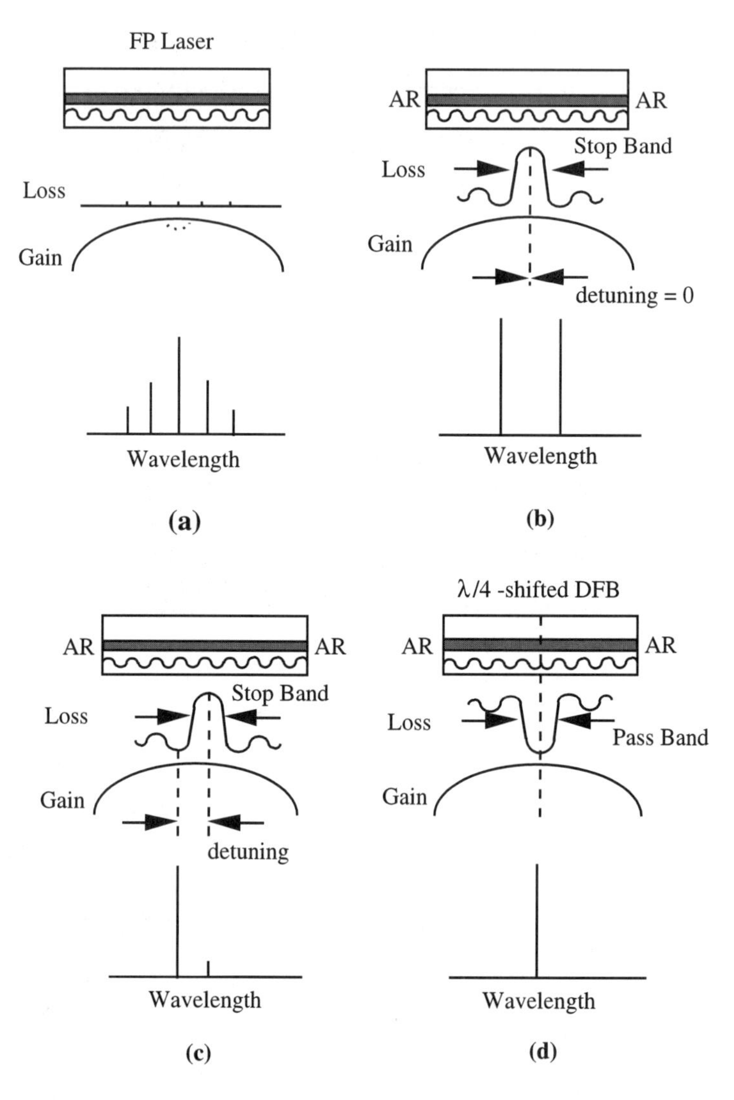

Figure 4.8: Basic structure, optical loss, optical gain, and the corresponding emitting wavelengths of (a) an FP laser; and a DFB laser with (b) zero detuning and (c) non-zero detuning between gain peak and the Bragg wavelength, and (d) a $\lambda/4$-shifted DFB laser.

As was shown in the derivations given in Section 3.3, with a distributed grating, a stopband will be formed which is centered at the Bragg wavelength (see Fig. 3.16). The bandwidth of the stopband depends on the coupling coefficient κ. A comparison of the spectral gain versus loss between a conventional FP laser and a DFB laser is given in Fig. 4.8 a–d. In Fig. 4.8b, the two modes with highest gain are just outside the stopband. This double-mode behavior, which ideally gives two identical

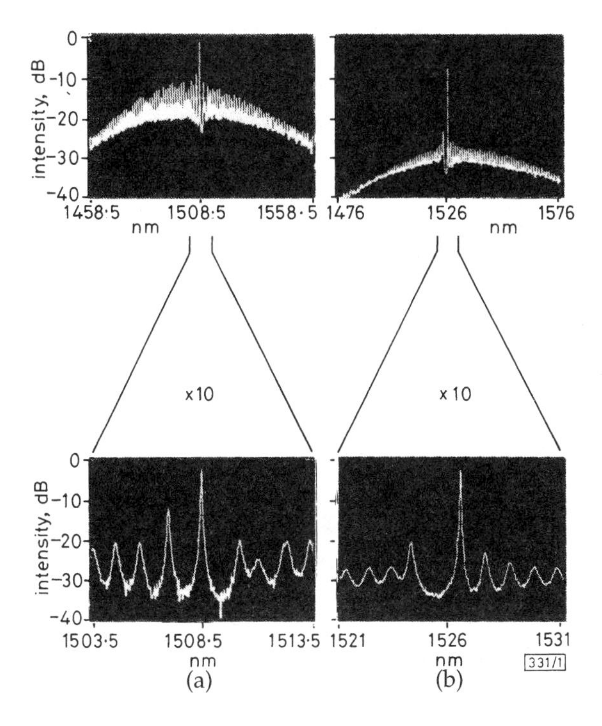

Figure 4.9: DFB lasers with two competing modes with power difference of (a) ~ 10 dB, and (b) ~18 dB. The stopband is about 1 or 2 nm in each case. After Ref. [10].

modes separated by about 1 nm, may or may not exist depending on the mirror asymmetries and/or grating phases. In practice, a random selection of DFB lasers will usually contain some devices with two dominant modes, as shown in Fig. 4.9a, where the power difference between the two modes is only 10 dB [10].

The rest of the devices can have essentially a dominant mode, as shown in Fig. 4.9b, in which the two modes can differ by about 18 dB. The stop band can be clearly seen in Fig. 4.9 to be as wide as 1–2 nm.

To ensure that only one longitudinal mode exists, the design in Fig. 4.8b can be modified to detune the Bragg wavelength from the gain peak, as shown in Fig. 4.8c, or a $\lambda/4$-shifted grating structure [11] can be used as shown in Fig. 4.8d. In Fig. 4.8c, the oscillating mode is purposely adjusted to be at the shorter wavelength side of the gain peak in order to obtain higher differential gain [12]. The operation principle a of the $\lambda/4$-shifted DFB laser is explained as follows.

As was shown in Fig. 3.16, the periodic grating acts as a reflector in its stopband ($|\delta| < |k|$) and has a set of transmission resonances in the passband ($|k| < |\delta|$). It is possible to achieve transmission within the stopband, if one spaces two periodic gratings by one-quarter wavelength ($\lambda_B/4$). This phenomenon could be understood as follows. First let us recall that the wavelength at the center of the stopband, λ_B, is equal to 2Λ [Eq. (3.14)], and that the spatial dependences of the forward wave $\bar{a}$ and

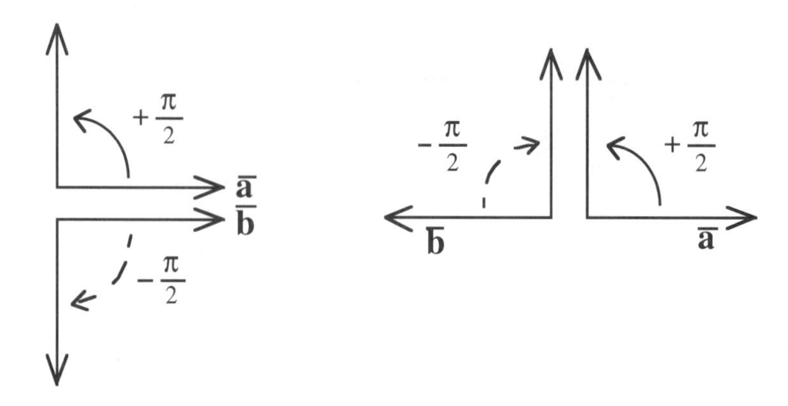

Figure 4.10: Phasor diagram showing the phase relationship between the coupled waves before and after $\lambda/4$-phase shifted grating. (a) The originally destructive waves (in the stop band) become constructive after the phase shift. (b) The originally constructive waves (in the transmission band) become destructive after the phase shift.

backward wave $\bar{b}$ are exp(−j(π/Λ)z) and exp(+j(π/Λ)z), respectively (from Eqs. (3.17) and (3.18)). For a $\lambda_B/4$ or $\Lambda/2$ phase shift, we will have a $-\pi/2$ and $+\pi/2$ phase shift on $\bar{a}$ and $\bar{b}$, respectively. Therefore, the originally constructively combined waves now become destructively combined, and the originally destructive waves (i.e., the "stopband") now become constructively combined, as illustrated in Fig. 4.10. Therefore, instead of obtaining a pair of modes outside the stopband, we can now obtain just one longitudinal mode inside the stopband, as shown in Fig. 4.8(d). Therefore, a $\lambda_B/4$-shifted DFB laser can exhibit a dominant single longitudinal mode with no concern about the competing modes that have such a severe deteriorating effect on optical fiber transmissions.

4.7 Multi-Quantum-Well Lasers

Most of the commercially available DFB lasers which can carry a large number of analog CATV channels are multi-quantum-well(MQW) lasers. The main difference between a QW laser and a conventional (bulk) semiconductor laser is mostly in the thickness of the active region. The active region is about 1000–3000 Å (0.1–0.3 μm) thick in the latter, while it is between 50 and 100 Å per well (corresponds to 7–15 atomic layers) in the former. Multiple (typically around 5) active layers of thickness 50–100 Å were used to improve the device performance, and such devices are called multi-quantum-well (MQW) lasers. The carrier motion normal to the wells in the active layer in these structures is restricted. Furthermore, as the thickness of each well L_z is comparable to the de Broglie wavelength ($\lambda \approx h/P$), the kinetic energy of the carriers moving in that direction is quantized (sometimes termed "quantum size

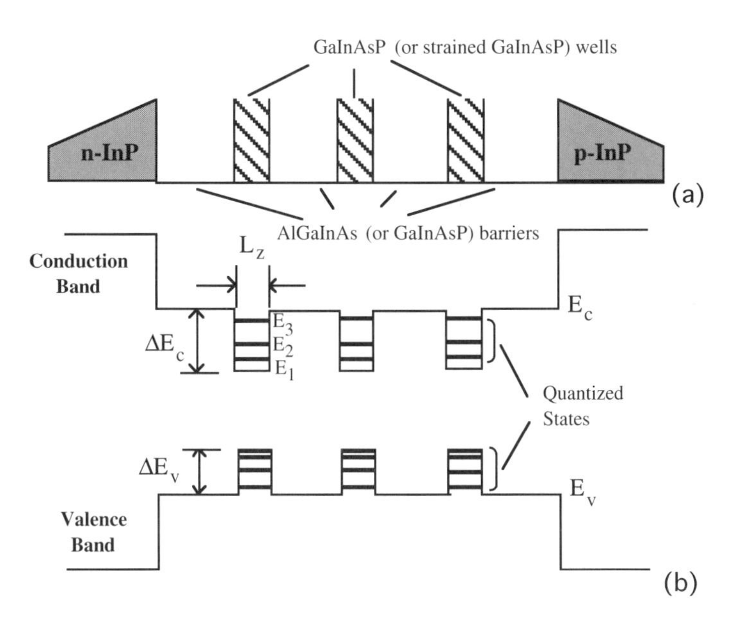

Figure 4.11: (a) AlGaInAs/InP (or GaInAsP/InP) multiple quantum wells. (b) Quantized states in wells ($\Delta E_c = 0.72\ \Delta E_g$ in AlGaInAs/InP materials, $\Delta E_c = 0.4\ \Delta E_g$ in GaInAsP/InP materials, and $\Delta E_c + \Delta E_v = \Delta E_g$).

effects"). As a result, the quantized carrier (electron or hole) energy levels in the wells (see Fig. 4.11b) are (assuming infinitely deep wells) [13]

$$E(n,k_x,k_y) = E_n + \frac{\hbar^2}{2m^*}\left[k_x^2 + k_y^2\right],\ n = 1,2,3,..., \tag{4.16}$$

where E_n (see Fig. 4.11b) is the nth confined particle (particle-in-a box) energy level for carrier motion normal to the well and is given by

$$E_n = \frac{\hbar^2}{2m^*}\left(\frac{n\pi}{L_z}\right)^2, \quad n = 1,2,3,.... \tag{4.17}$$

In the preceding equations, $\hbar$ is the Planck's constant ($h/2\pi$), m^* is the effective mass of the carrier, and k_x and k_y are the wave-vector components along the x and y directions, respectively.

The second term on the right-hand side of Eq. (4.16) represents a continuum of energy states along the x and y directions. As shown in Fig. 4.11 (this energy diagram should be compared with that of a bulk laser in Fig. 4.4), below the energy E_1 there are no allowed electron states. Then, as shown in Fig. 4.12, a subband with a constant density of permitted states per unit volume [13,14]

$$\rho(E) = \frac{m_n^*}{\pi\hbar^2 L_z} \tag{4.18}$$

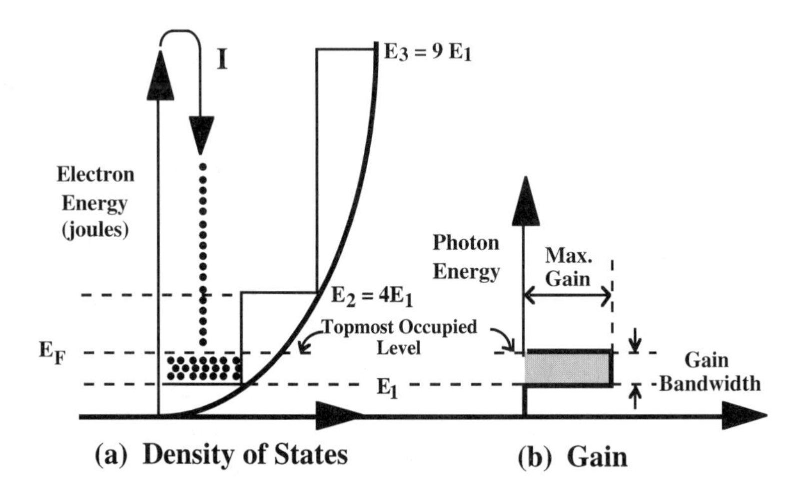

Figure 4.12: (a) Electron energy versus density of states for bulk lasers and quantum-well lasers, and the corresponding (b) photon energy versus gain. After Ref. [14] (© 1989 IEEE).

begins at E_1, followed by another subband (step) with a constant block of states appearing at E_2, etc. Note that, as shown in Fig. 4.12, $E_2 = 4E_1$, $E_3 = 9E_1$, etc., according to Eq. (4.17). The staircase-like function in Fig. 4.12 depicts the electron energy versus the density of available electron states. This is in contrast to the half-parabola, also shown in Fig. 4.12, that corresponds to the density of states of a bulk laser [2,13,14].

Note that the mean number of states actually occupied in an incremental energy range dE is given by $\rho(E)\bullet f(E)dE$, where $f(E)$ is the Fermi–Dirac probability of occupancy. The steps at E_1 and E_2 and beyond are the signature of the two-dimensional nature of the confined carriers. We can see that at a given current I all the electron states up to some uppermost level E_F (the quasi-Fermi level) are filled up.

Following the analogy given in Ref. [14], we can think of a quantum-well structure as having an electron fluid "sink" with a very shallow, high, and flat bottom, while the "sink" of a bulk laser diode has a deep, low, and round bottom. Therefore, much less injection current is needed to reach a certain energy level (equivalent to the water level in a sink) in the case of quantum-well lasers. In other words, a quantum-well laser has a group of electrons with nearly the same (and maximum) energy which can recombine with a group of holes, also with nearly the same (and maximum) energy. It follows that the maximum gain becomes available at very small injected currents. And here lies one of the main advantages of the QW laser—large gains at small currents, which implies that the threshold current of a QW laser is much lower than that of a bulk semiconductor laser. Therefore, as opposed to the *linear* relationship between the peak gain and injection carrier density N given earlier in Eq. (4.1), the peak optical gain of an MQW laser has an empirical *logarithmic* dependency on N [13],

$$g_p = g_o^{'}(1 + ln\frac{N}{N_o^{'}}) \tag{4.19}$$

where $N'_o = N_o e$ and is obtained at $g_p = 0$, and g'_o is obtained at $N = N'_o$.

Strained[1] MQW semiconductor lasers with a threshold current as low as 0.56 mA [15] and 1 mA [16] have been demonstrated for 1.3-μm InGaAsP/InP lasers at room temperature. Both MQW lasers used relatively short-cavity (100 and 200 μm, respectively) and optimized high-reflection coatings which are effective in achieving low threshold due to the reduction of cavity loss [17].

In addition to low threshold current, MQW lasers also exhibit the following features: high quantum efficiency and high output power because of the low losses and the excellent uniformity of the epitaxial QW material; narrow linewidth because of fewer spontaneous emissions; and less temperature dependence because a QW laser can be designed so that the carrier can be confined or recaptured during high-temperature operations [18,19].

The last feature mentioned is particularly important to applications such as return-channel lasers in applications such as fiber-in-the-loop and HFC systems. In those applications, the laser diode carrying the upstream signals is usually located in an outdoor enclosure. Therefore, "high-temperature" and "uncooled" laser diodes without the need of TE cooler, while maintaining high quantum efficiency and steady threshold current, are quite useful. Typical *L–I* curves of these lasers are illustrated in Fig. 4.13.

Figure 4.13a represents the performances of an uncooled 1.3-μm AlGaInAs/InP strained-layer quantum-well laser [18]. The laser shows excellent linearity and less than a 0.3-dB drop in differential quantum efficiency when the heat sink temperature changes from 25 to 100°C. This laser diode used AlGaInAs/InP material system which has a higher conduction band offset $\Delta E_c = 0.72\ \Delta E_g$ (see Fig. 4.11), instead of conventional GaInAsP/InP material system which has $\Delta E_c = 0.4\ \Delta E_g$ to provide a better carrier confinement under high-temperature operation. Figure 4.13b shows the performance of a conventional strained-quantum-well GaInAsP/InP laser under extremely high-temperature operation up to 170°C [19]. The laser used more wells (limited by the degradation of crystal quality of the wells due to the critical thickness) to improve the temperature characteristics, an effect which is due to the reduction of the gain saturation and threshold current density and the improvement of optical confinement in the well.

---

[1]Strained QWs use a material which has a different native lattice constant than the surrounding barrier material. If the QW's native lattice constant is larger than the surrounding lattice constant, the lattice is said to be under *compressive* strain. If the converse is true, the QW is under *tensile* strain.

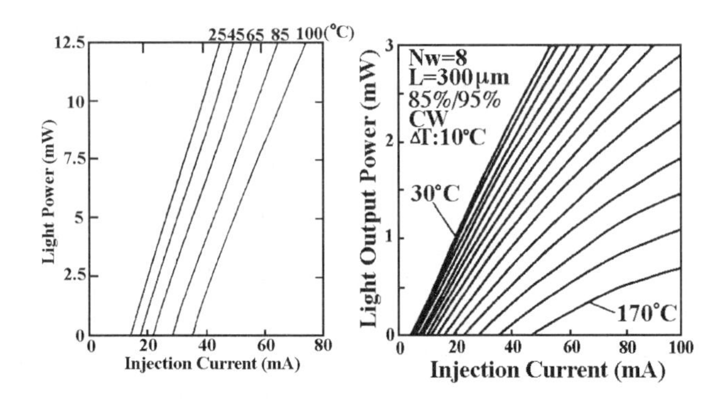

Figure 4.13: Light–current characteristics of "high-temperature" compressive-strained 300-μm-long 1.3-μm MQW lasers: (a) AlGaInAs/InP 5-well laser with 70% high reflection coating on the rear facet at various sink temperature After Ref. [18]., and (b) GaInAsP/InP 8-well laser with 85%/95% coatings with temperature from 30 to 170°C and steps of 10°C. After Ref. [19] (© 1994 IEEE).

4.8 Modulation Characteristics

So far we have only discussed the DC characteristics of a laser diode. In this section, we will examine the key modulation characteristics of semiconductor laser diodes. The understanding of these basic characteristics, including modal rate equations, small-signal modulation response, frequency chirp, and equivalent circuit models, can help an HFC system designer understand what critical parameters are controlling the laser linearity and noise levels.

4.8.1 Modal Rate Equations

Assuming a uniform distribution of injected carriers over the volume V of the active region, and assuming the photon density S_m (of each longitudinal mode) is constant within the active layer, then the time rates of change, $\frac{dN}{dt}$ and, $\frac{dS_m}{dt}$ of a multi-longitudinal-mode (MLM) laser, are given by the multimode rate equations [5]

$$\frac{dN}{dt} = \frac{I}{qV} - \frac{N}{\tau_n} - g_o\left(N - N_o\right)\left(1 - \varepsilon S_c\right)S_c - \sum_{m \neq 0} g_m (N - N_o)(1 - \varepsilon S_m)S_m \tag{4.20a}$$

$$\frac{dS_m}{dt} = \Gamma g_m\left(N - N_o\right)\left(1 - \varepsilon S_m\right)S_m - \frac{S_m}{\tau_p} + \Gamma\gamma\frac{N}{\tau_n} \quad , \tag{4.20b}$$

where the various parameters are defined as follows:

N: electron density (m^{-3})

S: photon density (m^{-3})

I: injection current (Amp)

N_o: transparent carrier density (m^{-3})

S_c,S_m: photon density of the (central mode) or (side mode m) (m^{-3})

V: volume of the active layer (m^3)

ε: gain compression coefficient (m^3)

τ_n: electron lifetime (ns)

Γ: optical mode confinement factor

τ_p: photon lifetime (ps)

γ: fraction of total spontaneous emission coupled into a cavity mode (~10^{-4})

g_o ($= v_g \bar{g}_o$): differential gain coefficient at central mode λ_o (m^3/s) [note that $\bar{g}_o$ was defined in Eq. (4.1), and v_g is the group velocity]

g_m: differential gain coefficient at λ_m (m^3/s)

Equation (4.20a) states that the time rate of change of carriers in the cavity is proportional to (1) the rate of injected carrier density, (2) the rate of loss due to spontaneous emission, (3) the rate of reduction in carriers due to stimulated emission of the central mode, and (4) the rate of reduction in carriers due to stimulated emission of the side modes. Equation (4.20b) states that the time rate of change of photon density is proportional to (1) the fraction Γ of photons due to stimulated emission which are confined in the active layer, (2) the rate of loss of photons [note that the photon lifetime τ_p can be viewed as the average time a photon spends in the cavity before it is lost by internal absorption or transmission through the facets, i.e., $(\tau_p)^{-1} = v_g(\alpha_{int} + \alpha_{mir})$, and (3) the rate of increase of photon density due to spontaneous emission in the active layer. The factor γ accounts for the fraction of the isotropically radiated spontaneous emission N/τ_n coupled into the mth longitudinal mode. Note the resemblance between Eq. (4.20b) and the simplified rate equation (4.10) which was given earlier, $G_m = \Gamma g_m(N - N_o)(1 - \varepsilon S_m)$, $\gamma_m = 1/\tau_p$, and $R_{sp}(\omega_m) = \gamma\Gamma N/\tau_n$.

Note that gain compression is included in the rate equations by the factor $(1 - \varepsilon S_m)$. Gain compression is due to a phenomenon called spectral hole burning (also called intraband gain saturation) and has an important implication for the linearity characteristics of a laser diode (see Problem 4.1). Spectral hole burning is a small localized gain reduction at the energy corresponding to the wavelength of an oscillating mode (see the dashed curve in Fig. 4.8a). It is caused by the fact that when stimulated emission rate is high, the localized reduction in the number of conduction-band states cannot be quickly refilled from the large pool of electrons in the conduction band. The refilling process does not occur instantaneously because it is a scattering process with electron–electron and electron–phonon collisions which has an intraband relaxation time on the order of 1 ps [20]. Considering just one oscillating mode (S_c), the reduction in gain due to spectral hole burning is roughly proportional to the net stimulated emission rate $g_o(N - N_o)S_c$ and the intraband relaxation time. Thus, the gain reduction can be written as

$$\Delta g \approx g_o\left(N - N_o\right) \bullet S_c \bullet \varepsilon, \tag{4.21}$$

where ε is a parameter which is proportional to intraband relaxation time. Therefore, the net gain becomes [5]

$$\begin{aligned} g &= g_o(N - N_o) - \Delta g \\ &= g_o(N - N_o)(1 - \varepsilon S_c). \end{aligned} \tag{4.22}$$

This is the term we used in Eq. (4.20). When S_c is large, the gain compression term $(1 - \varepsilon S_c)$ due to spectral hole burning reduces the gain for S_c. Note that the average photon density in the modal volume S_c is linearly proportional to the optical output power per facet P_o of the laser because

P_o = (energy of photon)(photon density)(volume of the optical mode)(escape rate of the photons per facet)

$$= (h\nu)(S_c)(V)(v_g \alpha_{mir} \eta_{facet}). \tag{4.23}$$

Therefore, the modulation of the bias current I results in modulation of S_c [Eq. (4.20)], and consequently the output optical power P_o.

For a single-longitudinal-mode (SLM) laser such as a DFB laser, the single-mode rate equations can be simplified as

$$\frac{dN}{dt} = \frac{I}{qV} - \frac{N}{\tau_n} - g_o\left(N - No\right)\left(1 - \varepsilon S\right)S \tag{4.24a}$$

$$\frac{dS}{dt} = \Gamma g_o\left(N - N_o\right)\left(1 - \varepsilon S\right)S - \frac{S}{\tau_p} + \Gamma\gamma\frac{N}{\tau_n}. \tag{4.24b}$$

Typical parameter values for a CATV DFB laser are V = width (1–2 μm) × height (~0.2 μm) × length (~300 μm), $\tau_p \approx$ 1–3 ps, $\tau_n \approx$ 1–3 ns, $\Gamma \approx$ 0.3–0.5, $\gamma \approx 10^{-4}$, $g_o \approx 1\text{–}3 \times 10^{-12}$ m^3/s, $N_o \approx 1\text{–}1.5 \times 10^{24}$ m^{-3}, and $\varepsilon \approx 1\text{–}5 \times 10^{-23}$ m^3. In order to obtain valid theoretical or numerical results which are based on these parameters, so that they can be experimentally verified, it is important to carefully extract meaningful laser parameters through various measurement techniques [21,22].

The rate equations for a single-quantum-well laser were used to show that the carrier transport across the separate confinement heterostructure (SCH) region (see Fig. 4.14) and the barriers can be critical in determining the modulation bandwidth [23]. The rate equations are

$$\frac{dN_b}{dt} = \frac{I}{qV_{SCH}} - \frac{N_b}{\tau_s} + \frac{N_w(V_w/V_{SCH})}{\tau_e} \tag{4.25a}$$

$$\frac{dN_w}{dt} = \frac{N_b(V_{SCH}/V_w)}{\tau_s} - \frac{N_w}{\tau_n} - \frac{N_w}{\tau_e} - g_o S(1 - \varepsilon S) \tag{4.25b}$$

$$\frac{dS}{dt} = \Gamma g_o S(1 - \varepsilon S) - \frac{S}{\tau_p}, \tag{4.25c}$$

where N_w and N_b are the carrier densities in the quantum well and the SCH layer, respectively; V_{SCH} and V_w are the volumes of the SCH and quantum well, respectively; τ_s is the carrier transport time across the SCH; and τ_e is the thermionic emission lifetime ($\propto exp(E_b/kT)$, where E_b is the effective barrier height, k is the Boltzmann constant, and T is the temperature in K). A closed-form solution of the QW laser modulation bandwidth using these equations is given in the next section.

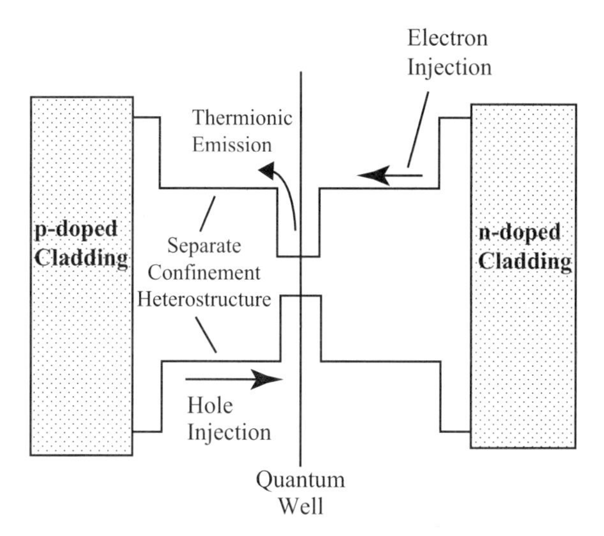

Figure 4.14: Schematic of a single-quantum-well laser with a separate confinement heterostructure used in the carrier transport model. After Ref. [23] (© 1992 IEEE).

4.8.2 Small-Signal Intensity Modulation Response and High-Speed Lasers

The small-signal modulation response is of practical interest because it is easy to measure and can provide useful information for the modulation dynamics of large-signal modulation. From the results of small-signal analysis, we will see what the controlling parameters are in making a high-speed laser diode. An intrinsically high-speed laser diode is essential for multichannel AM-VSB video signal transmission. The reason why transporting RF video channels below 1 GHz will require a laser diode with an intrinsically high bandwidth (i.e., not counting the package and chip parasitics) in excess of 8–10 GHz will become clear in Section 4.10.

Considering an SLM laser, we can obtain an analytical expression for the small-signal response by linearizing the rate equations (4.20a) and (4.20b). This can be done by first letting

$$
\begin{aligned}
N &= n_o + ne^{j\omega t} \\
I &= I_o + ie^{j\omega t} \\
S &= S_o + se^{j\omega t},
\end{aligned}
\tag{4.26}
$$

where ω is the radian frequency of the modulation signal. Equations (4.26) are substituted into the rate equations, and terms in $e^{-j\omega t}$ on both sides of the rate equations are equated, while products of two or more small-signal terms are neglected. This approach leads to the intensity modulation response given by (with small terms neglected) [5,24]

$$FR(\omega) = \frac{s(\omega)/S_o}{i(\omega)/I_o} = \frac{\omega_o^2}{\omega_o^2 - \omega^2 + j\omega\left\{(\gamma\Gamma I_{th}/qVS_o) + (S_o\varepsilon/\tau_p)\right\}}, \tag{4.27}$$

where I_{th} is the threshold current, and

$$\omega_o^2 = \frac{g_o S_o}{\tau_p}, \tag{4.28}$$

and ω_o is the so called laser resonance frequency. Equation (4.27) is a conventional second-order, low-pass transfer function with damping characteristics controlled by the magnitude of the $j\omega$ term in the denominator. For bias currents well above threshold, the term $(\gamma\Gamma I_{th}/qVS_o)$ can be neglected, and Eq. (4.27) becomes

$$FR(\omega) = \frac{\omega_o^2}{\omega_o^2 - \omega^2 + j\omega(S_o\varepsilon/\tau_p)}. \tag{4.29}$$

The corresponding damping constant is $\tau_p/(S_o\varepsilon)$. As illustrated in Fig. 4.15, the electrical power spectrum of this frequency response is

$$|FR(\omega)|^2 = \frac{\omega_o^4}{(\omega_o^2 - \omega^2)^2 + \omega^2(S_o\varepsilon/\tau_p)^2}. \tag{4.30}$$

From Eq. (4.30), it can be shown that

$$\omega_{3\text{-dB}} = \sqrt{1+\sqrt{2}}\,\omega_o \approx 1.55\omega_o. \tag{4.31}$$

Since from Eq. (4.28), $\omega_o \propto \sqrt{S_o} \propto \sqrt{P_o}$, we know that the useful laser modulation bandwidth is proportional to $\sqrt{P_o}$. The 3-dB bandwidth of $1.55\omega_o$ may be of use to baseband digital data transmission, but is not meaningful to HFC systems because all RF carriers must be kept away from the resonance peak to avoid the resonance-induced dynamic nonlinear distortions (see Section 4.10). Also, from Eq. (4.30),

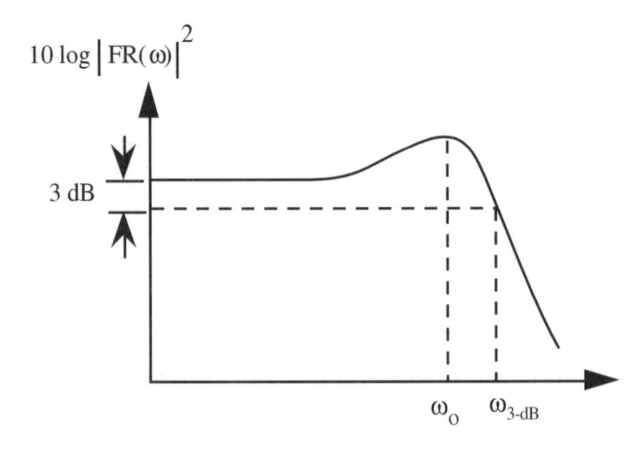

Figure 4.15: The electrical power spectrum of typical laser frequency response.

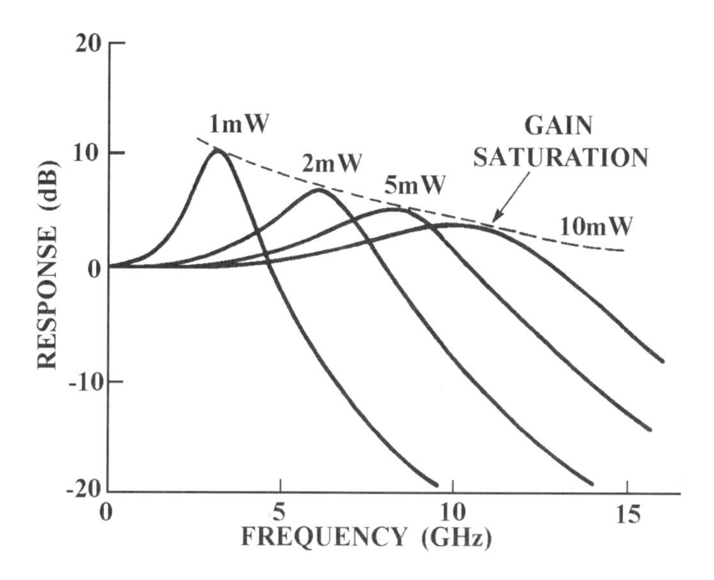

Figure 4.16: The intrinsic frequency response of a typical long-wavelength high-speed laser diode (without counting package and chip parasitics). After Ref. [25].

$|FR(\omega_0)|^2$ is inversely proportional to S_0 (note that ω_0^2 is proportional to S_0); hence, the resonance peak decreases with increasing output power. An example is illustrated in Fig. 4.16. We can see that the peak resonance frequency increases from about 3 to 10 GHz which approximately corresponds to $\sqrt{10\ mW/1\ mW}$. Furthermore, when the output is increased from 1 to 10 mW, the gain peak is decreased by about 6 dB (not quite reaching the theoretical value of ~10 dB).

It is known that, near its resonance frequency ω_0, a laser diode is highly nonlinear (see Section 4.10) and has a high intensity noise level (see Section 4.9). Therefore, it is always desirable to push this resonance frequency away from the operating frequency range, and a laser diode with a high modulation bandwidth is mandatory. But how can the laser diode bandwidth be increased? From Eq. (4.28), we can see that there are a number of approaches to increasing the modulation bandwidth: (1) increase the photon density by operating at higher power, or by using a laser with smaller active region volume (e.g., decrease the width of the active region); (2) increase the optical loss $(\tau_p)^{-1}$ by using a shorter cavity because $(\tau_p)^{-1} = v_g(\alpha_{int} + \frac{1}{2L} ln \frac{1}{R_1R_2})$; and (3) increase the differential gain coefficient g_0 by decreasing the temperature (Fig. 4.17) [26], by detuning the DFB Bragg wavelength toward the short-wavelength side of the gain peak [12], or by using quantum-well lasers.

By using a similar small-signal analysis, the resonance frequency of a single QW laser was obtained from Eqs. (4.25a–c) as [23]

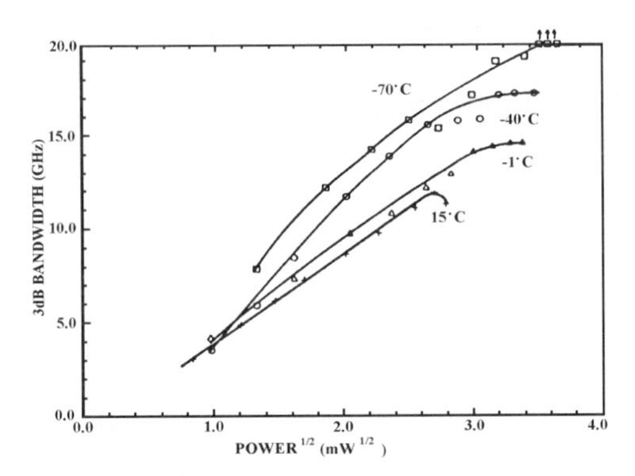

Figure 4.17: Bandwidth of constricted mesa laser against square root of output power, under various operating temperatures. Reprinted with permission from Ref. [26]. Copyright 1985 American Institute of Physics.

$$\omega_o^2 \approx \frac{(g_o / \chi) S_o}{\tau_p} \tag{4.32}$$

where $\chi = 1 + \tau_s / \tau_e$ is the transport factor. Therefore, long SCH (higher τ_s) can drastically decrease the modulation bandwidth of a QW laser [23].

4.8.3 Simultaneous Intensity and Frequency Modulations

The refractive index in the active region of a laser diode is changed as a result of an increment in the injected carrier density in the active layer, which in turn induces an increment change in the propagation constant. Therefore, amplitude modulation of the optical field is always accompanied by phase or frequency modulations (FM) [27]. This FM or "frequency chirp" has a profound effect on the transmission link performance, which will be covered in more details in Chapter 9. In this section, we will present the basic analytical tool for future system analysis.

Under a single-tone modulation at frequency f_m, the electrical field of the amplitude- and frequency-modulated optical wave can be represented as

$$E = E_o[1 + m\cos(\omega_m t)]^{\frac{1}{2}} \exp\{j\,[\Omega_o t + \beta\sin(\omega_m t)]\,\} \tag{4.33}$$

where $\Omega_0/2\pi$ is the optical carrier frequency, $\omega_m/2\pi$ is the modulating RF/microwave subcarrier (as opposed to the "main" optical carrier), and m and β are the intensity and frequency modulation indices, respectively. The parameter m is often given the name "optical modulation index" or abbreviated as OMI (see Fig. 5.13 for illustration). Assuming the modulating current level is small, we can approximate the IM-FM expression as an AM-FM expression (this approximation may not be valid for typical CATV systems because the total modulation index due to multichannel AM-VSB signals may reach 25 to 30%, but we want to use the approximation for illustration purpose):

$$E \approx E_o(1 + M\cos(\omega_m t)) \cdot \exp\{j\,[\Omega_o t + \beta\sin(\omega_m t)]\,\}. \tag{4.34}$$

From Bessel expansion, we have

$$ej^{\{\Omega_o t\, +\, \beta\sin(\omega_m t)\}} = \sum_{n\,=-\infty}^{\infty} J_n(\beta) \cdot e^{j[(\Omega_o\, +\, n\omega_m)t]} \tag{4.35}$$

Therefore, we know that the amplitudes of the central carrier at Ω_o, the first upper sideband at $\Omega_o + \Omega_m$, and the first lower sideband at $\Omega_o - \Omega_m$ are

$$\text{Central carrier:} \qquad E_o J_o(\beta) \tag{4.36a}$$

$$\text{Upper first sideband:} \left[J_1(\beta) + \frac{M}{2}\left\{J_2(\beta) + J_o(\beta)\right\}\right] \cdot E_o \tag{4.36b}$$

$$\text{Lower first sideband:} \left[-J_1(\beta) + \frac{M}{2}\left\{J_2(\beta) + J_o(\beta)\right\}\right] \cdot E_o \tag{4.36c}$$

An example of the FM sidebands of an intensity-modulated DFB laser is shown in Fig. 4.18. The measured results of the first upper and lower sidebands are clearly different, especially when the modulating current is large, as predicted by the preceding equations. Note that from the difference of the two sidebands, $2J_1(\beta)$, we can obtain the FM modulation index β, while from their amplitude average, $M/2 \bullet \{J_2(\beta) + J_0(\beta)\}$, and the known β, we can obtain the AM modulation depth M. It is clear from Fig. 4.18 that with a larger modulating signal power, the sidebands spread wider, which causes an effective linewidth broadening or wavelength "chirping." The same phenomenon can be illustrated using a coarse display resolution to show the effective linewidth broadening, as given in Fig. 4.19. These spectral broadening phenomena, when combined with optical fiber dispersions, can induce severe dispersion-induced nonlinear distortions (see Chapter 9).

4.8.4 Large-Signal Circuit Model

The small signal analysis of rate equations which was introduced in Section 4.8.2 is not applicable to a typical CATV laser under the modulation of multiple video channels. In addition, the small signal assumption may not be valid, since the total rms modulation index of >40 video channels may be in the range of 20–30%. In this case, it is convenient to model a laser diode by an equivalent large-signal circuit model [5,21,24], which is more convenient than solving the rate equations numerically. Such a large-signal circuit model has all the physical parameters in rate equations transformed into circuit parameters, and it can be used to predict dynamic nonlinear distortions in a laser diode [21]. A circuit model can be separated into two

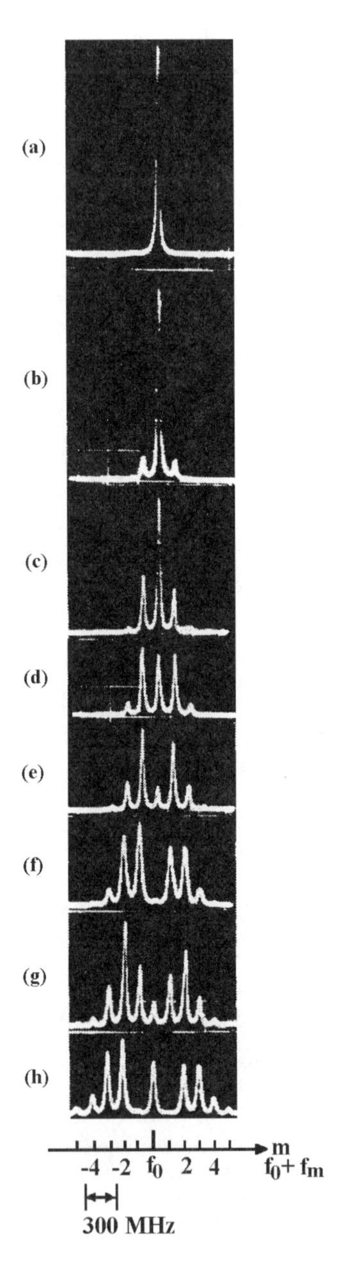

Fig. 4.18 Optical power spectra (measured by a Fabry–Perot interferometer) of a laser under 150-MHz modulation with different levels of modulating power (zero-to-peak modulating current amplitudes): (a) 0 mA, (b) 0.76 mA, (c) 1.4 mA, (d) 1.84 mA, (e) 2.24 mA, (f) 2.88 mA, (g) 3.44 mA, (h) 4 mA. After Ref. [28] (© 1982 IEEE).

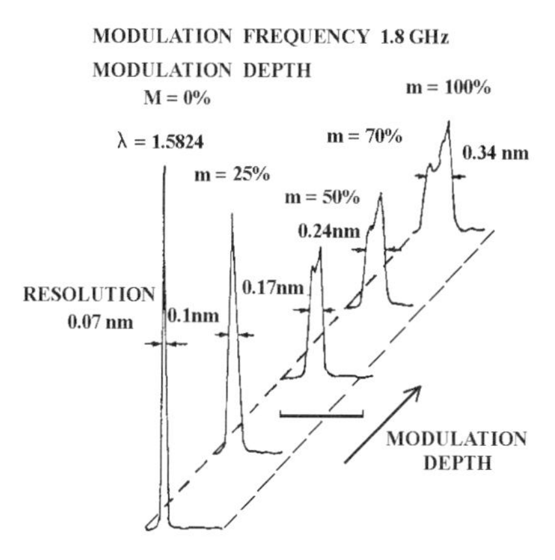

Figure 4.19: The broadening of a laser linewidth when the laser is modulated by a 1.8-GHz tone with different levels of modulation depth. After Ref. [25].

major parts, as shown in Fig. 4.20 [5]. The first part is the intrinsic laser diode, and the second part includes the package and chip parasitics. We will discuss these two parts separately.

4.8.4.1 Circuit Model of an Intrinsic Laser Diode In Eqs. (4.24a) and (4.24b), we define

$$I_{sp} = qV \bullet \frac{N}{\tau_n} \tag{4.37a}$$

$$I_{stim} = qV\Gamma S_n S' g_o (N - N_o)[1 - (\varepsilon \Gamma S_n) S'] \tag{4.37b}$$

$$R_{ph} = \frac{\tau_p}{qVS_n} \tag{4.37c}$$

$$C_{ph} = qVS_n \tag{4.37d}$$

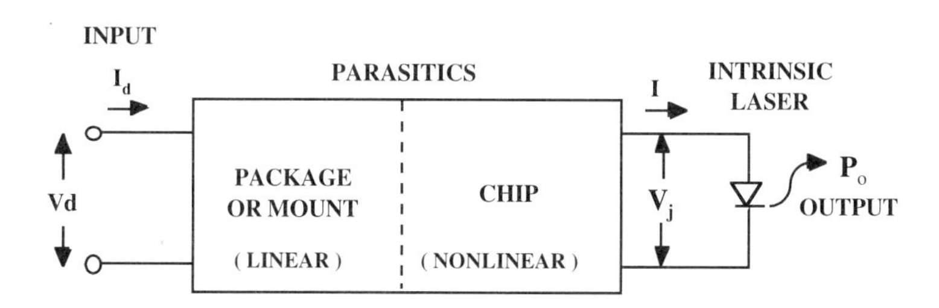

Figure 4.20: Illustration of package and chip parasitics of a laser diode package. The chip parasitics may include nonlinear elements such a diode which is used to model the leakage current. After Ref. [5].

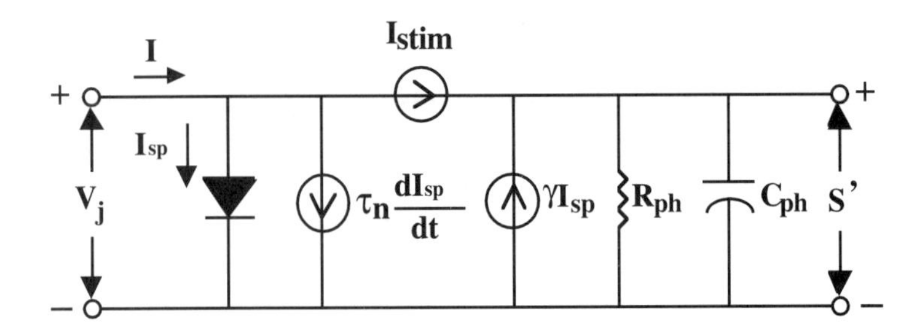

Figure 4.21: Equivalent circuit model of an intrinsic laser diode. After Ref. [24] (© 1984 IEEE).

where I_{sp} is the equivalent spontaneous emission current, I_{stim} is the equivalent stimulated emission current, $S' = S/\Gamma S_n$ is a normalized photon density, and S_n is a normalization constant to avoid divergences in numerical evaluation. Then Eqs. (4.24a) and (4.24b) become

$$\tau_n \frac{dI_{sp}}{dt} = I - I_{sp} - I_{stim} \tag{4.38a}$$

$$C_{ph} \frac{dS'}{dt} = I_{stim} - \frac{S'}{R_{ph}} + \gamma I_{sp} \tag{4.38b}$$

respectively. We can then obtain the large-signal circuit model as shown in Fig. 4.21.

4.8.4.2 Circuit Model of a Laser Module Parasitics As was shown in Fig. 4.20, the equivalent circuit model of a completely packaged laser diode should further consider the package parasitics and the laser chip parasitics. The package parasitics usually include a bond wire inductance and one or more stray capacitances to ground (these parasitics can be greatly eliminated if the laser drive circuit is monolithically integrated with the laser diode). These parasitics are represented by L_p, R_p, and C_p in Fig. 4.22. The chip parasitics, represented by R_S and C_S in Fig. 4.22, are stray capacitance and resistances associated with the semiconductor material sur-

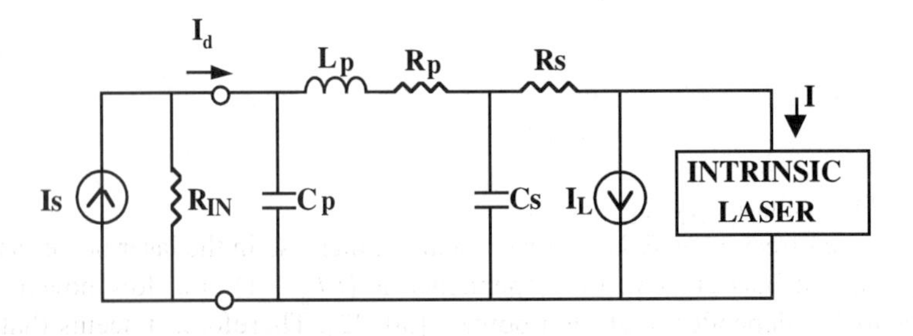

Figure 4.22: Equivalent circut model of the package and chip parasitics of a packaged laser diode. After Ref. [5].

rounding the active region. Chip capacitances are strongly structure dependent. The package and chip parasitics can normally be measured by using the reflection parameters S_{11}, which gives the corresponding equivalent circuit model by using the proper microwave computer-aided software [21]. The current source I_L represents DC leakage around the active region and causes severe second-order nonlinear distortions in the laser diode [29].

4.9 Laser Diode Noise

The output intensity, phase, and frequency of a laser diode exhibit fluctuations even when it is only DC-biased (without any AC signal modulation). These random fluctuations are caused mainly by the noncoherent spontaneous emission. The related noise terms can be analyzed by adding Langevin forces to the rate equations [2]. We shall not go into the details of the derivations, but shall summarize some of the important results and concepts.

4.9.1 Intensity Noise

The intensity noise of a laser diode is due to spontaneous events and quantum fluctuations in electron density. Intensity noise of a laser diode or a complete system is usually specified by a normalized quantity, relative intensity noise (RIN) [30–32], defined as

$$RIN = \frac{<i^2>}{I_p^2} \quad \text{(dB/Hz)}, \tag{4.39}$$

where $<i^2>$ is the noise current spectral density and I_p is the detected DC photocurrent. *RIN* noise has a frequency response very similar to that of the small-signal modulation response given in Eq. (4.30) [2,33]:

$$RIN(\omega) = \frac{A + B\omega^2}{(\omega_0^2 - \omega^2)^2 + \omega^2 (S_o \varepsilon / \tau_p)^2}. \tag{4.40}$$

Here, A and B are constants depending on various laser parameters. The intrinsic laser diode intensity noise has the same relaxation oscillation frequency and damping constant as that of the small-signal modulation frequency response given in Section 4.8.2. Consequently, sometimes it is more convenient to estimate the relaxation oscillation frequency of a laser diode by measuring its intensity noise spectrum instead of by directly modulating the laser diode, because the latter technique could be limited by the laser package parasitics.

At a given frequency, RIN decreases with an increase in the laser power as P^{-3} (or, in terms of bias current, the dependence is $(I/I_{th} - 1)^{-3}$) at low powers but changes to P^{-1} dependence at high powers [30–32]. Therefore, it seems that the laser should be biased as high as possible to simultaneously obtain high output power and low RIN. However, for a fixed RF or microwave driving signal power, when the bias current is increased, the optical modulation index is decreased (see Fig. 5.13), and consequently the carrier-to-noise ratio of the received RF signal is decreased (see Section 5.4). On the other hand, if a constant OMI is to be maintained, then the amplitude of the driving current must be increased, which could cause the laser or the RF/microwave amplifier operate in a severe nonlinear region. Hence, the optimum bias current depends on the trade-off between minimum RIN and maximum OMI. In addition, the optimum bias current may also depend on the linearity characteristics of the *L–I* curve, as will be discussed in Section 4.10.2.

In addition to the intrinsic laser diode itself, there are several other possible sources of intensity noise in an optical fiber system. In this section, we will limit our discussions only to the laser package module-related intensity noise, and will leave the system-related intensity noise to Chapter 9.

The intrinsic laser diode RIN can be severely degraded by the direct optical reflections (back into the laser cavity) from a near-end or far-end reflecting point which is within the coherence length of the laser. The near-end reflections could be due to the lenses or optical fiber inside the laser package, and the far-end reflections could be due to poor-quality fiber connectors [34].

Since the effect of an optical reflection depends on how coherent it is with the optical field inside the active region, the laser diode's coherent length plays an important role. It is defined as $L_c = (c/n)t_c$, where t_c is the coherent time of a laser diode, defined as $t_c = 1/(2\pi \bullet \Delta\upsilon)$, where $\Delta\upsilon$ is the linewidth of the laser diode (see Section 4.9.3). Because of the larger spontaneous emission noise of a FP laser as compared to a DFB laser, the correspondingly larger $\Delta\upsilon$ makes the coherent length of a FP laser shorter. Therefore, FP lasers usually are less sensitive to optical reflection than DFB lasers. On the other hand, the unavoidable optical FM modulation that we discussed in Section 4.8.3 increases the laser effective linewidth. Therefore, as the "effective" $\Delta\upsilon$ becomes larger for the case of a laser under direct intensity modulation than for that of a CW laser, the coherent length of a laser under modulation is decreased correspondingly. Consequently, an originally damaging reflecting interface at some distance may now be outside the coherent length and becomes harmless. This phenomenon can be seen in Fig. 4.23 in which case the optical reflection-induced spiky intensity noise is significantly suppressed when a large-signal modulation is applied to an FP laser without any optical isolator [34].

If coherent reflection-induced ripple in the small-signal frequency response of a laser diode can be observed, the frequency period of the ripple can be estimated from the long cavity formed by the laser active region and the external reflecting point: It is about $c/2n_f L$ or $c/2L$, depending on whether the reflecting point is from optical fiber connectors or from the lens inside the laser ($n_f \cong 1.45$ is the refractive index of the optical fiber). If the laser has a strong relaxation oscillation peak, reflection ripples can be observed most easily, as shown in Fig. 4.24. We can see that for noise frequencies far below the resonance frequency, the RIN was affected much less significantly. This phenomenon again stresses the important system design concept that the useful operation frequency range of a laser diode should be much smaller than the resonance frequency. It should be noted that the reflection will affect the laser device behavior more significantly only if the optical feedback is much greater than the amount of spontaneous emission [2]. A parameter which is used to quantize the amount of optical reflection power is given by the feedback power ratio (FPR), defined as FPR = $C_f \bullet \alpha_f \bullet R_{ext}$, where C_f is the coupling efficiency of light from the laser into the single-mode fiber, α_f is the roundtrip loss between the laser and the external reflection interface, and R_{ext} is the reflectivity of an external reflecting point.

An example of the significant effect of a near-end FPR on a laser RIN is shown in Fig. 4.25. This measured result shows that the FPR due to near-end reflections

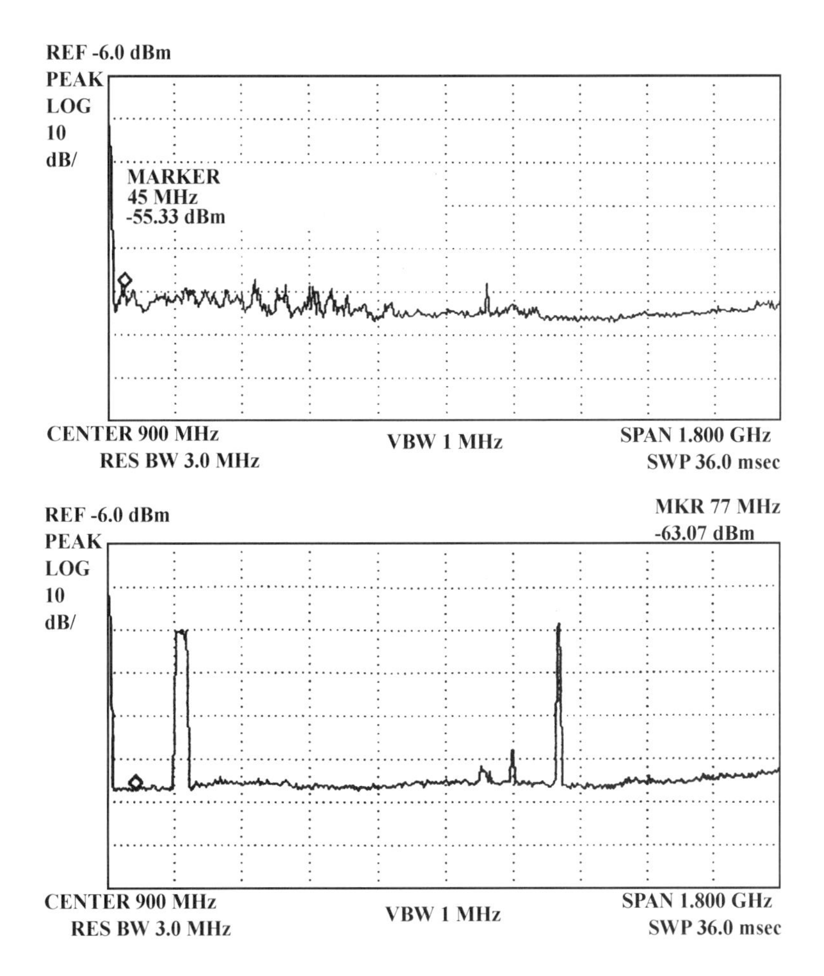

Figure 4.23 Intensity noise spectra of a Fabry-Perot laser diode without optical isolator showing the effect of optical reflections and signal modulation depths. In (b), there are signal channels around 200 MHz and a pilot tone at 1.2 GHz, with OMI/ch = 10%. (Courtesy of New Elite Technol, Inc.)

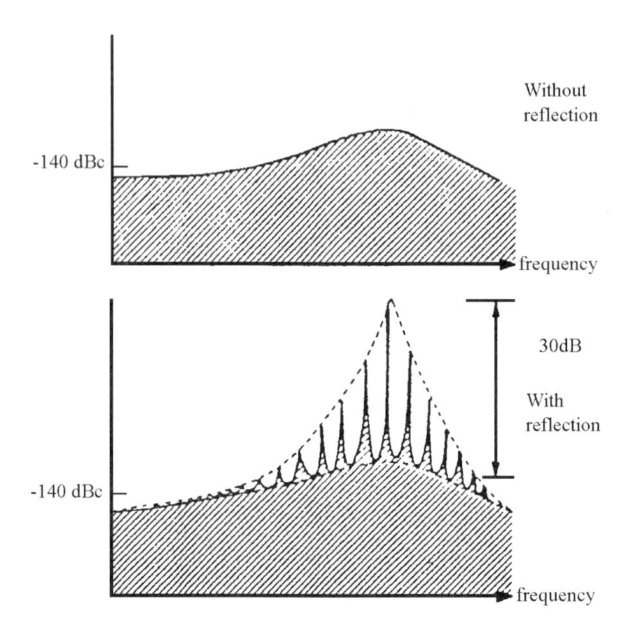

Figure 4.24: Comparison of the frequency response near the resonance frequency region without and with optical reflections. After Ref. [35].

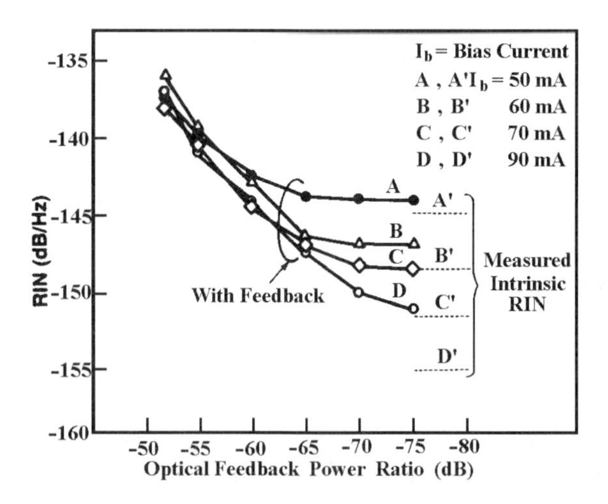

Figure 4.25: RIN versus optical feedback power ratio for a 1.54-μm DFB laser with and without an optical isolator. Curves A' to D' are the intrinsic laser RIN, and curves A to D are the RIN with some residual multiple reflection-induced intensity noise in the measurement setup. The threshold current of the laser is 20 mA. After Ref. [36] (© 1989 IEEE).

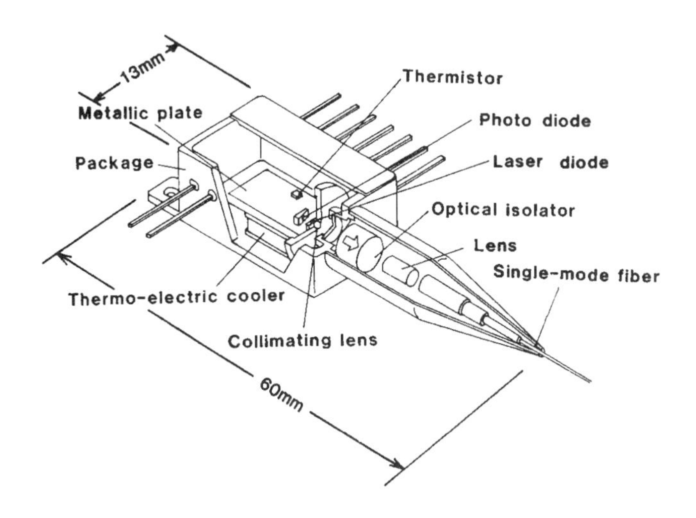

Figure 4.26: A typical DFB laser package. After Ref. [37] (© 1990 IEEE).

should be kept below –75 dB in order to obtain a RIN close to the intrinsic RIN value. This implies that an optical isolator with an isolation greater than 30 dB is a necessary component in the package to eliminate most of the external optical reflections (the reflectivity from a typical "good" optical fiber connector is around –45 dB, as was explained in Chapter 3). A multichannel AM-VSB CATV system will require a RIN in the range of –155 dB/Hz or below, as will become clear in Chapter 9.

Consequently, all of today's commercially available multichannel CATV DFB lasers have a built-in optical isolator, as can be seen in Fig. 4.26 (the thermoelectric cooler in that figure will be introduced at the end of this chapter) [37].

4.9.2 Mode-Partition Noise and Mode-Hopping Noise

So far we have only concerned ourselves with downstream CATV light sources which are DFB lasers with built-in optical isolators. However, based on economic considerations, return-path lasers are likely to be (uncooled) low-cost FP lasers [38] or uncooled DFB lasers without optical isolators [39]. In both types of lasers, mode-partition noise (MPN) and mode-hopping noise (MHN) can be observed.

MPN refers to the tendency for optical power to distribute itself between different optical modes of a laser in such a way that the power in individual modes fluctuates but the total power in all modes is relatively steady. A mode-resolved measured MPN of an AlGaAs laser is shown in Fig. 4.27 [40]. We can see that the intensity noise spectral density of the laser's dominant mode is essentially flat below a low-frequency region (in this "low-speed" laser case, it is about 50 MHz), and decreases at 6 dB/octave above that frequency region.

With a multiple-longitudinal-mode FP laser, fiber dispersion-enhanced MPN (EMPN) was measured to be (cyclically) varied significantly with temperature and

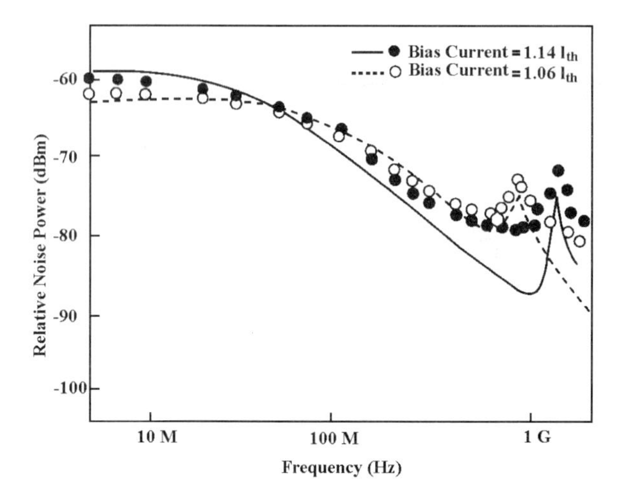

Fig. 4.27: Measured (mode-resolved) mode-partition noise in an AlGaAs laser. After Ref. [40] (© 1977 IEEE).

bias current [41]. An example of measured EMPN spectra for an unmodulated FP laser is shown in Fig. 4.28 [42].

The measured results can be explained using a simplified conceptual analysis as follows [43]. It was assumed that the laser had only two longitudinal modes; we can express the intensity noise of each mode at any frequency ω_n as (before transmission through optical fibers)

$$\begin{aligned} n_{p1} &= A_n \cos(\omega_n t + \phi) \\ n_{p2} &= -A_n \cos(\omega_n t + \phi), \end{aligned} \tag{4.41}$$

where the sum of the noise intensities is assumed to be zero (the MPN is constant when all modes are included), and ϕ represents a random phase angle.

In the presence of fiber dispersion, the two modes are delayed with respect to each other by τ, and Eq. (4.41) becomes

$$\begin{aligned} n_{p1} &= A_n \cos(\omega_n (t + \frac{\tau}{2}) + \phi) \\ n_{p2} &= -A_n \cos(\omega_n (t - \frac{\tau}{2}) + \phi). \end{aligned} \tag{4.42}$$

When these two terms are combined, the EMPN becomes

$$\begin{aligned} n_{p1}(t,\tau) + n_{p2}(t,\tau) &= 2A_n \sin(\omega_n \tau / 2)\sin(\omega_n t + \phi) \\ &\approx A_n \omega_n \tau \cdot \sin(\omega_n t + \phi), \qquad \omega_n \tau << 1. \end{aligned} \tag{4.43}$$

We can see that the EMPN is proportional to the noise frequency ω_n, the fiber dispersion τ, and the individual mode noise spectral noise density. Doubling any

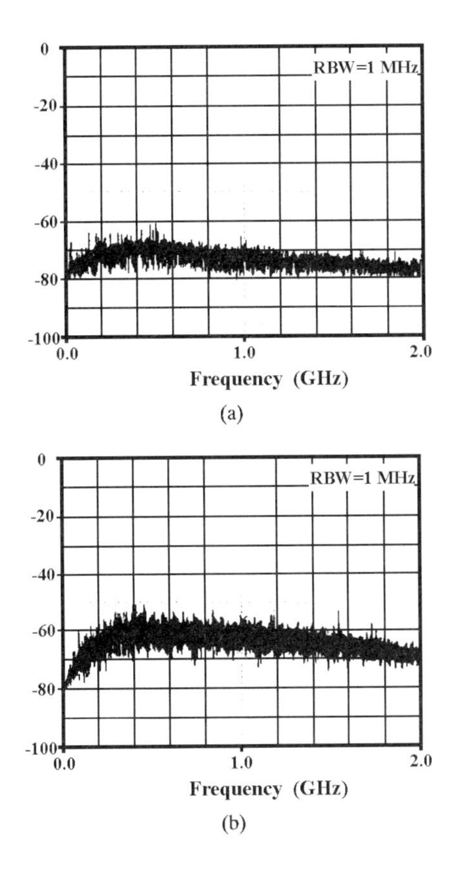

Figure 4.28: Mode-partition noise in FP lasers after transmission through (a) 6 km and (b) 20 km of single-mode fibers. Vertical scale in dBm. DC photocurrent is adjusted to 1 mA. The RF output from photodiode is amplified by a 20-dB amplifier in these measurement. After Ref. [42] (© 1993 IEEE).

term doubles noise intensity and increases EMPN spectral noise density by 6 dB at the photodetector output. Since the noise spectral density of dominant longitudinal mode (A_n) tends to be flat at low frequencies, then decrease at a rate of 6 dB/octave (recall Fig. 4.27), the resultant EMPN will rise at a 6 dB/octave rate at low frequencies, then become essentially flat until approaching resonance. This simple model predicts the measured EMPN spectra quite well, as can be observed in Fig. 4.28.

For DFB lasers without optical isolators which may have a few extremely small side modes accompanying the main mode, the intensity noise due to MPN has not yet been specifically measured. However, mode-resolved measurement technique shows that some DFB lasers could also have MPN, as shown in Fig. 4.29, where we can see both the averaged power spectrum and the time-resolved statistics [44]. From an optical spectrum analyzer which measures the average optical power, the

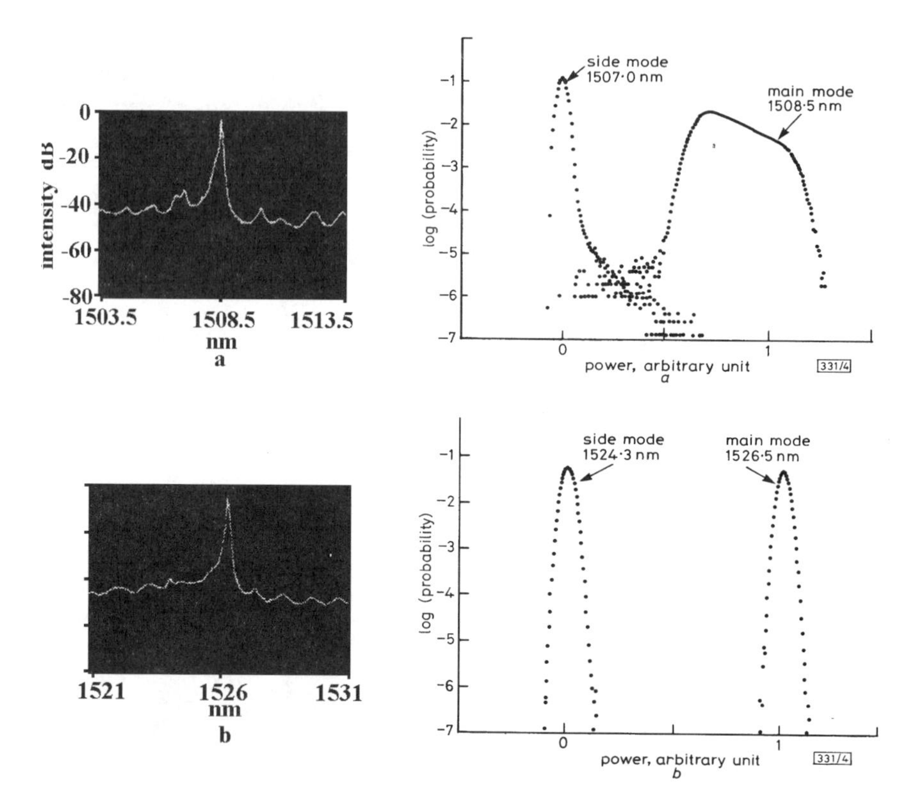

Figure 4.29: Measurement of time-resolved photon statistics of two DFB lasers, shown on the right, whose average power spectra appear to be single-mode, as shown on the left. The main mode of laser A has a high probability to jump into its side mode, while laser B has no mode partition. Reprinted with permission from Ref. [44]. Copyright 1988 American Institute of Physics.

side modes may appear to be 30 dB or more below the main mode. However, the instantaneous optical power may actually jump among these modes in such a way that instantaneously almost all of the power may jump into one of the side modes. Figure 4.29a represents a DFB laser with mode partition, and Fig. 4.29b represents a DFB without mode partition. In the former case, there is a probability up to 10^{-5} that the optical power may jump to the side mode at 1507 nm, even though the measured average optical power spectrum exhibits predominantly a single mode.

MHN is observed in FP lasers with multiple longitudinal modes or with nearly single longitudinal mode, when temperature or current varies, and such mode hopping causes significant noise at a frequency below several tens of megahertz [45]. The main difference between mode-hopping noise and mode-partition noise is that the former can be observed even when wavelength selectivity is eliminated, whereas the latter must be observed with wavelength-selective devices as was done in Figs. 4.27 and 4.29. The origin of MHN may be the random hopping of laser oscillation among several modes and the output power difference of those modes [45], and coupling effects among the modes [46]. Although this kind of noise only causes high RIN (< –130 dB/Hz) in the frequency region below a few tens of megahertz, the $f_1 \pm f_2$ type second-order nonlinearity can induce "skirt" noise around the multiple RF carriers [47]. Other low frequency laser RINs such as those due to asymmetric nonlinear gain [48] in a FP laser can also cause the "skirt" noise. Therefore, multi-longitudinal-mode FP lasers are not recommended for HFC system downstream transmissions where extremely low RIN (< –155 dB/Hz) is required, and must be carefully used for upstream signal transmission even though the system requirements may not be as stringent as those of downstream transmission.

4.9.3 Laser Phase Noise, Spectral Linewidth, and High-Power DFB Lasers

The spectral linewidth of a single longitudinal mode is determined by phase noise in the optical field. The phase noise is due to spontaneous emissions coupled into the lasing mode and fluctuations in the electron population. Both the amplitude and phase of the optical field are affected because of the coupled spontaneous photons which induce the refractive index changes Δn (= $\Delta n' - j\Delta n''$). An analysis [49] shows that the linewidth $\Delta\nu$ due to spontaneous emissions is given by

$$\Delta\nu \propto (1+\boldsymbol{a}^2)/P_O, \tag{4.44}$$

where $\boldsymbol{a}$ is the linewidth enhancement factor given by $\Delta n'/\Delta n''$ and is the amplitude–phase coupling parameter. Typical range of $\boldsymbol{a}$ is in the range 4 to 6 in bulk lasers, and ~2 in quantum-well lasers. Equation (4.44) indicates that the laser linewidth can be reduced by operating at high output power. Today's MQW DFB lasers which have been used in direct-modulation HFC systems have typical linewidths in the range of 1 to 10 MHz. In an external modulation system, CW laser linewidth has important effect on the stimulated-Brillouin-scattering-limited optical power into a single-mode fiber-based link, and on the multiple reflection-induced intensity noise, as will be discussed in Chapter 9. There has been growing interest in using a 1.55-μm high-power MQW DFB laser as the source for an external modulation system. This kind of DFB laser has a typical linewidth lower than ~3 MHz and has an output power in the range of 20 to 50 mW into a single-mode fiber.

4.10 Laser Physics and Structure Affecting Linearity Characteristics

Nonlinear distortions in laser diodes have three origins: (1) dynamic nonlinearity due to relaxation oscillation, (2) static nonlinearity due to light output power versus bias current (L–I) curve, and (3) static and dynamic nonlinearities due to overmodulation. Generally speaking, (1) dominates for modulating signals in the high-frequency region (e.g., ≥ 1 GHz), while (2) and (3) dominate for modulating signals in the low-frequency region (such as typical CATV bands). Most HFC laser diodes for transporting >60 channels of AM-VSB signals have their resonance frequencies beyond 8–10 GHz, and therefore the dominant nonlinear distortions usually come from the static nonlinearity (see Problem 4.3). That is why in recent years, most research effort has been devoted to the improvement of the L–I curve and the overmodulation-induced nonlinearities (see Chapter 9). In the following sections, we will only discuss the first and second types of laser nonlinearities, and will postpone the discussions of type 3 nonlinearities until Chapter 9.

4.10.1 Dynamic Nonlinearity in Semiconductor Laser Diodes

The dynamic nonlinearity due to relaxation oscillation in SLM semiconductor lasers can be characterized by the two-tone intermodulation products derived from the single-mode rate equations (4.24a) and (4.24b). Instead of using a small-signal single-tone analysis which was used to obtain a laser diode's small frequency response, we let the injection current be

$$I(t) = I_o + \delta I = I_o + (i_1 e^{j\omega_1 t} + i_2 e^{j\omega_2 t}) . \tag{4.45}$$

The photon density S and carrier density N in Eq. (4.24) also vary around their DC values as

$$S(t) = S_o + \delta S^1 + \delta S^2 + \delta S^3 + \ldots \tag{4.46}$$

$$N(t) = N_o + \delta n^1 + \delta n^2 + \delta n^3 + \ldots . \tag{4.47}$$

δS^1 and δn^1 are, respectively, the amplitudes of photon and carrier densities corresponding to the fundamental frequencies ω_1 and ω_2. δS^k and δn^k $(k \geq 2)$are respectively the kth-order amplitudes of photon and carrier densities and are caused by the dynamic nonlinearity. If we neglect terms of order >3, and substitute Eqs. (4.45)–(4.47) into Eq. (4.24), the photon and carrier densities of second and third order can be obtained. The results are given as [50,51]

$$\frac{S_{2f_1}}{S_{f_1}} = m \cdot FR(2f_1) \cdot \frac{f_1^2}{f_o^2} \tag{4.48}$$

$$\frac{S_{2f_1-f_2}}{S_{f_1}} = \frac{1}{2} m^2 \cdot FR(f_1) \cdot FR(2f_1)$$

$$\left[\left[\left\{(\frac{f_1}{f_o})^4 - \frac{f_1^2}{2f_o^2}\right\}^2 + (\frac{f_1}{f_o})^2 \left\{\frac{1}{4\pi f_o \tau_n} - (\frac{f_1}{f_o})^2 (2\pi f_o \tau_p + \frac{3}{4\pi f_o \tau_n} + \frac{3\varepsilon S_{f_1}}{2\pi f_o \tau_p})\right\}^2\right]\right]^{1/2}, \tag{4.49}$$

where m is the optical modulation index per channel given by S_{f_1}/S_o, FR(f) is the small-signal frequency response given by Eq. (4.29), and f_o is the resonance frequency given by Eq. (4.28). Since the photon density is proportional to the received

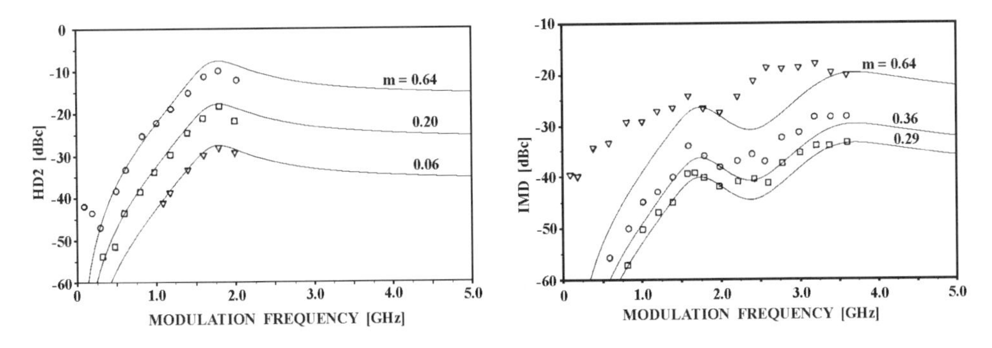

Figure 4.30: (a) Second-order harmonic and (b) third-order intermodulation distortions as a function of modulation frequency for different intensity modulation indices; — calculated; circles, triangles, and squares: measured. The laser resonance frequency is at 3.4 GHz. After Ref. [51] (© 1991 IEEE).

photocurrent, we have to square the results in Eqs. (4.48) and (4.49), and we obtain the ratio of second-order harmonic versus fundamental carrier, and the ratio of two-tone third-order intermodulation products on an RF/microwave spectrum analyzer as

$$\frac{HD_2}{C} \approx m^2 \cdot \left|FR(2f_1)\right|^2 \cdot \left(\frac{f_1}{f_o}\right)^4 \tag{4.50}$$

and

$$\frac{IMD_3}{C} \approx \frac{1}{4} m^4 \cdot \left|FR(f_1)\right|^2 \cdot \left|FR(2f_1)\right|^2$$
$$\left[\left\{(\frac{f_1}{f_o})^4 - \frac{f_1^2}{2f_o^2}\right\}^2 + (\frac{f_1}{f_o})^2 \left\{\frac{1}{4\pi f_o \tau_n} - (\frac{f_1}{f_o})^2 (2\pi f_o \tau_p + \frac{3}{4\pi f_o \tau_n} + \frac{3\varepsilon S_o}{2\pi f_o \tau_p})\right\}^2\right] \tag{4.51}$$

respectively. It is clear from Eqs. (4.50) and (4.51) that HD_2/C and IMD_3/C are proportional to m^2 and m^4, respectively. Both HD_2 and IMD_3 can become very small when the modulating frequency $f_1 << f_o$. This is the reason why we keep emphasizing that high laser resonance frequency (i.e., high f_o) is important to achieve the required low nonlinear distortions in a semiconductor laser. A good example illustrating this principle is a 1.3-μm thin-barrier, short-cavity MQW-DFB laser, which exhibits extremely low third-order intermodulation products and has a resonance frequency of about 14 GHz [52].

Calculated results using Eqs. (4.48) and (4.49) were compared with experimental results as shown in Fig. 4.30a and b, respectively [51]. We can see that the calculations agree quite well with the measurements in most cases. We also observe that the maximum of the second-order harmonics peaks at about half of the relaxation resonance frequency (f_0 / 2 = 1.7 GHz), as predicted by Eq. (4.50) through $|FR(2f_1)|^2$. The third-order intermodulation products exhibit peaks at 3.4 GHz and a subpeak at 1.7 GHz, which were expected in Eq. (4.51) through $|FR(f_1)|^2$ and $|FR(2f_1)|^2$, respectively. Note that for modulation frequencies $f_1 \ll f_0$, both HD_2 and IMD_3 drop at a rate proportional to $10 \log[(f_1/f_0)^4]$ or 40 dB/decade.

We also note that there are discrepancies between the calculated and measured IMD in Fig. 4.30b when the optical modulation index was 64% because the laser was constantly driven below threshold, and this overmodulation produces additional distortions in terms of clipping (see Chapter 9), which cannot be explained by the results of small-signal analysis. The discrepancies between the calculated and measured HD_2 in Fig. 4.30a at low modulating frequencies ($f < 250$ MHz, $m = 0.64$) were due to static nonlinearities such as current leakage [29] and nonlinear gain from spatial hole burning (SHB). Most of today's high-performance CATV lasers do not exhibit nonlinear distortions due to current leakage [37,53], and this fact is true even at high temperature [37]. Therefore, we will concentrate our attention on the effect of SHB on second-order nonlinear distortions in the following section.

4.10.2 Static Nonlinearities and Spatial Hole Burning in Semiconductor Laser Diodes

Besides SHB and leakage current, clipping due to overmodulation is also a type of static nonlinearity. It sets the fundamental limit which is independent of device physics, and will be discussed thoroughly in Chapter 9. Here we will concentrate our discussion only on SHB phenomena.

Early investigations on using DFB lasers for multi-channel AM-VSB video transmissions found that, surprisingly, most of the DFB lasers show higher second-order nonlinear distortions than FP lasers [37]. SHB [54] was found to be responsible for the worse linearity performance. SHB was caused by the nonuniform distribution of light intensity along the laser active layer axis, which is closely related to (1) the normalized coupling coefficient κL in a DFB laser cavity [37,53, 55,56], and (2) the facet reflectivity [55,56].

As shown in Fig. 4.31, assuming the injected carrier density is constant along the axis of the laser active layer, the light intensity near a certain location on the laser axis may become higher than that of other locations on the axis because of the nonoptimized κL, which in turn causes higher stimulated recombination carriers (more holes are burned out) and lower net carrier density near that location. This variation of photon density along the axis (e.g., z-direction) of an active layer can induce variation of carrier density according to a z-dependent form of Eq. 4.24a (under steady state and with the term $1 - \varepsilon S$ neglected),

$$N_n(z) = N_c - N_{st}(z) = \frac{\tau_n I}{eV} - \tau_n g(z) S(z), \tag{4.52}$$

where N_c, N_{st} and N_n are defined in Fig. 4.31. The first term on the right-hand side of Eq. (4.52) is due to the uniformly injected current (uniform along the active layer axis),

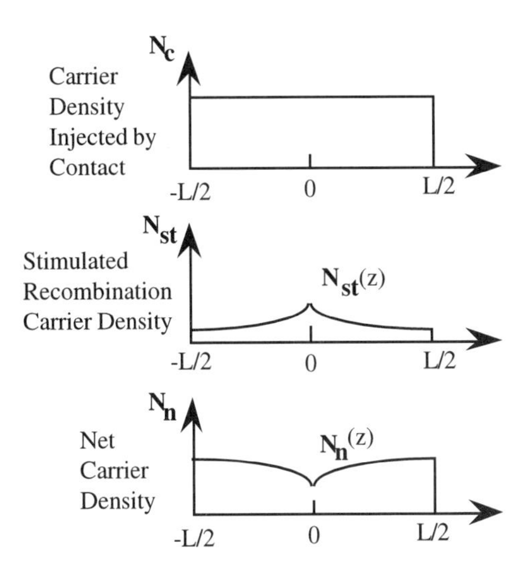

Figure 4.31: Carrier density distribution along a DFB laser. After Ref. [54] (© 1987 IEEE).

and the second term is due to the spatially dependent stimulated emission. The variation of carrier density in turn will affect the optical gain, since $g(z) = \bar{g}_0 (N_n(z) - N_0)$. The reduction of carrier density $N_n(z)$ causes gain compression $g(z)$, which is very similar to that caused by spectral hole-burning [recall Eq. (4.22)]. Therefore, one can add to the spectral hole burning-induced gain compression factor ε by a spatial hole burning-induced gain compression factor ε when using the rate equations to analyze nonlinear distortions [57]. It was calculated that the larger the factor ε or ε', the higher the second-order nonlinear distortions in the HFC operation frequency region. Therefore, analytically through the gain compression factor ε in rate equations, spatial hole burning introduces additional nonlinear distortions, although, strictly speaking, SHB should be a static phenomenon and should cause frequency-independent, instead of frequency-dependent, nonlinearities.

An actual example of the dependence of nonuniform axial light intensity distribution on κL for a DFB-PPIBH (P-substrate partially inverted buried-heterostructure) laser, with the reflectivities of front and rear facets at 1% and 80%, respectively, is shown in Fig. 4.32 [37]. The light intensity distribution becomes nonuniform when the normalized coupling coefficient κL between the grating and the laser active layer is too large or too small. Only when κL approaches unity can the uniform light distribution along the axis be obtained. When $\kappa L = 0.5$, the light intensity near the front facet is higher than that of the rear facet. The condition is reversed when $\kappa L = 2$. Figure 4.33 shows the relation between the measured second harmonic distortions and the standard deviation (σ_{opt}) of the optical intensity along the cavity [37]. It can be clearly seen that the more uniform the light intensity along the axis (lower σ_{opt}), the lower the second-order harmonic distortion.

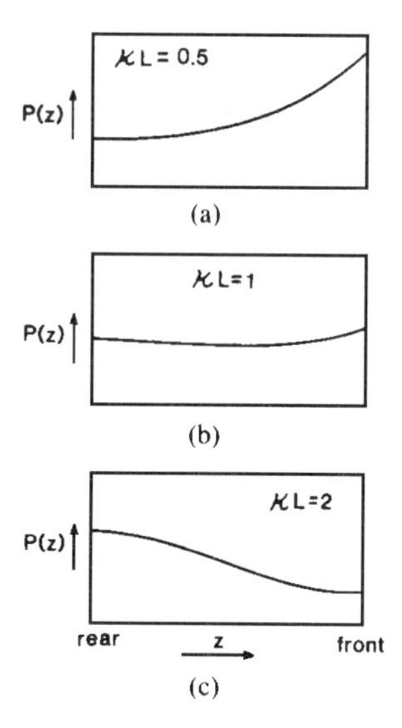

Figure 4.32: Optical intensity distribution in the cavity for DFB-LD with various κL. (a) $\kappa L = 0.5$; (b) $\kappa L = 1$; (c) $\kappa L = 2$. After Ref. [37] (© 1990 IEEE).

In another type of DFB laser (FBH-DFB) [53], the optimum value of κL was found to be ~0.7, as can be seen from the experimental results given in Figs. 4.34a and 4.34b. In this case, both $\kappa L = 0.5$ and 1.3 exhibit worse linearity performance than when $\kappa L = 0.7$.

A similar result was also obtained for the yield of DFB lasers which satisfy a certain CSO requirement, as shown in Fig. 4.35 [55]. This calculated result shows that the yield of conventional DFB lasers can only reach a maximum value of ~20%

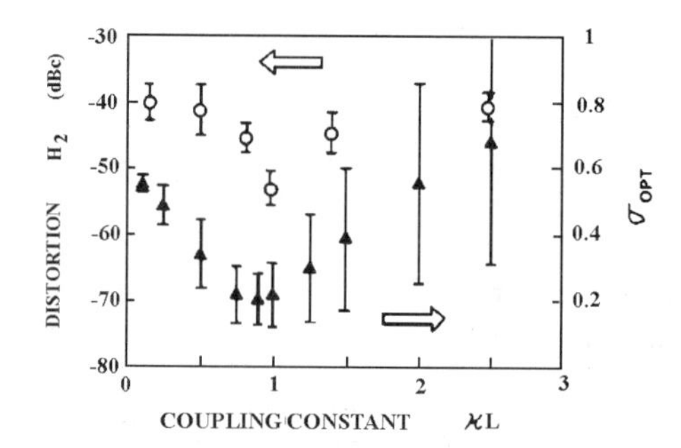

Figure 4.33: Coupling constant dependence of the second harmonic distortion H2. The calculated standard of the optical intensity in the cavity (σ_{opt} is also shown. After Ref. [37] (© 1990 IEEE).

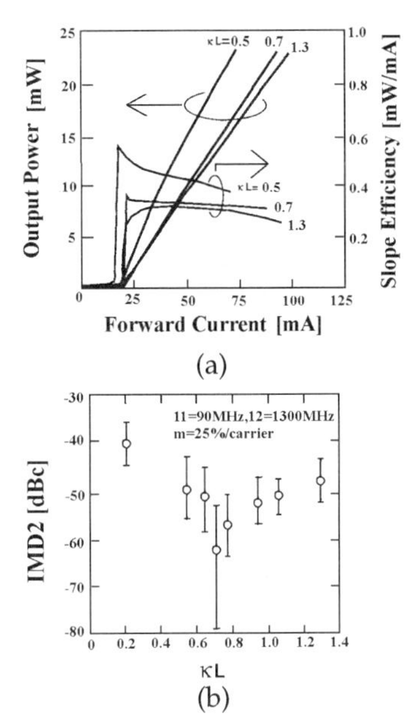

Figure 4.34: (a) L–I curve and slope efficiency dependence on coupling efficiency (κL). (b) Second-order intermodulation distortion (IMD2) versus coupling efficiency (κL). After Ref. [53] (© 1993 IEEE).

when κL is around 0.6–0.7 (for the case of 42 AM channels with OMI = 5% per channel). This is due to the nonuniform longitudinal electrical field distribution along the cavity and the grating phase uncertainty at the cleaved facets. It was also shown [55] that the maximum yield is obtained with a front facet reflectivity of about 4% is used, whereas the reflectivity of the rear facet has almost no effect on the yield. However, a high rear facet reflectivity was chosen to maximize the laser slope efficiency.

In a DFB laser with nonoptimized coupling coefficient κL, the degree of spatial hole burning depends on the light intensity along the laser cavity. The light

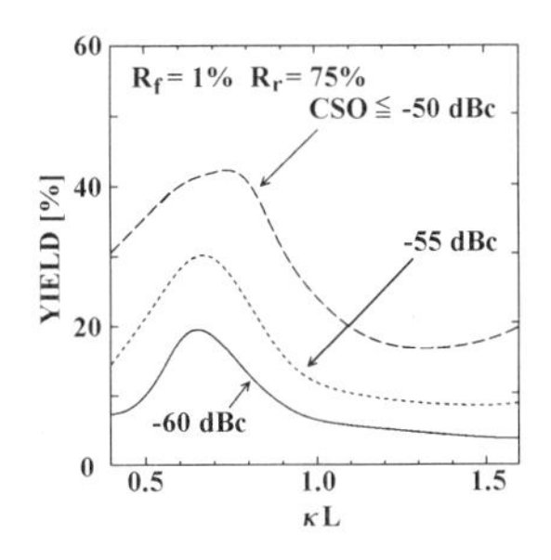

Figure 4.35: Calculated κL dependence of yield for DFB laser. After Ref. [55] (© 1994 IEEE).

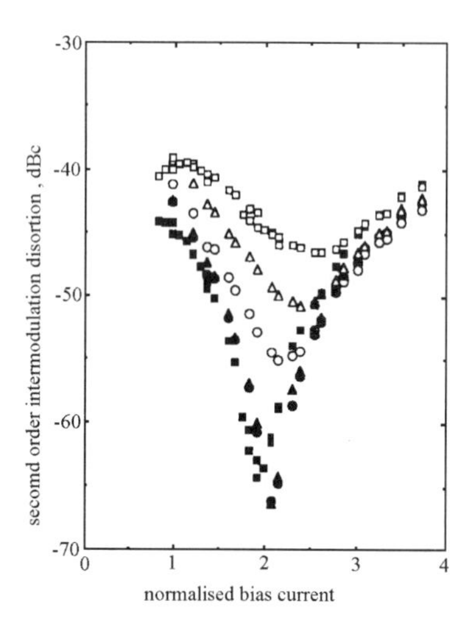

Figure 4.36: Second-order intermodulation distortions versus normalized bias current (I/I_{th}) for a typical DFB-LD with two modulating tones @ OMI/ch. = 0.25. (●: f = 109.25 – 55.25 MHz , O: f = 109.25 + 55.25 MHz, ▲: f = 163.25 - 109.25 MHz, △: f = 163.25 + 109.25 MHz, ■ : f = 355.25 - 301.25 MHz, □: f = 355.25 + 301.25 MHz). After Ref. [58].

intensity, on the other hand, is bias dependent. Therefore, observed bias-dependent second-order nonlinear distortions, shown in Fig. 4.36, were also attributed to spatial hole burning [58]. It can be seen in Fig. 4.36 that as the laser bias is increased above threshold current, the second-order intermod decreases correspondingly. However, when the bias current is increased above about two times the threshold current where spatial hole burning starts to occur, the second-order intermod starts to rise accordingly. In this bias region, the second-order intermod is essentially independent of the modulating frequencies. This phenomenon was successfully explained by assuming the photon lifetime, or equivalently, the threshold gain (recall that $1/\tau_p = v_g(\alpha_{int} + \alpha_{mir}) = v_g \Gamma g_{th}$), is modulated by the modulating signals when spatial hole burning takes place, that is, both $\tau_p(N)$ and $g_{th}(N)$ become functions of the carrier density N (= $N + \delta N$) when spatial hole burning occurs. Using the carrier-dependent photon lifetime and threshold gain, and following the same procedure as in the small-signal analysis in Section 4.10.1, one can derive the ratio of second harmonics and fundamental carrier. It was found that even a slight variation of threshold gain above threshold is sufficient to increase second-order nonlinear distortions.

Note that the nonlinear distortions caused by either leakage current or spatial hole burning are essentially second-order distortions. Consequently, most of the commercially available DFB lasers today have worse second-order nonlinear distortions than third-order nonlinear distortions.

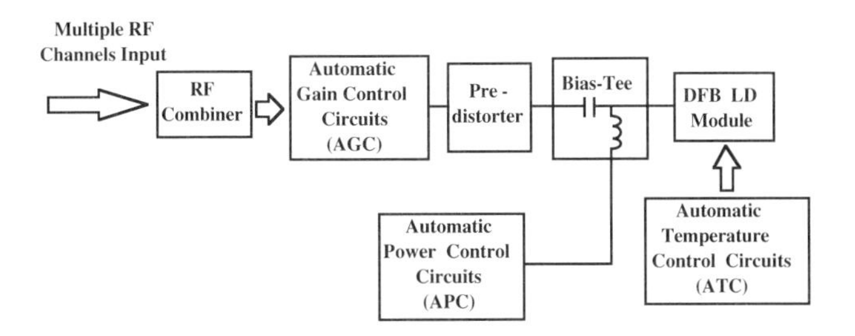

Figure 4.37: A complete CATV laser transmitter includes all the components shown above. The predistorter may or may not be included, depending on whether the DFB LD linearity needs improvement.

4.11 SCM DFB Laser Transmitter Design

A typical DFB laser diode package was previously shown in Fig. 4.26. We note that, in addition to the coupling lenses and an optical isolator, there are a TE cooler, a thermistor, and a monitoring photodiode. To achieve high power and high slope efficiency (the slope of the L–I curve), an antireflection (AR) coating (~1%) to the front facet and a high-reflection (HR) coating (~80%) to the rear facet are generally applied [37,53,56,57]. In this way, the slope efficiency of the DFB laser becomes 20% higher than that of a conventional DFB laser. To make a complete DFB transmitter, additional circuits such as automatic power control (APC), automatic temperature control (ATC), automatic gain control (AGC), and a predistortion circuit must be included, as shown in Fig. 4.37.

The predistortion circuit [59,60] in Fig. 4.37, located after the AGC circuits, is used to cancel the nonlinear distortions of the laser diode. A general configuration of the predistorter is shown in Fig. 4.38. An input signal feeds into a directional coupler and is split into a primary and two secondary electrical paths. The signal power on the primary path is much larger than those on the secondary paths. Both of the two secondary paths shown in Fig. 4.38 comprise a series of distortion generators, an amplitude adjustment block, a "tilt" or frequency adjustment block, and a phase adjustment block. Ideally, the fundamental frequency in both secondary paths is suppressed in the distortion generator by cancellation, filtering, or other means. The magnitude of the generated distortions is adjusted to be the same as that generated in the transmission device, while the phase of the generated distortions is adjusted to be opposite in sign to that generated in the transmission device.

A typical APC circuit is shown in Fig. 4.39, in which we see that the laser bias current I_C is controlled by I_b, while I_b is in turn controlled by the difference between node D, whose voltage is from a reference (V_{ref}) corresponding to a fixed output power, and node C, whose voltage is derived from the detected photocurrent in the monitoring photodiode. Therefore, any change in the output optical power will

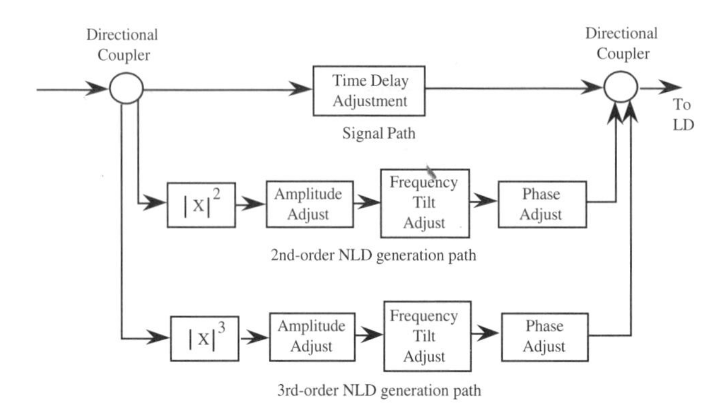

Figure 4.38: A general configuration for a predistorter which generates second- and third-order nonlinear distortions (NLDs). After Ref. [60].

change the monitored photocurrent, and consequently the bias current I_C will be changed to increase or decrease the output optical power.

A typical ATC circuit is shown in Fig. 4.40, in which we see that a thermistor RT (10 $k\Omega$ at 25°C) is placed in a balanced bridge with three other temperature-independent 10-$k\Omega$ resistors. When temperature rises above and decreases below 25°C, RT decreases and increases, respectively. Therefore, the bridge becomes unbalanced, which induces a current flow through the TE cooler to bring the laser temperature back to room temperature.

The AGC circuit in Fig. 4.37 is used to ensure that the modulating signal power is within a predefined range so as to ensure that the OMI per channel can be essentially fixed at an optimum value.

A TE cooler (TEC) is composed of an array of bulk *n*- and *p*-type semiconductor material connected electrically in series and thermally in parallel, as shown in Fig. 4.41a and b [61]. Current I_{TEC} passed through the device results in a temperature

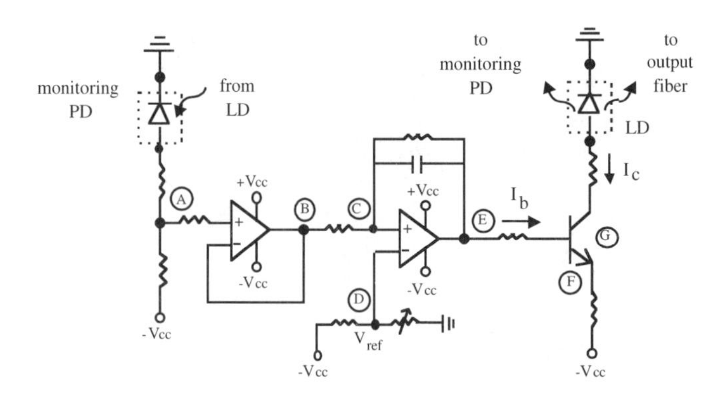

Figure 4.39: An example of an APC circuit.

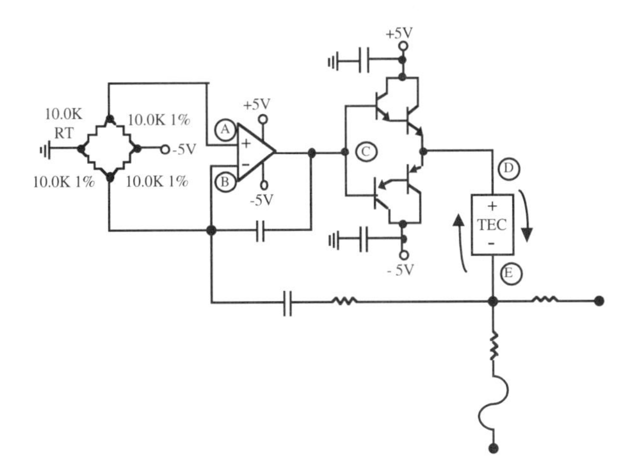

Figure 4.40: An example of an ATC circuit.

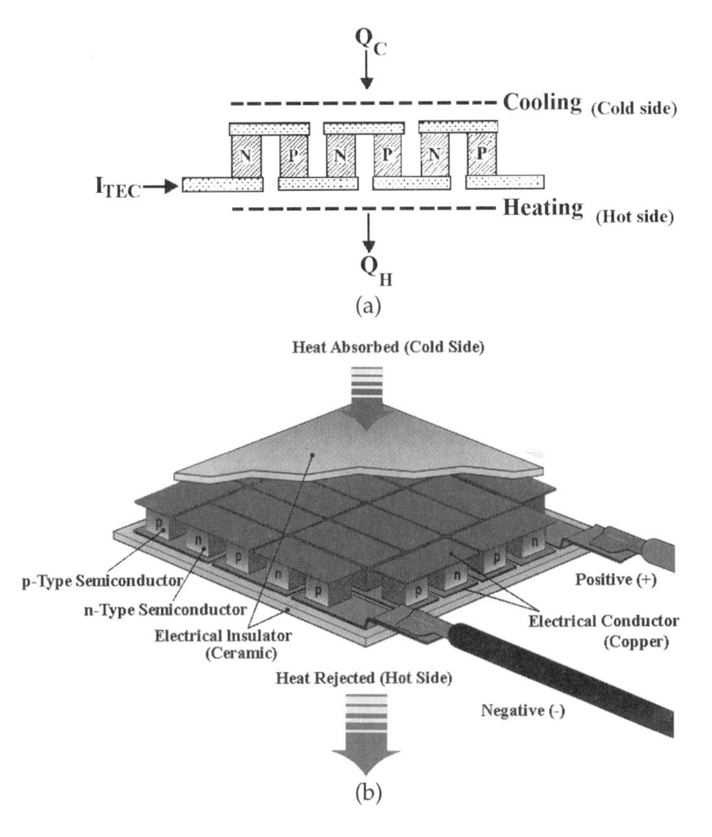

Figure 4.41: (a) Arrangement of semiconductor elements in a TEC. (After Ref. [61], with kind permission from Kluwer Academic Publishers.) (b) Cutaway drawing shows the construction of a TEC. (Courtesy of Melcor Corp.)

differential across the TEC, (called "peltier effect" [62]) leading to a removal of heat Q_C from the cold side to the hot side. The heat rejected at the hot side, Q_H, is the sum of Q_C plus the input power to the TEC, i.e., $Q_H = Q_C + I^2_{TEC} \bullet R$, where R is the total resistance across the series connected elements. To prevent the upper metal from "thermally shorting" the lower metal, the thermal conductivity of the semiconductor should be low; to minimize Joule self-heating, the electrical conductivity should be high.

Problems

4.1. For a DFB laser with the following parameters:

Table P4.1: Laser Parameters

Symbol	Definition	Value	Unit
q	Electron charge	1.6×10^{-19}	Coulomb
g_o	Differential gain	4.5×10^{-12}	m^3/s
N_0	Transparent carrier density	10^{24}	m^{-3}
ε	Gain saturation factor	3×10^{-23}	m^3
τ_p	Photon lifetime	1×10^{-12}	
τ_n	Electron lifetime	3×10^{-9}	
Γ	Optical confinement factor	0.34	—
m	Optical modulation index	0.04	—
γ	Spontaneous emission factor	10^{-4}	
V	Volume of active layer	1.8×10^{-16}	m^3

Assume a bias current I_b of 45 mA and 70 mA, use the large-circuit model (and SPICE simulation software) to find [Note: in the circuit model, let $I_{spon} = qVN/\tau_n = I_s \exp(qV_j/2kT)$ where I_s = 100 pA, and k is the Boltzman's constant] the following:

(1) Its small-signal frequency responses with (a) ε given in the table, and (b) ε equal to zero.

(2) What is the effect of ε on nonlinear distortions when the modulation signal is near the resonance frequency?

4.2. The modified rate equations for quantum-well lasers are given by Eqs. (4.25a)–(4.25c).

Following the same procedure as in Section 4.8.4.1, find the equivalent circuit model of a quantum-well laser.

4.3. Assume that product count can be used to estimate the resultant nonlinear distortions, so that we can use Eq. (2.2) to predict CTB and CSO in a multichannel CATV system. Now use Eqs. (4.50) and (4.51) to see what kind of CTB and CSO (for $f < 550$ MHz) can be obtained when a laser resonance frequency is 6 and 8 GHz (use the parameters given in Table P4.1, adjust the bias current to achieve the resonance peak at the desired frequencies, and adjust the amplitude of the modulation current so that $m = 0.04$/channel in an 80-AM-channel system). Use the relation

$$S_o \simeq \frac{\Gamma \tau_p}{qV}(I_b - I_{th}).$$

4.4. From Eqs. (4.24) and (4.26), derive Eqs. (4.27) and (4.29).

References

[1] N. K. Dutta, "Calculated absorption, emission, and gain in $In_{0.72}Ga_{0.28}As_{0.6}P_{0.4}$," J. Appl. Phys., vol. 51(12), pp.6095–6100, 1980.

[2] G. P. Agrawal, *Fiber-Optic Communication Systems*, Wiley, New York, 1993.

[3] G. P. Agrawal and N. K. Dutta, *Long-Wavelength Semiconductor Lasers*, Van Norstrand-Reinhold, Princeton, New Jersey, 1986.

[4] E. Kapon, "Semiconductor diode lasers: Theory and Techniques," in *Handbook of Solid-State Lasers*, Ed., K. Cheo, Dekker, New York, 1989.

[5] I. P. Kaminow and R. S. Tucker, "Mode-controlled semiconductor lasers," *in Guided-Wave Optoelectronics*, Ed., T. Tamir, 2nd Edition, Chapter 5, Springer-Verlag, Berlin, 1990.

[6] L. Coldren and S. W. Corzine, *Diode Lasers and Photonic Integrated Circuits*, Wiley, New York, 1995.

[7] D. A. Ackerman, P. A. Morton, G. E. Shtengel, M. S. Hybertsen, R. F. Kazarinov, T. Tanbun-Ek, and R. A. Logan, "Analysis of T_0 in 1.3 μm multiple-quantum-well and bulk active lasers," Appl. Phys. Lett., vol.66, pp.2613–2615, 1995.

[8] Y. Zou, J. S. Osinski, P. Grodzinski, P. D. Dapkus, W. C. Rideout, W. F. Sharfin, J. Schlafer, and F. D. Crawford, "Experimental study of auger recombination, gain and temperature sensitivity of 1.5 μm compressively strained semiconductor lasers," IEEE J. Quantum Electron., vol. 29, pp.1565–1575, 1993.

[9] Y. Suematsu, K. Kishino, S. Arai, and F. Koyama, "Dynamic single-mode semiconductor lasers with a distributed reflector," Chapter 4, in *Semiconductors and Semimetals, volume 2 Lightwave Communication Technology*, Vol. Ed. W. T. Tsang, Academic Press, San Diego, 1985.

[10] S. Sasaki, M. M. Choy, and N. K. Cheung, "Effects of dynamic spectral behavior and mode-partitioning of 1550 nm distributed feedback lasers on Gbit/s transmission systems," Electron. Lett., vol.24, pp.26–27, January 1988.

[11] S. Akiba, M. Usami, and K. Utaka, "1.5 μm λ/4-shifted InGaAsP/InP DFB lasers," IEEE J. Lightwave Technol., vol.5, pp.1564–1573, November 1987.

[12] H. Nishimoto, M. Yamaguchi, I. Mito, and K. Kobayashi, "High-frequency response for DFB LD due to wavelength detuning effect," IEEE J. Lightwave Technol., vol.5, pp.1399–1402, October 1987.

[13] S. L. Chuang, *Physics of Optoelectronic Devices*, Wiley, New York, 1995.

[14] A. Yariv, "Quantum well semiconductor lasers are taking over," IEEE Circuits Devices Magazine, pp.25–28, November 1989.

[15] K. Uomi, T. Tsuchiya, M. Komori, A. Oka, K. Shinoda, and A. Oishi, "Extremely low threshold (0.56 mA) operation in 1.3 μm InGaAsP/InP compressive-strained-MQW lasers," Electron. Lett., vol.30, pp.2037–2038, November 1994.

[16] H. Nobuhara, T. Inoue, T. Watanabe, K. Tanaka, T. Odagawa, T. Abe, and K. Wakao, "1.3 μm wavelength, low-threshold strained quantum well laser on p-type substrate," Electron. Lett., vol.30, pp.1292–1293, August 1994.

[17] H. Yamada, T. Terakado, Y. Sasaki, S. Takano, T. Tamura, T. Torikai, and T. Uji, "Extremely low operating current 1.3 μm MQW lasers at high temperature," *Tech. Digest, ECOC'92*, paper Mo.B3.3, pp.101–104, Berlin, 1992.

[18] C. E. Zah, R. Bhat, B. N. Pathak, F. Favire, W. Lin, M. C. Wang, N. C. Andreadakis, D. M. Hwang, M. A. Koza, T. P. Lee, Z. Ewang, D. Darby, D. Flanders, and J. J. Hsieh, "High-performance uncooled 1.3-μm AlGaInAs/InP strained-layer quantum-well lasers for subscriber loop applications," IEEE J. Quantum Electron., vol.30, pp.511–521, February, 1994.

[19] T. Namegaya, N. Matsumoto, N. Yamanaka, N. Iwai, H. Nakayama, and A. Kasukawa, "Effects of well number in 1.3-μm GaInAsP/InP GRIN-SCH strained-layer quantum-well lasers," IEEE J. Quantum Electron., vol.30, pp.578–583, February, 1994.

[20] M. Asada and Y. Suematsu, "Density-matrix theory of semiconductor lasers with relaxation broadening modal-gain and gain-suppression in semiconductor lasers," IEEE J. Quantum Electron., vol.QE-21, pp.434–442, 1985.

[21] W. I. Way, "Large signal nonlinear distortion prediction for a single-mode laser diode under microwave intensity modulation," IEEE J. Lightwave Technol., vol.LT-5, pp.305–315, March 1987.

[22] H. M. Salgado, J. M. Ferreira, and J. J. O'Reilly, "Extraction of semiconductor intrinsic laser parameters by intermodulation distortion analysis," IEEE Photon. Technol. Lett., vol.9, pp.1331–1333, October 1997.

[23] R. Nagarajan, M. Ishikawa, T. Fukushima, R. S. Geels, and J. Bowers, "High speed quantum-well lasers and carrier transport effects," IEEE J. Quantum Electron., vol.28, pp.1990–2007, October 1992.

[24] R.S. Tucker and I.P. Kaminow, "High frequency characteristics of directly modulated InGaAsP ridge waveguide and buried heterostructure lasers," *IEEE J. Lightwave Technol.*, vol.LT-2, pp.385–393, 1984.

[25] T. P. Lee, "Light sources for fiber systems," OFC'88 Tutorial.

[26] J. E. Bowers, B. R. Hemenway, A. H. Gnauck, T. J. Bridges, E. G. Burkhardt, "High-frequency constricted mesa lasers," Appl. Phys. Lett., vol.47, pp.78–80, 1985.

[27] K. Petermann, Laser Diode Modulation and Noise, Dordrecht Klumer Academic Publishers, 1988.

[28] S. Kobayashi, Y. Yamamoto, M. Ito, and T. Kimura, "Direct frequency modulation in AlGaAs semiconductor lasers," IEEE J. Quantum Electron., pp.582–595, April 1982.

[29] M. S. Lin, S. J. Wang, and N. K. Dutta, "Measurements and modeling of the harmonic distortion in InGaAsP distributed feedback lasers," IEEE J. Quantum Electron., vol.26, pp.998–1004, June 1990.

[30] Y. Yamamoto, "AM and FM quantum noise in semiconductor lasers—Part 1: Theoretical analysis," IEEE J. Quantum Electron., vol.QE-19, pp.34–46, 1984.

[31] Y. Yamamoto, "AM and FM quantum noise in semiconductor lasers—Part ll: Comparison of theoretical and experimental results for AlGaAs lasers," IEEE J. Quantum Electron., vol.QE-19, pp.47–58, 1984.

[32] K. Sato, "Intensity noise of semiconductor laser diodes in fiber-optic analog video transmission," IEEE J. Quantum Electron., vol. QE-19, pp.1380–1391, 1983.

[33] P. Hill, R. Olshansky, J. Schlafer, and W. Powazinik, "Reduction of relative intensity noise in 1.3 μm InGaAsP semiconductor lasers," Appl. Phys. Lett., vol.50(20), pp.1400–1402, May 1987.

[34] W. I. Way and M. M. Choy, "Optical feedback on linearity performance of 1.3 μm DFB and multimode lasers under microwave intensity modulation," IEEE J. Lightwave Technol., vol.6, pp.100–108, January 1988.

[35] K. Y. Lau, "Lasers for lightwave systems," Tutorial, Conf. Optical Fiber Commun., 1990.

[36] W. I. Way, "Subcarrier multiplexed lightwave system design considerations for subscriber loop applications," IEEE J. Lightwave Technol., vol.7, pp.1806–1818, November 1989.

[37] A. Takemoto, H. Wantanabe, Y. Nakajima, Y. Sakakibara, S. Kakimoto, J. Yamashita, T. Hatta, and Y. Miyake, "Distributed feedback laser diode and module for CATV systems," IEEE J. Selected Areas Commun., vol.8, pp.1359–1364, September 1990.

[38] S. L. Woodward and G. E. Bodeep, "Uncooled Fabry-Perot lasers for QPSK transmission," IEEE Photon. Technol. Lett., vol.7, pp.558–560, May 1995.

[39] S. L. Woodward, V. Swaminathan, G. E. Bodeep, and A. K. Singh, "Transmission of QPSK signals using unisolated DFB lasers, " IEEE Photon. Technol. Lett., vol.8, pp.127–129, January 1996; also, see T. Nakabayashi, K. Iwai, A. Miki, M. Murata, H. Kobayashi, M. Yoshimura, K. Yoshida, G. Sasaki, and T. Katsuyama, "Uncooled DFB laser modules operating with low distortion over a wide temperature range for a return path of CATV networks," Tech. Digest, Conf. Optical Fiber Commun., paper ThR3, 1997.

[40] T. Ito, S. Machida, K. Nawata, and T. Ikegami, "Intensity fluctuations in each longitudinal mode of a multimode AlGaAs laser," IEEE J. Quantum Electron., vol.13, pp.574–579, August 1977.

[41] G. J. Meslener, "Temperature dependence of mode distribution, intensity noise, and mode-partition noise in subcarrier multiplexed transmission systems," IEEE Photon. Technol. Lett., vol. 4, pp.939–941, August 1992.

[42] K. Y. Lau, C. M. Gee, T. R. Chen, N. Bar-Chaim, and I. Ury, "Signal-induced noise in fiber-optic links using directly modulated Fabry-Perot and distributed-feedback laser diodes," IEEE J. Lightwave Technol., vol.11, pp.1216–1225, July 1993.

[43] G. J. Meslener, "Mode-partition noise in microwave subcarrier transmission systems," IEEE J. Lightwave Technol., vol.12, pp.118–126, January 1994.

[44] M. M. Choy and P. L. Liu, "Origin of modulation-induced mode partition and Gb/s system performance of highly single-mode 1.5μm distributed feedback lasers," Appl. Phys. Lett., vol.52(21), pp.1762–1764, May 1988.

[45] N. Chinone, T. Kuroda, T. Ohtoshi, T. Takahashi, and T. Kajimura, "Mode hopping noise in index-guided semiconductor and its reduction by saturable absorbers," IEEE J. Quantum Electron., vol.21, pp.1264–1270, 1985.

[46] M. Yamada, "Theory of mode competition noise in semiconductor lasers," IEEE J. Quantum Electron., vol.22, pp.1052–1059, 1986.

[47] K. Y. Lau and H. Blauvelt, "Effect of low-frequency intensity noise on high-frequency direct modulation of semiconductor injection lasers," Appl. Phys. Lett., vol.52(9), pp.694–696, February 1988.

[48] X. Lu, C. B. Su, R. B. Lauer, G. J. Meslener, and L. W. Ulbricht, "Analysis of relative intensity noise in semiconductor lasers and its effect on subcarrier multiplexed lightwave systems," IEEE J. Lightwave Technol., vol.12, pp.1159–1166, July 1994.

[49] C.H. Henry, "Theory of the linewidth of semiconductor lasers," IEEE J. Quantum Electron., vol.QE-18, pp.259–264, 1982.

[50] J. Wang, M. K. Haldar, and F.V.C. Mendis, "Formula for two-tone third-order intermodulation distortion in semiconductor laser diodes," Electron. Lett., vol.29, pp.1341–1343, 1993.

[51] J. Helms, "Intermodulation and harmonic distortions of laser diodes with optical feedback," IEEE J. Lightwave Technol., vol.9, pp.1567–1575, November 1991.

[52] H. Yamada, T. Okuda, M. Shibutani, S. Tomida, T. Torikai, and T. Uji, "High modulation frequency low distortion 1.3 μm MQW-DFB-LDs for subcarrier multiplexed fiber-optic feeder systems," Electron. Lett., vol.29, pp.1944–1945, October 1993.

[53] H. Yonetani, I. Ushijima, T. Takada, and K. Shima, "Transmission characteristics of DFB laser modules for analog applications, " IEEE J. Lightwave Technol., vol.11, pp.147–153, January 1993.

[54] H. Soda, Y. Kotaki, H. Sudo, H. Ishikawa, S. Yamakoshi, and H. Imai, "Stability in single longitudinal mode operation in GaInAsP/InP phase-adjusted DFB lasers, " IEEE J. Quantum Electron., vol.23, pp.804–814, June 1987.

[55] T. Okuda, H. Yamda, T. Torikai, and T. Uji, "DFB laser intermodulation distortion analysis taking longitudinal electrical field distribution into account," IEEE Photon. Technol. Lett., vol.6, pp.27–30, January 1994.

[56] T. Okuda, H. Yamada, T. Torikai, and T. Uji, "Novel partially corrugated waveguide laser diode with low modulation distortion characteristics for subcarrier multiplexing," Electron. Lett., vol.30, pp.862–863, May 1994.

[57] C.Y. Kuo, "Fundamental nonlinear distortions in analog links with fiber amplifiers," IEEE J. Lightwave Technol., vol.11, pp.7–15, January 1993.

[58] H. Kawamura, K. Kamite, and H. Yonetani, "Effect of varying threshold gain on second-order intermodulation distortion in distributed feedback lasers," Electron. Lett., vol.26, no.20, pp.1720–1721, September 1990.

[59] H. A. Blauvelt and H. L. Loboda, "Predistorter for linearization of electronic and optical signals," U. S. Patent Number 4,992,754, February 1991.

[60] M. Nazarathy, C. H. Gall, and C. Y. Kuo, "Predistorter for high frequency optical communications devices," United States Patent Number 5,424,680, June 1995.

[61] P.W. Shumate, "Semiconductor laser transmitters," in *Optoelectronic Technology and Lightwave Communication Systems,* Ed. Chinlon Lin, Van Nostrand-Reinhold, Princeton, New Jersey, 1989.

[62] D. D. Pollock, Physics of Engineering Materials, chapter 7, Prentice Hall, New Jersey, 1990.

Chapter 5

Fundamentals of Optical *p-i-n* Diode and Optical Receiver for HFC Systems

In this chapter, the basic physics of a *p-i-n* diode and receiver design for subcarrier-multiplexed (SCM) systems, especially for CATV systems, will be covered. There are basically four requirements that must be met when designing an optical CATV receiver: (1) a wide bandwidth from 50 to 750 MHz (or 50 to 860 MHz for European CATV systems, and up to 1 GHz may be required in the near future) with a flat frequency response ($< \pm 0.75$ dB); (2) low thermal noise over the entire bandwidth ($<$ 7 to 8 pA/$\sqrt{Hz}$); (3) sufficient automatically controlled gain; and (4) high linearity for an input optical power level in the range of -3 to $+3$ dBm. The last requirement is due to the stringent requirements on >50 dB carrier-to-noise ratio (CNR), < -60 dBc CTB/CSO, for each of the >80 channels of analog AM-VSB signals. Note that the minimum received optical power for a CATV receiver is much higher than what is required in a typical baseband digital optical link (e.g., -25 to -30 dBm for a 2.5 Gb/s link). Note also that all of today's CATV receivers use *p-i-n* diodes as the photodetector, and avalanche photodetectors are rarely used because of their relatively small dynamic range and the high reverse bias voltage required (about -120 V as opposed to -15 V for a *p-i-n* photodetector).

In Section 5.1, we will review the basic physics, shot noise, and linearity characteristics associated with a *p-i-n* diode. In Section 5.2, basic receiver configuration, front-end design, related thermal noise, and bandwidth considerations are reviewed. In Section 5.3, state-of-the-art techniques to improve the linearity and bandwidth performance of an optical CATV receiver are reviewed. In Section 5.4, measurement techniques in characterizing the bandwidth and linearity performance of an optical receiver are discussed.

5.1 *P-i-n* Photodiode

A *p-i-n* photodiode, as shown in Fig. 5.1, is the most commonly used semiconductor photodetector. The device structure consists of *p* and *n* regions separated by a layer of lightly doped intrinsic semiconductor material. Therefore, this structure is referred to as a *p-i-n* photodiode. In normal operation, a sufficiently large reverse-bias voltage is applied across the device that the intrinsic region is fully depleted of carriers. Because the depleted region has a high resistance, most of the voltage drop is across this *i*-region. A *p-i-n* photodiode commonly used for long-wavelength HFC systems uses InGaAs for the middle layer and InP (or InGaAsP) for the surrounding *p*-type and *n*-type layers. Since the bandgaps of InP and InGaAs are 1.35 and 0.75 eV, respectively, the surrounding layers are transparent to, and the *i*-layer absorbs, incident light with a wavelength in the 1.3- to 1.6-μm region in which the photon energy is between 0.77 and 0.95 eV. The absorbed photon gives up its energy and generates free electron–hole pairs; the high electric field present in the depletion region causes the free carriers to separate and be collected across the reverse-biased junction. This gives rise to a current flow in an external circuit and is known as photocurrent (I_p).

Two important parameters of a *p-i-n* diode are its quantum efficiency (η) and its responsivity ($\mathcal{R}$). They are defined as follows:

$$\eta = (\text{number of photoelectrons})/(\text{number of incident photons})$$

$$\begin{aligned}\mathcal{R} &= (\text{detected photocurrent})/(\text{incident optical power})\\ &= (\eta q)/(h\nu)\\ &= (\eta\lambda)/1.24,\end{aligned}$$

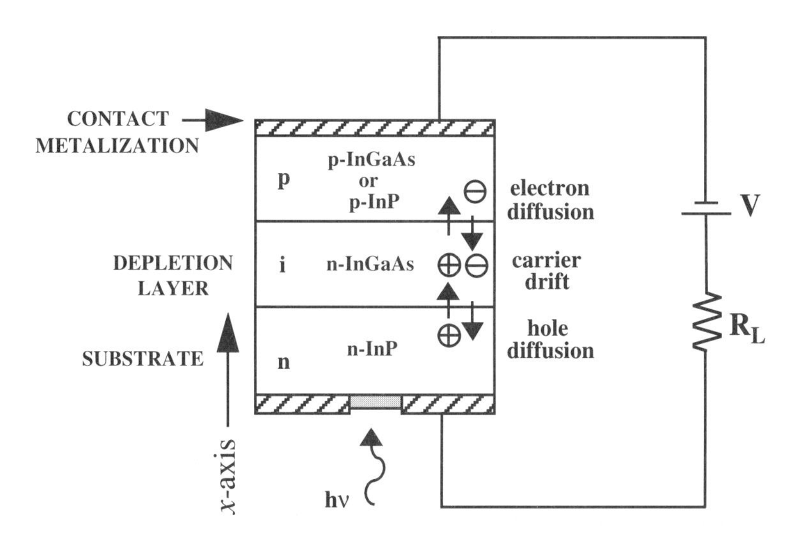

Figure 5.1: Schematic cross-section of a *p-i-n* diode and its external circuit.

where q is the electron charge, $h\nu$ is the photon energy, and λ is the light wavelength in micrometers. The quantum efficiency is affected by the photodiode facet reflectivity r, the absorption loss of the p- or n-layer, and the absorption efficiency of the depletion region, and can be expressed by

$$\eta = (1 - r)\, exp(-\alpha W_n)\, [1 - exp(-\alpha W_d)], \tag{5.1}$$

where α is the loss coefficient, W_n is the width of the n-layer (see Fig. 5.1), and W_d is the width of the depletion region. A typical quantum efficiency spectral response is shown in Fig. 5.2. It can be seen that Si detectors have a better quantum efficiency around the 0.7- to 0.8-μm region than Ge or InGaAs detectors in the 1.3- to 1.55-μm region, mainly because of its lower α in the 0.7- to 0.8-μm region. Ge detectors are not as popular as InGaAs detectors because the former requires high reverse bias voltages of around 30 V, whereas the latter generally requires only about 10 to 15 V. A Ge detector also has the disadvantage that it has a drastically reduced quantum efficiency at around 1.55 μm region, as can be seen in Fig. 5.2. Also note that the front facet of the *p-i-n* diode is usually anti-reflection-coated, and the coupling fiber angle-polished, not only to improve the quantum efficiency, but also to ensure a facet return loss less than about -40 dB ($r < \sim 0.0001$), as required by typical lightwave HFC systems (see Chapter 9).

The width of the depletion region, W_d, must be carefully controlled because it relates to not only the quantum efficiency, but also the modulation bandwidth and the linearity of the *p-i-n* diode. From Eq. (5.1), a wider W_d provides a higher quantum efficiency. In addition, when W_d is wider, the depletion layer capacitance of a *p-i-n* diode becomes smaller according to $C_d = \varepsilon A/W_d$, where ε is the semiconductor permittivity and A is the junction area. This in turn results in a higher modulation bandwidth which is given by $f_{max} = 1/(2\pi RC_d)$, where R is the combination of the load resistor and preamplifier input resistance (see Section 5.2).

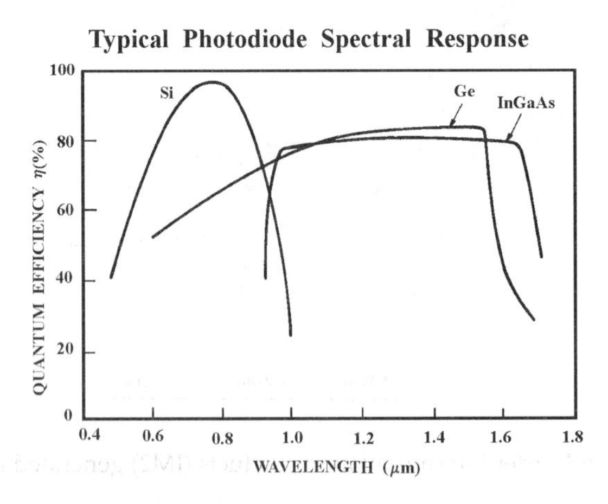

Figure 5.2: Spectral quantum efficiency response for Si, Ge, and InGaAs.

The effect of depletion width on the linearity performance of a *p-i-n* diode can first be illustrated through an actual measured results of an InGaAs *p-i-n* diode in a typical CATV receiver, as shown in Fig. 5.3. The second-order intermodulation (IM_2) performance as a function of the reverse junction bias voltage were obtained by using two tones at 135 and 189.25 MHz, each with a modulation index of 0.7 (see Section 5.4 for photoreceiver characterization techniques). It can be seen that with a reverse bias voltage below ~10 V, increasing the voltage which results in a wider W_d can quickly improve the linearity performance of the diode. However, when the reverse bias voltage is above ~10 V, the IM2 depends only slightly on the bias voltages. The strong bias-dependent nonlinear distortions (NLDs) with bias voltages < ~10 V are believed to be caused by the carrier velocity variations induced by photogenerated space-charge electric field [1,2], as can be explained from the three basic equations governing the *p-i-n* diode operation:

The Poisson equation:

$$\nabla \cdot E = \frac{q}{\varepsilon}(p - n + N_d - N_a). \tag{5.2}$$

The continuity equations:

$$\frac{\partial p(x,t)}{\partial t} = G(x,t) - R(x,t) - v_p \frac{\partial p}{\partial x} - p\frac{\partial v_p}{\partial x} - \frac{1}{q}\frac{\partial J_{p,diff}}{\partial x} \tag{5.3}$$

$$\frac{\partial n(x,t)}{\partial t} = G(x,t) - R(x,t) + v_n \frac{\partial n}{\partial x} + n\frac{\partial v_n}{\partial x} + \frac{1}{q}\frac{\partial J_{n,diff}}{\partial x}. \tag{5.4}$$

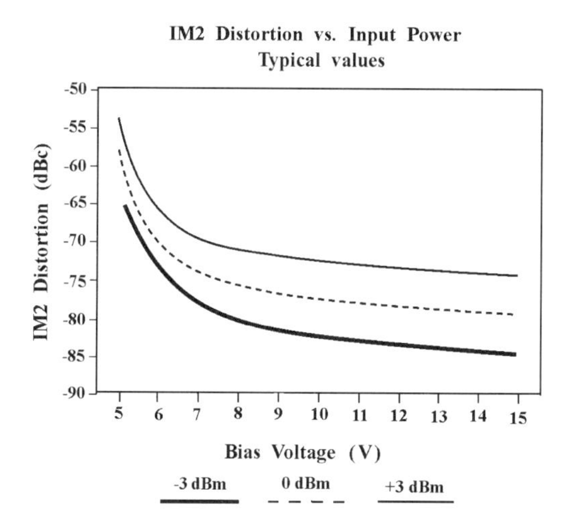

Figure 5.3: Second-order intermodulation products (IM2) generated from a *p-i-n* diode as a function of reverse bias voltage, for three different received optical power levels. OMI/channel = 0.7. (Courtesy of EPITAXX Inc.)

The current density equations:

$$J_p(x,t) = J_{p,drift} + J_{p,diff} = q \cdot p(x,t) \cdot \mathrm{v}_p(E) + J_{p,diff} \tag{5.5}$$

$$J_n(x,t) = J_{n,drift} + J_{n,diff} = q \cdot n(x,t) \cdot \mathrm{v}_n(E) + J_{n,diff}\,, \tag{5.6}$$

where

E = electrical field

$n(x,t)$ = free electron concentration

$p(x,t)$ = free hole concentration

x, t = position in the ternary (n-InGas) region (see Fig. 5.1) and the time

N_d, N_a = density of ionized donor dopants and acceptor dopants in the crystal

$G(x,t)$ = carrier generation rates

$R(x,t)$ = carrier recombination rates

$\mathrm{v}_p(E)$ = electric-field-dependent hole drift velocity

$\mathrm{v}_n(E)$ = electric-field-dependent electron drift velocity

$J_{p,drift}$, $J_{n,drift}$ = hole and electron drift current densities

$J_{p,diff}$, $J_{n,diff}$ = hole and electron diffusion current densities.

We can see that the continuity equations (5.3) and (5.4) are linear (with respect to p and n, respectively) only if the carrier velocities v_p and v_n are independent of the carrier densities, n and p. Unfortunately this is not true because the carrier velocities are related to the electrical field through the following two equations [and are therefore dependent on n and p through the Poisson equation (5.2)]:[1,2]

$$\mathrm{v}_n(E) = \frac{(\mu_n + \mathrm{v}_{nL}\beta|E|)E}{1 + \beta E^2} \tag{5.7}$$

and

$$\mathrm{v}_p(E) = \frac{\mu_p \mathrm{v}_{pL} E}{(\mathrm{v}_{pL}^{\gamma} + \mu_p^{\gamma} E^{\gamma})^{1/\gamma}} \tag{5.8}$$

where μ_n and μ_p are the electron and hole low-field mobilities, respectively; v_{nL} and v_{pL} are the high-field electron and hole velocities, respectively; β is a fitting parameter; and γ is a factor depending on temperature. Therefore, when the *p-i-n* diode is under illumination by a single-tone modulated light with carrier generation rates given by $G(x) = G_o(1 + m \cdot \sin \omega t)\exp(-\alpha(L-x))$ (L is the length of the depletion layer), the resultant $p(x,t)$ and $n(x,t)$ will contain harmonic nonlinear distortions. This will causes nonlinear distortions in $J_n(x,t)$ and $J_p(x,t)$ according to Eqs. (5.5) and (5.6). Finally, the total conduction current, which is given by

$I_c(t) = A \cdot (J_n(x,t) + J_p(x,t))$ where A is the junction area, will also contain nonlinear distortions.

The effect of space-charge electric field can be minimized by using longer W_d because when there is a *p-i-n* diode contact voltage change due to external circuit loading, a longer device can result in smaller change in the depletion-region electric field.

For high enough reverse bias voltage, such as $> \sim 10$ V in Fig. 5.3, the carrier velocities gradually saturates with increasing electric field, and another nonlinear mechanism dominated by absorption in *undepleted* semiconductor regions can start to play a major role [2]. This mechanism is essentially bias-independent (since the reverse bias voltage only affects the depletion region electric field).

It was also experimentally observed that for low to medium reverse bias voltages, the NLDs are frequency-dependent , while at higher applied voltages, the NLD is nearly independent of frequencies [2]. For very low modulating frequencies (e.g., below a few hundred megahertz), the nonlinear distortions in a *p-i-n* photodetector are essentially caused by "static" nonlinearities such as nonlinear relation between I_p and the incident light power, or a variation in responsivity with detector voltage.

We also note that the modulation bandwidth of a *p-i-n* diode may be limited by the transit time—the time it takes the photogenerated carriers to drift through the depletion region. Therefore, the width of the depletion region W_d cannot be too large. Fortunately, the required modulation bandwidth of a *p-i-n* diode in an HFC system is only about 2 GHz, which can be easily achieved even with a W_d as wide as a few micrometers.

So far it seems that wide W_d is advantageous to high quantum efficiency, high linearity, and high bandwidth. However, one must be cautioned by the negative effect in increasing W_d, that is, the dark current. A dark current is the output photocurrent of a diode even when there is no incident light and is dominated by the thermal generation of carriers in the depletion region at bias voltages < 20 V [3]. Figure 5.4 shows an example of a reverse bias voltage- and temperature-dependent dark current of a CATV *p-i-n* diode. We can see that when the reverse bias voltage is increased, W_d is increased, and consequently the dark current is increased. Note that the dark current can increase by two orders of magnitude when the temperature is increased from 25 to 85°C.

From the preceding discussions, we know that the *p-i-n* diode depletion width W_d is determined by finding a trade-off among high quantum efficiency, high linearity (large W_d), and low dark current (small W_d).

A typical *p-i-n* diode used for CATV lightwave systems has a responsivity of 0.8–0.9, a modulation bandwidth of around 2 GHz, and a two-tone second-order intermodulation product of -75 dBc for both tones with modulation index of 70% at an input optical power of 1 mW.

5.1.1 Shot Noise

Shot noise [4] associated with a *p-i-n* diode is a quantum noise which is caused by the random (unpredictable) number of electron–hole pairs produced within the

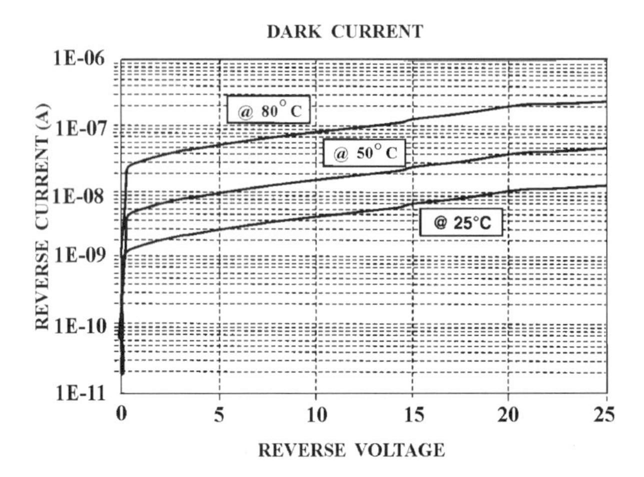

Figure 5.4: Dark current as a function of reverse bias voltage for three different operation temperature conditions. Courtesy of Epitaxx Inc.

photodetector. If we assume that the probability of a single electron–hole pair generation event in a very small time interval Δt is proportional to Δt, the probability of more than one event in this interval is negligible, and the events are independent, we can then use the Poisson statistics to describe the probability that exactly n electron–hole pairs (per unit of interval) will be produced is given by

$$f(n) = S_o^n \exp(-S_o) /n!, \ (n = 0, 1, 2, \ldots) \tag{5.9}$$

where S_o is the average total number of received photons over a time interval (O, T) and is equal to $E/h\nu$, where E is the average received photon energy. We can see that even though E or S_o is known in advance, we may get anywhere from zero to an infinite number of electron–hole pairs produced. The average of this statistical distribution is S_o or $E/h\nu$ pairs. The shot noise associated with a received optical signal can be illustrated in Fig. 5.5. The received optical pulse, shown in Fig. 5.5a, illuminates a detector which produces electron–hole pairs in response. We cannot predict exactly how many electron–hole pairs will be produced, or at what times (Fig. 5.5b). We assume that the detector produces a photocurrent in response to each pair, having the shape shown in Fig. 5.5c. Figure 5.5d shows the composite response from all of the photocurrents and the shot noise fluctuations can be observed.

Analytically, let us consider the photocurrent $i(t)$ which is made up of a very large number of individual electron–hole pairs $h(t-t_i)$, which occur at random times t_i. An observation of $i(t)$ during a period T will yield

$$i_T(t) = \sum_{i=1}^{S_O} h(t-t_i), \quad 0 \le t \le T, \tag{5.10}$$

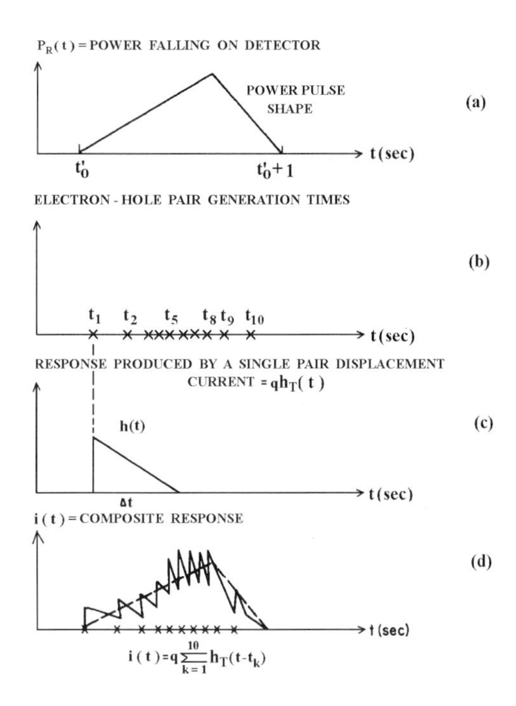

Figure 5.5: (a) Received optical pulse shape, (b) times at which pairs are produced, (c) individual hole–electron pair displacement current response waveform, (d) composite response resulting from all hole–electron pairs. After Ref. [5].

Note that since Δt in Fig. 5.5 is very small, $h(t)$ is almost like a delta function, and from the properties of the electron, we have

$$\int_{-\infty}^{\infty} h(t)dt = \int_{0}^{\Delta t} h(t)dt = q, \tag{5.11}$$

where q is the electron charge. The Fourier transform of $i_T(t)$ is given by

$$I_T(w) = \sum_{i=1}^{S_o} H_i(w), \tag{5.12}$$

where $Hi(\omega)$ is the Fourier transform of $h(t-t_i)$, that is $H(\omega)exp(-i\omega t_i)$. We then have

$$\begin{aligned} \left|I_T(\omega)\right|^2 &= \left|H(\omega)\right|^2 \sum_{i=1}^{S_o}\sum_{j=1}^{S_o} e^{-i\omega(t_i-t_j)} \\ &= \left|H(\omega)\right|^2 [S_o + \sum_{i\neq j,}\sum_{j=1}^{S_o} e^{-i\omega(t_i-t_j)}] \, . \end{aligned} \tag{5.13}$$

If we take the ensemble average of both sides, the second term on the right-hand side can be neglected in comparison to S_O since the times t_i are random. This results in

$$E[|I_T(\omega)|^2] = S_O \cdot E[|H(\omega)|^2] \tag{5.14}$$

where $E[\bullet]$ represents ensemble averaging. Therefore, the single-sided current spectral density of the photocurrent $i_T(t)$ is given by [6]

$$<i_{sh}^2> = \lim_{T\to\infty} \frac{2E[|I_T(\omega)|^2]}{T} \ (\text{A}^2/\text{Hz}). \tag{5.15}$$

We also note that $H(\omega)$ is given by

$$H(\omega) = \int_0^{\Delta t} h(t) e^{-i\omega t} dt \approx q, \tag{5.16}$$

where we have used Eq. (5.11) and the approximation is valid when $\omega \bullet \Delta t << 1$. From Eqs. (5.14) and (5.15), and by assuming a 100% quantum efficiency, we finally obtain the spectral current density of the shot noise as

$$\begin{aligned} <i_{sh}^2> &= 2 \cdot (S_O/T) \cdot q^2 = 2 \cdot (S_O/T \cdot q) \cdot q \\ &= 2 \cdot I_p \cdot q \end{aligned} \quad (\text{A}^2/\text{Hz}). \tag{5.17}$$

The dark current can also be included to give the total shot noise current density as

$$<i_{sh}^2> = 2q(I_p + I_d) \quad (\text{A}^2/\text{Hz}). \tag{5.18}$$

5.2 Basic Receiver Configurations, Front-End Design, and Related Thermal Noise

5.2.1 Basic Configuration

Optical receivers can be classified into two major categories: digital receiver (for baseband digital signals) and analog receiver (for signals on RF/microwave carriers, i.e., for subcarrier-multiplexed systems). Their functional block diagrams are shown in Figs. 5.6a and b, respectively. The major difference between the two types of receivers lies in the fact that digital receivers must have a very good low-frequency response in addition to high-frequency response (maintaining flat up to about 70%

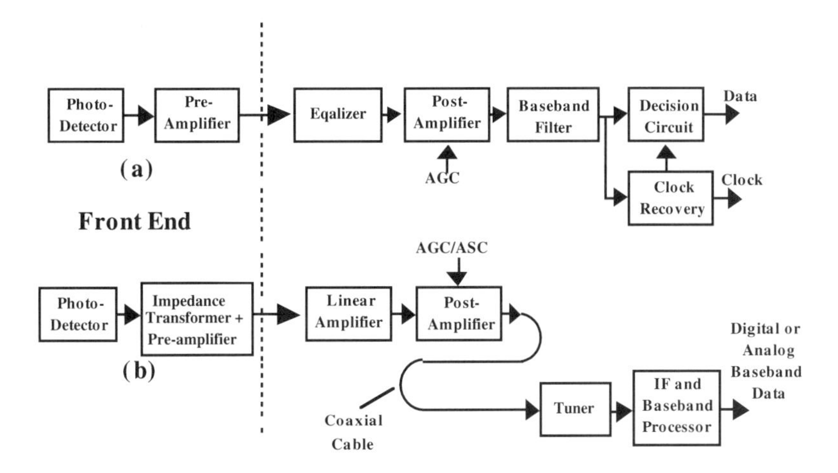

Figure 5.6: Basic block diagrams of a baseband digital (a) and SCM (b) optical receiver. (AGC: automatic gain control. ASC: automatic slope control.)

of the bit rate) because a pseudorandom non-return-to-zero (NRZ) signal consists of frequency components all the way down to R_b/L_p, where R_b is the data bit-rate and L_p is the pattern length. For example, a 622 Mb/s data with a pattern length of 2^{23}-1 may require a receiver with low-frequency response flat all the way down to about 80 Hz. Therefore, a DC-coupled receiver must be designed. An SCM receiver, however, only has to take care of the frequency range where the RF carriers occupy, such as 5–40 and 50–750(860) MHz for the upstream and downstream HFC systems, respectively.

Another major difference is that SCM receivers must have an excellent linearity to meet the stringent requirements on the CTB, CSO, and CNR of AM-VSB video signals. Therefore, the front-end preamplifier design considerations are quite different for the two types of receivers.

5.2.2 Front-End Design and Related Thermal Noise

In addition to the shot noise which we discussed in Section 5.1.1, another major noise source in a typical optical fiber receiver is the receiver thermal noise. Receiver thermal noise exists even if there is no incident optical power at the receiver. Although the effective receiver thermal noise depends critically on the front-end design (e.g., preamplifier), the sole source of thermal noise comes from resistors. A resistor has free electrons that have some random motion if the resistor has a temperature above absolute zero. This random motion causes a noise voltage to be generated at the terminals of the resistor. The resistor can be modeled by an equivalent circuit which consists of a noiseless resistor in parallel with a noise current source.

From quantum mechanics, it can be shown that the two-sided power spectral density $< i_{th}^2(f) >$ corresponding to the current source is [7]

$$\left\langle i_{th}^2(f) \right\rangle_{two-sided} = \frac{2}{R} \frac{h|v|}{\exp(\frac{h|v|}{kT}) - 1} \quad (A^2/Hz), \tag{5.19}$$

where v is the frequency, h is Planck's constant, and k is the Boltzmann constant. At room temperature kT is approximately 4×10^{-21} *J*. At 10^9 Hz hv is approximately 10^{-24} *J*. Thus, we see that for classical radio frequencies, kT is much greater than hv. In this case, Eq. (5.19) reduces to

$$< i_{th}^2 >_{two-sided} \cong 2\,k\,T/R \qquad (A^2/Hz). \tag{5.20}$$

Note, however, that at 3×10^{14} Hz (1 μm wavelength), hv is approximately 2×10^{-19} J. In this case hv is approximately 50 times kT. Therefore, at optical wavelengths shorter than about 2 μm, thermal noise is typically negligible. At those wavelengths, it is the quantum noise which ultimately limits the system performance.

The open-circuit one-sided (positive frequency) current spectral density is then given by

$$< i_{th}^2 > = 4\,k\,T/R \qquad (A^2/Hz). \tag{5.21}$$

When an amplifier with a noise figure of F is connected directly to a *p-i-n* diode, as shown in Fig. 5.7, we have the noise current spectral density at the output of the amplifier as

$$< i_{th}^2 > = 4\,k\,T\,F/R \qquad (A^2/Hz). \tag{5.22}$$

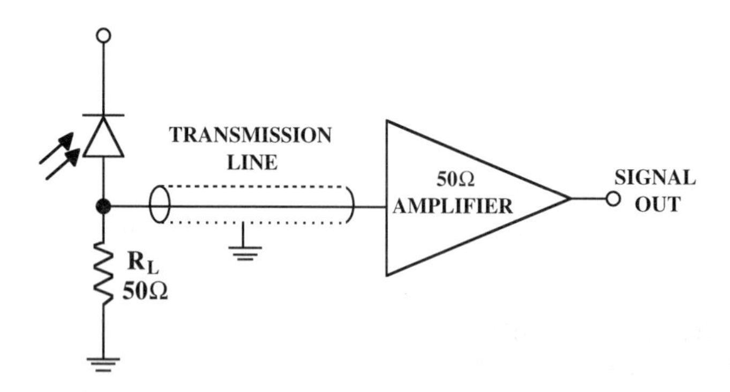

Figure 5.7: A simple optical receiver using an amplifier with a 50-ohm input impedance and a noise figure F.

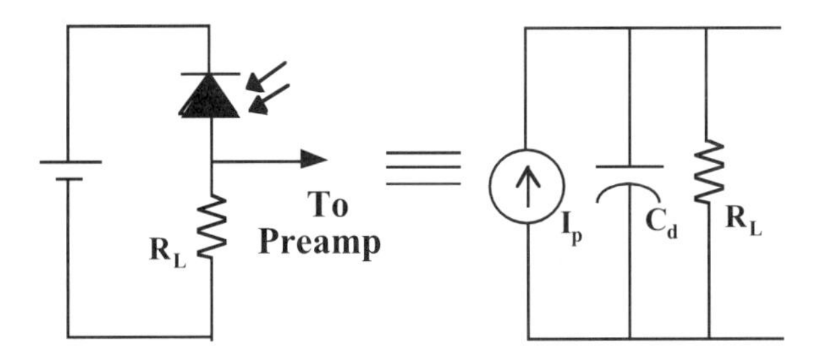

Figure 5.8: Equivalent circuit of a *p-i-n* diode and its load resistor.

Having learned that the major source of thermal noise comes from resistors, we can now take a look of the equivalent circuit of a *p-i-n* diode and its associated bias circuit, as shown in Fig. 5.8, where a *p-i-n* diode is modeled as a current source in parallel with its depletion capacitance C_d (today's InGaAs *p-i-n* diode chips for CATV applications are available with shunt capacitance values as low as 0.3 pF). Since we already know that the thermal noise current density generated by the load resistor is, from Eq. (5.21), $4kT/R_L$, the best signal-to-noise ratio (SNR) that can be achieved is when R_L approaches infinity. A receiver with a large R_L is called a *high-impedance* receiver. However, there is a trade-off between the high SNR and the available modulation bandwidth which is proportional to $1/(2\pi R_L C_T)$. Here C_T is the total capacitance of the depletion capacitance and the stray capacitance C_a (which is in parallel with C_d) associated with the preamplifier input impedance. Today's FET amplifier can have an input capacitance as low as 0.6 pF. Packaging parasitics can further increase the value of C_T (e.g., up to about 1.3 pF [8]. To increase the bandwidth, a high-impedance receiver may require an equalizer, as illustrated in Fig. 5.9.

The pole-zero matching between front amplifiers and the equalizer, however, is not a straightforward task. The high-impedance receiver, even with an equalizer built in, still has the major disadvantage that the enhanced amplitude of the low-frequency signal components at front-end input causes early saturation of the amplifier; therefore, its linearity at high input optical power is very poor.

The other extreme case is a *low-impedance* receiver in which the load resistor R_L is equal to 50 or 75 ohms. A readily available 50- or 75-ohm amplifier can follow immediately after the *p-i-n* diode, as was shown in Fig. 5.7. The advantages of this configuration are (1) many commercially available linear and high-gain amplifiers can readily be used, and (2) wide bandwidth is easily available. Its biggest disadvantage, however, is its high noise level, which is equal to $4kTF/R_L$ [Eq. (5.22)] where R_L is 50 or 75 ohms.

Having learned the disadvantages of the previous two front-end designs, we can arrive at a compromise design by decreasing the effective load resistance with the configuration shown in Fig. 5.10, which forms the *transimpedance* receiver

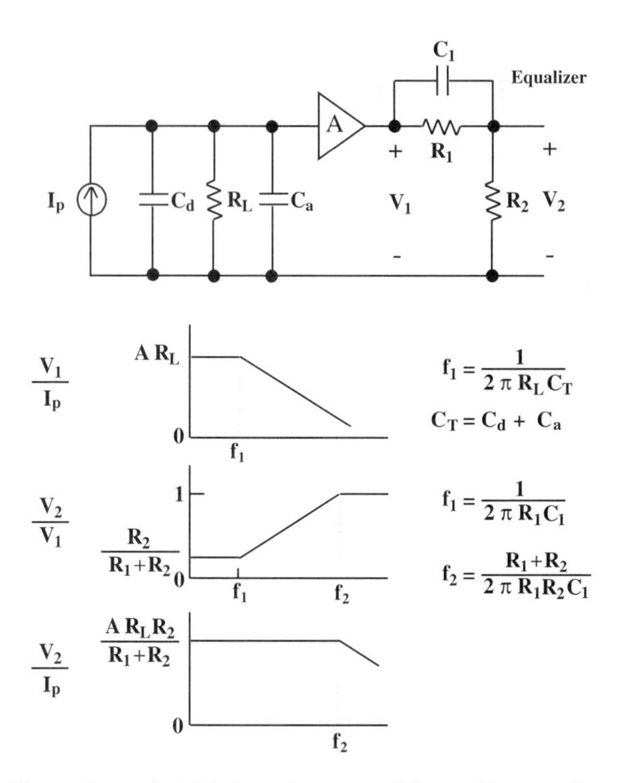

Figure 5.9: Illustrations of a high-impedance amplifier and its equalizer. (After Ref. [9], with kind permission from Kluwer Academic Publishers.)

configuration. The advantages of a transimpedance receiver are that it has (1) a better dynamic range than high-impedance receiver, (2) a better sensitivity than that of low-impedance receiver, and (3) no need to match the poles and zeros between equalizer and front end. On the other hand, a transimpedance receiver has a higher noise level than a high-impedance receiver, and it has limited bandwidth due to the feedback resistor and the associated shunt capacitance. Later in Section 5.3, we will find that a modified transimpedance front end has been adopted in most wideband CATV optical receiver design.

In designing a high- or trans-impedance amplifier, we can use either a GaAs MESFET or a silicon bipolar-junction-transistor (BJT) amplifier. In the case of an FET amplifier, the equivalent input thermal noise current variance per unit bandwidth is given by [9,10]

$$< i_{th}^2 > /\Delta f = 2q\, I_L\, I_2 + (4kT/R_f)\, I_2 + 4kT\,\Gamma\, (2\,\pi\, C_T)^2\, (\Delta f)^2\, I_3 / g_m, \quad (5.23)$$

where I_2 and I_3 are the Personick integrals given in Table 5.1 (solve Problem 5.3), for digital NRZ, RZ, and analog signals, respectively (in our case of subcarrier-multiplexed lightwave systems, we use the I_2 and I_3 values for analog signals); Δf is

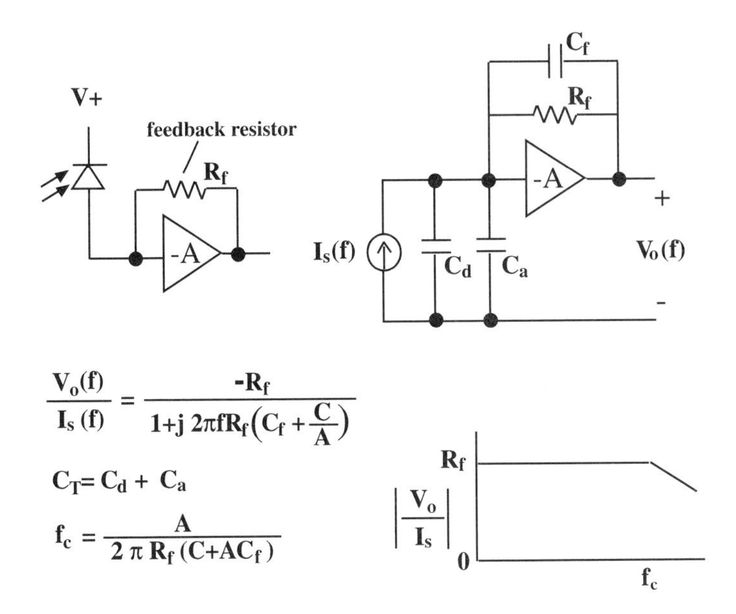

Figure 5.10: Transimpedance front-end design. (After Ref. [9], with kind permission from Kluwer Academic Publishers.)

the bandpass filter 3-dB bandwidth, I_L is the FET leakage current, R_f is the feedback or load resistor, g_m is the FET transconductance, C_T is the total input capacitance, and Γ is the FET channel noise factor, which is about 1.75 for a GaAs MESFETs. In Eq. (5.23), the first term is the shot noise associated with the FET gate leakage current, the second term is the thermal noise due to the load resistor, and the third term is the thermal noise associated with the FET conductance. In an ideal FET preamplifier, only the third term dominates, and we see that the limiting circuit noise for a FET varies as Δf^3 and also as C_T^2 / g_m. The optimized FET preamplifier figure of merit is thus given by

$$\text{Figure of merit (FET)} = g_m / C_T^2 . \tag{5.24}$$

We can also express the figure of merit in terms of f_T , the FET short-circuit common-source current–gain–bandwidth product, given by $f_T = g_m / (2 \pi C_a)$, where C_a is the sum of the gate-source and gate-drain capacitances. For an optimum

Table 5.1: Values of Personick integrals (weighting constants).

Weighting constants	Raised cosine NRZ	Raised cosine RZ (50% duty cycle)	Analog
I_2	0.562	0.403	1
I_3	0.0868	0.0361	⅓

designed gate width, we have $C_T = 2\,C_a$ [10]. Now if we substitute $C_T = 2\,C_a$ together with the relation $g_m = 2\,\pi\,C_a f_T$ into the third term of Eq. (5.23), we obtain

$$< i_{th}^2 >_{FET,\,optim.} = (32/3)\,\pi\,k_T\,\Gamma\,C_a\,(\Delta f)^3\,/\,f_T. \tag{5.25}$$

Therefore, the optimized FET preamplifier figure of merit can also be expressed by

$$Figure\ of\ merit\ (optimized\ FET) = f_T\,/\,C_a. \tag{5.26}$$

From Eqs. (5.24) and (5.26), we know that we should choose FETs with as large a transconductance and f_T as possible.

In the case of BJT amplifiers, the input noise-current variance per unit bandwidth for an optimized BJT preamplifier is given by [10]

$$< i_{th}^2 > /\,\Delta f = (4kT/R_f) + 4\pi\,kT\,C_0 f_T\,\Psi(f)\,/\,\beta + 4\,\pi\,k_T\,C_0\,\Delta f^2\,[1 + \Psi(f)\,]^2 \,/\,(3\,\Psi(f)\,f_T) + 4kT\,r_{bb'}\,[2\,\pi\,(C_d + C_s)]^2\,\Delta f^2/3, \tag{5.27}$$

where C_d is the detector capacitance, C_s is the stray capacitance including portion of the transistor input capacitance associated with leads and package, C_0 is the total input capacitance for zero bias current (including C_d, C_s, and transistor base-emitter and base-collector capacitances), $r_{bb'}$ is the base spreading resistance, β is the small-signal current gain, f_T is the short-circuit common-emitter current–gain–bandwidth product at high collector current (where the diffusion capacitance dominates), and $\Psi(f)$ can be approximated by $(\Delta f\,/\,f_T)\bullet\,[\beta\,/3]^{1/2}$. The figure of merit of a bipolar front end can be given by

$$Figure\ of\ merit\ (bipolar) = \beta^{\,1/2}\,/\,C_0. \tag{5.28}$$

Therefore, in a BJT front-end design, we prefer using a bipolar transistor with high small-signal current gain and low total input capacitance (at zero bias current).

Note that the figures of merit of FETs and bipolars which we obtained in Eqs. (5.24), (5.26), and (5.28) are all based on SNR considerations. These considerations are quite enough for SCM–QPSK or QAM signals whose SNR and linearity requirements are both low. However, for multichannel AM-VSB signals, the linearity performance of an SCM receiver must also be considered, as will be covered in the next section.

5.3 State-of-the-Art Optical CATV Receiver Design

Generally speaking, the design considerations of an optical CATV receiver can be divided into two parts: (a) the interface between a *p-i-n* diode and the following preamplifier, and (b) the preamplifier.

The most critical part of (a) is to minimize the total stray capacitance to ground from any element in the interface circuitry and any preamplifier input capacitance that is translated back across the *p-i-n* diode. One approach to reach this goal is to integrate the photodiode, the interface, and the preamplifier into a single hybrid assembly module [8]. Usually, an impedance transformer (with some impedance-matching circuitry) is used as the interface, as shown in Fig. 5.11. This interface can be used in an optical CATV receiver because the low-frequency response (say, below 50 MHz) is not important, and the high-frequency response is only up to <1 GHz.

In Fig. 5.11, the choice of the load resistor value should be based on the following considerations: (1) the CNR is proportional to $<i_{sig}^2>/(4kT/R_L)$, (2) the bandwidth is inversely proportional to $2\pi(R_L//(N^2 \bullet 75))C$, where C includes the depletion region capacitance and other stray capacitances at the front end and N is the turn ratio of the transformer ("//" represents resistors in parallel), (3) R_L and $N^2 \bullet 75$ should not be too far different for maximum power transfer, and (4) the preamplifier after the transformer must be operated in its linear range. Most of today's low-cost CATV receivers use these design guidelines and can achieve an equivalent input noise current density in the range between 5 and 8 $pA/\sqrt{Hz}$ and a maximum bandwidth of 750 or 860 MHz.

As for preamplifier design, we should consider the following factors [11]:

1. *Push–pull versus single-ended:* Conventional transimpedance FET or BJT amplifiers are susceptible to second-order NLDs when they are used for the amplification of broadband signals which include a high number of video carriers. This problem can be alleviated by using a dual matched amplifier configuration coupled in a push–pull relationship (see Section 2.2.2). This precludes the use of single-ended designs in favor of push–pull where an additional broadband second order improvement of between 15 and 20 dB can be achieved.
2. *Circuit configuration:* Conventional optical receivers typically used transimpedance amplifiers which have two stages. The first transimpedance stage comprises an FET or BJT in a common-source or common-emitter configuration. The output of the transimpedance is then buffered by a second stage used for impedance matching and power transfer to a output power combiner. A problem inherent in the common source (emitter) configuration is the "Miller effect," where the amplifier

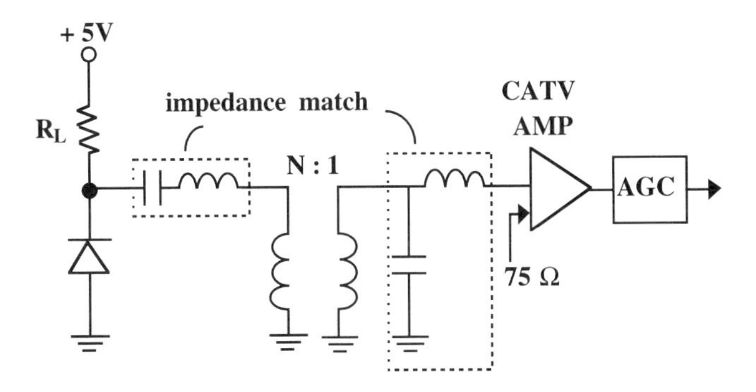

Figure 5.11: Single-ended optical receiver with an input impedance transformer.

stage amplifies the feedback capacitance of the device by its voltage gain and thereby reduce the high-frequency response of the amplifier. Because most devices configured for voltage gains produce such effects, it is difficult to increase the bandwidth of the amplifier. To solve this problem, a circuit configuration of common source (emitter) with common gate (base) stages, termed the *cascode* topology, is used to maximize bandwidth. It was also found that less amplifier noise is produced with this dual stage configuration than with the configuration using cascade common source (emitter) stages [12].

3. *Reduce power consumption and parasitics*. Note in particular that if the incident light is not launched into the central region of a *p-i-n* diode due to misalignment, the *p-i-n* diode bandwidth can be severely reduced [11].

A block diagram of an optical CATV receiver using a push–pull "cascode" configuration is shown in Fig. 5.12 [12].

5.4 Carrier-to-Noise Ratio at the Receiver

In subcarrier multiplexed systems, an additional important noise term, the intensity noise that we discussed in Chapter 4, must also be considered. Now we have the analytical expressions for shot noise, receiver thermal noise, and in addition, the intensity noise due to the laser diode and due to the system. Shot noise and laser intensity noise exist only when there is incident light, and intensity noise depends on both the

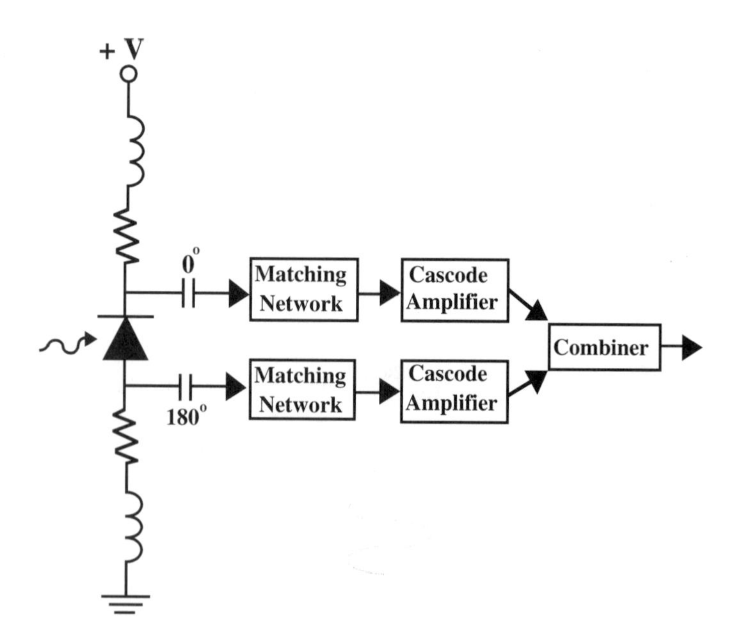

Figure 5.12: Push–pull optical receiver with cascode amplifiers. After Ref. [12].

light source characteristics and other system factors which may be further affected by the modulation index in a laser diode. We can therefore write down the expression for the CNR in an *RF* channel with noise bandwidth Δf (most commonly 4.2 MHz in a U.S. HFC system):

$$\mathrm{CNR} = \frac{\langle i_s^2 \rangle}{\left[\langle i_{sh}^2 \rangle + \langle i_{th}^2 \rangle + \langle i_{RIN}^2 \rangle\right] \bullet \Delta f} \tag{5.29}$$

where again $\langle i_s^2 \rangle$ is the average received carrier power, $\langle i_{RIN}^2 \rangle$ is given by $RIN \bullet I_p^2$, and $\langle i_{sh}^2 \rangle$ is given by Eq. (5.18); the value of $\langle i_{th}^2 \rangle$ depends on what kind of preamplifiers are used. For an RF signal modulation on laser diode with a modulation index m (defined in Fig. 5.13), we have the average received carrier power for this particular modulation frequency as $\frac{1}{2} m^2 I_p^2$. Therefore, using Eqs. (4.39), (5.18), and (5.29) we can have

$$\mathrm{CNR}^{-1} = \frac{4q(I_p + I_d) \bullet \Delta f}{m^2 I_p^2} + \frac{2\langle i_{th}^2 \rangle \bullet \Delta f}{m^2 I_p^2} + \frac{2\mathrm{RIN} \bullet I_p^2 \bullet \Delta f}{m^2 I_p^2}. \tag{5.30}$$

From this equation, we can see that if the system noise is dominated by RIN noise, the system CNR is independent of the received photocurrent I_p (or equivalently, the received optical power); if the system noise is dominated by shot noise, the system CNR depends linearly on the received photocurrent I_p; and if the system noise is dominated by receiver thermal noise, the system CNR depends on I_p^2 (practice Problem 5.1).

Note that some manufacturers of optical receivers use noise-equivalent power (*NEP*) to quantify receiver noise. *NEP* is defined as the minimum optical power per unit bandwidth required to generate an SNR of 1, that is, $NEP \bullet \mathcal{R} = \sqrt{\langle i_{th}^2 \rangle}$ or $NEP = h\nu \sqrt{\langle i_{th}^2 \rangle} / \eta q$.

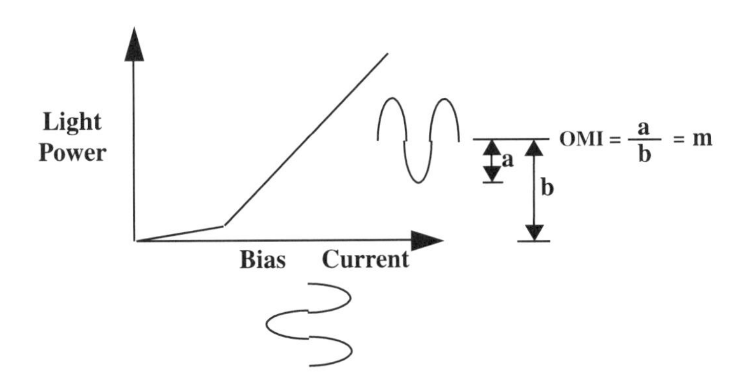

Figure 5.13: Defintion of optical modulation index (OMI).

5.5 Bandwidth and Linearity Characterization Techniques for Optical Receivers

Typical optical receivers for HFC systems require a flat frequency response from about 50 to 750 or 860 MHz, and the frequency response ripple has to be within ±0.5 or ±0.75 dB over the entire bandwidth. In addition, the optical receiver should not add any more CTB or CSO to that generated by a laser transmitter. Therefore, when characterizing the frequency response and linearity of an optical receiver, it is very important to make sure that its frequency ripples and nonlinearities can be distinguished from those of a laser transmitter. In the following two subsections, we will introduce such characterization techniques.

5.5.1 Bandwidth Measurement Techniques

The simplest method is to use a well-calibrated optical transmitter which has a wide bandwidth (typically >3 GHz) and small frequency response ripples (<±0.3 dB). The 750- or 860-MHz bandwidth and the frequency response of an optical CATV receiver can then be easily characterized. Such well-calibrated optical transmitters are commercially available.

The second method is to use two heterodyned DFB lasers, as shown in Fig. 5.14 [13]. One of the lasers is temperature tuned (~12 GHz /°C in [13]) so that the beat frequency of the two lasers (after the photodiode) can be tuned from DC to over 100 GHz while maintaining a high SNR (~40 dB). The linewidths of the two lasers should be kept as narrow as possible to increase the SNR.

The third method, which is available today as commercial product, is to use the amplified spontaneous emission (ASE) noise (see Chapter 6 for definitions) from an

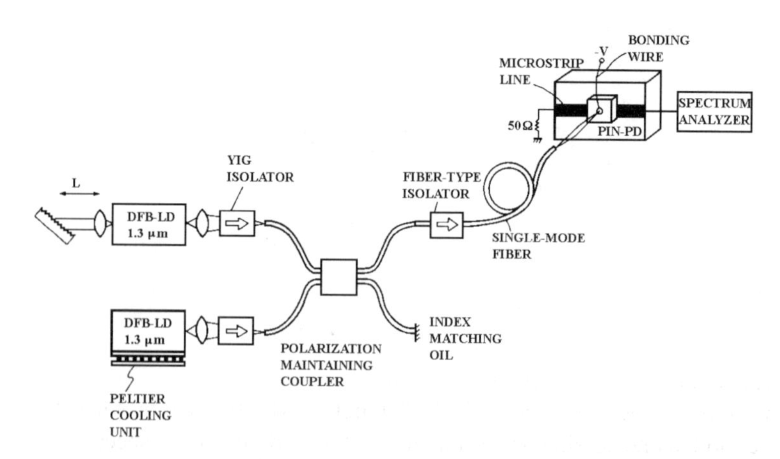

Figure 5.14: A heterodyne technique for measuring optical receiver bandwidth. The external cavity laser is used to reduce the laser linewidth and improve the measurement SNR. After Ref. [13] (© 1989 IEEE).

optical amplifier [14,15], practically an erbium-doped fiber amplifier [15]. The ASE noise beats with itself at the photodetector and results in a white noise with an extremely wide bandwidth. However, care must be taken to eliminate a significant portion of the ASE noise by using an optical filter, so that the SNR of the measured frequency response can be increased [15].

Other methods, such as applying time-domain short pulse technique [13,16], and a delayed self-homodyne method [17], have been proposed. But both methods have their measurement uncertainties: The former is limited by not knowing the exact shape of the optical pulse, and the latter is limited by not knowing exactly the power spectrum of a frequency-modulated laser diode.

5.5.2 Linearity Measurement Techniques

Before the successful development of highly linear laser diodes, most of the measured nonlinear distortions generated from a paired optical transmitter and receiver can be attributed entirely to the transmitter. However, with the development of highly linear MQW DFB lasers with >80-AM-VSB-channel capacity, it becomes important to separately evaluate the linearity performance of an optical receiver. Direct single-tone or two-tone modulation of a laser transmitter may no longer be appropriate because the measured results may include the NLDs of the laser.

In measuring the linearity of a photodetector, one can measure the single-tone harmonics and/or two-tone intermodulation products. In the former case, a typical test setup is to use heterodyning of two narrow-linewidth offset-phase-locked solid-state lasers [2] or two wavelength-stabilized narrow-linewidth DFB lasers. Tuning the wavelength of one laser (through phase-locked loop or temperature) allows the beat note between the two lasers to sweep from DC to very high frequencies. In the latter case, one can use two DFB lasers with wavelengths sufficiently apart (to avoid any undesired beat note), and with one of them directly modulated by a tone at frequency 1 and the other modulated by a tone at frequency 2 [18]. At the receiver output, any frequency components that are the sum or difference of those two frequencies must come from the photodetector.

Problems

5.1. In a multi-AM-channel optical link, the laser transmitter has the following parameters: RIN $= -155$ dB/Hz, optical fiber output $= 10$ mW, OMI/ch $= 0.04$; the corresponding optical receiver has a noise spectral density $= 7$ pA/ $\sqrt{Hz}$; and the optical attenuation loss of single-mode fibers is 0.35 dB/km. Assuming the entire optical link uses only two connectors with 0.2 dB loss each, what is the maximum allowable distance to achieve CNR/channel $= 51$ dB? Within what distance ranges do RIN noise, shot noise, and thermal noise dominate, respectively?

5.2. How can optical modulation index (OMI) be measured? Why do we use OMI instead of electrical current modulation index based on $I_a/(I_b\text{-}I_{th})$? (I_a is the peak modulating current, I_b is the bias current, and I_{th} is the threshold current.)

5.3. Refer to Refs. [19] and [10]; derive Eqs. (5.23) and (5.27), and confirm Table 5.1.

5.4 If the S_{22} parameter of a packaged *p-i-n* diode is measured and the result is shown in Fig. P5.1, use a standard computer-aided-design software such as Libra to find the equivalent circuit of the packaged *p-i-n* diode. Your circuit model should include a bonding wire from the *p-i-n* diode to the output port.

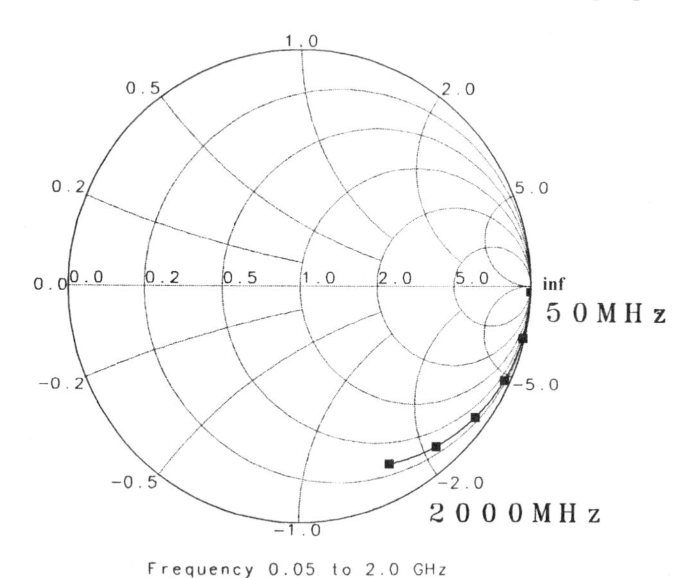

Figure P5.1: Measured S_{22} parameter of a packaged p-i-n diode. The frequency range of the measurement is between 50 and 2000 MHz.

References

[1] M. Dentan and B. D. Cremoux, "Numerical simulation of the nonlinear response of a p-i-n photodiode under high illumination," IEEE J. Lightwave Technol., vol.8, pp.1137–1144, August 1990.

[2] K. J. Williams, R. D. Esman, and M. Daenais, "Nonlinearities in *p-i-n* microwave photodetectors," IEEE J. Lightwave Technol., vol.14, pp.84–96, January 1996.

[3] J.C. Campbell, Photodetectors for long-wavelength lightwave systems, in *Optoelectronic Technology and Lightwave Communication Systems*, Ed., C. Lin, Chapter 14, Van Norstrand-Reinhold, Princeton, New Jersey, 1989.

[4] A. Yariv, *Introduction to Optical Electronics*, 4th Edition, McGraw-Hill, New York, 1994.

[5] S. D. Personick, *Fiber Optics Technology and Applications*, Plenum Press, New York and London, 1988.

[6] P. Z. Peeble, Jr., *Probability, Random Variables, and Random Signal Principles*, 3rd Edition, McGraw-Hill, New York, 1993.

[7] A. B. Carlson, *Communication Systems: An Introduction to Signals and Noise in Electrical Engineering*, McGraw-Hill, New York, 1986.

[8] D. C. Brayton and N. M. Lau, "High frequency linear amplifier assembly," U.S. Patent 5,142,239, August 1992.

[9] T. V. Muoi, "Optical receivers," in *Optoelectronic Technology and Lightwave Communications Systems*, Ed., Chinlon Lin, Chapter 16, Van Nostrand-Reinhold, Princeton, New Jersey, 1989.

[10] B. L. Kasper and J. C. Campbell, "Multigigabit-per-second avalanche photodiode lightwave receivers," IEEE J. Lightwave Technol., vol.LT-5, pp.1351–1364, 1987.

[11] E. J. Johnson, D. Brayton, and P. Holmes, "Physical considerations for the design of a linear opto-to-RF converter for analog fiber optics," Proc. 43rd Electron. Components Technol. Conf. pp.957–967, 1993.

[12] J. Skrobko, "Push–pull optical receiver with cascode amplifiers," U.S. Patent 5,347,389, September 1994.

[13] S. Kawanishi, A. Takada, and M. Saruwatari, "Wideband frequency-response measurement of optical receivers using optical heterodyne detection," IEEE J. Lightwave Technol., vol.7, pp.92–98, 1989.

[14] E. Eichen, J. Schlafer, W. Rideout, and J. McCabe, "Wide-bandwidth receiver/photodetector frequency response measurements using amplified spontaneous emission form a semiconductor optical amplifier," IEEE J. Lightwave Technol., vol.8, pp.912–916, 1990.

[15] D. M. Baney, W. V. Sorin, and S. A. Newton, "High-frequency photodiode characterization using a filtered intensity noise technique," IEEE Photon. Technol. Lett., vol.6, pp.1258–1260, October 1994.

[16] C. A. Burrus, J. E. Bowers, and R. S. Tucker, "Improved very-high-speed packaged InGaAs *p-i-n* punch-through photodiode," Electron. Lett., vol.21, pp.262–263, 1985.

[17] J. Wang, U. Kruger, B. Schwarz, and K. Petermann, "Measurement of frequency response of photoreceivers using self-homodyne method," Electron. Lett., vol.25, pp.722–723, May 1989.

[18] B. Kanack, "Measurement of intermodulation distortion in optical diodes," Tech. Digest, Int. Microwave Symp. vol.1, pp.61–64, 1995.

[19] R. G. Smith and S. D. Personick, "Receiver design for optical fiber communication systems," in *Semiconductor Devices for Optical Communications*, Chapter 4, Springer-Verlag, New York, 1980.

Chapter 6

Optical Amplifiers

Since the early 1990s, optical amplifiers, especially erbium-doped-fiber amplifiers (EDFAs), have revolutionized telecommunication system design and implementations. Long-distance terrestrial and undersea cable systems have already started to massively deploy EDFAs. HFC systems, however, have only begun to pick up momentum in using EDFAs as booster amplifiers (or power amplifiers). The driving force comes from the high optical power budget required for (1) replacing multiple direct-modulation DFB laser transmitters, (2) adding redundant-ring-based supertrunks to interconnect headends and hubs over distances greater than those used in the distribution paths (see Fig. 2.20), and (3) power splitting in redundant ring networks. For economic reasons, booster amplifiers, located right after an optical transmitter (Fig. 6.1) will be used in the early stages because their cost can be shared by all customers in the same CATV serving area, and because minimum environmental and supervision factors must be considered. In-line amplifiers (Fig. 6.1) may gradually be used if environmental factors such as temperature extremes are taken into design considerations and if cost-effective supervisory configurations can be incorporated to monitor when and where each EDFA fails. It should be mentioned, however, that in-line amplifiers can possibly be implemented by remote pumping so that the active pump lasers can be kept indoor. Preamplifiers (Fig. 6.1) will probably never be used in HFC systems because their dedication to only a single end user is too costly.

EDFAs can only amplify light signals in the 1.55-μm region. A problem associated with this wavelength constraint is that tens of thousands of 1.3-μm transmitters have been installed and may be continuously used for a long time, and EDFAs cannot be used for these *existing* transmitters. This implies that future deeper fiber penetration may still require some kind of 1.3-μm optical amplifier. The other problem associated with 1.55 μm EDFAs is that the existing 1.3-μm zero-dispersion single-mode fibers present a high fiber dispersion of 17–20 ps/nm/km in the 1.55-μm signal region, which can cause severe nonlinear distortions and intensity noise in direct-

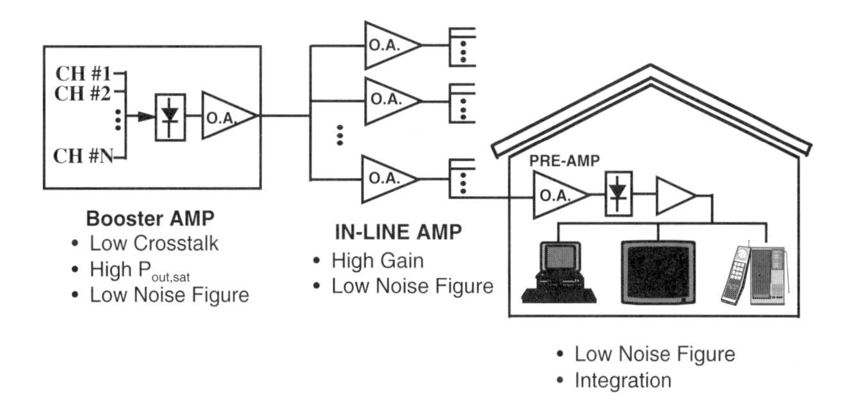

Figure 6.1: Locations of optical amplifier in the subscriber loop. The network architecture can be star, double-star, or ring.

modulation subcarrier-multiplexed (SCM) lightwave systems even for a short distance less than several kilometers (see Section 9.4). Therefore, most of today's commercially available 1.55 μm transmitters are external-modulator-based. However, external-modulation systems generally are more expensive, are less reliable, and require high driving voltages (e.g., $LiNbO_3$ modulators; see Chapter 7). Observing these constraints, we can see that there is still a great need for the development of cost-effective and reliable 1.3-μm optical amplifiers, despite the slow progress in this research and development area.

Potential 1.3-μm optical amplifiers for HFC systems include praseodymium-doped fluoride fiber amplifiers (PDFFAs), Raman fiber amplifiers (RFAs), and semiconductor optical amplifiers (SOAs). The major problem with PDFFAs and RFAs is that their poor pumping efficiency still needs to be resolved. SOAs were developed successfully in the late 1980s, but were quickly overwhelmed by EDFAs because of several system disadvantages, such as higher coupling loss from input/output fiber to SOA waveguide, high noise figure (typically >6 dB), polarization dependence, and most importantly, their inherent short carrier lifetime (<1 ns), which results in pulse distortions in digital systems, high nonlinear distortions in analog systems, and four-wave mixing in wavelength-division-multiplexing (WDM) systems. Therefore, SOAs have thus far received little attention for HFC system applications. Despite these problems, research and development efforts are being continued toward practical applications of SOAs, as will be briefly discussed in Section 6.2.

Note that all of the practically used optical amplifiers belong to the "traveling-wave" type which is just an optical gain medium without feedback, and optical amplification takes place when either electrical or optical pumping is applied to achieve the necessary population inversion. SOAs use electrical bias current for pumping, while EDFAs, PDFFAs, and RFAs use optical pumping.

6.1 Rare-Earth-Doped Fiber Amplifiers—EDFAs, PDFFAs, and RFAs

6.1.1 Basic EDFA Configuration

A complete module of EDFA can be constructed as shown in Fig. 6.2. The key components include a piece of erbium-doped fiber (trivalent erbium ions, Er^{3+}, incorporated into the core of a silica glass fiber, at an erbium concentration typically around 100–1000 ppm); one or more pump lasers; one or more wavelength-division multiplexers (WDMs) to combine the signal and the pump laser wavelengths; and one or more optical isolators to prevent system degradation because of optical reflections or self-saturation because of amplified spontaneous noise (see Section 6.1.3). But if bidirectional amplification is needed, the optical isolators must be eliminated unless a special arrangement is implemented [1,2]. Depending on the relative propagation direction between the input signal and the pump lasers, we have three different arrangements of the pump lasers as shown in Fig. 6.2: (a) forward or co-propagating pumping, (b) backward or counter-propagating pumping, and (c) bidirectional pumping. For insufficient pumping power, forward pumping can achieve a better noise-figure performance than backward pumping (Section 6.1.6.2). However, for the case when the remnant pump power is significant, forward pumping configuration will need either an optical filter or a fiber grating [3] to reject the remnant pump and prevent the optical receiver from being saturated or overwhelmed by shot noise, whereas backward pumping may not need an optical bandpass filter (when the ASE noise is heavily suppressed due to large optical input signals).

Another point to consider in comparing forward and backward pumping is that the noise figure of an amplifier should include the loss of the WDM at the amplifier input in a forward pumping configuration (see Section 6.1.6); if pump power is suf-

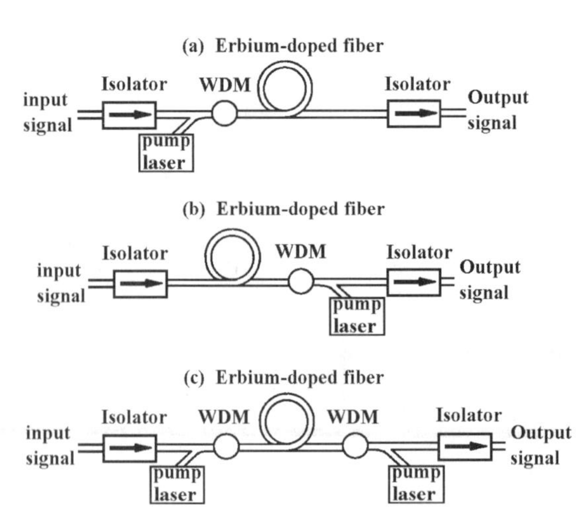

Figure 6.2: Arrangement of passive components and pump lasers in an EDFA module.

ficiently strong to invert the population in the entire amplifier length, the backward pumping option is likely to give the best overall system performance.

To achieve a high output power or high gain, it is often necessary to use bidirectional pumping in a longer erbium-doped fiber (EDF). However, when both pumps have similar wavelengths, it was noted that the remnant pump power from one laser can change the spectral shape and power level of the other laser; therefore, an optical isolator [4] or fiber grating [5] associated with each laser may be necessary. Further consideration will be given to power conversion efficiency and noise figure of each configuration in Sections 6.1.5 and 6.1.6.

6.1.2 Energy-Level Diagrams

6.1.2.1 Three-Level and Two-Level EDFAs Figure 6.3 shows the energy-level diagram (which includes ground-state absorption and excited-state absorption) of erbium ions in silica fibers. We can see that many different pump laser wavelengths can be used to achieve population inversion. These wavelengths include 514, 532, 670, 800, 980, and 1480 nm. Figure 6.4 shows the ground-state absorption spectrum of an EDF with the peaks at the corresponding wavelengths.

The energy diagram in Fig. 6.3[7] indicates that erbium ions are first excited to a high energy state (level 3) by absorbing the pump photons. The erbium ions then stay at that level for an extremely short time of about 10^{-12} and quickly relax to a lower-energy excited state (level 2). There the erbium ions stay for a relatively long time of about 10 ms for silica- and fluoride-based glasses (for tellurite-based EDFA, it is about 4 ms), and then go back to the ground state (level 1) through stimulated emission. The incident light is amplified through stimulated emission, and the amplified signal wavelength is around 1.530–1.560 nm, which can be observed from the stimulated emission cross-section curves shown in Fig. 6.5. By adding aluminum

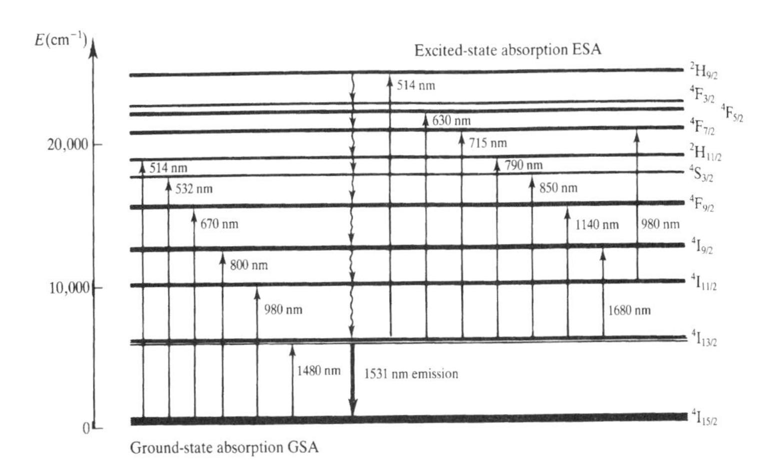

Figure 6.3: Erbium energy-level diagram of erbium ions in silica fibers (including ground-state absorption and excited-state absorption). After Ref. 7.

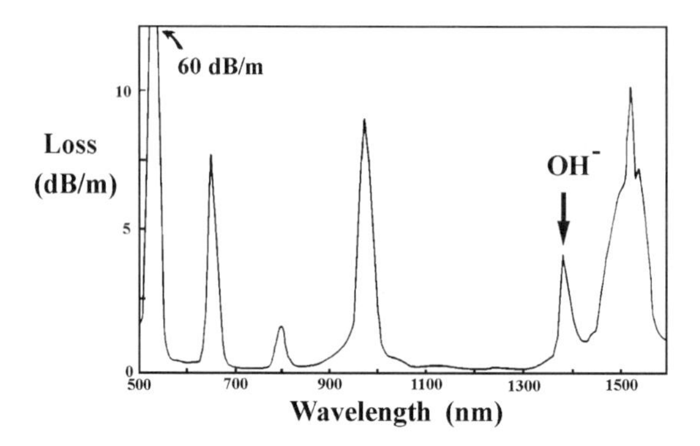

Figure 6.4: Absorption spectrum of silica fiber doped with Er^{3+} ions. After Ref. 6.

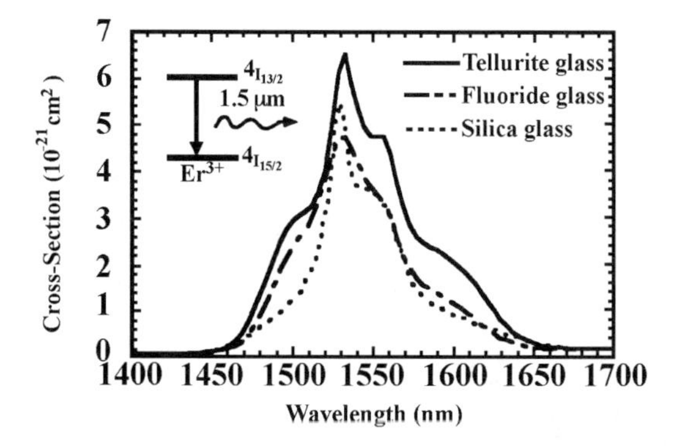

Figure 6.5: The stimulated emission cross-section for Er^{3+} in tellurite, fluoride and silica-based glasses. After Ref. [8].

together with erbium ions in silica-based glass (sometimes termed "aluminosilicate Er^{3+} : glass"), one can significantly increase the gain bandwidth (3-dB gain bandwidth of about 30 nm). Note also that alumina codoping in silica glass makes it possible to incorporate nearly 10 times more rare earth ions than germania codoping [7]. Furthermore, it was recently found that tellurite glass has 1.2–1.6 times larger emission cross-sections than those in fluoride and silica-based glasses in the 1530–1580 nm range, and has 2 times larger emission cross-sections (around 1600 nm). Therefore, a tellurite-based EDFAs can achieve a gain exceeding 20 dB over 80-nm gain bandwidth from 1530–1610 nm (see Section 6.1.9.1) [8].

The reason why the gain spectrum of aluminosilicate Er^{3+}: glass can cover a relatively wide bandwidth of about 30 nm is because each energy level shown in Fig. 6.3 actually consists of multiple Stark splitting[1] levels, as magnified in Fig. 6.6.

---

[1]Stark splitting of laser transitions is induced by Stark effect [7], caused by the host's crystalline electric field.

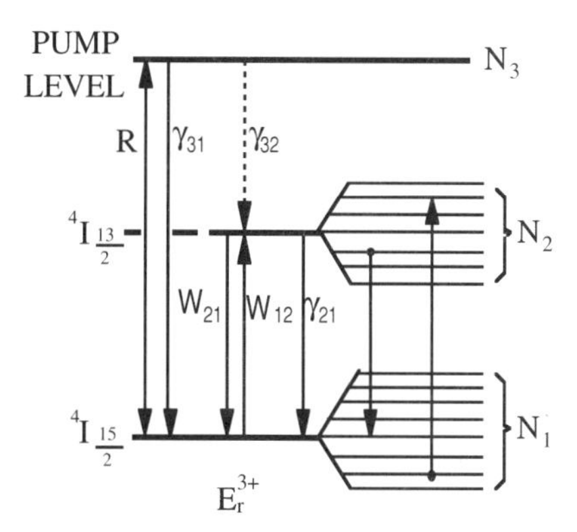

Figure 6.6: Energy diagrams of a three-level system. The dashed line represents nonradiative transition; the solid lines represent radiative transitions (the symbols R, W, and γ are defined in Section 6.1.3). After Ref. [9] (© 1989 IEEE).

When the lower and upper energy levels are split into manifolds of m_1 and m_2 sublevels, respectively, the transition spectrum corresponding to the two main levels, made of the superposition of $m_1 \times m_2$ possible transitions, is effectively broadened. The three-level energy system shown in Fig. 6.6 can be used to model the dynamics in EDFAs with 980 pump lasers. In a three-level system, the ground state is also the lower lasing level, that is, the pump and the signal transition share the $^4I_{15/2}$ ground level. Therefore, with the three-level erbium system, there is an optimum length for a given amount of pump power, beyond which the fiber begins to reabsorb signals and suppress the gain from the initial section of fiber. In a three-level system, some of the absorbed power is used to bleach the ground state (to make the amplifier transparent first) and cannot contribute to the actual gain.

As shown in Fig. 6.3, when the pump laser wavelength is around 1480 nm, an EDFA can be modeled as a two-level system. In this case, the pump and the signal transitions share both the $^4I_{15/2}$ ground level and the $^4I_{13/2}$ upper level. The Er^{3+} ion is excited directly into the second laser level, and the stimulated emission of pump (from $^4I_{13/2}$ to $^4I_{15/2}$) can increase the ground-state population; therefore, the inversion is limited. In this case, we say that the pump emission cross-section $\sigma_{pe} \neq 0$.

When the pump wavelength is around 800 nm, an EDFA is modeled as a three-level system with excited-state absorption (ESA).[2] Because of the ESA, the pump

[2] The existence of pump ESA means that the pump energy is absorbed from an excited energy level 2 (instead of the ground level) because of the existence of another level whose energy gap with level 2 happens to closely match the pump photon energy. Pump ESA exists near 790 and 850 nm, and is therefore important to the 800-nm pump band. However, 980-nm and 1480-nm pump ESAs could also exist at very high pump power levels [10,11].

efficiency and noise figure at 800 nm are generally poor. The ESA can be physically observed in a 800-nm-pumped erbium-doped fiber as a glowing green color when reabsorption from $^4I_{13/2}$ to $^2H_{11/2}$ occurs (refer to Fig. 6.3).

6.1.2.2 Er/Yb EDFA through Cooperative Energy Transfer The energy diagram in Fig. 6.6 concerns isolated rare earth ions, which is true for the case of small rare earth concentration with large enough interionic distance to prevent any interaction. However, when the concentration is increased above a certain level, the interionic distance is reduced such that ion–ion coupling is possible, resulting in the so-called "cooperative energy transfer" (CET) [7]. The commercially available Er/Yb-codoped fiber amplifier takes advantage of the CET principle whose energy diagram is shown in Fig. 6.7a. Er/Yb codoped fibers broadens the choice of pump source wavelength by using a "sensitizer," Yb (Ytterbium), which has a broad absorption band between 800 and 1060 nm (see the absorption spectrum given in Fig. 6.7b). The "sensitizer" or "donor" transfers its energy to the "acceptor" (Er), which eventually emits fluorescence. The advantage of using Yb as a sensitizer is using the commercially available high-power GaAs laser-diode array around 808 nm as the pump laser, as will be described in Section 6.1.8.2.

6.1.2.3 Four-Level PDFFA PDFFAs and NDFAs (neodymium-doped fiber amplifiers) are four-level systems. In such systems, if the ground state is considered level 1, the emitted signal transition during stimulated emission is between level 3 and

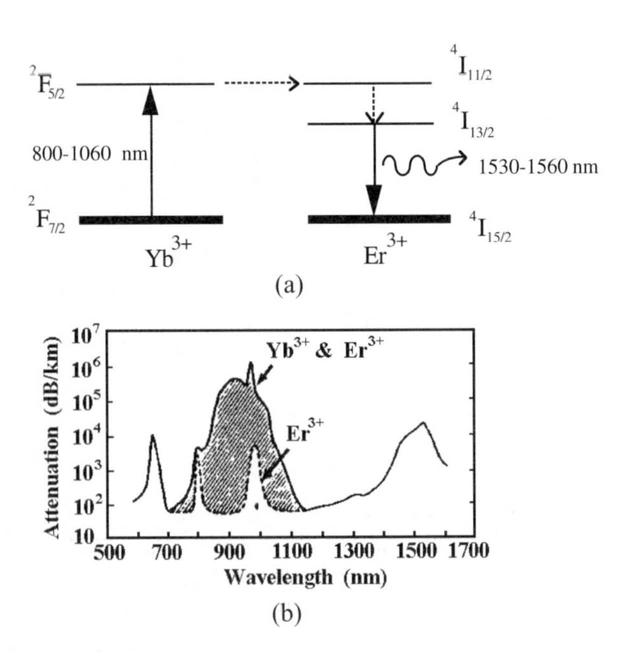

Figure 6.7: Principle of Er/Yb codoping. (a) Energy level diagram, (b) absorption spectrum. After Ref. [12].

level 2, and does not go back to level 1. Therefore, there is essentially no background loss, and positive gain occurs as soon as a single atom is in the excited state. Consequently, there is no bleaching effect (i.e., zero threshold pump power), and most of the pump is absorbed along the amplifier length, that is, $P_p^{abs} \approx P_p^{in}$.

The energy diagram and absorption spectrum Pr^{3+} in a standard ZrF_4-based fluoride glass, ZBLAN ($ZrF_4 - BaF_2 - LaF_3 - AlF_3 - NaF$), are shown in Figs. 6.8a and b, respectively. It can be seen that amplification at wavelength around 1.31 μm is achieved by establishing a population inversion between the $^1G_4$ and $^3H_5$ levels of triply ionized praseodymium. This inversion can be achieved by pumping at wavelengths anywhere between 950 and 1050 nm, with the peak absorption around 1017 nm [13]. Because of the existence of excited-state absorption (between $^1G_4$ and $^1D_2$ levels) and ground-state absorption (between $^3F_4$ and $^3H_4$ levels), the pumping efficiency can only reach about 0.65 dB/mW [14].

6.1.3 The Rate Equations

As mentioned in the previous section, an EDFA without excited-state absorption can be considered as a three-level energy system, as shown in Fig. 6.6, with level 1 as the ground state, level 2 as the metastable level, and level 3 as the pump level. The

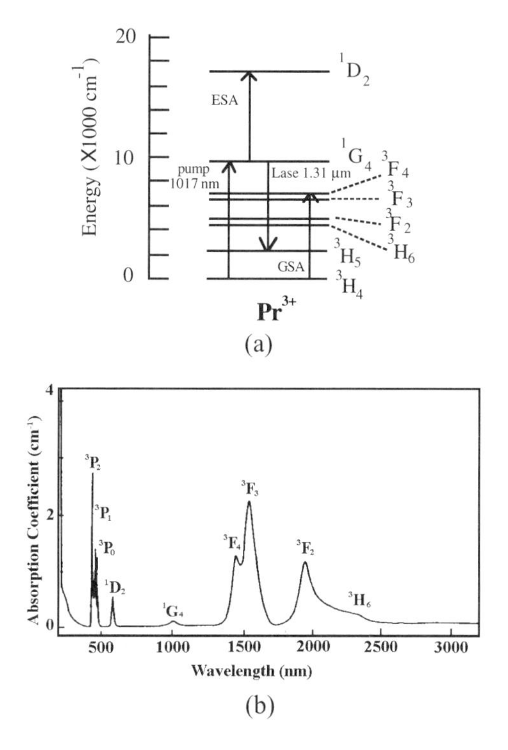

Figure 6.8: Energy-level diagram (a) and absorption spectrum (b) of Pr^{3+} in ZBLAN. After Ref. [13].

rate equations of the populations in each level can be obtained from the matrix notation [15]

$$\begin{bmatrix} \frac{dN_1}{dt} \\ \frac{dN_2}{dt} \\ \frac{dN_3}{dt} \end{bmatrix} = \begin{bmatrix} -(W_{12}+R_{13}) & W_{21}+\gamma_{21} & \gamma_{31}+R_{31} \\ W_{12} & -(W_{21}+\gamma_{21}) & \gamma_{32} \\ R_{13} & 0 & -(\gamma_{32}+\gamma_{31}+R_{31}) \end{bmatrix} \begin{bmatrix} N_1 \\ N_2 \\ N_3 \end{bmatrix}, \quad (6.1)$$

where

$N_i (i = 1,2,3)$ = the population in each energy level

$W_{ij}(i, j = 1,2)$ = stimulated emission (absorption) rate from level 2 (1) to level 1 (2)

$$= \frac{P_s \cdot \sigma_{se}}{h\nu_s \cdot A_s} \psi_s(r) \text{ or } (\frac{P_s \cdot \sigma_{sa}}{h\nu_s \cdot A_s}) \psi_s(r) \quad (6.2)$$

$$R_{13} = \text{pumping rate} = \frac{P_p \cdot \sigma_{pa}}{h\nu_p \cdot A_p} \psi_p(r) \quad (6.3)$$

$$R_{31} = \text{pump emission rate} = \frac{P_p \cdot \sigma_{pe}}{h\nu_p \cdot A_p} \psi_p(r) \quad (6.4)$$

$\gamma_{ij}(i, j = 1,2,3, i > j)$ = spontaneous emission rates from level i to level j.

Note that A_s and A_p are the cross-sectional areas of the signal and pump modes inside the fiber, respectively; σ_{se}, and σ_{sa}, and σ_{pa} are the signal emission, signal absorption, and pump absorption cross-sections (in cm^2), respectively (a diagram illustrating signal emission and absorption cross-sections is shown in Fig. 6.9 [7]); the pump emission cross section σ_{pe} is zero as from level 3 the decay is essentially nonradiative; P_p and P_s are the pump and signal power confined in A_p and A_s, respectively; and $\psi_{p,s}(r) \approx \exp(-r^2/\omega_{p,s}^2)$ is the pump or signal Gaussian envelope determined by the fiber core radius r and the power mode radius $\omega_{p,s}$ at λ_p or λ_s. The factor $\psi_p(r)$ suggests that in order to achieve maximum signal gain, the Er-doping distribution should be confined to the center region of the fiber core, or that the pump intensity should be increased near the center of the core.

To simplify Eq. (6.1), we further make the following two assumptions:

1. $N_3 \approx 0$, which means that ions do not reside in level 3 for a long time, but rapidly relax to level 2.
2. $\gamma_{21} << \gamma_{32}$, which implies that the ions remain in level 2 for a considerable length of time. Indeed, the fluorescence lifetime $\gamma_{21} = 1/\tau \approx 10$ ms for silica- and fluoride-based glasses [16, 65].

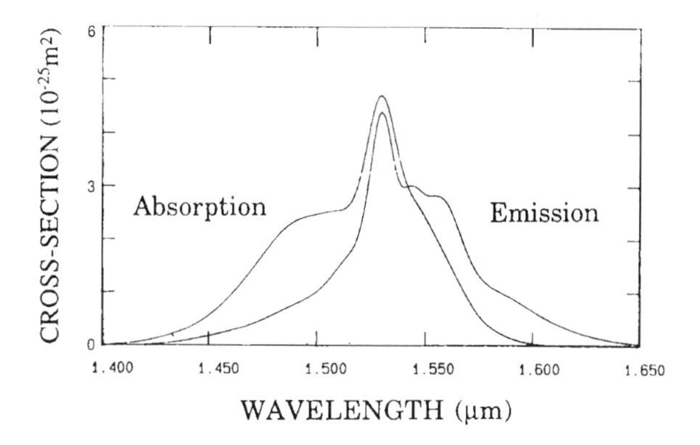

Figure 6.9: Emission and absorption cross-sections (σ_e and σ_a) for aluminogermanosilicate fibers. After Ref. [7].

With these assumptions, Eq. (6.1) becomes

$$\frac{dN_1}{dt} = -(W_{12} + R_{13})N_1 + (W_{21} + \gamma_{21})N_2 \tag{6.5a}$$

$$\frac{dN_2}{dt} = W_{12}N_1 - (W_{21} + \gamma_{21})N_2 . \tag{6.5b}$$

Considering steady-state conditions, and let the sum of the populations in levels 1 and 2 be equal to the total ion density N_t ($\approx N_1 + N_2$, with $N_3 = 0$), we obtain

$$N_1 = N_t \frac{1 + W_{21} \cdot \tau}{1 + (W_{12} + W_{21})\tau + R_{13}\tau} \tag{6.6}$$

$$N_2 = N_t \frac{R_{13}\tau + W_{12} \cdot \tau}{1 + (W_{12} + W_{21})\tau + R_{13}\tau} . \tag{6.7}$$

So far, we have not discussed the fiber-length-dependent distribution of the pump power, the signal power, and the amplified spontaneous emission power. These dependencies can be neglected in semiconductor lasers or amplifiers because their cavity lengths are only a few hundred micrometers. The length of a discrete EDF, however, can range from a few meters to a few tens of meters. Therefore, the spatial evolution of the pump, signal, and ASE power in the fiber must be considered [17],

$$\frac{dP_p^{\pm}}{dz} = \mp P_p^{\pm}\Gamma_p(\sigma_{pa}N_1 - \sigma_{pe}N_2) \mp \alpha_p P_p^{\pm} \tag{6.8}$$

$$\frac{dP_s}{dz} = P_s\Gamma_s(\sigma_{se}N_2 - \sigma_{sa}N_1) - \alpha_s P_s \tag{6.9}$$

$$\frac{dP_{ase}^{\pm}}{dz} = \pm P_{ase}^{\pm}\Gamma_s(\sigma_{se}N_2 - \sigma_{sa}N_1) \pm 2\sigma_{se}N_2\Gamma_s \cdot h\nu_s \cdot \Delta\nu \mp \alpha_s P_{ase}^{\pm}, \quad (6.10)$$

where we have neglected the transverse power variations, as erbium doping is confined near the center of the fiber core in most practical cases. The superscript + designates pump and ASE co-propagating with the signal, and – when they counter-propagate to the signal; Γ_s and Γ_p are the signal-to-core and pump-to-core overlap factors, respectively.

The second term in Eq. (6.10) is the ASE power produced in the amplifier per unit length within an optical filter bandwidth $\Delta\nu$ for both polarization states (so that a factor of 2 is included). The loss terms $\alpha_{s,p}$ represent internal loss of the fiber for signal and pump wavelengths, respectively, which can be neglected for nondistributed amplifiers with typical length < 10–50 m. Note, once again, that for three-level systems (e.g., when using 980-nm pumps), $\sigma_{pe} = 0$, as there is negligible stimulated emission at the pump wavelength. However, for two-level systems (e.g., when using 1480-nm pumps), $\sigma_{pe} \neq 0$. In Eq. (6.10), if we neglect $\alpha_{s,p}$ and let the gain coefficient $g = \Gamma_s(\sigma_{se}N_2 - \sigma_{se}N_1)$ be independent of the z, that is, assuming the amplifier is uniformly pumped by a sufficiently large pump, then we can rewrite Eq. (6.10) as

$$\frac{1}{g}\frac{dP_{ase}^{\pm}}{dz} = \pm P_{ase}^{\pm} \pm 2n_{sp} \cdot h\nu_s \cdot \Delta\nu, \quad (6.11)$$

where we have defined the population inversion factor n_{sp} as

$$n_{sp} = \frac{\eta_s N_2}{\eta_s N_2 - N_1}, \quad (6.12)$$

where $\eta_s = \sigma_{se}/\sigma_{sa}$. The solution of Eq. (6.11), for a length of L, is

$$P_{ase}(L) = e^{gL}P_{ase}(0) + g \cdot \int_{z=0}^{L} e^{g(L-z)}\left[2n_{sp}h\nu \cdot \Delta\nu\right]dz, \quad (6.13)$$

and when the input noise $P_{ase}(0) = 0$, which is true for a single-stage amplifier, we have the power of the amplified spontaneous emission noise from an amplifier with a length of L as

$$P_{ase}(L) = 2n_{sp}(G-1)h\nu \cdot \Delta\nu, \quad (6.14a)$$

where $G = e^{gL}$ is the amplifier gain. This is a formula that will be frequently used in calculating ASE power. But we should remember that the assumption we made to obtain Eq. (6.14a) is that the entire length of the amplifier is uniformly pumped. Note also that Eq. (6.14a) can be written in terms of the number of amplified spontaneous emission photons S_{ase} as

$$S_{ase} = 2n_{sp}(G-1). \quad (6.14b)$$

If we substitute Eqs. (6.6) and (6.7) into Eq. (6.9) (neglect the loss due to attenuation), we obtain

$$\frac{dP_s}{dz} = P_s\Gamma_s N_t \frac{\left[\sigma_{se}(R_{13}\ \tau + W_{12}\tau) - \sigma_{sa}(1 + W_{21}\tau)\right]}{1 + (W_{12} + W_{21})\tau + R_{13}\tau}. \tag{6.15}$$

Following the same approach, a similar expression for P_p can be obtained. It turns out that when the pump power approaches infinity (so that $R_{13} \to \infty$), we have

$$\frac{dP_s}{dz} \approx P_s\Gamma_s N_t \sigma_{se}, \qquad P_p \to \infty, \tag{6.16}$$

which implies that at complete population inversion, the signal gain coefficient is proportional to $\Gamma_s N_t \sigma_{se}$, and the signal emission cross-section plays a key role. The maximum unsaturated gain for a three-level system EDFA with length L is then given by

$$G(P_p \to \infty) = \exp\{\Gamma_s N_t \sigma_e L\}. \tag{6.17}$$

To obtain a complete picture of $G = P_s(L)/P_s(0)$ as a function of pump power and EDF length, we have to resort to Eqs. (6.2)–(6.4), (6.15), and the known emission and absorption cross-sections (recall that when the transverse power variations are neglected, i.e., $\psi_{p,s}(r) \approx 1$, we see that $W_{12} \approx W_{21}$ is a function of P_s, and R_{13} is a function of P_p). We can numerically solve Eq. (6.15) for G as a function of pump power (L fixed), or solve for G or output power against L (P_p fixed). Typical results are shown in Figs. 6.10 and 6.11, respectively [15]. Figure 6.10 shows that for a

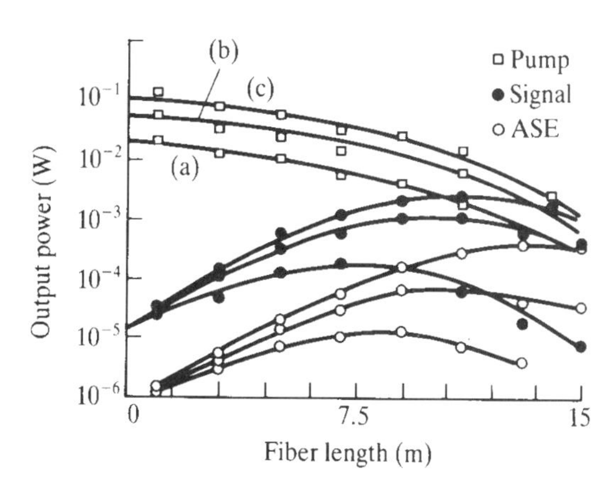

Figure 6.10: Experimental characterization of EDFA: (❑) pump, (●) signal, and (❍) ASE powers versus length, for three different input pump powers: $P_p(0)$ = (a) 20 mW, (b) 55 mW, and (c) 100 mW. The input signal power is $P_s(0) = 15\ \mu$w. After Ref. [15].

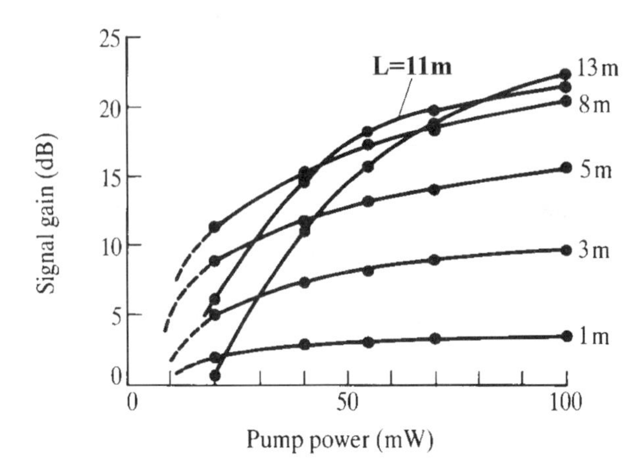

Figure 6.11: Signal gain at λ = 1.531 μm versus input pump power at λ = 514 nm for different fiber lengths. After Ref. [15].

given pump power, there is an optimum fiber length to achieve the maximum output signal. The higher the maximum output signal and the pump power, the longer the optimum fiber length is. As the length exceeds the optimum length, the pump power is not strong enough to excite the ground populations in the latter portion of the amplifier, and therefore the amplified signal is reabsorbed in this portion. Figure 6.11 shows that for a given amplifier length L, the amplifier gain initially increases exponentially with the pump power, but the increase becomes much smaller when the pump power exceeds a "knee." The commonly mentioned "pump efficiency" is typically taken as the slope at this knee (dB/mW). Note that there are significant threshold pump powers for longer lengths of EDFs (see also the discussion of Fig. 6.16).

The spatial evolutions of P_p and P_{ase} can also be obtained numerically by substituting Eqs. (6.6) and (6.7) into (6.8) and (with $\psi_{p,s}(r) \approx 1$) (6.10), respectively.

The derivations in this section, so far, have been on three-level systems, such as 980-nm pumped systems. Suppose we consider a two-level system, such as a 1480-nm pumped system. The preceding derivations can be modified as follows: (i) $R_{13} = 0$ because the pump goes directly to level 2 in systems when using 1480 nm pumps; (ii) the stimulated emission (absorption) rate from level 2 (1) to level 1 (2) $W_{ij}(i,j = 1,2)$ should be modified as [7]

$$W_{ij}(i,j=1,2) = \frac{P_S \cdot \sigma_{se}}{h\nu_S \cdot A_S} + \frac{P_p \cdot \sigma_{pe}}{h\nu_p \cdot A_p} \text{ or } \left(\frac{P_S \cdot \sigma_{sa}}{h\nu_S \cdot A_S} + \frac{P_p \cdot \sigma_{pa}}{h\nu_p \cdot A_p}\right) \tag{6.18}$$

to include the fact that the pump and the signal transitions share both the $^4I_{15/2}$ ground level and the $^4I_{13/2}$ upper level. An expression similar to Eq. (6.15) can be obtained, and when the pump power approaches infinity, we have [7]

$$\frac{dP_s}{dz} \approx P_s \Gamma_s N_t \sigma_{se} \frac{1-(\eta_p/\eta_s)}{1+\eta_p}, \tag{6.19}$$

where $\eta_s = \sigma_{se}/\sigma_{sa}$ and $\eta_p = \sigma_{pe}/\sigma_{pa}$. Therefore, the maximum unsaturated gain for a two-level system EDFA with length L is given by

$$G(P_p \to \infty) = \exp\left\{\Gamma_s N_t \sigma_{se} L \frac{1-(\eta_p/\eta_s)}{1+\eta_p}\right\}. \tag{6.20}$$

From Eqs. (6.17) and (6.20), it appears that the maximum amplifier gain can be increased without limit as long as the EDF fiber length can be increased and the required pump power can be provided. In fact, however, there are three additional limiting factors which prevent the amplifier gain from being increased indefinitely. The first is *amplifier self-saturation*, the second is *laser oscillation*, and the third is Rayleigh double scattering. *Amplifier self-saturation* occurs when the gain is so high that stimulated emission by ASE is enhanced to the point where it eventually competes with the pumping rate. This competition occurs in regions located near the fiber ends where the ASE power is the highest. In these regions, the population inversion is reduced and the amplifier gain saturates in the absence of any input signal. Laser oscillation occurs when the amplifier gain is high enough to compensate for the reflectivities of any reflecting points in the path of the ASE signal. The ASE is reflected back and forth between the reflecting points, which provide optical feedback, and therefore the device can operate as a laser. The last factor, Rayleigh double scattering, will be discussed in Sections 6.1.9.2 and 9.3.1.

6.1.4 Gain Spectra and Homogeneous Broadening

The gain spectrum of an EDFA is proportional to $\sigma_{se}(\lambda)N_2 - \sigma_{sa}(\lambda)N_1$. With three-level pumping, the spectrum has two envelopes represented by the cross-section line shapes $\sigma_{sa}(\lambda)$ and $\sigma_{se}(\lambda)$. In the case of two-level pumping, the two envelopes are $\sigma_{sa}(\lambda)$ and $\sigma_{se}(\lambda)[1 - \eta_p/\eta_s(\lambda)]/(1 + \eta_p)$. Figure 6.12 shows the gain and loss spectra of an aluminosilicate Er^{3+}: glass for different levels of laser pumping power [18]. Typically, when there is no pump applied, the Er/Al fiber can absorb 2.5 to 6 dB/m power at 1530 nm,[3] depending on the specific EDF parameters. It should be noted that as the relative pumping power increases, it is the longer wavelength region (e.g., >1550 nm) which becomes transparent first; the shorter wavelength region (e.g., near 1530 nm) becomes transparent later. This effect is because of the fact that the overlap between absorption and emission cross-sections is maximum around 1530 nm (see Fig. 6.9), which minimizes the net gain $N_2\sigma_{se} - N_1\sigma_{sa}$ around that wavelength. Therefore, the EDFA can be viewed as a continuous superimposition of laser systems, with pure three-level behavior at short wavelengths and pure

[3] From specifications of commercially available EDFs.

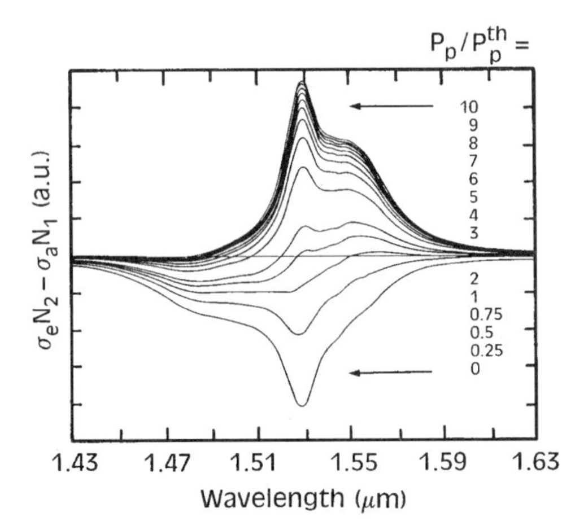

Figure 6.12: Spectral gain and loss versus pump power (P_{th} is the threshold pump power). After Ref. [18] (© 1990 IEEE).

four-level behavior at long wavelengths [7]. Consequently, if there is not sufficient pump power available, operating in the longer wavelength region is beneficial in terms of lower noise figure because of less ground-state absorption.

The gain spectrum shown earlier is under the condition of no input signal or with small input signal. From a system designer's point of view, it is very important to know the basic characteristics of the EDFA gain spectra under gain saturation, that is, when the input signals (either single or WDM channels) are large.

Before we examine the gain spectra of a saturated EDFA, we must first understand a phenomenon called "homogeneous broadening." Homogeneous broadening originates from the fact that the Stark sublevels have small energy gaps and are therefore strongly coupled by the effect of thermalization. Such a strong coupling causes the overall laser line to exhibit homogeneous saturation $\Delta\lambda_{hom}$ characteristics. An example of a measured homogeneous transition linewidth is shown in Fig. 6.13 [19]. The spectra were measured at the output of a GeO_2 EDFA when the input signal is a strong saturating signal near 1.535 μm. The output signal was eliminated by a polarizer whose polarization is orthogonal to that of the output signal. We can clearly see that the "hole" width $\Delta\lambda_{hole}$ (= $2\Delta\lambda_{hom}$) is increased when the temperature increases. As a matter of fact, at room temperature (300 K), $\Delta\lambda_{hole}$ is as high as high 8 nm for GeO_2 EDF, and 23 nm for Al_2O_3 EDF [20]. The essentially homogeneous saturation behavior across the whole gain bandwidth of aluminosilicate Er-doped fibers is because of the effects of fast Stark manifold thermalization and strong spectral overlap between Stark components. Therefore, although adding aluminum in erbium-doped fibers can increase the gain spectral bandwidth significantly, the negative effect is that, at room temperature, there is a homogeneous saturation characteristic across a spectrum as wide as 23 nm. This means that when a strong input signal saturates an Al_2O_3 EDFA, the gain spectrum near the saturating signal is affected, that is, the

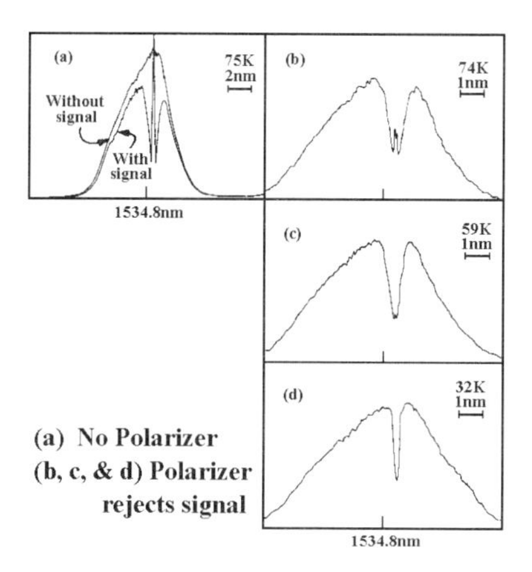

Figure 6.13: (a) Unpolarized output spectra recorded at T = 75 K with and without signal. (b), (c), (d) Polarized output spectra recorded with polarization direction normal to that of the saturating signal for temperatures 74, 59, and 32 K, respectively. After Ref. [19] (© 1990 IEEE).

originally flat gain spectrum becomes "tilted," as shown in Fig. 6.14. We can see that the upper curve is the small-signal gain spectrum (or the amplified spontaneous emission noise), and the lower curve is when a 1552-nm DFB laser with a large input power is launched into the EDFA. It is clear that because of the gain saturation at a wavelength of 1552 nm, the gain in all other wavelengths is compressed. The affected wavelength range apparently is much wider than 23 nm. The resultant "gain tilt" effect has a strong impact on AM video lightwave transmission systems, as will

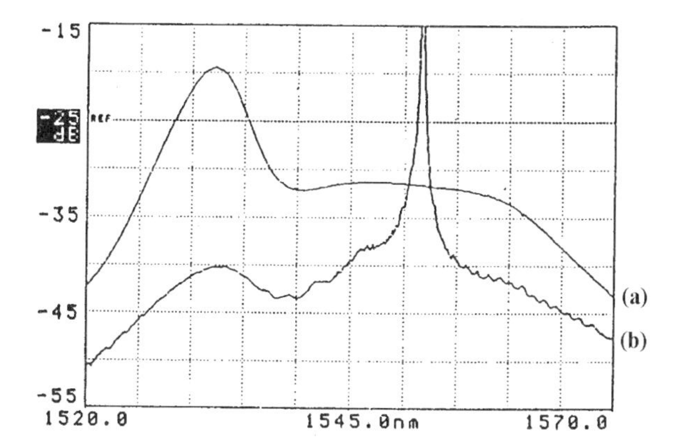

Figure 6.14: Optical spectra of the amplified spontaneous emission noise when there is no input signal (a), and when there is an input signal with a large optical power of 1 mW (b), which saturates the amplifier.

be discussed in Section 9.3.3. The homogeneous saturation characteristics could also mean when several wavelength-division-multiplexed lasers occupying a range ~ 23 nm are amplified by the same EDFA with aluminum codopant, gain saturation in one laser can cause gain saturation in another laser.

The reason why the gain compression is the largest around $\lambda = 1531$ nm is because when there is a change of manifold populations $\Delta N_1 = \Delta N_2$ induced by saturation, the local gain coefficient decreases according to $\Delta g = \sigma_{se}(\lambda)\Delta N_2 - \sigma_{sa}(\lambda)\Delta N_1 = [\sigma_{se}(\lambda) + \sigma_{sa}(\lambda)]\Delta N_2$, which is largest at the wavelength where the sum of the cross-sections is maximum (see Fig. 6.9).

6.1.5 Pump Lasers

6.1.5.1 Semiconductor Pump Lasers and Their Power/Wavelength Stabilization

In selecting pump power sources for practical system applications, we should consider the following factors: (1) compactness and reliability, (2) reasonably high dB/mW gain coefficient and freedom from pump excited-state absorption (ESA), (3) confinement of Er-doping with respect to the pump intensity distribution, (4) low noise figure, and (5) high power-conversion efficiency (PCE). Factor 2 is particularly important for in-line and preamplifiers, whereas factor (5) is particularly important for booster amplifiers.

Compact and reliable semiconductor laser diodes are considered the most practical pump sources for communication systems. In this respect, 1480-nm, 980-nm, and 800-nm pump bands are good candidates. Strained-layer 1480-nm GaInAsP-MQW lasers with CW powers near 250 mW [21,22] and 980-nm strain-layer InGaAs MQW laser diodes with up to 320-mW single-transverse mode operation [23] have been reported. Since a single spatial mode output from a traditional Fabry–Perot laser exceeding ~300 mW is very difficult, therefore, a new device structure based on a monolithic MOPA (master oscillator power amplifier) has been developed to increase the single-mode power available from a diode laser (up to ~1.2 W) [24], the structure of which is shown in Fig. 6.15. However, the reliability of MOPAs is still under investigation at this stage.

It is worth mentioning that even though the reliability of 1480-nm pump diodes was proved in the early 1990s, packaging technologies to eliminate packaging-induced failures in 980-nm pump diodes were not developed until 1994 [25].

When an EDFA is used as an in-line amplifier or preamplifier, the signal input power is so low that the amplifier operates in the unsaturated region. The design goal is then to minimize the pump power needed for a given gain. To characterize a specific EDFA's performance with respect to this design criterion, the gain coefficient is introduced, defined as the *peak* ratio between small-signal gain in decibels and launched pump power in milliwatts. Figure 6.16 summarizes the highest dB/mW gain coefficients experimentally reported [7]. We can see that, for the 980-nm pump band, the record value can reach 11 dB/mW [26]. For the 1480-nm pump band, the best result was 6.3 dB/mW [27] and is about 60% of that obtained from the 980-nm pump band. For the 800-nm pump band, we can see that the best obtain-

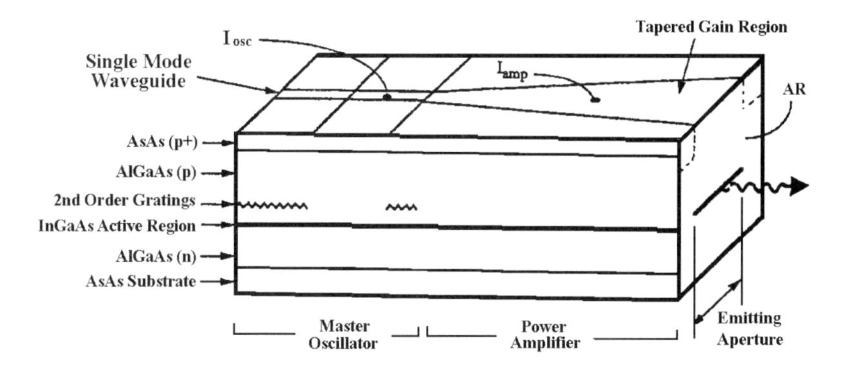

Figure 6.15: Schematic of a monolithic MOPA chip. After Ref. [24].

able gain coefficient is very low, about 1.3 dB/mW. The main reason why the gain coefficient for 980-nm pumping is higher than for 1480-nm pumping is that the maximum obtainable population inversion is only 0.7 for the latter, whereas a complete inversion is possible for the former.

The quick pump power decay at some distance from the center of the EDF core could result in reduced medium inversion and even signal loss. This problem can be solved by increasing the fiber numerical aperture (NA), by confining the doped Er ions to the central core region [28], and by increasing the pump intensity near the center of the core. The latter can be achieved by having single-mode propagation of the pump wave in the EDF, by carefully designing the cutoff wavelengths of EDFs

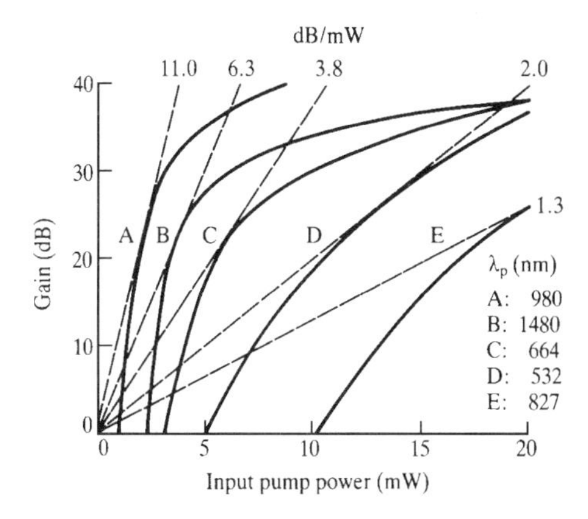

Figure 6.16: Representative record experimental EDFA gain versus input pump power characteristics reported using different pump bands near 532, 820, 665, 980, and 1480 nm. After Ref. [7].

for 980 and 1480 nm, respectively. Commercially available EDF cutoff wavelengths for 980- and 1480-nm pumps fall in the range of 850–950 nm. It should be noted that the fiber NA cannot be increased without limit, because the gain efficiency and noise characteristics of the amplifier are affected by the excess background loss at the pump and signal wavelength, as will be discussed in Section 6.1.9.2.

Because of their full-inversion characteristics under high pump power, 980-nm laser diodes also exhibit an inherent low noise figure [18,29], which approaches the quantum limit of 3 dB (see Section 6.1.6.2), even under deep saturation conditions [29]. The 1480-nm pump diodes, however, because of their incomplete inversion characteristics, can only reach a best noise figure of about 4 dB [30].

Power conversion efficiency (PCE) is of particular importance for booster amplifiers, in which case the maximum available output power must be as high as possible. The definition of PCE is

$$PCE = \frac{P_s^{out} - P_s^{in}}{P_p^{in}}, \tag{6.21}$$

where P_p^{in}, P_s^{in}, P_s^{out} are the input pump, input signal, and output signal power levels, respectively. Generally speaking, $P_p^{in} >> P_s^{in}$, and the maximum PCE becomes

$$PCE_{\max} = \frac{\lambda_p}{\lambda_s}. \tag{6.22}$$

Therefore, we see that the highest PCEs are achievable for the longest pump wavelengths, which corresponds to $\lambda_p \approx 1480$ nm. The other figure of merit which readers may frequently encounter in published literature is the quantum conversion efficiency (QCE) defined by

$$\boldsymbol{QCE} = \frac{\varphi_s^{out} - \varphi_s^{in}}{\varphi_p^{in}} = \frac{\lambda_s}{\lambda_p}\boldsymbol{PCE}, \tag{6.23}$$

where φ_p^{in}, φ_s^{in}, φ_s^{out} are the photon fluxes ($= \boldsymbol{P_p^{in}/h\nu_p, P_s^{in}/h\nu_s, P_s^{out}/h\nu_s}$) of input pump, input signal, and output signal, respectively. Therefore, the ideal QCE is 100%, which represents the case when all input pump photons are converted to output signal photons (assuming $\varphi_s^{out} >> \varphi_s^{in}$). Typical examples are PCE = 75.6% (QCE = 79.4%) for $\lambda_p \approx 1475$ nm (with P_s^{out} = 140 mW and P_p^{in} = 185 mW) [31], and PCE = 56% (QCE = 89%) for $\lambda_p \approx 980$ nm (with P_s^{out} = 78 mW and P_p^{in} = 135mW) [28].

It was observed that backward pumping yields the highest PCE [32]. The observed difference between the forward and backward pumping configurations for the PCE can be mainly attributed to the effect of amplifier self-saturation [33].

Many high-power 1480- and 980-nm pump lasers are multi-longitudinal-mode Fabry–Perot lasers, and they generally occupy a broad spectral range of 15 to 25 nm, as can be seen in Figs. 6.17a. We can guess that the optimum pump wavelength for a 980-nm laser exists around the peak of the ground-state absorption (~979 nm).

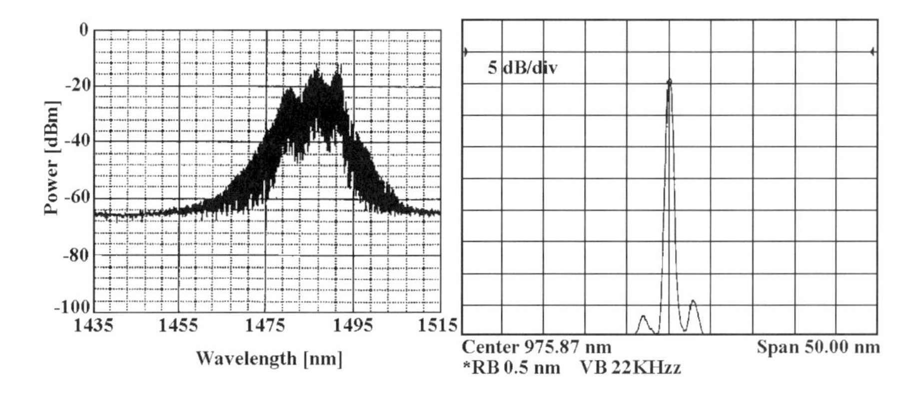

Figure 6.17: Optical spectra of (a) 1480-nm and (b) 980-nm pump lasers. The 980-nm laser in (b) is grating stabilized.

In the case of two-level pumping, however (1480-nm pump band), we note that when λ_p is detuned toward λ_s, the required pump power must be increased because the signal stimulated emission will increase the ground-state population; when λ_p is detuned toward shorter wavelengths, the pump absorption coefficient decreases, and the required pump power must also be increased. Therefore, there exists an optimum pump wavelength region (~1480 nm) to minimize the required pump power. A question then arises as to how exact the pump wavelength must be in the optimum region of these two pump bands in order to minimize the required pump power. Quite a number of investigations have been reported with regard to this question [34–40]. The general observations are as follows: (1) At low pump power levels, the EDFA gain is very sensitive to the pump wavelength variation, whereas in the heavy-gain-saturated regime (e.g., 10-dB gain compression), both the EDFA gain and noise figures are not sensitive to the pump power and wavelength variation; (2) for typical pump power levels applicable for high-gain in-line amplifiers, the tolerances on pump wavelength are significantly tighter around 980 nm (typically $\Delta\lambda_p \approx \pm 8$ to ± 9 nm) than around 1480 nm (typically $\Delta\lambda_p \approx \pm 20$ nm); and (3) increasing the NA (say, from 0.15 to 0.25) of the EDF can significantly improves the tolerance of pump wavelength variations (but increasing NA may have other negative effects; see Section 6.1.9.2).

Now we know that semiconductor pump lasers may require good spectral stability, especially the 980-nm lasers. However, high-power semiconductor pump lasers exhibit severe wavelength and power-level fluctuations, because of mode hopping (see Section 4.9.2), spurious optical reflections [5,41,42], and remnant pump power injection [4], which in turn degrade EDFA system performance. To solve this problem, optical feedback from fiber Bragg gratings has been used to stabilize both pump wavelength and output power levels in 980-nm laser diodes [5, 41,42].

Figure 6.18 shows an example of pump wavelength stabilization. Without a grating reflector, the laser spectra exhibited strong mode hopping over the

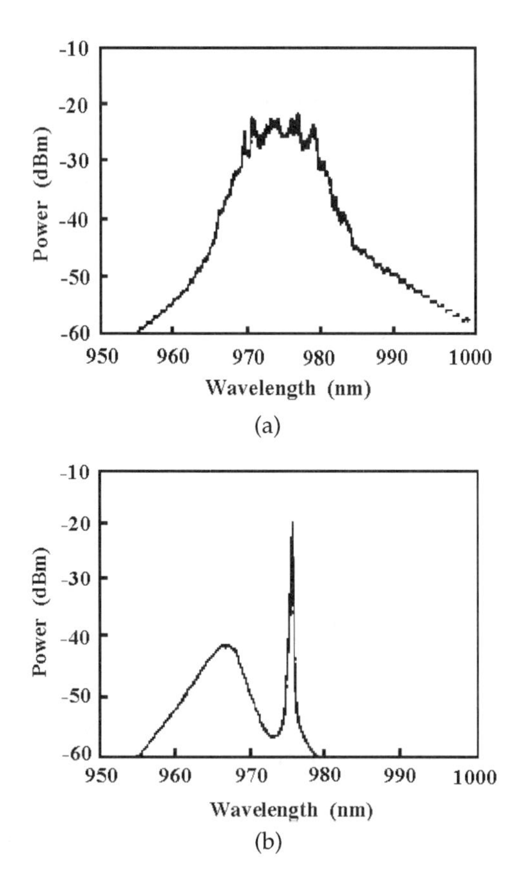

Figure 6.18: Pump lasers wavelength-stabilized to 975 nm using optical feedback from a fiber grating. Figures show laser output spectra with 0.2-nm resolution. (a) Single laser, no fiber grating, and (b) single laser, fiber grating. After Ref. [42] (© 1994 IEEE).

wavelength range 965–980-nm; when a grating with reflectivity greater than about 1% is spliced, a stabilized output wavelength at 975 nm is obtained [42].

Low-frequency intensity fluctuations in the free-running pump laser power due to mode hopping can also be greatly reduced, as shown in Fig. 6.19 [5], by using fiber gratings which force the laser to operate on multiple longitudinal modes (the laser spectrum consists of tens of external cavity modes centered near the peak of the fiber grating). The large number of modes and their lack of coherence averages out the low-frequency power fluctuations due to mode hopping.

Now we have seen in Figs. 6.18 and 6.19 that both wavelength and power of a pump laser are stabilized. This is mainly because the *incoherent* optical feedback provided by the fiber grating makes the laser enter the "coherence-collapse" regime [43], which is characterized by a substantial broadening of the laser diode modes and a large reduction in the coherence length of the laser.

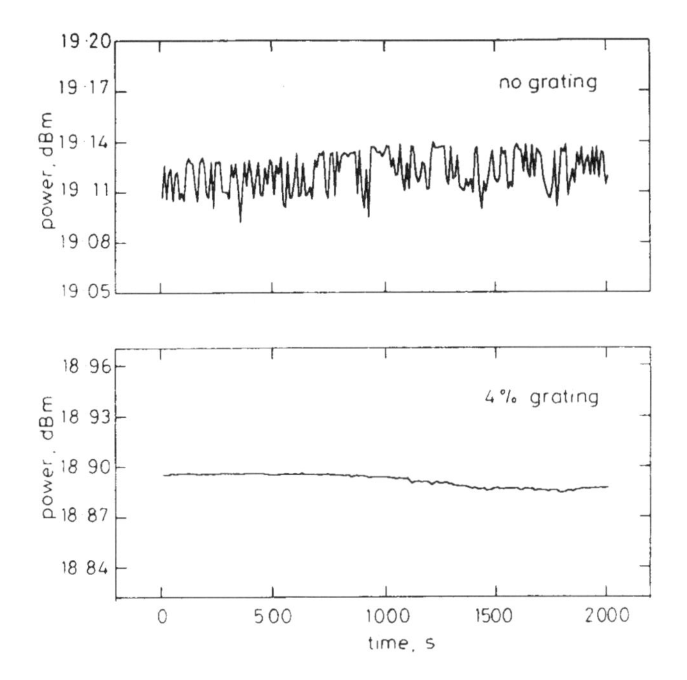

Figure 6.19: Effect of fiber grating on power fluctuations of 980-nm diode laser. After Ref. [5].

Operation of a grating-stabilized 980-nm pump laser at high powers without a thermoelectric cooler has also been demonstrated [44]. This is made possible by the low temperature dependence of the grating which keeps the laser wavelength sufficiently constant over a wide temperature range. The peak wavelength of the laser remains locked to the Bragg wavelength of the grating over a 75°C variation in outside temperature. Although the ASE peak becomes more prominent at the temperature extremes, more than 90% of the power remains within ±1 nm of the grating wavelength at all times.

Grating-stabilized 980-nm laser diodes operating in the coherence-collapse regime are now commercially available.

6.1.5.2 Diode-pumped Solid-State Lasers and High-Power Cladding-Pumped Fiber Lasers The pump powers currently available from laser diodes are limited by intrinsic material properties (i.e., facet damage) and are not likely to increase significantly beyond 200 mW with reliable operation. Power scaling by polarization-multiplexing multiple lasers (see Fig. 6.31) could encounter pump laser diode crosstalk [2] and is limited to a total of four laser diodes (two for forward pumping and the other two for backward pumping). Therefore, for extremely high pump power levels, one would have to resort to the well-developed, arbitrarily scalable high-power pump GaAlAs laser diode arrays (with output of several watts, easily).

GaAlAs phase-array laser diodes (around 808 nm) can be used to pump solid-state crystals (Nd:YAG or Nd:YLF), termed a diode-pumped solid-state (DPSS) laser, or to pump Nd-doped fibers, to generate a very high output power. Experimental results include diode-pumped Nd:YLF lasers with a dual line output at 1.047/1.053 μm (power coupled into Er/Yb-codoped fiber was 900 mW) [45]; and diode-pumped Nd-doped fiber lasers operating at 1.053 μm with an output power up to about 680 mW [46]. Note that DPSS lasers are used to pump Er/Yb-codoped fibers (Section 6.1.2.2) to achieve very high-power booster amplifiers, whose performance will be summarized in Section 6.1.8.2.

Another "brightness-conversion" approach which can utilize the scalable high output power from GaAlAs laser-diode arrays is the Nd^{3+} or Yb^{3+} cladding-pumped fiber lasers [47,48]. In this approach, a single-mode fiber laser with a double-cladding configuration was employed, as shown in Fig. 6.20, where we can see a large multimode guiding region for the diode pump light and a rare-earth-doped single-mode core. As opposed to a conventional single-mode fiber which has a small core diameter of less than 10 μm and a low numerical aperture (e.g., NA = 0.11), the multimode rectangular core (the "pump cladding") has a very wide dimension (360 $\times$ 20 μm^2) as shown in the figure, and a large numerical aperture (e.g., NA = 0.59). This large numerical aperture provided by the cladding makes it possible to couple the pump light into the inner cladding (alignment tolerances to pump are tens of micrometers), from where it is subsequently gradually absorbed in the core. When feedback elements such as fiber Bragg gratings are written into this innermost core codoped with germania, as shown in Fig. 6.20, high laser output power can be available from the

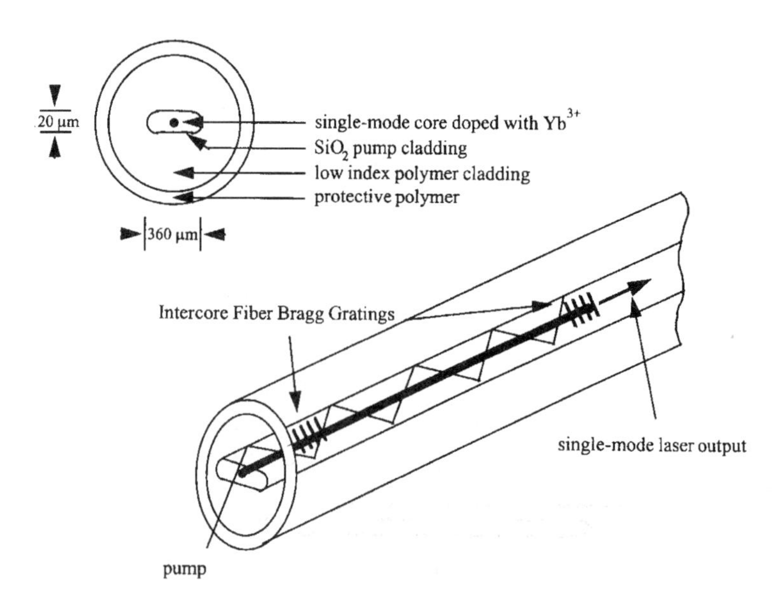

Figure 6.20: Schematic diagram of cladding-pumped fiber laser with fiber gratings. After Ref. [47].

single-mode core. The use of fiber Bragg gratings is necessary to narrow and stabilize the high-power fiber lasers with single-mode fiber output [48]. Output power as high as 30 W has been obtained from a diode-pumped Nd^{3+} cladding-pumped fiber laser [47], and such lasers are commercially available with output of 9 W.

6.1.6 Amplifier Noise

With an optical amplifier introduced in the system, new noise terms such as signal-spontaneous beat noise, spontaneous-spontaneous beat noise, and multiple reflection-induced inteferometric intensity noise (MRIN) will be generated. A simple illustration of signal-spontaneous beat noise and spontaneous-spontaneous beat noise is given in Fig. 6.21a. Both the beat noise terms are apparently generated by the square-law photodiode and will be included in a complete system CNR analysis in Section 6.1.6.1. MRIN is generated by reflections on both sides of the amplifier if no isolators are used, as shown in Fig. 6.21b, or by double Rayleigh backscattering inside the EDF. We will leave the discussions of the former to Chapter 9, and the latter to Section 6.1.9.2.

In the limit of high pump powers, we can obtain the minimum population inversion factor from Eqs. (6.6), (6.7), (6.12), and (6.18) as

$$n_{sp}^{\min}(\lambda_p, \lambda_s) = \frac{1}{1 - \dfrac{\eta_p}{\eta_s}} = \frac{1}{1 - \dfrac{\sigma_{pe}\sigma_{sa}}{\sigma_{pa}\sigma_{se}}} \quad (\text{high } P_p). \tag{6.24}$$

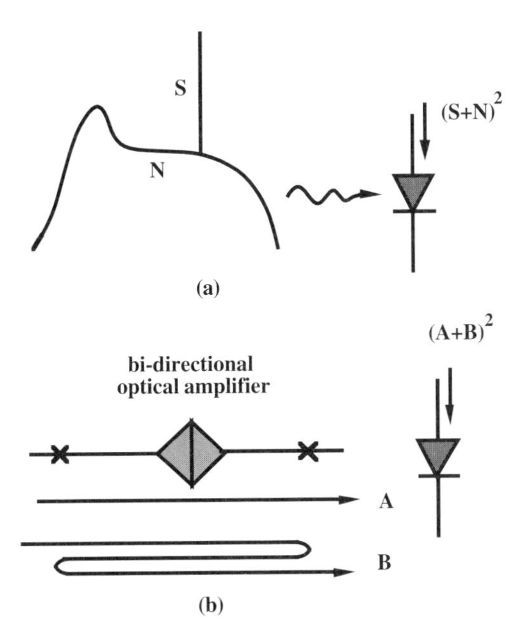

Figure 6.21: Illustrations of (a) signal-spontaneous beat (2S·N) and spontaneous-spontaneous beat noise (N^2), and (b) amplified interferometric noise (2A·B).

Therefore, we can see that the minimum population inversion factor depends on the emission and absorption cross-sections at both pump and signal wavelengths. For a three-level system, when the pump power is very high, $\sigma_{pe} \to 0$, and $n_{sp}^{min} \approx 1$. For a two-level system, however, $\sigma_{pe} \neq 0$, and $n_{sp}^{min} > 1$. Note that η_s is always larger than η_p, as can be seen in Fig. 6.22 for the example of an alumino-germanosilicate Er:glass [7]. In this example, $1.9 > \eta_s > 0.85$ in the signal wavelength region of 1.53–1.57 μm, while $< \eta_p \approx 0.3$ in the pump wavelength region of 1.46–1.49 μm. The best n_{sp} we can obtain in the 1.48-μm pump wavelength region is about 1.18 (equivalent to a noise figure of ~3.8 dB) when the pump and signal wavelengths are 1.46 and 1.57 μm, respectively. Therefore, we can understand why the noise figure of a 980-nm-pumped amplifier is always better than that of a 1480-nm-pumped amplifier. The dependence of n_{sp}^{min} on different signal wavelengths (1.52–1.57 μm) and pump wavelengths (1.45–1.50 μm) is shown in Fig. 6.23. We can see that the closer the pump wavelength is to the amplified signal wavelength, the larger is η_p, and the worse is the population inversion parameter. Also, we see that n_{sp}^{min} is lower in the longer wavelength region, whose physics is consistent with what we have observed in Fig. 6.12.

6.1.6.1 SCM System CNR Calculations, Including Signal-Spontaneous and Spontaneous-Spontaneous Beat Noise In analyzing the SNR or CNR in an intensity-modulated optical fiber system which employs traveling-wave optical amplifiers, we must consider all possible noise terms,

$$SNR\ (CNR) = \frac{\langle i_s^2 \rangle}{(\langle i_{RIN}^2 \rangle + \langle i_{shot}^2 \rangle + \langle i_{th}^2 \rangle + \langle i_{sig-sp}^2 \rangle + \langle i_{sp-sp}^2 \rangle) \cdot B_e}, \tag{6.25}$$

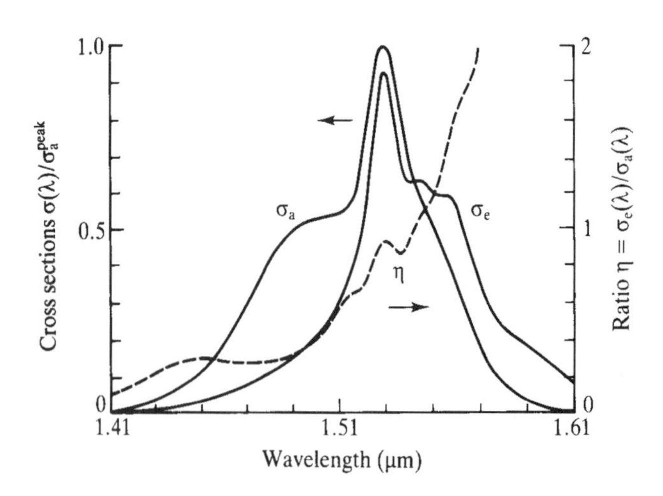

Figure 6.22: Absorption and emission cross-sections $\sigma_{a,e}$ and the corresponding ratio η, where η is given by $\eta_p = \sigma_{pe}/\sigma_{pa}$ or $\eta_s = \sigma_{se}/\sigma_{sa}$ depending on wavelength (for a typical alumino-germanosilicate Er-doped fiber). After Ref. [7].

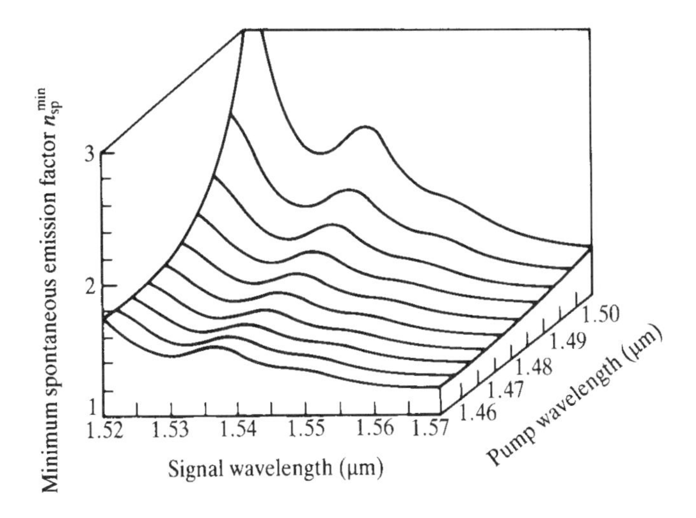

Figure 6.23: Minimum population inversion factor n_{sp}^{min} versus signal wavelengths from 1.52 to 1.57 μm, and pump wavelengths from 1.46 to 1.50 μm. After Ref. [7].

where $<i_s^2> = \frac{1}{2}m^2I_p^2$ = the mean-square signal photocurrent (recall that m is the optical modulation index per RF channel, I_p is the DC photocurrent, and $<i_{RIN}^2>$, $<i_{shot}^2>$, and $<i_{th}^2>$ are the equivalent noise current densities for RIN, shot noise and receiver thermal noise, respectively, which were all introduced in Chapter 5). Now with an optical amplifier added to the system, the two major additional noise terms are the signal-spontaneous beat noise, $<i_{sig-sp}^2>$, and the spontaneous-spontaneous beat noise, $<i_{sig-sp}^2>$. For a single-stage optical amplifier shown in Fig. 6.24a, we let the average amplified signal power and ASE power be given by P_s and P_{ase} ($= 2n_{sp}h\nu(G-1)\cdot\Delta\nu$), respectively. In Fig. 6.24a, L_f is the loss factor between the amplifier and the photodetector which could be because of transmission loss or power splitting loss, and $0< L_f <1$.

Following the signal and noise flows in Fig. 6.24a, we have

$$I_p = \frac{q\eta}{h\nu}\cdot P_{in}c_1G\cdot L_f \tag{6.26}$$

$$I_{ase} = \frac{q\eta}{h\nu}P_{ase}L_f = \frac{q\eta}{h\nu}2n_{sp}h\nu(G-1)\cdot\Delta\nu\cdot L_f \tag{6.27}$$

$$\langle i_s^2\rangle = \frac{1}{2}m^2I_p^2 \tag{6.28}$$

$$\langle i_{RIN}^2\rangle = RIN\cdot I_p^2 \tag{6.29}$$

$$\langle i_{shot}^2\rangle = 2q\cdot(I_p + I_{ase}) \tag{6.30}$$

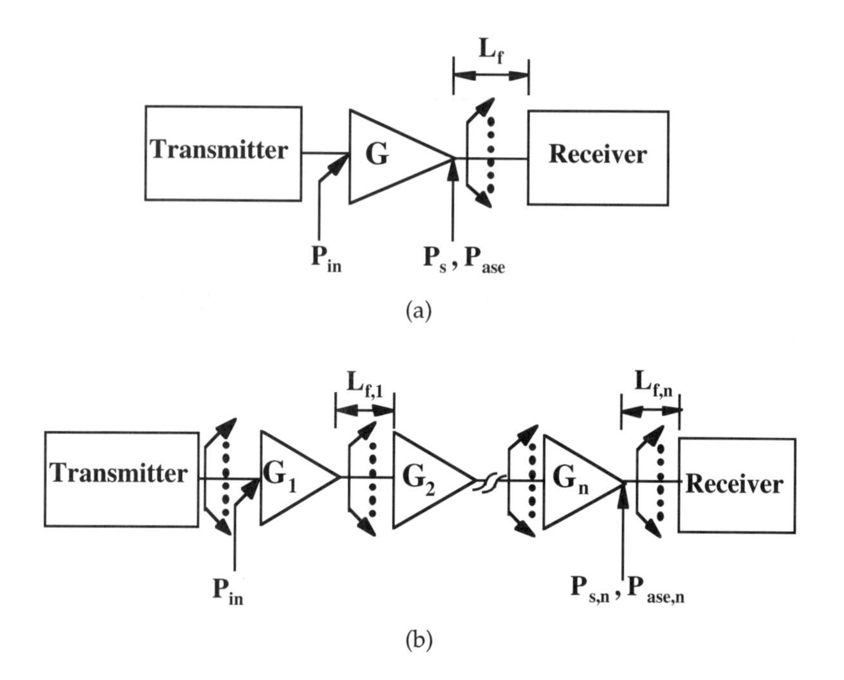

Figure 6.24: (a) Single and (b) cascaded optical amplifiers. An optical filter may be integrated in each amplifier.

$$\left\langle i_{sig-sp}^{2} \right\rangle = 2 \cdot \{2I_p \cdot \frac{I_{ase}}{2} / \Delta v\} \tag{6.31}$$

$$\left\langle i_{sp-sp}^{2} \right\rangle = 2\left\{ \left(\frac{I_{ase}}{2} \right)^2 + \left(\frac{I_{ase}}{2} \right)^2 \right\} / \Delta v. \tag{6.32}$$

In Eq. (6.26), c_l and P_{in} represent the input coupling loss and input power to the amplifier, respectively. Note that I_{ase} in Eq. (6.27) includes both polarizations, and therefore a factor of 2 is included. $<i^2_{sig-sp}>$ is a beat product between the signal and the ASE noise *in the same polarization* [that is why we divide I_{ase} by 2 in Eq. (6.31) to consider only one polarization], and the extra multiplication factor of 2 in Eq. (6.31) is because of the fact that we are considering single-sided noise spectral density. (Note that the shot noise derived in Section 5.1.1, the thermal noise derived in Section 5.2.2, and the RIN noise given by Eq. (4.39), are all single-sided spectral noise densities.) The same considerations apply to Eq. (6.32), where the first and second terms represent spontaneous-spontaneous beat noise in horizontal and vertical polarizations, respectively, and the extra multiplication factor of 2 is for single-sided spectral noise density. It is important to note that the signal-spontaneous beat noise is independent of the optical bandwidth $\Delta\nu$ [note that I_{ase} is dependent on $\Delta\nu$, and that is why we place $\Delta\nu$, in the denominator of Eq. (6.31) to eliminate that dependence], which implies that optical filters cannot be used to eliminate signal-spontaneous noise. This is because there is always a beat noise between signal and ASE which falls into a bandpass (or baseband) filter in a receiver, and one cannot eliminate this beat noise without eliminating the desired signal at the same time. On the other hand, $<i^2_{sp-sp}>$ is a result of beating among ASE components and is proportional to $\Delta\nu$, which implies that the narrower the optical filter is, the less the spontaneous-spontaneous beat noise is. Note also that there are two shot noise components; one is because of the signal, $2qI_p$, and the other is because of the ASE noise, $2qI_{ase}$. When the amplifier gain is high (e.g., with small input signal), shot noise components have much lower power levels than beat noise components because the former is proportional to G while the latter is proportional to $\sim G^2$. This can be seen more clearly as follows:

Signal shot noise:

$$\begin{aligned}\left\langle i^2_{shot,sig}\right\rangle &= 2q(\frac{q\eta}{h\nu}P_{in}\cdot c_1 GL_f) \\ &= \frac{2q^2\eta}{h\nu}P_{in}\cdot c_1 GL_f. \end{aligned} \tag{6.33}$$

ASE shot noise:

$$\begin{aligned}\left\langle i^2_{shot,ase}\right\rangle &= 2q(\frac{q\eta}{h\nu}2n_{sp}h\nu(G-1)\cdot\Delta\nu\cdot L_f) \\ &= 4q^2\eta\cdot n_{sp}(G-1)\cdot\Delta\nu\cdot L_f. \end{aligned} \tag{6.34}$$

Signal-spontaneous beat noise:

$$\begin{aligned}\left\langle i^2_{sig-sp}\right\rangle &= 2(\frac{q\eta}{h\nu}P_{in}c_1\cdot G\cdot L_f)(2q\eta\; n_{sp}(G-1)\cdot\Delta\nu\cdot L_f)/\Delta\nu \\ &= 4\frac{q^2\eta^2}{h\nu}n_{sp}c_1 P_{in}\cdot G(G-1)\cdot L_f{}^2. \end{aligned} \tag{6.35}$$

Spontaneous-spontaneous beat noise:

$$\begin{aligned}\left\langle i^2_{sp-sp}\right\rangle &= (2q\eta\; n_{sp}(G-1)\cdot L_f)^2\cdot\Delta\nu \\ &= 4q^2\eta^2\cdot n^2_{sp}(G-1)^2\cdot L^2_f\cdot\Delta\nu. \end{aligned} \tag{6.36}$$

An example showing the relative amplitudes of the four noise terms in Eqs. (6.33)–(6.36) or $L_f = 1$ is given in Fig. 6.25. A typical EDFA gain versus output signal power is shown in Fig. 6.25a, where we can see that the maximum output signal power is about 16 dBm. In Fig. 6.25b, we see that the ASE shot noise is the smallest term which can be neglected. The signal shot noise can also be neglected because in the small P_{in} region, it is less than both the spontaneous-spontaneous and signal-spontaneous beat noise, while in the large P_{in} region, it is less than the RIN noise (which may not be true if $L_f << 1$, see Eqs. (6.33) and (6.35)). Among the three major noise terms, $<i^2_{sig-sp}>$ is proportional to $G^2(P_{in}) \cdot P_{in}$ [see Eq. (6.35)], $<i^2_{RIN}>$ is proportional to $G^2(P_{in}) \cdot P^2_{in}$ [see Eqs. (6.26) and (6.29)], and $<i^2_{sp-sp}>$ is proportional to $G^2(P_{in})$. When the amplifier is operated in a linear regime, $G(P_{in})$ remains essentially constant. In this regime, as P_{in} increases, $<i^2_{sig-sp}>$ increases linearly with P_{in}, $<i^2_{RIN}>$ increases in proportion to P^2_{in}, and $<i^2_{sig-sp}>$ remains constant. However, when the amplifier is operated in the gain-saturation region, all these three noise terms are affected by the saturated gain G. As shown in Fig. 6.25b, once P_{out} increases above about 13 dBm, both the increasing P_{in} and the decreasing G^2 contribute to vary $<i^2_{sig-sp}>$. Therefore, $<i^2_{sig-sp}>$ first reaches a peak value because of the increase of P_{in}, and then decreases quickly corresponding to the degree of gain saturation. On the other hand, $<i^2_{RIN}>$ increases steadily as the P_{in} and the received optical power increase steadily, even when the amplifier is saturated. Therefore, when the amplified output signal is high, $<i^2_{sig-sp}>$ and $<i^2_{RIN}>$ become comparable, and the latter can easily exceed the former when RIN noise is high (we only show two RIN values of –150 and –160 dB/Hz in Fig. 6.25b). Therefore, the measured signal-spontaneous beat noise must be carefully calibrated to obtain the *intrinsic* amplifier noise figure.

Let us now examine SNR (or CNR) because of signal-shot noise, signal-spontaneous beat noise, and spontaneous-spontaneous beat noise terms in more detail. From Eqs. (6.28), (6.33), (6.35), and (6.36), we have

$$\text{CNR}^{-1} \text{ (due to signal shot noise only)} = \frac{4h\nu}{m^2 \eta \cdot c_1 P_{in} G L_f} \tag{6.37}$$

$$\text{CNR}^{-1} \text{ (due to sig-sp beat noise only)} = \frac{8n_{sp} h\nu}{m^2 c_1 P_{in}} (1 - \frac{1}{G}) \tag{6.38}$$

$$\text{CNR}^{-1} \text{ (due to sp-sp beat noise only)} = \frac{8n_{sp}^2 \cdot h^2\nu^2 \cdot \Delta\nu}{m^2 c_1^2 P_{in}^2} \left(1 - \frac{1}{G}\right)^2. \tag{6.39}$$

It is clear that these CNRs depend critically on the amplifier input optical power P_{in}. An important concept which Eqs. (6.37)–(6.39) bring forward is that optical loss *before* an optical amplifier can severely degrade the resultant CNR, whereas optical loss *after* an optical amplifier (that is, L_f in Fig. 6.24) does not affect the CNR (because signal and beat noise power all depend on L_f^2). Therefore, from the standpoint of signal-spontaneous beat noise, booster amplifiers could offer better CNR than in-line and preamplifiers, because the input optical power in the former case is higher. However, one must note that, when a booster amplifier is operated under deep gain saturation conditions, the system's RIN may dominate over beat noise terms, as was shown in Fig. 6.25b [29]. Therefore, if RIN is high, when P_{in} becomes very large, the total intensity noise and consequently the *effective* noise figure may actually increase (see Section 6.1.6.3).

In the case of n-concatenated amplifiers, we can still use the preceding formulas, except that P_s and P_{ase} in Fig. 6.24a should be replaced by the signal and ASE noise power of the last-stage amplifier, that is, $P_{s,n}$ and $P_{ase,n}$ in Fig. 6.24b, respec-

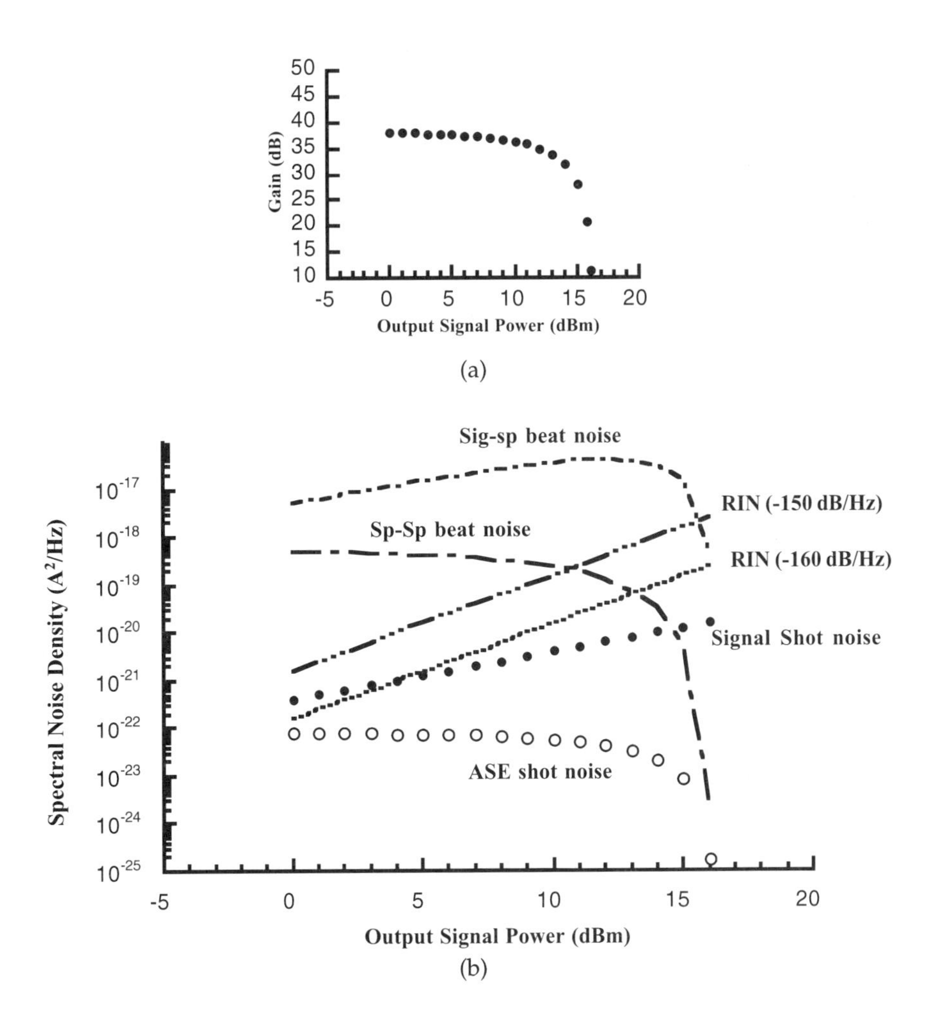

Figure 6.25: Typical gain versus output signal power of an EDFA (a), and the corresponding spectral noise densities of various noise terms (b). In the calculations, we have assumed $n_{sp} = 1$, the input coupling loss to the amplifier is 0.5 dB, the detector quantum efficiency is 100%, $\lambda_c = 1550$ nm, $L_f = 1$, and the optical filter bandwidth $\Delta\upsilon = 1$ nm.

tively. Also, L_f should be replaced by $L_{f,n}$. The two key parameters $P_{s,n}$ and $P_{ase,n}$ can be obtained by iterative calculations [17]:

$$P_{s,n} = P_{s,n-1} \cdot L_{f,\,n-1} \cdot G_n \tag{6.40}$$

$$P_{ase,n} = P_{ase,n-1} \cdot L_{f,\,n-1} \cdot G_n + 2n_{sp}h\nu \cdot (G_n - 1) \cdot \Delta\upsilon. \tag{6.41}$$

If we let $L_{f,\,n-1} \cdot G_n = 1$, that is, all amplifier gain just compensates the transmission or splitting loss before the amplifier, and let $G_n = G$ for all n, then we see that $P_{s,n}$

remains the same at each amplifier output and $P_{ase,n}$ grows linearly with n. From Eqs. (6.31) and (6.32), this implies that signal-spontaneous beat noise is proportional to n, while the spontaneous-spontaneous beat noise is proportional to n^2. Naturally, one may have to install optical bandpass filters between optical amplifiers to eliminate most of the ASE noise, not only so that the gains of the following amplifiers will not be wasted on ASE noise, but also so that the spontaneous-spontaneous beat noise can be minimized.

6.1.6.2 Amplifier Noise Figure The basic definition of an optical amplifier's noise figure F_o is given by

$$F_o = \frac{(SNR)_{in}}{(SNR)_{out}} = \frac{(CNR)_{in}}{(CNR)_{out}}, \tag{6.42a}$$

or, in terms of log scale,

$$\mathrm{NF_o} = 10 \cdot \log \frac{(SNR)_{in}}{(SNR)_{out}} = 10 \cdot \log \frac{(CNR)_{in}}{(CNR)_{out}}, \tag{6.42b}$$

where the subscripts "in" and "out" represent the signal-to-noise (carrier-to-noise) ratios, measured *after* a photodiode, without and with an optical amplifier, respectively. We assume that P_{in} of an optical amplifier is sufficiently large that $(\mathrm{SNR})_{\mathrm{in}}$ is *shot-noise-limited*, and $(\mathrm{SNR})_{\mathrm{out}}$ is signal-spontaneous-beat-noise- and signal-shot-noise-limited. In this case, we shall prove that the ideal amplifier noise figure is 3 dB.

Overall $(\mathrm{CNR})_{\mathrm{out}}$ is derived from Eq. (6.38) (signal-spontaneous-beat-noise-dominated) and Eq. (6.37) (signal-shot-noise-dominated),

$$\begin{aligned}(CNR)_{out} &= (CNR_{sig-shot}^{-1} + CNR_{sig-sp}^{-1})^{-1} \\ &= \left(\frac{4h\nu}{m^2 \eta \cdot c_1 P_{in} G L_f} + \frac{8 n_{sp} h\nu}{m^2 c_1 P_{in}} (1 - \frac{1}{G}) \right)^{-1},\end{aligned} \tag{6.43}$$

while the shot-noise-limited $(CNR)_{in}$ of an SCM channel is given by

$$(CNR)_{in} = \frac{\frac{1}{2} m^2 I_p^2}{2qI_p} = \frac{m^2 \eta P_{in} L_f}{4h\nu}. \tag{6.44}$$

From Eqs. (6.42)–(6.44), we obtain

$$F_o = \frac{1}{c_1 G} + \frac{\eta L_f}{c_1} (\frac{G-1}{G}) 2n_{sp}. \tag{6.45}$$

If we let the detector quantum efficiency $\eta = 1$, optical amplifier input coupling loss $c_1 = 1$, and assume no loss between the amplifier and the detector ($L_f = 1$), we obtain an ideal optical amplifier noise figure as

$$F_o = \left(\frac{G-1}{G}\right) \cdot 2n_{sp} + \frac{1}{G} = 2n'_{sp} + \frac{1}{G}, \tag{6.46}$$

where $n'_{sp} = n_{sp}(1 - \frac{1}{G})$. When $G >> 1$ and the population inversion parameter is ideal, that is, $n_{sp} = 1$, we obtain the ideal optical amplifier noise figure to be 2, or 3 dB. This ideal noise figure performance has been achieved by 980-nm-pumped optical amplifiers. Note that from Eq. (6.14b), Eq. (6.46) can also be written as [7]

$$F_o = \frac{S_{ase} + 1}{G}. \tag{6.47}$$

The effect of cascading amplifiers on SNR can also be examined from the viewpoint of "effective noise figure." We give an example in Fig. 6.26, where two amplifiers are in cascade. We can treat the link which consists of G_1, $L_{f,1}$, and G_2 as a black box, and find its effective noise figure as follows. The total ASE photon number out of this black box is given by

$$\begin{aligned} S_{ase} &= 2n_{sp,1}(G_1 - 1) \cdot L_{f,1} \cdot G_2 + 2n_{sp,2}(G_2 - 1) \\ &= 2n'_{sp,1} G_T + 2n'_{sp,2} G_T \frac{1}{G_1 L_{f,1}} \end{aligned} \tag{6.48}$$

where

$$\begin{aligned} n'_{sp,1} &= n_{sp,1}\left(1 - \frac{1}{G_1}\right) \\ n'_{sp,2} &= n_{sp,2}\left(1 - \frac{1}{G_2}\right). \end{aligned} \tag{6.49}$$

In Eq. (6.48), the first term is the ASE photon number generated from the first amplifier and subsequently amplified by the second amplifier; the second term is the ASE photon number of the second amplifier; and $G_T = G_1 L_{f,1} G_2$ is the total gain of

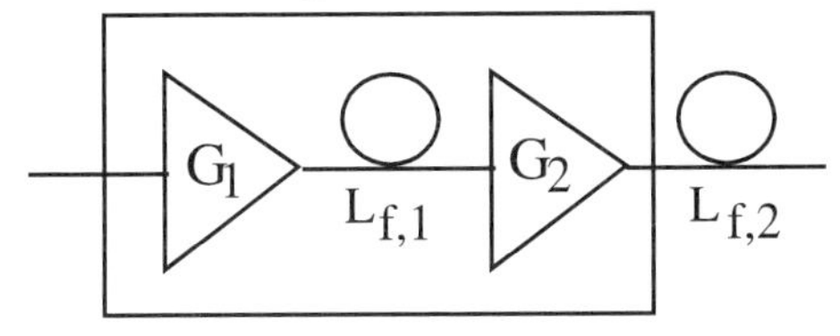

Figure 6.26: Optical amplifiers in cascade. We can treat G_1, $L_{f,1}$, and G_2 as a black box when deriving the effective noise figure of the amplifiers in cascade.

the cascaded amplifier (in the black box). According to Eqs. (6.47) and (6.48), the optical noise figure of this cascaded amplifier is given by

$$F_{o,T} = \frac{S_{ase}+1}{G_T} = 2n'_{sp,1} + 2n'_{sp,2}\frac{1}{G_1 L_{f,1}} + \frac{1}{G_T}. \tag{6.50}$$

If we assume $G_T >> 1$, then Eq. (6.50) becomes

$$F_{o,T} = 2n'_{sp,1} + \frac{2n'_{sp,2}}{G_1 L_{f,1}} = F'_{o,1} + \frac{F'_{o,2}}{G_1 L_{f,1}}, \tag{6.51}$$

where $F'_{o,i} = 2n'_{sp,i}$ $(i = 1,2)$. Therefore, we see that the cascaded amplifier noise figure is slightly different from the case of cascaded *electric* amplifiers in Chapter 2, that is, $\boldsymbol{F_e = F_1 + (F_2 - 1)/(G_1 L_1)}$. From Eq. (6.51), if $G_1 L_{f,1} = 1$ and $\boldsymbol{n'_{sp,1} = n'_{sp,2}}$, the effective noise figure is seen to be twice of that of a single optical amplifier. This is consistent with the previous observation that signal-spontaneous beat noise increases linearly with the number of optical amplifiers (note that in defining the amplifier noise figure the spontaneous-spontaneous beat noise is neglected).

Equation (6.51) states that the noise figure of the first-stage optical amplifier must be low, and its gain must be high, in order for the entire link (which consists of cascaded amplifiers with interstage loss) to have a low effective noise figure. If the first stage has a high noise figure, the over link could have a poor noise figure performance even if the following stages have high gain and low noise figures. This fact can be extended to the case of a single EDFA, which is assumed to be composed of many small cascaded pieces of EDFAs. Now if the pump power is not high enough to fully invert the ion populations in the entire EDF, we can model the amplifier as a cascade of a high-gain amplifier followed by a power attenuator for the forward-pumping case, as shown in Fig. 6.27a. The power attenuator represents the end portion of the EDF where the pump is too weak to excite the ground state population. For the backward pumping case, the amplifier can be modeled as shown in Fig. 6.27b, where the order

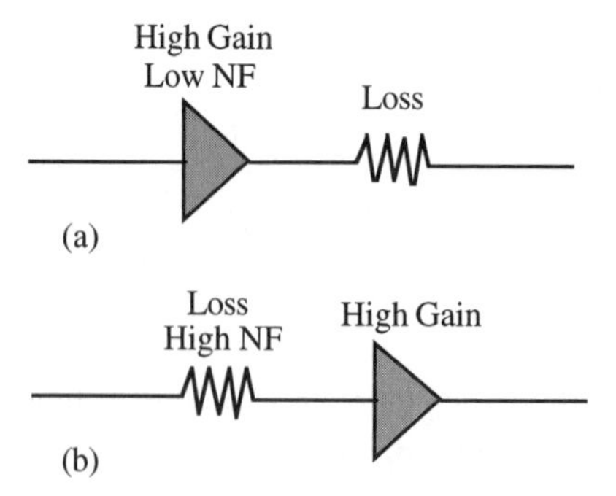

Figure 6.27: An analogy to electrical amplification and attenuation of a forward-pumped (a) and backward-pumped (b) EDFA when the pump power is not sufficient to fully invert the populations in the entire EDF.

of the high-gain amplifier and the attenuator is reversed. It is obvious, either from Eq. (6.51) or from Eq. (2.4) for cascaded electrical amplifiers, that configuration (a) provides a better noise figure performance than configuration (b). Therefore, when the pump power is not strong enough, or when the fiber length is not optimized, it is preferable to have forward pumping rather than backward pumping.

6.1.6.3 Measurement of Amplifier Noise Figure For SCM system applications, optical amplifier noise figure can be measured in terms of optical or electrical methods. The optical method uses an optical power meter and/or an optical spectrum analyzer. It is simple to implement and is accurate in the small-signal regime. The electrical method measures the intensity noise on a spectrum analyzer after a photodetector[4] and is a more accurate method in the large signal regime, that is, when the amplifier is operated in the saturation region. Generally speaking, the electrical method is more complicated, but can directly measure the intensity noise level in a real SCM system. Both the optical and electrical methods are introduced as follows.

Optical Method for Noise Figure Measurement. The simplest optical method is to take advantage of Eq. (6.14a), that is, $P_{ase}(L) = 2n_{sp}(G - 1)h\nu \cdot \Delta\nu$, in the small-signal regime. Recall from Eq. (6.42) that the amplifier noise figure can be approximated by $2n_{sp}$ when the amplifier gain $G >> 1$. Therefore, if we can measure the amplifier gain G (when the input signal is small) and P_{ase} after an optical filter (which has a bandwidth $\Delta\nu$ centered at $\lambda_c = c/\nu$) with the input signal turned off, we can easily calculate $2n_{sp}$.

When the input signal power is increased, the ASE power will be compressed as the gain saturates. In order to measure this compressed ASE, we cannot use the previous method because once the signal is turned off, the ASE goes back to its uncompressed level. Under this condition, the compressed ASE power level can possibly be measured by nulling the signal using a polarization controller and/or a polarizer, as shown in Fig. 6.28.

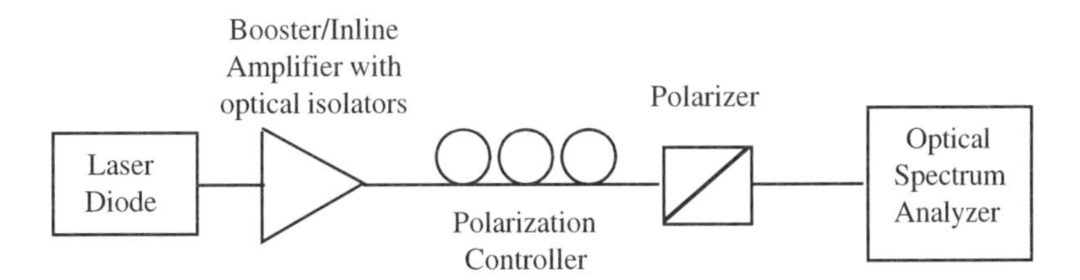

Figure 6.28: Optical method: the polarization nulling technique for optical amplifier noise figure measurement. After Ref. [50].

[4]Another electrical method termed "time-domain extinction measurement," which is more suitable for digital optical fiber systems, can be found in [49].

The polarization nulling technique usually shows large and unsatisfactory residual spectral components of the amplified signal even with the use of a polarizer with a 40-dB polarization extinction ratio. Furthermore, when the input signal power is increased even further, the polarization-nulling method may overreport the amplifier noise figure by a phenomenon called polarization hole burning (PHB). PHB occurs because the measured "orthogonal" ASE noise may not be compressed as much as the "parallel" ASE noise which has the same polarization state as the signal. Since it is the "parallel" ASE noise that generates signal-spontaneous beat noise, the measured "orthogonal" ASE noise may not give a precise indication of the actual signal-spontaneous beat noise. Therefore, in this highly saturated operation condition, the optical method may not be very reliable.

The optical method is especially unreliable when we realize the fact that frequency-dependent multiple-reflection-induced interferometric noise (MRIN) may exist (see Sections 6.1.9.2 and 9.3.1) after a square-law photodetector.

Electrical Method for Noise Figure Measurement. This method, although more tedious, is particularly useful for measuring the noise figure of a booster amplifier. Besides measuring signal-spontaneous beat noise, this method can also measure the total RIN noise because of a laser transmitter and the interferometric phenomenon (that is, multiple reflections), and it can faithfully reflect the total RIN noise in a real SCM system.

In using this method, we first have to measure $G(P_{in})$ for a specific amplifier, as was shown in Fig. 6.25a. Next, we have to measure the total RIN noise after the photodiode. As was shown in Fig. 6.25b, when the intensity noise is dominated by signal-spontaneous beat noise, the noise level will first increase and then decrease as the input signal to the amplifier increases. However, after a certain input signal level, the total intensity noise may start to increase again, and when this happens, according to Fig. 6.25b, we know we may have a significant noise contribution from RIN or MRIN. Therefore, we can derive the amplifier intrinsic noise figure $2n_{sp}$ when the signal-spontaneous beat noise is dominant, and derive the *effective* amplifier noise figure in the deep saturation region when RIN or MRIN is dominant. This *effective* amplifier noise figure can be expressed [based on Eqs. (6.29) and (6.35)] as

$$F_{o,eff} = 2n_{sp} = \frac{c_1 P_{in}}{2h\nu} RIN. \tag{6.52}$$

Another interesting phenomenon which was observed in the measurement of a deeply gain-saturated amplifier noise figure is that in certain regions of P_{in}, there may exist a dip of the noise figure value at the onset of amplifier saturation [51,52], as illustrated in Fig. 6.29. This decrease of noise figure occurs especially in high-gain amplifiers and arises from suppression of backward ASE by the high forward-traveling signal power. The backward-traveling ASE degrades the noise figure by depleting the inversion at the beginning of the fiber which gives rise to higher signal-spontaneous beat noise. Because the ASE power never becomes large enough

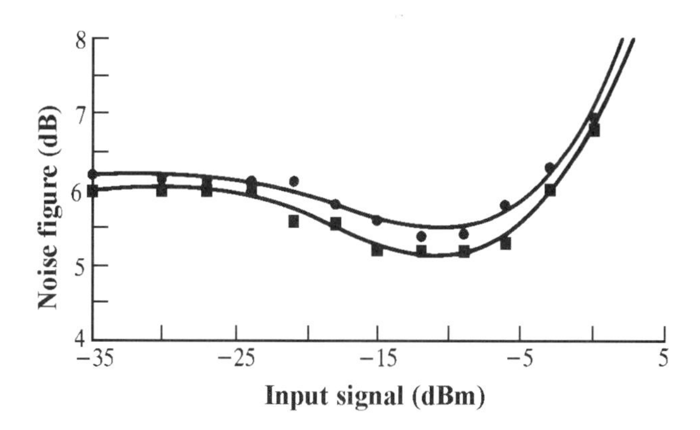

Figure 6.29: Example of noise figure behavior in a both-end 1.48-mm-pumped amplifier. The signal is at 1550 nm. After Ref. [51] (©1989 IEEE).

to saturate low-gain amplifiers, this effect is observed only for amplifiers with higher gains.

6.1.7 EDFA Gain Dynamics and Linearity

The key advantages of EDFAs for SCM system applications is that, because of the long lifetime (~10 ms) of the $^4I_{13/2}$ first excited state of the erbium ion, the dynamic response of the erbium ions is much slower than the modulating signals (remember that all CATV channels have frequencies >54 MHz). No CSO or CTB degradations have been observed even in a deeply saturated EDFA.

An experimental result illustrating the slow gain dynamics effect on EDFA is shown in Fig. 6.30 [53]. Two P_{out} versus P_{in} curves are shown for the cases when the input light is CW and modulated by a 1-MHz tone, respectively. We can see that once the input light changes with a rate much faster than the inverse of the carrier lifetime (~100 Hz), the output light is linearly proportional to the input light, whereas output power saturation can be observed when the input light is unmodulated.

6.1.8 Design Considerations for Booster Amplifiers

EDFAs with high output power greater than +20 dBm are desirable to overcome the large splitting and transmission losses in HFC systems. Therefore, in the following two subsections, we will review techniques for building high-power booster amplifiers and the associated high-power lasers.

6.1.8.1 High-Power EDFAs Using 980- and/or 1480-nm Semiconductor Pumps
When using purely semiconductor pump lasers, several schemes can be considered to increase the output power of the booster amplifier: (1) using multiple

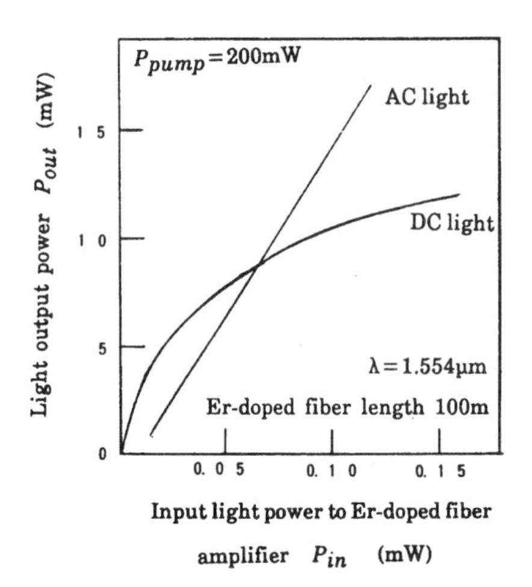

Figure 6.30: Under CW conditions, when the input light power is increased, the output light power is gradually saturated. Under modulation of a 1-MHz tone and the same level of input light power, the output light power is linearly increased without saturation. After Ref. [53] (© 1993 IEEE).

polarization-multiplexed 1480-nm laser diodes and bidirectional pumping (Fig. 6.31 [54]) to provide a total of 545 mW pump power in Er-doped fiber, and a signal output power of +25.2 dBm; (2) using a high-brightness 980-nm semiconductor monolithic master-oscillator power amplifier (M-MOPA), previously shown in Fig. 6.15, which can reach 1.2 W into a single-mode fiber [24] to pump erbium-doped fibers, and an EDFA with an output power up to +24.8 dBm at 1.55 μm has been reached by using this M-MOPA [55]; and (3) using two to four >100mW fiber-grating stabilized 980-nm pumps and a ~980 WDM multiplexer with a channel spacing > 2.0 nm.

Scheme 1 requires the use of polarization-preserving diode pigtails and bulky polarization-combining optics, which had a high insertion loss and introduced non-negligible reflections in the amplifier system. Scheme 2 is promising if more reliability data can be demonstrated on the 980-nm M-MOPA. Scheme 3 can possibly be implemented by using commercially available high-power 980-nm pump lasers.

Note that for high-gain amplifiers under conditions of heavy saturation, such as will be used for power amplifiers, there is no difference in PCE among the three pumping configurations given in Section 6.1.1 [52]. However, from the consideration of low noise figure, the higher the forward pump (using a 980-nm pump), the better.

6.1.8.2 High-Power EDFAs Based on Diode-Pumped Solid-State Lasers or Diode-Pumped Fiber Lasers The first few booster EDFAs based on diode-pumped solid-state (DPSS) lasers used a GaAs LD-array-pumped Nd:YLF or Nd:YAG laser to pump Er/Yb-codoped fiber to reach an output power greater than +20 dBm: (1) using a miniature 1064-nm diode-pumped Nd:YAG laser to pump an Er/Yb/phospho-

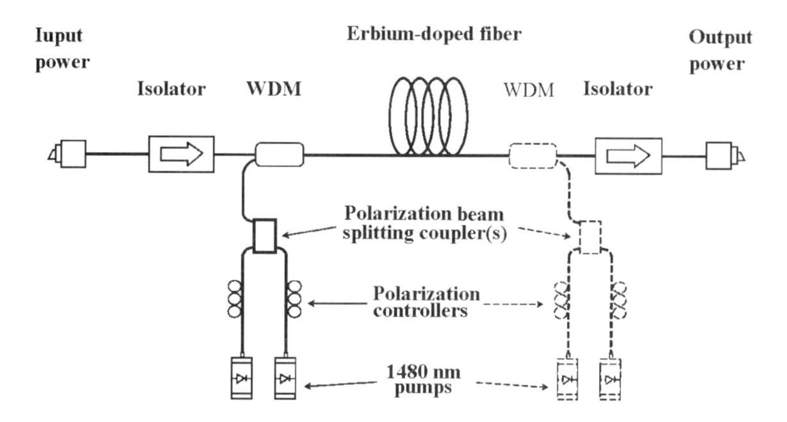

Figure 6.31: High-power EDFA configuration which uses four polarization-multiplexed 1480-nm pump lasers. After Ref. [54].

rus codoped silica fiber (390 mW of coupled pump power to give an output power of +21 dBm [56] and 365 mW pump power to give +21.6 dBm output power [57]), as illustrated in Fig. 6.32 [58]; or (2) using a single counter-propagating 1047/1053-nm diode-pumped Nd:YLF laser to pump an Er/Yb-codoped fiber (900 mW of coupled pump power to offer an output power of +24.6 dBm [45]).

Note that schemes (1) and (2) used Er/Yb/phosphorus codoped fibers. Er/Yb codoped fibers broaden the choice of pump source wavelength by using a "sensitizer," Yb, which has a broad absorption band between 800 and 1080 nm (see Section 6.1.2.2). We can see that 1064-, 1053-, and 1047-nm diode-pumped laser

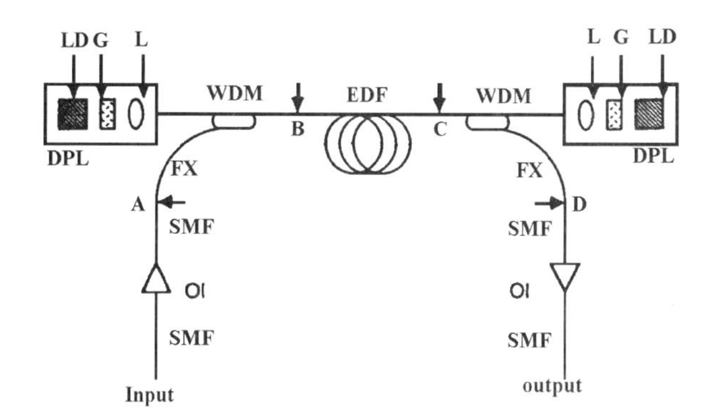

Figure 6.32: Experimental setup for the codoped Er/Yb DPSS-pumped optical amplifier. The components are as follows: DPL, diode-pumped laser; LD, AlGaAs laser diode array; G, Nd:YAG gain medium; L, lens, WDM, wavelength division multiplexing coupler; EDF, Er/Yb codoped fiber; OI, optical isolator; FX, Corning Flexcore 1064 single-mode fiber; SMF, Corning SMF–28 fiber. After Ref. [58] (© 1993 IEEE).

outputs which were used in both schemes all fall into that wavelength range. Er/Yb codoped fiber exhibits excited-state absorption with peaks at 970 and 1120 nm, which can degrade the pump efficiency. Therefore, pump sources between 1047 and 1064 nm are preferred because no ESA exists.

As another approach to obtaining high-power amplifiers, an Er/Yb booster amplifier pumped by a single diode-pumped cladding-pumped laser has yielded an output power of +31.8 dBm at 1550-nm with a 0-dBm input signal [57,59]. When a three-stage Er/Yb amplifier was constructed, with a cladding-pumped laser pumping each stage, an output of +36.3 dBm was obtained, for a total pump power of 11.5 W at 1060 nm [60].

As a last note before we leave this section, the PCE in booster EDFAs can be optimized by increasing the NA and confining the Er^{3+} ions to the central region of the core, and by choosing the optimum cutoff wavelengths for 980-nm or 1480-nm pump bands [31].

6.1.9 Gain-Flattened EDFAs and Double Rayleigh Backscattering

Looking back at the recent history of developing EDFAs, we see that the initial focus in EDFA optimization was simply on maximizing the gain per milliwatt of pump power. Then this gave way to designs to achieve high-PCE and low-noise figure. Recently, however, because of the requirement for flat amplifier gain for dense-wavelength-division-multiplexing (DWDM) and/or analog CATV systems, various broadband gain-flattening techniques have been proposed. Furthermore, because of the stringent requirement of CNR in AM-CATV systems, design considerations for minimizing the double Rayleigh backscattering (DRB)-induced RIN noise (which increases the effective amplifier noise figure according to Eq. (6.52)) have been discussed in a number of papers. In the following two sections, we will briefly review the research and development status of these two important topics: broadband gain-flattening and DRB-induced RIN minimization.

6.1.9.1 EDFAs with Broad and Flat Gain Flat amplifier gain in a wavelength range of a few nanometers can be very helpful in reducing the CSO in an AM-CATV link caused by amplifier gain tilt and laser frequency chirping (see Section 9.3.3). Flat amplifier gain in a wavelength range of a few tens of nanometers are needed for DWDM systems which could transport both analog and digital signals (see Section 9.5). The proposed techniques to broaden and flatten the EDFA gain include using static gain equalizers such as a long-period grating [61] to reach a 3-dB gain bandwidth of 47 nm (with a flat gain within 1 dB over 40 nm) [62] and 52 nm [63], or using dynamic gain equalizers such as acouto-optic tunable filters to reach a gain flatness of <0.7 dB over 35-nm [64]; using Tellurite-based EDFA and gain equalizers to reach a gain exceeding 20 dB over 80 nm (1530–1610 nm) and a flat gain within 1.5 dB in 1535–1610 nm [8, 65]; using fluoride-based EDFA [66] to reach a gain bandwidth (flat within 1 dB) of 28nm (1532–1560nm) [67]; using 1.55 and 1.58 μm band EDFAs in parallel configuration [68-70] with a 3-dB bandwidth of 68 nm [70]; and using a partially gain-flattened EDFFA and a distributed Raman amplifier to achieve a 3-dB bandwidth of 75nm (1531–1506 nm) [71].

Examples of the signal gain spectra of a tellurite-based EDFA (for various input signal powers) and an EDFA with a parallel pump configuration are shown in Fig. 6.33a and 6.33b, respectively.

6.1.9.2 Double Rayleigh Backscattering Rayleigh scattering is because of the inhomogeneities of the fiber refractive index along the transmission path and therefore is an intrinsic, unavoidable loss mechanism which depends on the signal wavelength and various codopants. When light propagates along an optical fiber, part of the scattered light is recaptured by the fiber and propagates in both directions. In a relatively short piece of EDF (a few tens of meters), amplified signal double Rayleigh scattering (SDRS) has been shown to affect the gain [72,73] and noise figure [74,75] of an EDFA. The undesirable high Rayleigh backscattering coefficient depends on the following EDF parameters: high NA [72,73], high aluminum concentration [72], and low erbium concentration with longer EDF length [73,74]. However, to obtain a better gain coefficient in EDFA, the conditions of the preceding parameters are actually desirable. Therefore, there exist optimum values of these EDF parameters [74].

The coherent and incoherent mixing of the directly and doubly backscattered fraction of the signal light at the photodetector can severely degrade the amplifier noise figure because of the increased interferometric intensity noise [75], as explained in more detail in Section 9.3.1.

6.1.10 1.3-μm PDFFA Amplifiers

As discussed earlier in the beginning of this chapter, there are now three 1.3-μm optical amplifiers to choose from: the praseodymium-doped fluoride fiber amplifier (PDFFA) [13,76,77], the semiconductor optical amplifier (SOA), and the Raman amplifier. In this section, we will concentrate on reviewing the current status of PDFFA development. We can see from Table 6.1 that the state-of-the-art PDFFA generally requires a very high pump power of about 300–500 mW in order to reach

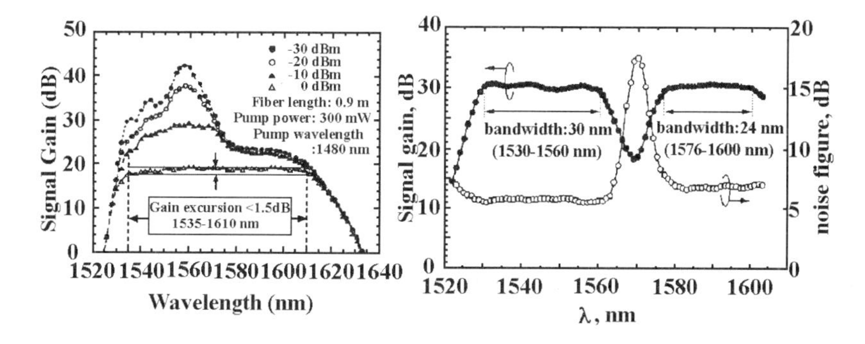

Figure 6.33: Gain spectra of a tellurite-based EDFA (for various input powers) (a) After Ref. [8] and a EDFA with a parallel pump configuration (b). After Ref. [69]

Table 6.1: Parameters for state-of-the-art PDFFAs.

Company	**NTT [78]**	**NTT [79]**	**SDL&Lucent [80]**	**Hoya [14]**
Year	1993	1995	1996	1996
Pump λ (μm)	1.017 (laser diode)	1.047 (Nd:YLF)	1.013 (M-MOPA)	1.004 (Ti:Sapphire)
Pump power (mW)	124	550	380/470	300
Signal λ (μm)	1.30	1.302	1.310	1.311
Small-signal gain (dB)	24	20	26 @380 mW	28
Max. output power (dBm)	9	19.2	17.1 @470 mW	16
Noise figure (dB)	—	6	—	—
Gain coeff. (dB/mW)	0.4	—	0.11	0.65

a reasonable output power level and/or an appreciable small-signal gain. The poor pump efficiency is because of the short fluorescence lifetime in Pr^{3+}-doped fibers (~110 μs [81]). A typical PDFFA module is supplied with one or two powerful Nd:YLF lasers at 1.047 μm or four laser diodes at 1.017 μm.

The efficiency of PDFFAs can be improved by (a) using a low-phonon-energy glass as a host in order to realize a longer fluorescence lifetime and (b) increasing the normalized index difference Δ. In the former case, use of Pr^{3+}-doped sulfide-based chalcogenide glasses (Ga:La:S glasses) was proposed, and the quantum efficiency was measured to be 58% with up to 80% predicted [82,83]. Both (a) and (b) were used in a InF_3/GaF_3-based fiber obtained with $\Delta = 6.1\%$ and a double-pass configuration to achieve a high gain coefficient of 0.65 dB/mW [14].

In addition to the disadvantage of poor pump efficiency, a special packaging technique of inserting a high-NA silica fiber (with relative refractive index of ~3.5%) between the PDF and an input or output coupler should be implemented [78,84]. This inserted fiber is used to reduce mode-field mismatching loss between the PDF and the fiber coupler. While using this high-NA silica fiber, a thermally diffused expanded core technique [85] or fusion splicing [80] was used at the junction of the high-NA silica fiber and the fiber coupler, and a tilted V-groove connection [78] or butt-coupled joint [80] was used at the junction of the high-NA silica fiber and the PDF. The splicing loss between the PDF and fiber coupler, and the reflection at the tilted V-groove connection, were 0.4 dB and <-60 dB, respectively [79].

The noise figure of PDFFAs was found to increase from ~5 to ~13 dB when the signal wavelength increased from 1280 to 1340 nm [86,87], mainly because of the ground-state absorption between $^3H_4$ and $^3F_4$ levels (see Fig. 6.8). It was also shown that a significantly RIN-noise-enhanced noise figure can be observed because of the splice points between the silica fiber and the PDFF [88] and because of amplified Rayleigh backscattering [87].

It should be noted that the relatively short fluorescence lifetime of ~110 μs does not cause any CTB or CSO degradations, and this fact was experimentally verified in an externally modulated lightwave CATV system [88].

6.1.11 Raman Fiber Amplifiers

The Raman fiber amplifier (RFA) is the only silica-based fiber amplifier for 1.3-μm use. Because this amplifier uses silica fiber as its amplification medium, it can be easily spliced to standard silica fiber with somewhat higher reliability than that obtained with a PDFFA. Since its gain efficiency is even lower than that of a PDFFA (only up to 0.1 dB/mW [89]), it must use a high-power Nd-doped fiber laser with over 1 W output power as the pump source [90]. This high power consumption could pose a serious problem for practical applications.

Nevertheless, it is worthwhile to understand the basic physics and the state-of-the-art performance of Raman fiber amplifiers [91]. The Raman effect takes place at high powers in a nonlinear optical medium. It consists of an inelastic photon scattering induced by elementary lattice excitations (that is, phonons) of the medium. The photon–phonon interaction gives rise to sidebands (Stokes lines) which are shifted from the original frequency by an amount equal to the phonon frequency (Stokes shift). In other words, incident light at a specific frequency can generate frequency-shifted Stokes waves. When the pump power exceeds a threshold value, the Stokes waves build up almost exponentially, which is the stimulated Raman scattering (SRS) phenomenon. A suitable choice of nonlinear medium and pump frequency can give a Stokes line tuned to the frequency of the signal to be amplified. Figure 6.34 shows five Stokes lines when a pump wavelength near 1.06 μm with 840 mW was applied [92]. We can see that to generate a Stokes line around 1.3 μm, a pump source with a wavelength around 1.24 μm is needed. However, no pump sources capable of emitting ~1 W around 1.24 μm were available, so Raman effects at higher orders had to be exploited and were experimentally demonstrated [89,90,93,94]. Two schemes have been used to amplify signals at 1.31 μm in silica fibers. The first scheme is to generate the third Stokes order at 1.24 μm using a cascaded Raman laser pumped by a cladding-pumped neodymium laser operating at 1064 nm (1064 nm provides gain for 1115 nm, an 1115-nm laser provides gain for 1175 nm, and a 1175-nm laser provides gain for 1240 nm), as shown in Fig. 6.35a, and subsequently pump a separate section of germanosilicate fibers to amplify at 1.3 μm, as shown in Fig. 6.35b. The second scheme, the multistage Raman ring amplifier shown in Fig. 6.35c, uses a neodymium cladding-pumped fiber laser at 1060 nm which is injected into the upper half of the ring amplifier in order to pump two cascaded Raman lasers that lase between short-period fiber Bragg gratings at 1115 and 1175 nm. The 1175-nm laser provides gain for a 1240-nm pump laser that lases in a clockwise fashion about the ring. The 1.3-μm signal propagates in the lower half of the ring through a two-stage amplifier with an interstage isolator. The purpose of the interstage optical isolator is reduce

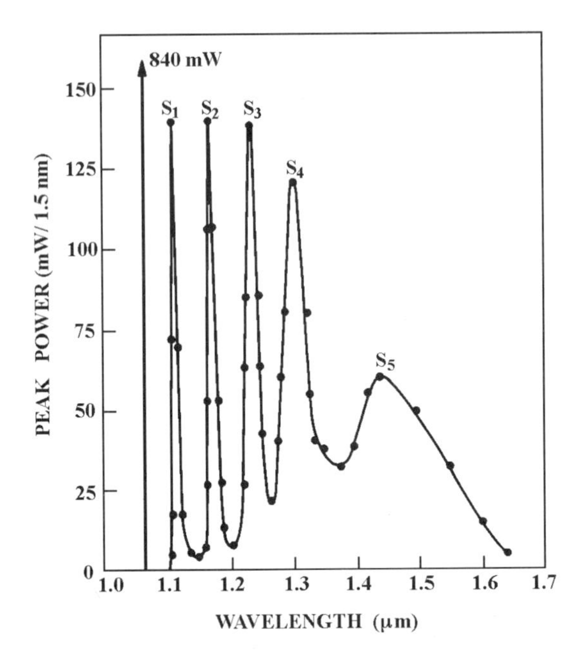

Figure 6.34: SRS spectrum showing generation of five Stokes lines S1 to S5 by using 1.06-μm pump pulses. Vertical line shows the pump output. After Ref. [92] (© 1978 IEEE).

the effects of Rayleigh double scattering discussed in Section 6.1.9.2. Note that the Raman amplifiers in Figs. 6.35b and c used a total of 7- and 2.1-km optical fibers, respectively. Both Figs. 6.35b and c use counter-propagating pumps to buffer the Raman gain from intensity fluctuations in the pump light. The Raman ring amplifier, which provides a 1.3-μm output power of +23 dBm, was tested in a 77-channel AM-VSB CATV system with CNR > 48 dB, CSO <-59 dBc, and CTB <-70 dBc [94].

6.2 Semiconductor Optical Amplifiers

The 1.3-μm semiconductor optical amplifiers (SOAs) are the most compact amplifiers with the lowest power consumption when compared to PDFFAs or Raman amplifiers. However, there are two major problems in using SOAs for SCM systems. The first is poor SOA linearity performance owing to its short carrier lifetime or gain response time, which is in the range of several hundred picoseconds, much faster than that of an EDFA. The second is the poor SOA noise figure performance (typically 6 to 10 dB) owing to the large input coupling loss in the range of 2 dB (from single-mode fiber to a semiconductor waveguide) and the residual SOA facet-reflectivities-induced MRIN noise. In the following, we discuss SOA linearity performance and the techniques to improve it.

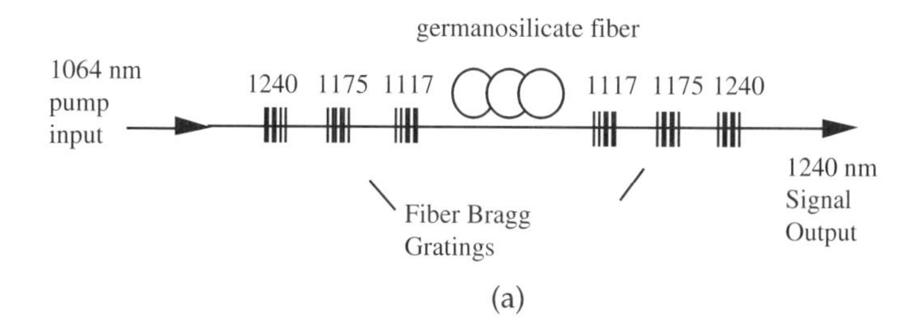

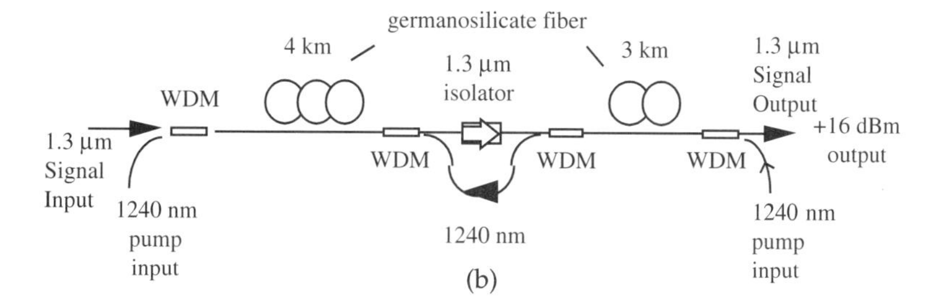

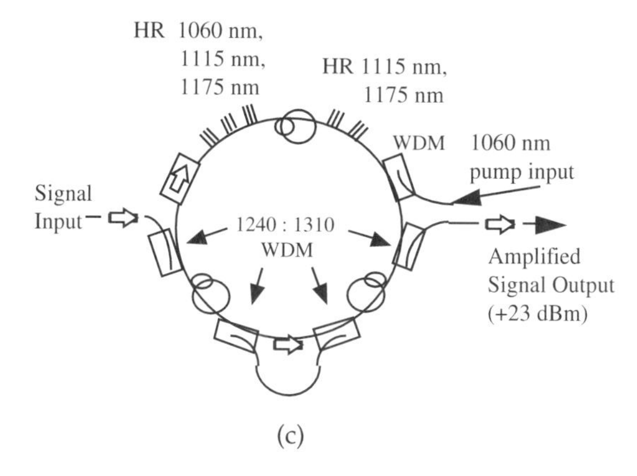

Figure 6.35: Schematic illustrations of (a) a third-order (1240-nm) cascaded Raman laser (after Ref. [90]); (b) a two-stage Raman amplifier with a counter-propagating pump (after Ref. [93]); (c) a multistage Raman amplifier with an intracavity, counter-propagating pump. After Ref. [94].

Since gain response time of the population inversion (~300 ps) is faster than signal time variations (1 to 20 ns for CATV signals), if we consider an amplifier gain (G) versus input signal power (P_{in}) curve, a small variation in P_{in} can cause a corresponding G variation, that is, the population inversion is modulated by the signal. In this case, severe signal distortions occur because, for example, low levels are more amplified than high levels.

Electrical and optical SOA linearization techniques have been proposed. The former includes feedforward [95] and bias-current feedback [96] techniques. The latter includes gain-clamping [97] and external-light-injection [98] techniques. SCM system experiments were carried out on the latter two optical linearization techniques and have attracted significant interest.

The basic concept of gain-clamping by laser oscillation in an SOA has been successfully demonstrated in various laboratories for both digital and analog optical fiber systems. The principle consists of pinning the population inversion by forcing the SOA to oscillate at a wavelength located far away from the useful spectral gain window, as shown in Fig. 6.36. This lasing is enforced by introducing a wavelength-selective feedback, such as from a distributed Bragg reflector (as wavelength-selective mirrors) built into the semiconductor optical amplifier. The optical amplifier gain is clamped when the carrier concentration is fixed by the laser oscillation. The reason why the lasing wavelength must be far away from the SOA gain peak is to avoid large unsaturated gain reduction. Note, however, that when the input signal power increases beyond a certain level, the amplified signal light depletes the carrier concentration to such an extent that the laser oscillation cannot be sustained, and the gain drops steeply; nonlinear distortions therefore increase sharply.

The basic concept of external light injection is to increase the SOA stimulated emission rate, which in turn results in decreased carrier lifetime, and consequently increase the corresponding SOA saturation output power [99,100]. The wavelength of the pumping light must also be at the edge of the SOA gain spectrum to avoid large unsaturated gain reduction, similar to the case of gain clamping.

It should be noted that, despite all the effort devoted to SOA linearizations, SOA linearity performance still cannot meet the standard 60- or 80-channel CATV system requirement, but it can easily meet the linearity requirement for multichannel M-QAM signal transmissions [98]. Therefore, in the near future when all AM-

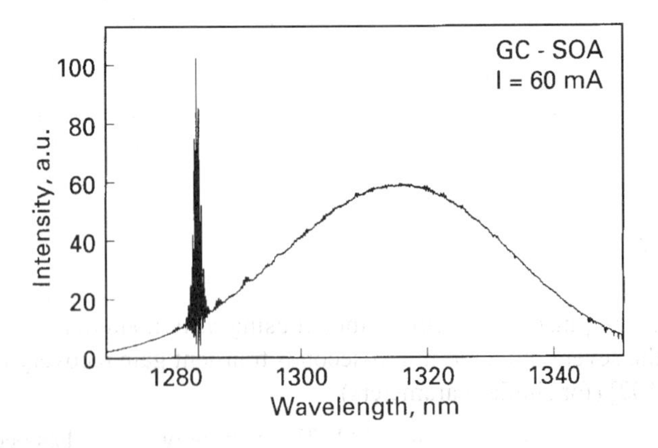

Figure 6.36: Amplified spontaneous emission spectrum of a gain-clamped MQW SOA measured at $I = 60$ mA, that is, below the lasing threshold. After Ref. [97] (© 1996 IEEE).

VSB analog signals are replaced by digital M-QAM signals, SOAs can potentially be used to help upgrade fiber penetration for those 1.3-μm optical CATV transmitters in use today.

Problems

6.1 From Eqs. (6.37) to (6.39), we can see that if optical amplifiers are installed in an SCM system, the resultant CNR per channel depends critically on the input optical power (P_{in}) to the amplifiers. Suppose we have a multichannel AM-CATV system with OMI/channel = 0.035 and the laser wavelength is 1550 nm. The noise figure and input coupling loss of each EDFA are 5 and 0.5 dB, respectively. The received optical power is 0 dBm. Answer the following questions:

(a) For a single-stage EDFA, what is the minimum P_{in} that must be used to meet the requirement of CNR = 51 dB?

(b) For three cascaded stages of EDFAs, what must be the minimum P_{in} to each EDFA? Here we assume that the EDFA inter-stage loss is equal to the amplifier gain.

(c) For a 2.5 Gb/s non-return-to-zero (NRZ) signal, the signal-spontaneous and spontaneous-spontaneous limited SNR is given by $\frac{I_1^2}{2I_1 I_{ase} B_e/B_o}$ and $\frac{I_1^2}{I_{ase}^2 \cdot B_e/B_o}$, respectively, where $I_1 = 2q\eta P_s L_f/h\nu$, P_s is the average signal power measured at the output of an EDFA, and all other terms have been defined in section 6.1.6.1. Assume the minimum SNR for an NRZ signal to achieve a bit-error-rate of 10^{-9} is 16 dB, and $B_e = 0.7 \times (2.5 \times 10^9)$ Hz, calculate the minimum required P_{in} for single-stage and three-cascaded-stage EDFAs. Assume the digital receiver sensitivity at BER = 10^{-9} is given by –35 dBm.

(d) From the results obtained from (a)-(c), what kind of system problems do you think we may encounter if we want to use WDM and EDFAs to transport both multichannel AM and 2.5 Gb/s band NRZ signals?

6.2 The relatively long fluorescence lifetime (~10 ms) of an EDFA prevents a direct modulation of a pump laser at high bit rates. However, it is beneficial to modulate the pump laser with a low bit rate data to transport some supervisory telemetry or data signals. Based on the rate equations given in Eqs. (6.5a) and (6.5b), find a guideline to estimate the maximum useful modulating data rates for a pump laser.

6.3 First use Ref. [101] to derive the equivalent circuit model of an EDFA. Then send a square pulse to this circuit model using SPICE simulation. See if you can see the several hundred micro-seconds transient gain recovery time given in Ref. [102] (for similar parameters).

6.4 The EDFA model presented in Ref. [103] is very popular and has been used in a few commercial EDFA products to predict amplifier performances. Study that paper and see if you can repeat the results given in Fig. 8 therein.

References

[1] J. M. Delavaux, C. R. Giles, S. W. Granlund, C. D. Chen, "Repeated bi-directional 10-Gbit/s 240-km fiber transmission experiment Tech. Digest, Conf. Optical Fiber Commun., paper TuD5, 1996.

[2] D. J. DiGiovanni, S. Plains, and C. R. Giles, "Multi-stage optical amplifier," U.S. Patent 5,115,338, May 1992.

[3] M. C. Sfarries, C. M. Ragdale, and D. C. J. Reid, "Broadband chirped fibre Bragg filters for pump rejection and recycling in erbium-doped fibre amplifiers," Electron. Lett., vol.28, pp.487–489, February 1992.

[4] T. Kimura, T. Aikiyo, and Y. Shirasaka, "The 980 nm pump laser module with built-in isolator," Optical Amplifiers and Their Applications, Trends in Optics and Photonics (TOPS), volume 5, pp.164–167, 1996.

[5] B. F. Ventrudo, G. A. Rogers, G. S. Lick, D. Hargreaves, and T. N. Demayo, "Wavelength and intensity stabilisation of 980-nm diode lasers coupled to fiber Bragg gratings," Electron. Lett., vol.30, pp.2147–2149, December 1994.

[6] D. Payne and R. I. Laming, "Optical Fiber Amplifiers," OFC'90 Tutorial, pp.331–354, January 1990.

[7] E. Desurvire, *Erbium-doped Fiber Amplifiers*, Wiley, New York, 1994.

[8] A. Mori, Y. Ohishi, M. Yamada, H. Ono, Y. Nishida, K. Oikawa, and S. Sudo, "1.5 μm broadband amplification by tellurite-based EDFAs," Post-deadline paper PD-1, Optical Fiber Communication Conference, Dallas, Texas, 1997.

[9] E. Desurvire and J. R. Simpson, "Amplification of spontaneous emission in erbium-doped single-mode fibers," J. Lightwave Technol., vol.7, pp.835–845, May 1989.

[10] M. G. Seats, P. A. Krug, G. R. Atkins, S. C. Guy, and S. B. Poole, "Non-linear excited state absorption in Er^{3+} -doped fibre with high power 980-nm pumping," Proc. Topical Meeting Optical Amplifiers Appl. paper WD2, p.48, 1991.

[11] P. Blixt, J. Nilsson, J. Babonas, and B. Jaskorzynska, "Excited-state absorption at 1.5 μm in Er^{3+} -doped fiber amplifiers," Proc. Topical Meeting Optical Amplifiers Appl., paper WE2, p.63, 1992.

[12] S. G. Grubb, "High power fiber amplifiers and lasers," OFC'96 Tutorial, pp.241–260, February 1996.

[13] Y. Ohishi, T. Kanamori, M. Shimizu, M. Yamada, Y. Terunuma, J. Temmyo, M. Wada, and S. Sudo, "Praseodymium-doped fiber amplifiers at 1.3 μm," IEICE Trans. Commun., vol.E77-B, pp.421–440, 1994.

[14] E. Ishikawa, H. Yanagita, K. Itoh, and H. Aoki, "Amplification at 1.3 μm in Pr^{3+}-doped InF_3/GaF_3-based fibers," OSA TOPS Optical Amplifiers and Their Applications, Vol.5, pp.116–118, 1996.

[15] E. Desurvire, J. R. Simpson, and P. C. Becker, "High-gain erbium-doped traveling-wave fiber amplifier," Optics Lett., vol.12, pp.888–890, November 1986.

[16] E. Desurvire and J. R. Simpson, "Amplification of spontaneous emission in erbium-doped single mode fibers," IEEE J. Lightwave Technol., vol.5, p.835, May 1989.

[17] C. R. Giles and E. Desurvire, "Propagation of signal and noise in concatenated erbium-doped fiber optical amplifiers, "IEEE J. Lightwave Technol., Vol.9, No.2. pp.147–154, February 1991.

[18] E. Desurvire, "Spectral noise figure of Er^{3+}-doped fiber amplifiers," IEEE Photon. Technol. Lett., vol.2, no.3, pp.208–210, March 1990.

[19] J. L. Zyskind, E. Desurvire, J. W. Sulhoff, and D. DiGiovanni, "Determination of homogeneous linewidth by spectral gain hole-burning in an erbium-doped fiber amplifier with GeO2-SiO2 core," IEEE Photon. Technol. Lett., vol.2, pp.869–871, December 1990.

[20] E. Desurvire, J. L. Zyskind, and J. R. Simpson, "Spectral gain hole-burning at 1.53 mm in erbium-doped fiber amplifiers," IEEE Photon. Technol. Lett., vol.2, pp.246–248, April 1990.

[21] H. Kamei, M. Yoshimura, H. Kobayashi, N. Tatoh, and H. Hayashi, "High-power operation of 1.48 μm GaInAsP/GaInAsP strained-layer multiple quantum well lasers," in Proc. Topical Meeting Optical Amplifiers Appl., paper WC3, p.30, 1991.

[22] H. Asano, S. Takano, M. Kawaradani, M. Kitamura, and I. Mito, "1.48 μm high-power InGaAs/InGaAsP MQW LD's for Er-doped fiber amplifiers," IEEE Photon. Technol. Lett., vol.3, pp.415–417, May 1991.

[23] H. A. Zarem, J. Paslaski, M. Mittlestein, J. Ungar, and I. Ury, "High-power fiber coupled strained layer InGaAs lasers emitting at 980-nm," Proc. Topical Meeting Optical Amplifers Appl., postdeadline paper PdP4, 1991.

[24] X. F. Shum, R. Parke, G. Harnagel, R. Lang, and D. Welch, "1.2 W single-mode-fiber-coupled MOPA at 980 nm," Tech. Digest, Conf. Optical Fiber Commun., postdeadline paper PD13, 1995.

[25] J. A. Sharps, "Reliability of hermetically packaged 980 nm diode lasers," Proc. IEEE Laser Electro-Optics Soc. (LEOS) Annu. Meeting, paper DL1.1, 1994.

[26] M. Shimizu, M. Yamada, H. Horiguchi, T. Takeshita, and M. Okayasu, "Erbium-doped fiber amplifiers with an extremely high gain coefficient of 11.0 dB/mW," Electron. Lett., vol.26, pp.1641–1643, 1990.

[27] T. Kashiwada, M. Shigematsu, T. Kougo, H. Kanamori, and M. Nishimura, "Erbium-doped fiber amplifier pumped at 1.48 μm with extremely high efficiency," IEEE Photon. Technol. Lett., vol.3, pp.1991–1993, August 1991.

[28] B. Pedersen, M. L. Dakss, B. A. Thompson, W. J. Miniscalco, T. Wei, and L. J. Andrews, "Experimental and theoretical analysis of efficient erbium-doped fiber power amplifiers," IEEE Photon. Technol. Lett., vol.3, pp.1085–1087, December 1991.

[29] W. I. Way, A. C. Von Lehman, M. J. Andrejco, M. A. Saifi, and C. Lin, "Noise figure of a gain-saturated erbium-doped fiber amplifier pumped at 980 nm," Tech. Digest, Optical Amplifiers and Their Applications, pp.134–137, 1990.

[30] M. Yamada, M. Shimizu, M. Okayasu, T. Takeshita, M. Horiguchi, Y. Tachikawa, and E. Sugita, "Noise characteristics of Er^{3+}-doped fiber amplifiers pumped by 0.98 and 1.48 μm laser diodes," IEEE Photon. Technol. Lett., vol.2, pp.205–207, 1990.

[31] J. F. Massicott, R. Wyatt, B. J. Ainslie, and S. P. Craig-Ryan, "Efficient, high power, high gain, Er^{3+} doped silica fibre amplifiers," Electron. Lett., vol.26, p.1038, 1990.

[32] R. I. Laming, J. E. Townsend, D. N. Payne, F. Meli, G. Grasso, and E. J. Tarbox, "High-power erbium-doped fiber amplifiers operating in the saturated regime," IEEE Photon. Technol. Lett., vol.3, p.253, 1991.

[33] E. Desurvire, "Analysis of gain difference between forward- and backward-pumped erbium-doped fiber amplifiers in the saturation regime," IEEE Photon. Technol. Lett., vol.4, pp.711–713, 1992.

[34] J. L. Zyskind, C. R. Giles, E. Desurvire, and J. R. Simpson, "Optimum pump wavelength in the $^4I_{15/2}$–$^4I_{13/2}$ absorption band for efficient Er^{3+}-doped fiber amplifiers," IEEE Photon. Technol. Lett., vol.1, p.428, December 1989.

[35] Y. Kimura, K. Suzuki, and M. Nakazawa, "Pump wavelength dependence of the gain factor in 1.48 μm pumped Er^{3+}-doped fiber amplifiers," Appl. Phys. Lett., vol.56, no.17, p.1611, 1990.

[36] K. Suzuki, Y. Kimura, and M. Nakazawa, "Pumping wavelength dependence on gain factor of 0.98 μm pumped Er^{3+} fiber amplifiers," Appl. Phys. Lett., vol.55, no.25, p.2573, 1989.

[37] P. C. Becker, A. Lidgard, J. R. Simpson, and N. A. Olsson, "Erbium-doped fiber amplifier pumped in the 950–1000 nm region," IEEE Photon. Technol. Lett., vol.2, p.35, January 1990.

[38] B. Pedeersen, J. Chirravuuri, and W. J. Miniscalco, "Gain and noise penalty for detuned 980-nm pumping of erbium-doped fiber power amplifiers," IEEE Photon. Technol. Lett., vol.4, pp.351–353, April 1992.

[39] R. G. Smart, J. L. Zyskind, J. W. Sulhoff, and D. J. DiGiovanni, "Dependence of performance of saturated in-line erbium-doped fiber amplifiers on pump wavelength around 1480 nm," IEEE Photon. Technol. Lett., vol.5, pp.523–525, May 1993.

[40] R. G. Smart, J. L. Zyskind, and D. J. DiGiovanni, "Experimental comparison of 980 nm and 1480 nm-pumped saturated in-line erbium-doped fiber amplifiers suitable for long-haul soliton transmission systems," IEEE Photon. Technol. Lett., vol.5, pp.770–773, July 1993.

[41] C. R. Giles, T. Erdigan, and V. Mizhahi, "Reflection-induced changes in the optical spectra of 980-nm QW lasers," IEEE Photon. Technol. Lett., vol.6, pp.903–906, August 1994.

[42] C. R. Giles, T. Erdigan, and V. Mizhahi, "Simultaneous wavelength-stabilization of 980-nm pump lasers," IEEE Photon. Technol. Lett., vol.6, pp.907–909, August 1994.

[43] R. W. Tkach, and A. R. Chraplyvy, "Regimes of feedback effects in 1.5 μm distributed feedback laser," J. Lightwave Technol., vol.4, pp.1655–1661, 1986.

[44] D. Hargreaves, G. S. Lick, and B. F. Ventrudo, "High-power 980 nm pump module operating without a thermoelectric cooler," Tech. Digest, Conf. on Optical Fiber Commun., paper ThG3, 1996.

[45] S. G. Grubb, W. H. Humer, R. S. Cannon, S. W. Vendetta, K. L. Sweeney, P. A. Leilabady, M. R. Keur, J. G. Kwasegroch, T. C. Munks, and D. W. Anthon, "+24.6 dBm output power Er/Yb codoped optical amplifier pumped by diode-pumped Nd:YLF laser," Electron. Lett., vol.28, pp.1275–1276, June 1992.

[46] J. D. Minelli, R. I. Laming, J. E. Townsend, W. L. Barnes, E. R. Taylor, K. P. Jedrezejewski, and D. N. Payne, "High gain power amplifier tandem-pumped by a 3 W multistripe diode," Tech. Digest, Conf. Optical Fiber Commun., paper TuG2, 1992.

[47] H. Zellmer, A. Willamowski, A. Tunnermann, H. Welling, S. Unger, V. Reichel, H. Muller, J. Kirchhof, and P. Albers, "High-power cw neodymium-doped fiber laser operating at 9.2 W with high beam quality," Optics Lett., vol.20, pp.578–580, 1995. Also, H. Zellmer, A. Tunnermann, H. Welling, and V. Reichel, "Double-clad fiber laser with 30 W output power," OSA TOPS, vol. 16, Optical Amplifiers and Their Applications, pp. 137–140, 1997.

[48] J. L. Archambault and S. G. Grubb, "Fiber gratings in lasers and amplifiers," J. Lightwave Technol., vol.15, pp.1378–1390, August 1997.

[49] K. Bertilsson, P. A. Andrekson, and B. E. Olsson, "Noise figure of erbium doped fiber amplifiers in the saturated regime," IEEE Photon. Technol. Lett., vol.6, pp.199–201, February 1994.

[50] J. Aspell, J. F. Federici, B. M. Nyman, D. L. Wilson, and D. S. Shenk, "Accurate noise figure measurements of erbium-doped fiber amplifiers in saturation conditions," Tech. Digest, Conf. Optical Fiber Commun., paper ThA4, 1992.

[51] J. F. Marcerou, H. Fevrier, J. Hervo, and J. Auge, "Noise characteristics of the EDFA in gain saturation regimes," Proc. OSA Topical Meeting Optical Amplifiers and Their Applications, paper ThE1–1, 1991.

[52] R. G. Smith, J. L. Zyskind, J. W. Sulhoff, and D. J. DiGiovanni, "An investigation of the noise figure and conversion efficiency of 0.98 μm pumped erbium-doped fiber amplifiers under saturated conditions," IEEE Photon. Technol. Lett., vol.4, pp.1261–1264, November 1992.

[53] E. Yoneda, K. Suto, K. Kikushima, and H. Yoshinaga, "All-fiber video distribution (AFVD) systems using SCM and EDFA techniques," IEEE J. Lightwave Technol., vol.11, pp.128–137, 1993.

[54] P. Bousselet, R. Meilleur, A. Coquelin, P. Garabedian, and J. L. Beylat, "+25.2-dBm output power from an Er-doped fiber amplifier with 1.48-μm SMQW laser-diode modules," Tech. Digest, Conf. Optical Fiber Commun., paper TuJ2, 1995.

[55] S. Sanders, F. Shum, R. J. Lang, J. D. Ralston, D. G. Mehuys, R. G. Waarts, and D. F. Welch, "Fiber-coupled M-MOPA laser diode pumping a high-power erbium-doped fiber amplifier," Tech. Digest, Conf. Optical Fiber Commun., paper TuG5, 1996.

[56] J. E. Townsend, W. L. Barnes, K. P. Jedrzejewski, and S. G. Grubb, "Yb^{3+} sensitised Er^{3+}-doped silica optical fiber with ultrahigh transfer efficiency and gain," Electron. Lett., vol.27, pp.1958–1959, October 1991.

[57] S. G. Grubb, W. F. Humer, R. S. Cannon, T. H. Windhorn, S. W. Vendetta, K. L. Sweeney, P. A. Lelabady, W. L. Barnes, K. P. Jedrezejewski, and J. E. Townsend, "+21 dBm erbium power amplifier pumped by a diode-pumped Nd:YAG laser," IEEE Photon. Technol. Lett., vol.4, pp.553–555, June 1992.

[58] S. G. Grubb, P. A. Leilabady, and D. E. Frymyer, "Solid-state laser pumping of 1.5-μm optical amplifiers and sources for lightwave video transmission," J. Lightwave Technol., vol.11, pp.27–33, January 1993.

[59] S. G. Grubb, "High power diode-pumped fiber lasers and amplifiers," Tech. Digest, Conf. Optical Fiber Commun., paper TuJ1, 1995.

[60] S. G. Grubb, D. J. DiGiovanni, J. R. Simpson, W. Y. Cheung, S. Sanders, D. F. Welch, and B. Rockney, "Ultrahigh power diode-pumped 1.5 μm fiber amplifiers," Tech. Digest, Conf. Optical Fiber Commun., paper TuG4, 1996.

[61] A. M. Vensarkar, P. J. Lemaire, J. B. Judkins, V. Bhatia, T. Erdogan, and J. E. Sipe, "Long-period fiber gratings as band-rejection filters," J. Lightwave Technol., vol. 14 pp.58–65, January 1996.

[62] P. F. Wysocki, J. B. Judkins, R. Espondola, M. Andrejco, A. M. Vengsarkar, and K. Walker, "Erbium-doped fiber amplifier flattened beyond 40 nm using long-period gratings," Tech. Digest, Conf. Optical Fiber Commun., postdeadline paper PD2, 1997.

[63] H. Masuda, S. Kawai, K. -I. Suzuki, and K. Aida, "Wideband, gain-flattened, erbium-doped fiber amplifiers 3 dB bandwidth of 50 nm," Electron. Lett., vol.33, pp.1070–1071, 1997.

[64] H. S. Kim, S. H. Yun, H. K. Kim, N. Park, and B. Y. Kim, "Dynamic gain equalization of erbium-doped fiber amplifier with all-fiber acousto-optic tunable filters," Tech. digest, Conf. Optical Fiber Commun., paper WG4, 1998.

[65] A. Mori, Y. Ohishi, M. Yamada, H. Ono, and S. Sudo, "Broadband amplification characteristics of tellurite-based EDFAs," Proc. of European Conference on Optical Commun., paper We2C.4, 1997.

[66] S. Artuigarud, M. Chbat, P. Noushi, F. Chiquet, D. Bayart, L. Hamon, A. Pitel, F. Goudeseune, P. Bousselet, and J. L. Beylat, "Transmission of 16 × 10 Gbit/s channels spacing 24 nm over 531 km of conventional single-mode fiber using 7 in-line fluoride-based EDFAs," Tech. Digest, Conf. Optical Fiber Commun., postdeadline paper PD27, 1996.

[67] M. Yamada, Y. Ohishi, T. Kanamori, H. Ono, S. Sudo, and M. Shimizu, "Low-noise and gain-flattened fluoride-based Er^{3+}-doped fibre amplifier pumped by 0.97 μm laser diode," Electron. Lett., vol.33, pp.809–810, April 1997.

[68] J. F. Massicott, J. R. Armitage, R. Wyatt, B. J. Ainslie, S. P. Craig-Ryan, "High gain, broadband, 1.6 μm Er^{3+}-doped silica fibre amplifier," Electron. Lett., vol.26, pp.1645–1646, 1990.

[69] M. Yamada, H. Ono, T. Kanamori, S. Sudo, and Y. Ohishi, "Broadband and gain-flattened amplifier composed of a 1.55 μm-band and a 1.58-μm-band Er^{3+}-doped fibre amplifier in a parallel configuration," Electron. Lett., vol.33, pp.710–711, April 1997.

[70] Y. Sun, J. W. Shlhoff, A. K. Srivastava, J. L. Zyskind, C. Wolf, T. A. Strasser, J. R. Pedrazzani, J. B. Judkins, R. P. Esopindola, and J. Zhou, "Ultra wide band erbium-doped silica fiber amplifier with 80 nm of bandwidth," Tech. Digest, Optical Amplifiers and Their Applications, postdeadline paper PD2, 1997.

[71] H. Masuda, S. Kawai, K.-I. Suzuki, and K. Aida, "Ultrawide 75-nm 3-dB gain-band optical amplification with erbium-doped fluoride fiber amplifiers and distributed Raman amplifiers," IEEE Photon. Technol. Lett., vol. 10, pp. 516–518, April 1998.

[72] S. L. Hansen, K. Dybdal, and C. C. Larsen, "Gain limit in erbium-doped fiber amplifiers because of internal Rayleigh backscattering," IEEE Photon. Technol. Lett., vol.4, pp.559–561, June 1992.

[73] M. N. Zervas and R. I. Laming, "Rayleigh scattering effect on the gain efficiency and noise of erbium-doped fiber amplifiers," IEEE J. Quantum Electron., vol. 31, pp.468–471, March 1995.

[74] S. Shikii, A. Sasaki, T. Sakai, and A. Wada, "Influence of Rayleigh backscattering on noise characteristics of Er-doped optical fiber amplifier for analog video

transmission systems," OSA TOPS on Optical Amplifiers and Their Applications, vol.5, pp.152–155, 1996.

[75] F. W. Willems, J. C. van der Plaats, and D. J. DiGiovanni, "EDFA noise figure degradation caused by amplified signal double Rayleigh scattering in erbium-doped fibres," Electron. Lett., vol.30, pp.645–646, 1994.

[76] T. J. Whitley, "A review of recent system demonstrations incorporating 1.3 μm Praseodymium-doped fluoride fiber amplifiers," J. Lightwave Technol., vol.13, no.5, pp.744–759, May 1995.

[77] M. Potenza, B. Sordo, and T. Tambosso, "Second window optical amplifiers: comparison of different pumping wavelengths and fibers," Opt. Fiber Technol., vol.1, no.4, pp.322–326, 1995.

[78] M. Yamada, M. Shimizu, Y. Ohishi, T. Kanamori, S. Sudo, and J. Temmyo, "One-LD-pumped Pr^{3+}-doped fluoride fibre amplifier module with signal gain of 23 dB," Electron. Lett., vol.29, pp.1950–1951, October 1993.

[79] M. Yamada, M. Shimizu, H. Yoshinaga, K. Kikushima, T. Kanamori, Y. Ohishi, Y. Terunuma, K. Oikawa, and S. Sudo, "Low-noise Pr^{3+}-doped fluoride fiber amplifier," Electron. Lett., vol.31, pp.806–807, May 1995.

[80] S. Sanders, K. Dzurko, R. Parke, S. O'Brien, D. F. Welch, S. G. Grubb, G. Nykolak, and P.C. Becker, "Praseodymium doped fiber amplifiers (PDFAs) pumped by monolithic master oscillator power amplifier (M-MOPA) laser diodes," Electron. Lett., vol.32, pp.343–344, February 1996.

[81] M. Karasek, "Analysis of gain dynamics in Pr^{3+}-doped fluoride fiber amplifiers," IEEE Photon. Technol. Lett., vol.7, pp.299–301, March 1995.

[82] D. W. Hewak, J. A. Medeiros Neto, B. Samson, R. S. Brown, K. P. Jedrzejewski, J. Wang, E. Taylor, R. I. Laming, G. Wylangowski, and D. N. Payne, "Quantum-efficiency of praseodymium doped Ga:La:S glass for 1.3 μm optical fibre amplifiers," IEEE Photon. Technol. Lett., vol.6, pp.609–612, May 1994.

[83] J. A. M. Neto, E. R. Taylor, B. N. Samson, J. Wang, D. W. Hewak, R. I. Laming, D. N. Payne, E. Tarbos, P. D. Maton, G. M. Roba, B. E. Kinsman, and R. Hanney, "The application of Ga:La:S-based glass for optical amplification at 1.3 μm," J. Non-Cryst. Sol., vol.184, pp.292–296, 1995.

[84] J. Temmyo, M. Sugo, T. Nishiya, T. Tamamura, M. Yamada, S. Sudo, F. Yamamoto, F. Bilodeau, and K. O. Hill, "Low noise characteristics of an AM-VSB video signal transmission using a PDFA pumped by InGaAs semiconductor lasers through fibre Bragg gratings," Electron. Lett., vol.32, pp.1910–1911, September 1996.

[85] H. Hanafusa, M. Horiguchi, and J. Noda, "Thermally-diffused expanded core fibre for low-cost and inexpensive photonic components," Electron. Lett., vol.27, pp.1968–1969, 1991.

[86] Y. Ohishi, M. Yamada, T. Kanamori, Y. Terunuma, and S. Sudo, "Analysis of gain characteristics of bi-directionally pumped praseodymium-doped fiber amplifiers," IEEE Photon. Technol. Lett., vol.8, pp.512–514, April 1996.

[87] M. Yamada, M. Shimizu, T. Kanamori, Y. Ohishi, Y. Terunuma, K. Oikawa, H. Yoshinaga, K. Kikushima, Y. Miyamoto, and S. Sudo, "Low-noise and high-power Pr^{3+}-doped fluoride fiber amplifier," IEEE Photon. Technol. Lett., vol.7, pp.869-871, August 1995.

[88] K. Kikushima, M. Yamada, M. Shimizu, and J. Temmyo, "Distortion and noise properties of a praseodymium-doped fluoride fiber amplifier in 1.3 μm AM-SCM video transmission systems," IEEE Photon. Technol. Lett., vol.6, pp.440–442, March 1994.

[89] E. M. Dianove, A. A. Abramove, M. M. Bubnov, A. M. Prokhorov, A. V. Shipulin, G. G. Devjatykh, A. N. Guryanov, and V. F. Khopin, "30 dB gain Raman amplifier at 1.3 μm in low-loss high GeO2-doped silica fibres," Electron. Lett., vol.31, pp.1057–1058, June 1995.

[90] S. G. Grubb, T. Erdogan, V. Mizrahi, T. Strasser, W. Y. Cheung, W. A. Reed, P. J. Lemaire, A. E. Miller, S. G. Kosinski, G. Nykolak, and P. C. Becker, "1.3 μm cascaded Raman amplifier in germanosilicate fibers," Tech. Digest, Optical Amplifiers and Their Applications, paper PD3, 1993.

[91] G. P. Agrawal, *Nonlinear Fiber Optics*, Academic Press, San Diego, 1995.

[92] L. G. Cohen and C. Lin, "A universal fiber-optic (UFO) measurement system based on a Near-IR fiber Raman laser," IEEE J. Quantum Electron., QE–14, p.855, 1978.

[93] A. J. Stenz, S. G. Grubb, C. E. Headley, J. R. Simpson, T. Strasser, and N. Park, "Raman amplifier with improved system performance," Tech. Digest, Conf. Optical Fiber Commun., paper TuD3, 1996.

[94] A. J. Stenz, T. Nielsen, S. G. Grubb, T. Strasser, and J. R. Pedrazzani, "Raman ring amplifier at 1.3 μm with analog-grade noise performance and an output of 23 dBm," Tech. Digest, Conf. Optical Fiber Commun., postdeadline paper PD16, 1996.

[95] A. A. M. Saleh, R. M. Jopson, and T. E. Darcie, "Compensation of nonlinearity in semiconductor optical amplifiers," Electron. Lett., vol.24, pp.950–952, 1988.

[96] J. A. Constable, I. H. White, A. N. Coles, and D. G. Cunningham, "Reduction of harmonic distortion and noise in a semiconductor optical amplifier using bias current feedback," Electron. Lett., vol.29, pp.2042–2044, 1993.

[97] L. F. Tiemeijer, G. N. van den Hoven, P. J. A. Thijs, T. van Dongen, J. J. M. Binsma, and E. J. Jansen, "1310-nm DBR-type MQW gain-clamped semiconductor optical amplifiers with AM-CATV-grade linearity," IEEE Photon. Technol.

Lett., vol.8, pp.1453–1455, November 1996; also, V. G. Mutalik, G. van den Hoven, and L. Tiemeijer, "Analog performance of 1310-nm gain-clamped semiconductor optical amplifiers," Tech. Digest, Conf. Optical Fiber Commun., paper ThG4, 1997.

[98] C. Tai, S. L. Tzeng, H. C. Chang, and W. I. Way, "Reduction of saturation induced nonlinear distortion in MQW semiconductor optical amplifier using light injection and its application in multi-channel M-QAM signal transmission systems," IEEE Photon. Technol. Lett., vol.10, pp.609–611, April 1998.

[99] M. Yoshino and K. Inoue, "Improvement of saturation output power in a semiconductor laser amplifier through pumping light injection," IEEE Photon. Technol. Lett., vol.8, pp.58–59, 1996.

[100] R. J. Manning and D. A. O. Davies, "Three-wavelength device for all-optical signal processing," Optics Lett., vol.19, pp.889–891, 1994.

[101] A. Bononi, L. A. Rusch, and L. Tancevski, "Simple dynamic model of fibre amplifiers and equivalent electrical circuit," Electron. Lett., vol.33, pp.1887–1888, October 1997.

[102] C. R. Giles, E. Desurvire, and J. R. Simpson, "Transient gain and cross talk in erbium-doped fiber amplifiers," Optics Lett., vol.14, pp.880–882, August 1989.

[103] C. R. Giles and E. Desurvire, "Modeling erbium-doped fiber amplifiers," IEEE J. Lightwave Technol., vol.9, pp.271–283, February 1991.

Chapter 7

$LiNbO_3$ External Modulator-Based CATV Lightwave Transmitter

In HFC systems, direct modulation of a laser diode is not the only transmission technique. As a matter of fact, there are many system problems associated with frequency chirping of a 1.55-μm directly modulated laser diode in combination with the large fiber dispersion (~17 ps/nm/km) of a conventional single-mode fiber with zero dispersion at 1.31 μm (see Chapter 9). These problems can severely limit the transmission link distance. An alternative method which can be used to overcome these problems is to launch a high-power CW light into an external electro-optic $LiNbO_3$ modulator, and apply the RF modulations to the electrodes of the external modulator. Using this approach, and with the help of erbium-doped fiber amplifiers (EDFAs), optical fiber links which transported >79 channels of AM-VSB video channels over a distance of >100 km have been demonstrated [1,2].

Figure 7.1 illustrates two 1.3-μm and one 1.55-μm commercially available lightwave CATV systems. We can see that two of them are external modulation systems. The 1.3-μm external modulation systems are facing direct competition with DFB laser-based direct modulation systems, because the latter can achieve comparable output power ($\geq$15 mW) with lower cost. On the other hand, the 1.55-μm external modulation system can achieve a very high power budget and transmission distance by taking advantage of EDFAs and the low attenuation loss at 1.55 μm (~0.22 dB/km). Consequently, a major external modulator manufacturer has recently announced that it will no longer produce 1.3 μm $LiNbO_3$ modulators.

In a 1.3-μm external-modulation-based transmitter, the high-power CW optical source usually is a diode-pumped Nd:YAG laser at 1319 nm which has extremely narrow linewidth (<200 KHz) [3] and low-intensity noise (< -165 dB/Hz) [4]. On the other hand, the successful development of 1.55-μm external modulation systems

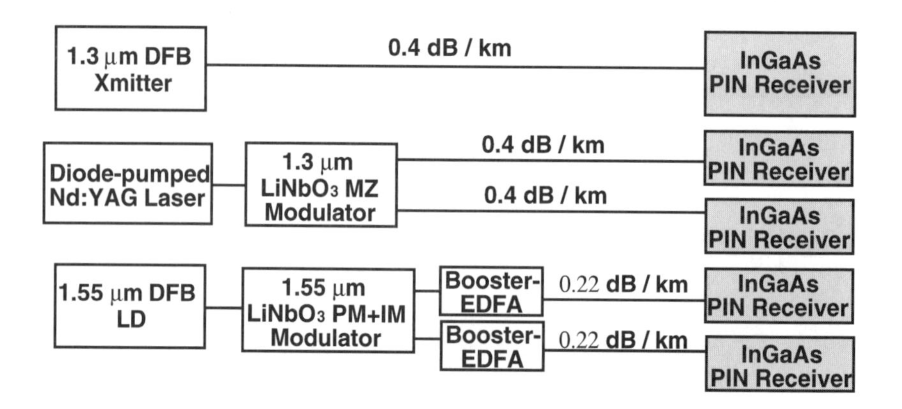

Figure 7.1: Three types of commercially available lightwave CATV links.

is partly due to the emerging high-power (>20 mW) and low-intensity-noise (~ -160 dB/Hz) multi-quantum-well (MQW) DFB lasers [5,6].

In this chapter, we will go over several fundamental operation principles of $LiNbO_3$ external modulators and the related high power optical sources.

7.1 Basic Material and Fabrication Techniques

$LiNbO_3$ is the most commonly used substrate material because of its large optical electro-optic coefficient and the fact that lithium niobate is commercially available in large substrates (3- to 4-inch wafers). The large electro-optic coefficient is important to produce large changes of refractive index for a fixed applied electrical field strength. In addition, the index change necessary to produce waveguiding is achieved by locally doping the crystal, and the resulting guides have low propagation losses (less than a few tenths of a decibel per centimeter) [7].

Optical waveguides are delineated on the $LiNbO_3$ wafer surface using standard photolithography techniques. The two dominant methods for fabricating optical waveguides in $LiNbO_3$ are annealed proton exchange (APE) and titanium-indiffused. Both methods can produce waveguides with comparable properties such as mode size, propagation loss, and fiber–waveguide coupling efficiency (~0.35 to 0.5 dB/facet) [7,8]. There are two main differences between the two fabrication processes. First, APE only increases the extraordinary index, while the ordinary index actually decreases slightly. This means that APE guides will only support light of a single polarization (with >50 dB polarization extinction ratio), which can be advantageous because one could avoid the problem of having different modulation efficiencies for different polarization modes in waveguides. However, it can be disadvantageous because the input light polarization state must be aligned with that the waveguide. Second, APE can produce a rather large refractive index change (5% versus 1% for Ti-indiffusion), and this is attractive for use in devices where one wants guides with small-radius bends.

The Ti-indiffused technique was well developed in the 1980s, whereas APE technique became mature only a few years ago. Accelerated aging and advanced diffusion modeling predict only minute changes in APE waveguide characteristics (including insertion loss, half-wave voltage, Y-branch split ratio, and on/off extinction) over 13 years when they are operated at 125°C [9].

7.2 Basic Operation Principles of $LiNbO_3$ Amplitude Modulators

The geometry of most commonly used external modulators is shown in Fig. 7.2, where we can see the schematic cross sections of embedded strip channel waveguides. A voltage V applied to the electrodes placed alongside or over the waveguide, as shown in Figs. 7.2a and b, respectively, creates an internal electric field of approximate magnitude $|E| = V/G_p$, G_p being the width of the electrode gap. The waveguide is fabricated by either the APE or the Ti-indiffusion technique mentioned earlier. In order to utilize the largest electro-optic coefficient, the electrical field is generally applied along the z-direction (E_z), and the electro-optically induced index change for light polarized along the z-axis is [7]

$$\Delta n = -n^3 r_{33} E_z / 2, \tag{7.1}$$

where r_{33} is the largest electro-optic tensor and n is either the ordinary or extraordinary refractive-index value. We can clearly see that the electrode arrangement in Fig. 7.2 is to utilize E_z. The electrode orientation relative to the waveguide needed to generate E_z depends upon the orientation of the crystal used. The orientation is frequently specified by the "cut"—the direction perpendicular to the flat surface on which the waveguide is fabricated. In Figs. 7.2a and b, we have X-cut and Z-cut $LiNbO_3$, respectively. If we use a correction factor Γ to account for the fact that the electrical and optical field may not overlap completely, the effective electro-optically induced index change can be written as

$$\Delta n \approx -n^3 r_{33} \Gamma (V/(2\, G_p), \tag{7.2}$$

where we have used $E_z \approx V/G_p$. The total phase shift resulting from Δn over the interaction length L can be written as

$$\Delta\beta L = 2\, \pi\, \Delta n\, L / \lambda = -\pi\, n^3 r_{33} \Gamma\, V L / (G_p \lambda). \tag{7.3}$$

For a predetermined phase shift, the voltage–length product (V•L) required for modulation is thus proportional to G_p/Γ. Therefore, for a given electro-optic material and wavelength, one should try to minimize V•L (the voltage-length product) by optimizing the lateral geometric parameter G_p/Γ. However, it should be noted that G_p cannot be too small; otherwise, the electrode capacitance may be too large, which in turn decreases the modulation bandwidth and increases the amount of DC drift (see Sec. 7.4 for more details).

Both the driving voltage and modulation bandwidth scale inversely with the waveguide length. This means that a shorter device is preferred if a high modulation

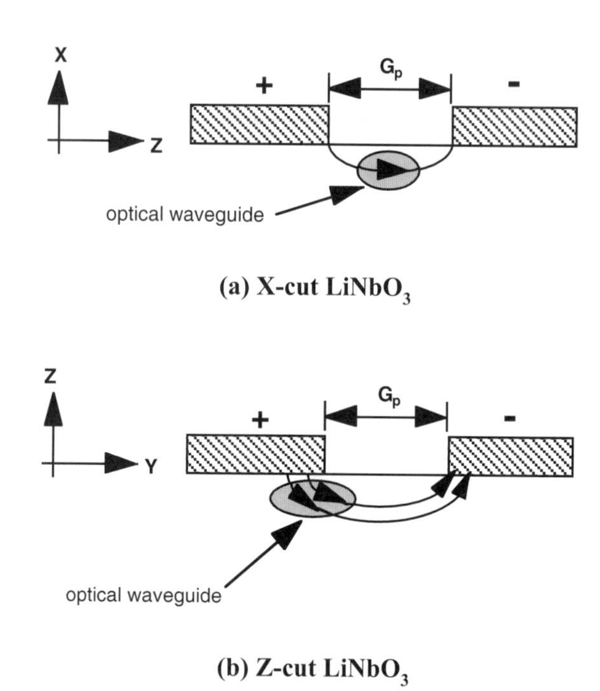

Figure 7.2: Schematic cross-sections of an embedded optical waveguide and the associated electrodes for (a) X-cut Y-propagating (or Y-cut X-propagating) and (b) Z-cut $LiNbO_3$. The electrode orientation relative to the waveguide needed to generate E_z depends upon the orientation of the crystal used. The orientation is frequently specified by the "cut"—the direction perpendicular to the flat surface on which the waveguide is fabricated.

bandwidth is more important, while a longer device is preferred if low driving voltage is more important. Therefore, an appropriate modulator figure of merit is the voltage/bandwidth ratio (V/f_{3-dB}). This figure of merit allows for meaningful comparison of modulators of different lengths and types. Since the required bandwidth in an HFC system is typically lower than 1 GHz, driving voltage reduction is actually a more critical parameter. Currently, typical driving voltage to turn the light through an external modulator from on to off (V_π) is about 3–4 V.

External modulators can be separated into two categories based on the electrode configurations. They are the *lumped* circuit-element modulator and the *traveling-wave* modulator. The basic structure of the lumped circuit-element phase modulator is illustrated in Fig. 7.3a. The modulation is fed from the input transmission line, which terminates at the pair of electrodes in parallel with a matched load. The pair of electrodes acts like a capacitor at the end of the transmission line. The bandwidth of this type of modulator is mainly limited by the *RC* time constant, which is the product of the load resistance and the electrode capacitance. Capacitance *C* can be reduced either by decreasing the electrode length or by increasing the electrode gap.

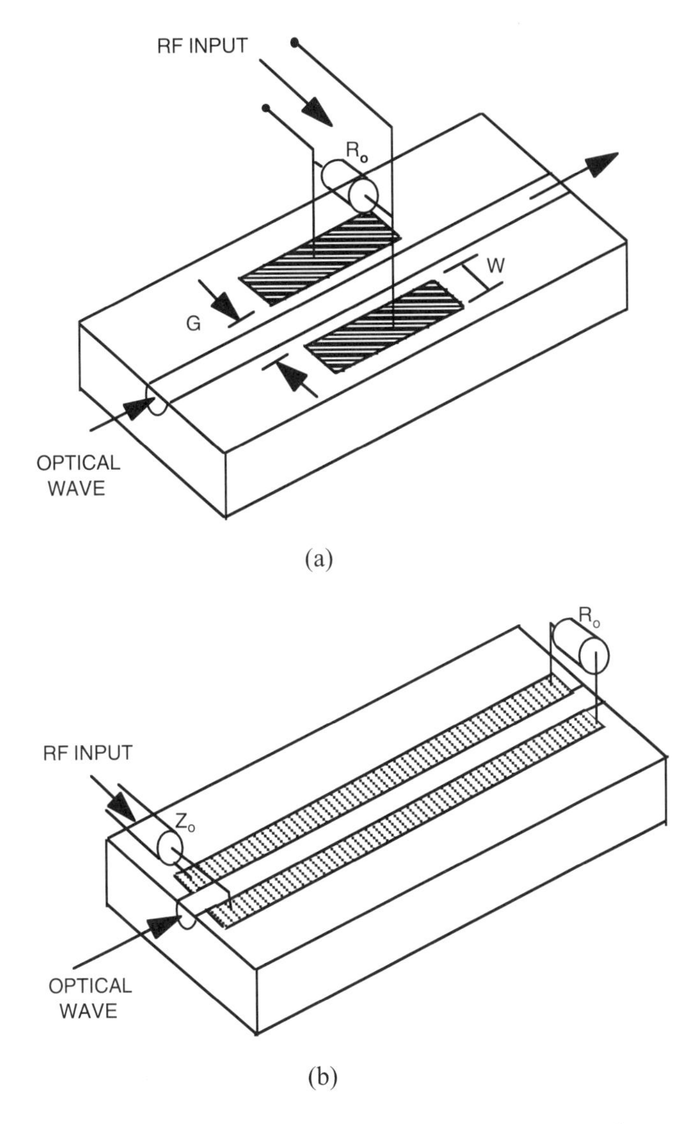

Figure 7.3: Schematic illustration of (a) lumped-element electrodes and (b) traveling-wave electrodes.

However, both these strategies also increase the drive voltage: the first because of the fixed *VL* product, and the second because it results in a weaker electrical field. One possible solution is to match the electrodes to the signal source using matching circuitry, which is possible for bandwidth below about 1 GHz.

The limitation due to the lumped-element electrodes can be significantly alleviated for traveling-wave modulators. The basic structure of a traveling-wave phase modulator is shown in Fig. 7.3b. The electrode (with greatly increased metal thickness) is constructed along the length of the optical guide in such a way that it is a part of a strip transmission line (and acts as a transmission line). The characteristic impedance of this transmission line is designed to match that of the RF system. The end of this electrode is terminated in a matched load. The modulation signal is launched into this transmission line in such a way that the signal travels along with the propagating optical wave inside the guide. The interaction between the propagating optical wave and the traveling RF/microwave signal along the length of the modulator results in efficient phase modulation of the optical wave, and the bandwidth is no longer limited by the *RC* time constant. Thus, the traveling-wave can achieve larger bandwidth with much longer electrodes, providing more efficient modulation. The bandwidth of the traveling-wave modulator is limited by the velocity mismatch between the propagating optical wave and the traveling RF/microwave signal.

So far, only optical phase modulators have been mentioned. In HFC systems, however, mainly amplitude modulators are used. There are several ways to build an amplitude modulator. Three of the most popular are Mach–Zehnder interferometer (MZI) modulators (Fig. 7.4), directional coupler modulators (DCMs) (Fig. 7.5), and Y-fed coupler modulators (modified 1 × 2 directional couplers) (Fig. 7.6). In a Mach–Zehnder interferometer modulator, shown in Fig. 7.4, a single input strip waveguide is split into two arms by a Y-junction 3 dB-splitter. These two arms are then recombined by a second Y-junction 3-dB combiner to form a single output strip guide. Traveling-wave electrode structures are laid down along one or both of the interferometer arms. The modulation signal traveling along these transmission lines induces a phase shift between the two arms. The phase modulated optical wave will then recombine at the output Y-junction either constructively or destructively, resulting in amplitude modulation at the output guide. Analytically, we can have the modulated optical power as

$$\begin{aligned} P_{out} &= (P_{in} L_a/4) \cdot | \exp[j(\omega t - \beta_1 L + \phi)] + \exp[j(\omega t - \beta_2 L)] |^2 \\ &= (P_{in} L_a) \cdot \cos^2 [(\Delta\beta L - \phi)/2], \end{aligned} \tag{7.4}$$

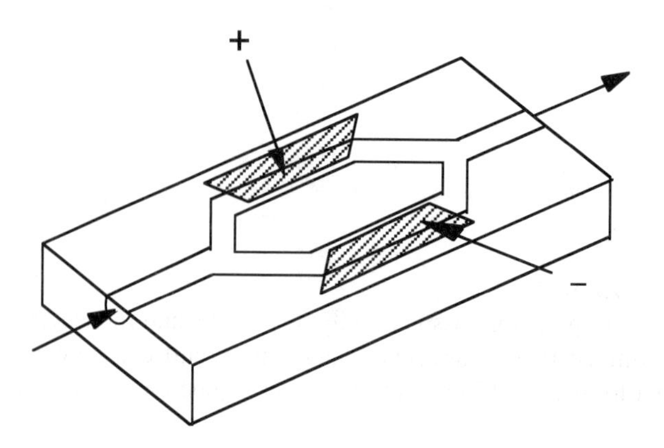

Figure 7.4: Mach–Zehnder interferometer (MZI) modulator.

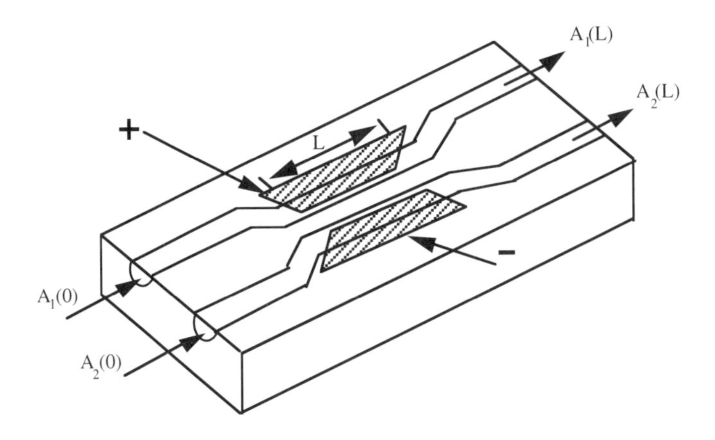

Figure 7.5: Directional coupler modulator (DCM).

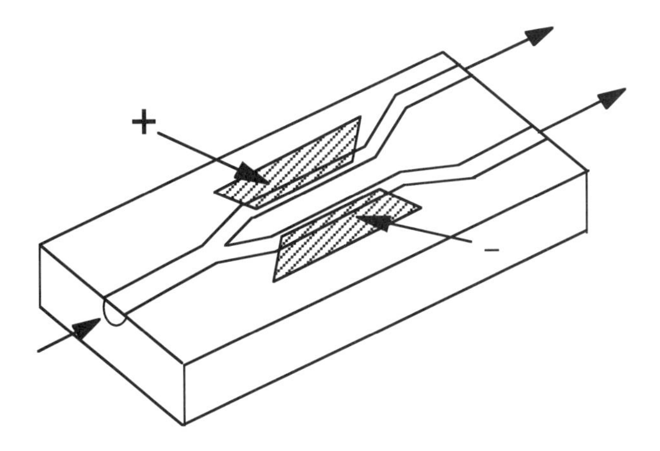

Figure 7.6: Y-fed coupler modulator.

where P_{in} is the average input optical power, L_a is the loss factor of the modulator (to be elaborated in the following paragraph), ω is the optical carrier frequency, ϕ is the static bias phase shift between the two arms, and $\Delta\beta L = (\beta_1 - \beta_2)L$ is the applied voltage-induced phase difference between the two arms.

The loss factors L_a in an external modulator include coupling loss, propagation loss (including scattering and absorption losses), electrode loading loss and, in the case of MZI modulator, the inherent 3-dB waveguide loss at the combiner Y-junction. The propagation loss is usually negligible, and the electrode loading loss can be minimized if the impedance-matching circuit can be carefully designed. Thus, except for the 3-dB loss in the combiner Y-junction of an MZI modulator, the coupling losses dominate the modulator loss factor. Assuming perfect alignment between the fiber and Ti:$LiNbO_3$ waveguide, the two contributions to fiber–waveguide coupling loss are reflection (or Fresnel loss) and the loss caused by mismatch between the fiber and waveguide modes. The former can be greatly reduced by using antireflection coatings on slanted lithium niobate end faces. The loss resulting from mismatch between the fiber and waveguide modes is in general the principal source of fiber–waveguide coupling loss. In order to have an optimum fiber–waveguide coupling between a 1.3–1.55 μm single-mode fiber and the modulator waveguide, a guide with transverse dimensions of about 8 to 9 μm and a waveguide index difference of $(n_s - n_g)/n_s \approx 0.4\%$ (where n_s is the substrate refractive index and n_g is the refractive index of the embedded guiding region) can be used [7]. Typical coupling loss per facet of commercial devices is around 0.5 dB. Note that significant coupling loss may be incurred if there is polarization mismatch between the launched light and the mode that the waveguide can support (as in the case of APE waveguide). Therefore, typically, there is a polarization-maintaining fiber between the light source and the modulator, or the polarized source light can be launched into the waveguide directly without using optical fibers [10]. One last factor, which is often carelessly neglected, is the additional 3-dB loss due to the fact that the bias point must be at the inflection point of the cosine curve given in Eq. (7.4). This bias point is usually chosen for subcarrier multiplexed lightwave systems based on linearity considerations (see Section 7.3). Therefore, the total loss of a typical MZI modulator under normal operation condition (for subcarrier multiplexed lightwave systems) is about $3 + 3 + 0.5 + 0.5 = 7$ dB.

The inherent 3-dB loss through the MZI modulator at the output Y-junction can be briefly explained as shown in Fig. 7.7, assuming light is incident from just one of the two input ports on the left. This light is assumed to be the superposition of two supermodes which are supported by the *composite* guide on the left-hand end. The symmetric supermode gradually converts into the single guide mode with an output power which is 50% of the input power, while the antisymmetric supermode cannot be converted into a symmetric distribution by a symmetrically tapered Y-junction and hence cannot emerge as a guided mode at all. Instead, this component must be converted into antisymmetric *radiation* modes. Excitation from the left-hand side with a single input therefore results in an automatic loss of 50% of the power to radiation. This fact is not immediately apparent and should therefore be taken into consideration carefully [11].

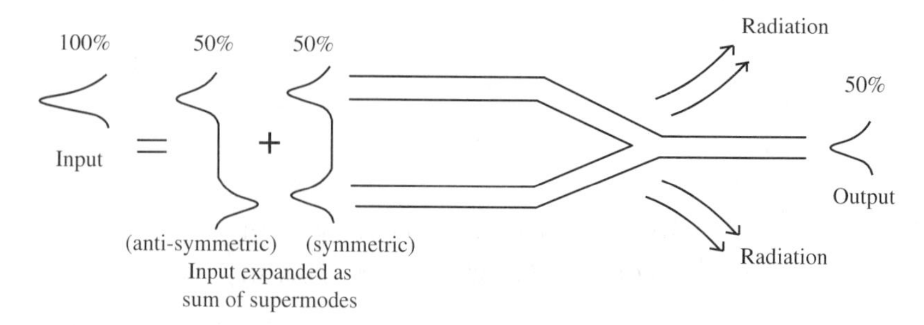

Figure 7.7: Excitation of a Y-junction at the left-hand end. After Ref. [11].

Coming back to Eq. (7.4), we can rewrite the equation as

$$P_{out} = \frac{P_{in}L_a}{2}\left[1+\cos\left(\frac{\pi V}{V_\pi} - \phi\right)\right], \tag{7.5}$$

where we have let $\Delta\beta L = \pi V/V_\pi$. V_π is the voltage required to achieve a π optical phase shift and is equal to the required peak-to-peak driving voltage to achieve 100% OMI. Practically speaking, V_π should be as small as possible. Judging from Eq. (7.3), we can see that a device with a longer length and smaller waveguide separation gives smaller V_π.

In a directional coupler modulator (DCM), as shown in Fig. 7.5, parts of the two optical waveguides are placed in close proximity to each other, resulting in coupling of modes between the two guides. The electric field is applied in such a way that there is a change in propagation constant in one or both of the waveguide channels. This will affect the amount of light coupling between the two guides and result in amplitude modulation at the output. The analysis of mode coupling in a directional coupler can be done by using the coupled-mode equations [11]

$$\begin{aligned} \frac{dA_1}{dz} + j\kappa A_2 \exp(-j\Delta\beta z) = 0 \\ \frac{dA_2}{dz} + j\kappa A_1 \exp(+j\Delta\beta z) = 0, \end{aligned} \tag{7.6}$$

where $A_1(z)$ and $A_2(z)$ are the local amplitudes of the modes in the two waveguides, respectively; κ is the coupling coefficient; and $\Delta\beta$ is the difference of the propagation constants in the two coupled waveguides. $\Delta\beta$ is proportional to the applied voltage (therefore, the electrode structure in Fig. 7.5 is sometimes called a pair of $\Delta\beta$ electrodes). In the case when $\Delta\beta = 0$, let us assume that light with unit modal amplitude is launched into guide 1 only (see Fig. 7.5). Then the expressions for the two outputs of the directional coupler are, from Eq. (7.6), $A_1(L) = \cos\kappa L$ and $A_2(L) = -j\sin\kappa L$, where L is electrode length. Therefore, for an arbitrary electrode length z, the output power at port 1 and port 2 are $P_1 = \cos^2\kappa z$ and $P_2 = \sin^2\kappa z$. Thus, the power distribution between the guides is an oscillatory function of z and the total power $\cos^2\kappa z + \sin^2\kappa z = 1$ is conserved by the device. This condition is the same as that of an optical fiber coupler (Sec. 3.2). We also observe that a complete power transfer occurs at $\kappa z = \pi/2$, $3\pi/2$, $5\pi/2$, etc., from which we define a *coupling length* given by $\ell = \pi/2\kappa$. Similarly, if a light with unit modal amplitude is launched into guide 2 only, the two output light amplitudes are $A_1(L) = -j\sin\kappa L$ and $A_2(L) = \cos\kappa L$. Therefore, when both input ports 1 and 2 have input lights $A_1(0)$ and $A_2(0)$, we can write

$$\begin{pmatrix} A_1(L) \\ A_2(L) \end{pmatrix} = \begin{pmatrix} \cos\kappa L & -j\sin\kappa L \\ -j\sin\kappa L & \cos\kappa L \end{pmatrix}\begin{pmatrix} A_1(0) \\ A_2(0) \end{pmatrix}, \tag{7.7}$$

and the 2×2 matrix is the transfer function matrix of a directional coupler with $\Delta\beta = 0$.

For the case when $\Delta\beta$ is not zero, assume again that light with unit modal amplitude is launched into guide 1 only. The power output from guide 2 can be derived from Eq. (7.6) and is given by [11,12] (see Problem 7.1)

$$P_2 = \frac{\sin^2\left(\kappa L\sqrt{1+(\delta/\kappa)^2}\right)}{1+(\delta/\kappa)^2}, \tag{7.8}$$

where $\delta = \Delta\beta/2$. We note that when $\delta = 0$, as in the previous case, we should have $\kappa L = \pi/2, 3\pi/2, 5\pi/2$, etc., to get the maximum output power. However, to minimize δ, which is proportional to the driving voltage, we choose $\kappa L = \pi/2$. Figure 7.8 shows this $\text{sinc}^2 x$-like function versus $\Delta\beta L$. Comparing this function to the transfer function of an MZI modulator [Eq. (7.4) with $\phi = 0$] shown in the same figure, we can see that the directional coupler requires $\sqrt{3}$ times higher voltage to switch the modulator light from on to off (or vice versa). This can also be obtained from Eq. (7.8) by letting $\kappa L\sqrt{1+(\delta/\kappa)^2} = \pi$ for the first zero to occur which gives $\Delta\beta L = \pi\sqrt{3}$. This driving voltage, $\sqrt{3}$ times higher than that of an MZI modulator, is considered disadvantageous for HFC systems (signal bandwidth < 1 GHz), even though for the same electrode length, a DCM could have a larger modulation bandwidth [8].

Following a similar analytical procedure as for a directional coupler, we can also obtain the modulation transfer characteristics of the Y-fed coupler modulator shown in Fig. 7.6, given by [12,13] (see Problem 7.1), using

$$P_2 = \frac{1}{2}\left[1 - \frac{2xy}{r^2}\sin^2\left(\frac{\pi r}{2}\right)\right], \tag{7.9}$$

where $y = L/\ell$, $x = \Delta\beta L/\pi$, and $r^2 = x^2 + y^2$. When the normalized length $y = 1/\sqrt{2}$ the optical power can be switched between the two outputs with normalized voltages $\Delta\beta L = \pi\sqrt{2}$ (change from $\Delta\beta L = -\pi/\sqrt{2}$ to $\pi/\sqrt{2}$ in Fig. 7.6), which is lower than that of a conventional directional coupler, but higher than that of an MZI modulator. A plot of the calculated P_2 against $\Delta\beta L$ is also shown in Fig. 7.8 so that it can be compared to those of the other two types of modulators. It can be clearly seen that the MZI modulator requires the lowest voltage swing to produce 100% modulation depth of the optical intensity. This is one of the reasons why almost all of the commercially available devices are based on the MZI operation principle.

All three types of amplitude modulators just discussed require two parallel waveguides. Therefore, their electrode configuration is usually designed such that the applied electric field in each waveguide are in opposite directions, as shown in Fig. 7.9. This is the so called "push–pull" electrode placement. This configuration produces a change of Δn in one waveguide and $-\Delta n$ in the other. Consequently, the driving voltage can be reduced by a factor of 2.

A type of commercially available MZI modulator called a balanced bridge interferometer (BBI) modulator, is built by replacing the output combiner Y-junction in an MZI modulator with an optical directional coupler [14]. The schematic diagram

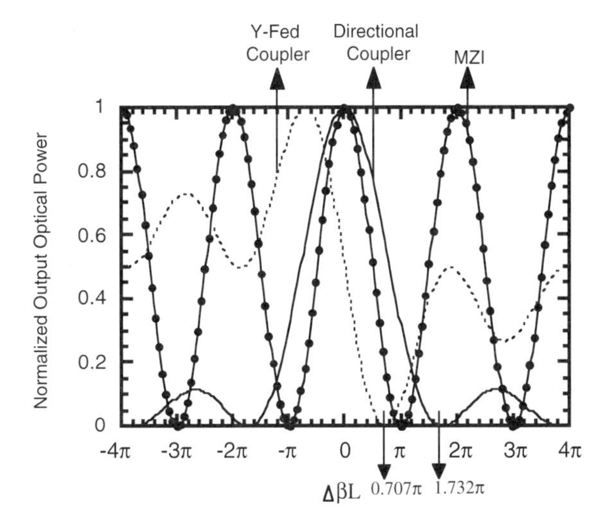

Figure 7.8: Comparison of the output optical power as a function of $\Delta\beta L$ for MZIs, directional couplers, and Y-fed couplers.

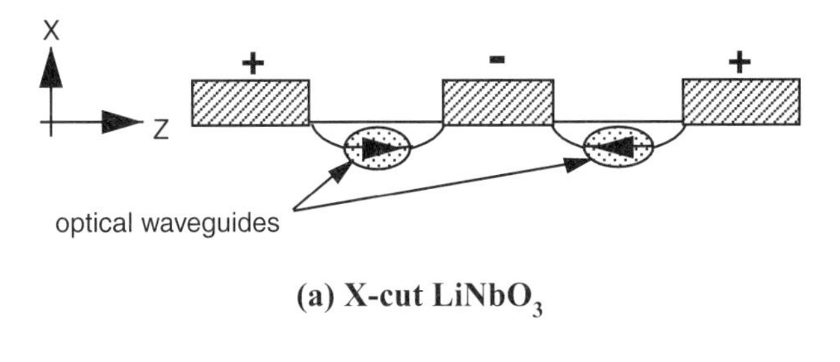

(a) X-cut $LiNbO_3$

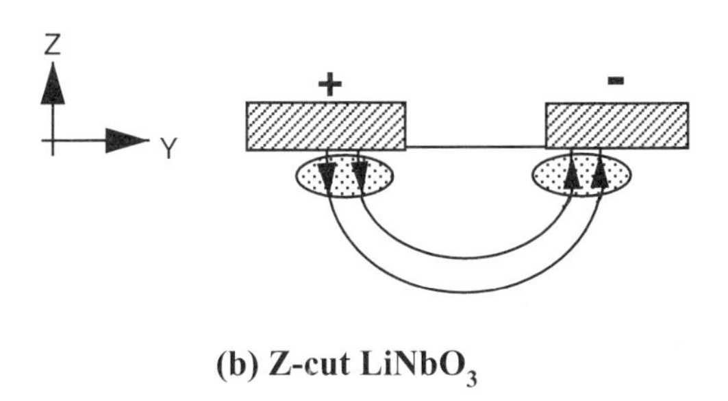

(b) Z-cut $LiNbO_3$

Figure 7.9: Push–pull electrode placement for utilization of r_{33} on $LiNbO_3$.

of a BBI modulator (Fig. 7.10) is shown in Fig. 7.10b [15]. Let the electrical fields in the upper and lower branches after the input Y-splitter and at the output of the two-stage modulator be given by

$$\begin{pmatrix} A_{in,1} \\ A_{in,2} \end{pmatrix} \text{ and } \begin{pmatrix} A_{out,1} \\ A_{out,2} \end{pmatrix},$$

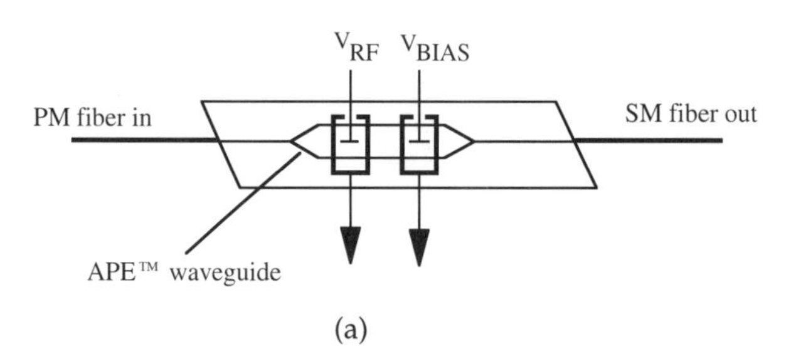

(a)

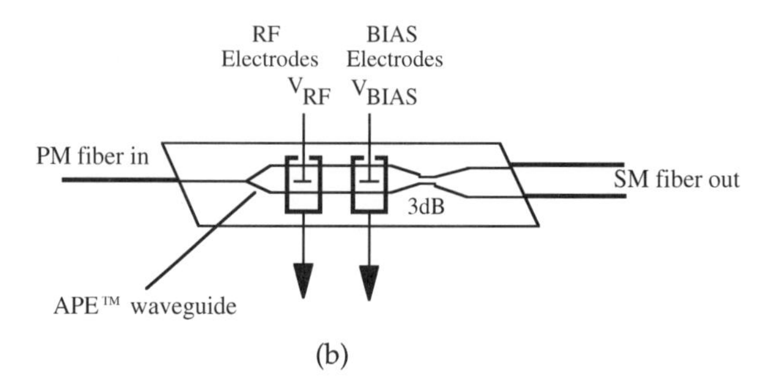

(b)

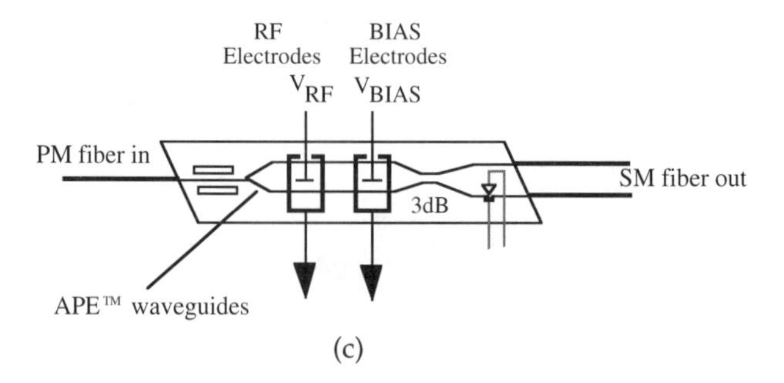

(c)

Figure 7.10: Evolutionary stages of commercially available MZI-based intensity modulators: (a) conventional MZI, (b) MZI modulator with output Y-junction replaced by a 2×2 directional coupler, and (c) MZI modulator with a phase modulator and an on-chip photodetector added to (b). The function of the added phase modulator is briefly explained in Section 7.6. After Ref. [15].

respectively. They are related by

$$\begin{pmatrix} A_{out,1} \\ A_{out,2} \end{pmatrix} = L_a \overline{M}_{DC} \overline{M}_{PM} \begin{pmatrix} A_{in,1} \\ A_{in,2} \end{pmatrix}, \tag{7.10}$$

where L_a is the loss factor considered in Eq. (7.4), but without counting the output Y-junction 3-dB loss, and $\overline{M}_{PM}$ is the matrix transfer function of the MZI modulator (after the Y-splitter), given by

$$\overline{M}_{PM} = \begin{pmatrix} \exp\left(j\pi V/2V_\pi + j\phi/2\right) & 0 \\ 0 & \exp\left(-j\pi V/2V_\pi - j\phi/2\right) \end{pmatrix}, \tag{7.11}$$

where V_π and ϕ have been defined in Eqs (7.4) and (7.5). $\overline{M}_{DC}$ is the transfer matrix of the output directional coupler, which can be expressed as

$$\overline{M}_{DC} = \begin{pmatrix} \cos\gamma & -j\sin\gamma \\ -j\sin\gamma & \cos\gamma \end{pmatrix}, \tag{7.12}$$

where γ is equal to κL given in Eq. (7.7). After the signal is detected by a receiving photodetector, we obtain the output optical power for the upper-branch output as (assuming

$$\begin{pmatrix} A_1(0) \\ A_2(0) \end{pmatrix} = P_{in} \begin{pmatrix} 1/\sqrt{2} \\ 1/\sqrt{2} \end{pmatrix}$$

for normalized input optical power)

$$P_{out,1} = \frac{P_{in} L_a}{2}\left[1 - \sin\left(\frac{\pi V}{V_\pi} + \phi\right)\sin 2\gamma\right]. \tag{7.13}$$

If we now design the output directional coupler such that $\gamma = 45°$, that is, equal power splitting, we can finally obtain the output power for the upper branch as

$$P_{out,1} = \frac{P_{in} L_a}{2}\left[1 - \sin\left(\frac{\pi V}{V_\pi} + \phi\right)\right]. \tag{7.14}$$

This is basically the same result as in Eq. (7.5), considering the adjustable static bias phase shift ϕ. Following the same derivation, the output power for the lower branch is given by

$$P_{out,2} = \frac{P_{in} L_a}{2}\left[1 + \sin\left(\frac{\pi V}{V_\pi} + \phi\right)\right]. \tag{7.15}$$

From Eqs. (7.14) and (7.15), we can see that the outputs from the two arms are 180° out of phase, and the total output power of the two arms become twice as high as that of the conventional MZI modulator. Hence, the progression from MZI modulator to BBI modulator allows two outputs (doubling the power) delivered to an HFC network. However, in practical conditions, the two output ports may not be exactly

180° out of phase, and therefore the optimum bias for one port may not be the optimum for the other port. In this case, we can either let only one of the ports be biased at the optimum condition, or we have to find a compromised bias point for both outputs while having degraded linearity performance.

7.3 Linearity Characteristics of MZI/ BBI, DCM, and Y-Fed Coupler Modulators

The fundamental linearity characteristics of three types of amplitude modulators can be compared by expanding their transfer functions, given in Eqs. (7.14), (7.8), and (7.9), in terms of Taylor series [12] as

$$f(\frac{V}{V_s}) = f(V_b) + \frac{df}{dV}\Big|_{V_b} V_{RF} + \frac{1}{2!}\frac{d^2 f}{dV^2}\Big|_{V_b} V_{RF}^2 + ... \tag{7.16}$$

$$= c_0 + c_1 V_{RF} + c_2 V_{RF}^2 + c_3 V_{RF}^3 + ...,$$

where $V = V_b + V_{RF}$, V_b is the bias voltage, V_{RF} is the modulating voltage, and V_s is the minimum swing voltage required to produce 100% modulation depth of the optical intensity. The results presented by Halemane and Korotky [12] are repeated in Table 7.1.

From the Table, we can see that all three types of modulators have nearly identical two-tone third-order intermodulation products (~ −74 dBc at 4% OMI). There are subtle differences , however. For example, biasing to zero second-harmonic point in DCM or Y-fed coupler modulators does not eliminate higher-order even harmonics [16]. If higher than third-order nonlinear distortions can be neglected, the distortion characteristics of all three types of modulators are similar, and other considerations may

Table 7.1: The coefficients in Taylor series expansion for the three types of modulators for the bias point corresponding to maximally linear response c_2=0, and the third-order intermodulation distortions relative to the power at the carrier frequency for 4% optical modulation depth. (After Ref. [12])

Modulator type	Bias point V/V_s	c_0	c_1	c_2	c_3	P_{IM3} (dBc) at OMI = 4%
MZI	0.5	0.5	−1.57	0	2.58	−74.0
DC	0.439	0.536	−1.65	0	3.23	−72.0
Y-Fed	0	0.5	−1.61	0	2.89	−73.6

influence the preferred structure. These considerations include (1) the RF efficiency of the driving voltage, (2) the manufacture yield when considering nonideal devices with intrinsic phase mismatch between the coupled arms of the directional coupler and Y-fed coupler, or the uncoupled arms of the MZI [17], (3) the relative sensitivity to photorefractive effects [18–21], and (4) the residual chirp in the modulators [22]. Considering point 1, we have seen that MZI modulators require the lowest swing voltage to reach 100% OMI. Considering point 2, Howerton *et al.* [17] pointed out that the sensitivity of the coupling coefficient κ in DCM or Y-fed modulators during fabrication process makes it difficult to achieve a theoretical linearity performance, and a higher yield would therefore be expected from MZI modulators. Considering point 3, experiments on Ti:$LiNbO_3$ directional couplers have shown that at optical power levels of 20 mW within a waveguide at 1.3 μm, the photorefractive phenomenon can act to modify the transfer function, $f(V)$, from its ideal form [20], whereas with a careful fabrication process, Ti-indiffused $LiNbO_3$ interferometric modulators can be made with photorefractive sensitivity low enough for most practical applications (at 1320 nm, there are no optical transmission or extinction ratio problems even at 400 mW [23]). As for point 4, MZI modulators can exhibit much lower residual chirp than directional coupler-type modulators [22].

As a side point, the photorefractive effect was a concern in interfacing modulators with a high-power (~100 mW) 1.3-μm Nd:YAG laser. However, later it was found that, for 1.55-μm systems, the optical power can be increased by a booster EDFA after the output of the external modulator; therefore, the laser power at the modulator input does not have to be as high as that in 1.3-μm systems, and consequently photorefractive effects can be significantly reduced. It can be argued that the introduction of a booster EDFA would degrade the signal-to-noise ratio because of the $\geq$3 dB EDFA noise figure, (see Section 6.1.6). Therefore, if one wants to avoid using an EDFA booster amplifier, a high-power 1.55-μm source is still required, and eventually photorefractive effects will have to be considered.

Coming back to the four considerations just discussed, we can see why most of the commercially available modulators for analog CATV applications are MZI/BBI modulators, despite the fact that there have been quite a number of proposed structures to linearize the DCM performance [16,24–26]. Therefore, in the following, we will only concentrate on analyzing the linearity performance of MZI/BBI and linearized-MZI/BBI modulators.

To analyze MZI/BBI modulator linearity performance, we start with Eqs. (7.14) and (7.15), and combine them as

$$P_{out} = (P_{in}L_a / 2)[1 \pm \sin(\frac{\pi V}{V_\pi} + \phi)]. \tag{7.17}$$

Since MZI/BBI modulators are normally biased at their inflection points, as shown in Fig. 7.11, Eq. (7.17) can be expanded at $V = 0$ (assuming initial phase $\phi = 0$) in terms of Taylor series, as was done in Eq. (7.16). Note that if the transfer function is written in terms of Eq. (7.5), the Taylor series should be expanded with respect to $V = V_\pi/2$. By keeping the first three terms in the expansion, we obtain

$$P_{out} \cong (P_{in}L_a/2)\cdot[1 \pm (\frac{\pi V}{V_\pi}) \mp \frac{(\pi V/V_\pi)^3}{6}]. \tag{7.18}$$

As can be seen, ideally there is no even-order distortion with the MZI or BBI modulator when biased at an inflection point. This is contrary to the linearity characteristics of laser diodes—the main distortions caused by spatial hole burning in a DFB laser (see Sec. 4.10.2) are second-order, whereas third-order distortions are generally small. This contrast in linearity characteristics between external modulators and DFB laser diodes was experimentally verified [27].

We can examine the fundamental CTB and CNR in a multichannel AM-VSB system by using a modulator described by Eq. (7.17) as follows. In Eq. (7.17), we first let $V = \sum_{i=1}^{N} A\sin(\omega_i t + \psi_i)$, $\phi = 0$, $\chi = \pi A/V_\pi$, and ψ_i is the random phase of each channel. Next, we use Bessel function expansions to obtain

$$\begin{aligned}\sin(\frac{\pi V}{V_\pi}) &= \sin(\chi\cdot\sum_{i=1}^{N}\sin(\omega_i t+\psi_i)) \\ &= \sum_{n_1=-\infty}^{\infty}\sum_{n_2\cdots}\cdots\sum_{n_N} J_{n_1}(\chi)\cdot J_{n_2}(\chi)\ldots J_{n_N}(\chi)\cdot \\ &\quad \sin\{n_1(\omega_1 t+\psi_1)+n_2(\omega_2 t+\psi_2)+\ldots+n_N(\omega_N t+\psi_N)\},\end{aligned} \tag{7.19}$$

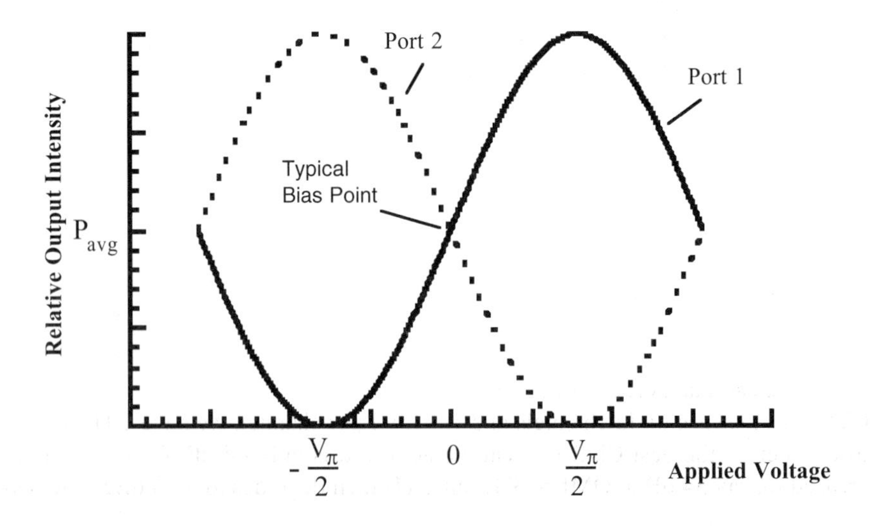

Figure 7.11: Modulation transfer function of an MZI or BBI.

where J_n is the nth-order Bessel function of the first kind. The fundamental signal photocurrent is given by letting $n_i = \pm 1$, $n_{j,\, j \neq i} = 0$ in Eq. (7.19), that is, $2J_1(\chi)[J_0(\chi)]^{N-1}$, and the average *electrical* signal power is

$$\text{Fundamental: } 2J_1^2(\chi)\cdot[J_0(\chi)]^{2N-2}. \tag{7.20}$$

The second-order terms are zero because $\sum n_i =$ odd in Eq. (7.19). In calculating the third-order intermodulation products, we choose terms of the form $\omega_i + \omega_j - \omega_k$ (terms of the form $2\,\omega_i - \omega_j$ are 6 dB smaller than the former; see Section 2.3.1). Therefore, the third-order intermod photocurrent is given by letting $n_i = n_j = \pm 1$, $n_k = \pm 1$, $n_{m,\, m\neq i\neq j\neq k} = 0$, that is, $2J_1^2(x)J_{-1}(x)\bullet[J_0(x)]^{N-3}$, and the average electrical third-order intermod power is (note $J_{-1}(x) = -J_1(x)$)

$$2J_1^6(\chi)\cdot[J_0(\chi)]^{3N-6}\cdot N_{CTB}, \tag{7.21}$$

where N_{CTB} is the product count of CTB in a particular channel which has been introduced in Section 2.3.1, by assuming all nonlinear distortions are due to static nonlinearity. The power ratio of carrier to CTB is then given by

$$\frac{C}{CTB} = \frac{2J_1^2(\chi)[J_0(\chi)]^{2N-2}}{N_{CTB}\cdot\{2J_1^6(\chi)\cdot[J_0(\chi)]^{2N-6}\}} \approx \frac{1}{PC\cdot(\frac{\chi}{2})^4}, \tag{7.22}$$

where we have used the approximations $J_1(\chi)/J_0(\chi) \cong \chi/2$ for $\chi << 1$. If we consider $N_{CTB} = 1$, that is; only one set of triple-tone beat, we have $CTB = -68$ dBc. Since the two-tone third-order intermodulation product is 6 dB lower than this value, we have $P_{IM3} = -74$ dBc, which is consistent with the value given in Table 7.1.

To calculate CNR per channel, we first observe that the OMI per channel is given by $2J_1(\chi)[J_0(\chi)]^{N-1}$, which can be approximated by letting $J_0(\chi) \approx 1$ to obtain χ. Therefore, if we assume the RIN noise of the high power source is negligible, we obtain (from Eq. (5.29))

$$CNR \approx \frac{\frac{1}{2}\left[\chi\right]^2 I_p^2}{\left[2qI_p + \left\langle i_{th}^2\right\rangle\right]\cdot BW}, \tag{7.23}$$

where the received photocurrent I_p is equal to $(\boldsymbol{R}\, P_{in}\, L_a)/2$. From Eqs. (7.22) and (7.23), if we let $I_p = 1$ mA, $<i_{th}^2> = 8$ pA/√Hz, and BW = 4.2 MHz, we can obtain the results shown in Fig. 7.12.

It is clear that there is a trade-off between the maximum CNR and the minimum CTB. For the case of 42 channels and 1% OMI per channel, we have CTB of −64 dBc, whereas the best CNR that can be achieved is only 44 dB. CNR can be increased to, say 54 dB at OMI = 3%, but CTB then degrades to −45 dBc. The performance of the 80-channel case is even worse. Therefore, we see that without applying linearization techniques, MZI/BBI modulators cannot be used in practical

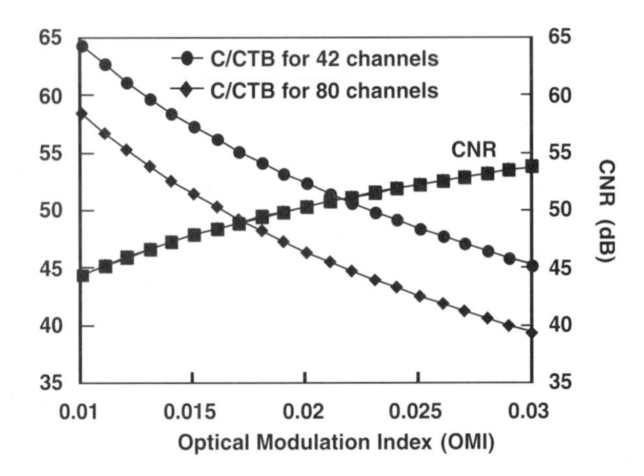

Figure 7.12: Calculated worst-case C/CTB and CNR versus OMI for 42-channel and 80-channel cases. Parameters used in the calculation are I_p = 1 mA, $<i_{th}^2>$ = 8 pA/$\sqrt{\text{Hz}}$, BW = 4.2 MHz, and N_{CTB} = 585 and 2295 for the two cases, respectively.

AM-VSB lightwave CATV systems. Therefore, in the next section, we discuss various linearization techniques that have been applied to MZI/BBI modulators.

7.4 Linearization Techniques

Linearization techniques which have been applied to external modulators are used to minimize third-order distortions (note that the predistortion circuit for a laser diode, as described in Section 4.10, was used to minimize mainly second-order distortions, although it can also suppress third-order distortions). Before we explain the operation principle of several linearization techniques, we must ensure that the bias point at a quadrature point is stabilized; otherwise, unexpected high second-order distortions may be incurred. The bias point may drift because of temperature variations and stress, especially when nonzero DC bias is applied to the electrode, as illustrated by curve (a) of Fig. 7.13.

We can see that if the DC bias does not have a feedback adjustment, the output optical power changes from 0 to 0.9 mW monotonically in a time interval of 72 hours. If a DC feedback loop is implemented to automatically adjust the DC bias and maintain the bias point at the inflection point, then the output power can be maintained at a constant value. However, since the DC voltage is increased or decreased monotonically in order to track the DC drift, it may exceed the maximum voltage of a power supply after some time. Therefore, a reset of the bias voltage may become necessary once in a while. To avoid this problem, current commercial devices are made such that their quadrature point occurs at zero DC bias, although this condition

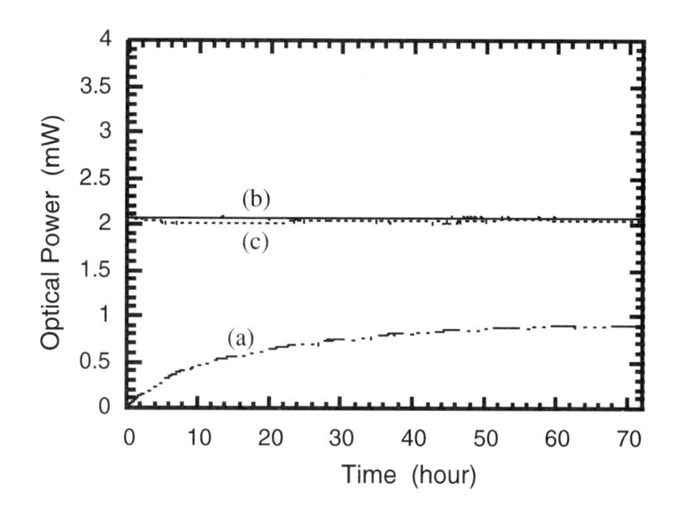

Figure 7.13: Measured results of a commercial $LiNbO_3$ MZI (APE) modulator. When a DC voltage is applied to the modulator electrode (without feedback bias control) such that the modulator output is 0 mW, after 72 hours it drifts to ~0.9 mW (a). However, if no DC bias is applied, the output power remains constant with (b) or without (c) feedback bias control.

cannot always be met practically. As shown in Fig. 7.13, curves (b) and (c), the modulator output power maintains a constant value with or without DC feedback control.

DC drift or bias instability is due to the capacitance effects of silicon dioxide buffer layers (which separate the $Ti:LiNbO_3$ waveguides from the metal electrodes), usually used on Z-cut devices [28], and due to humidity [29]. DC drift in $Ti:LiNbO_3$ optical waveguide devices can be significantly reduced by adding an ion-exchange process before the deposition of the buffer layers in device fabrication [30]. It was also reported that DC drift can be greatly reduced by taking care with the interface conditions between the substrate and the buffer layer during the fabrication process through the use of low-temperature sputtering and prolonged annealing to reduce oxygen defects at the interface and buffer layer [31]. Another approach is to make a device whose quadrature point occurs at zero DC bias [32], so that no accumulated charges can induce bias drift.

The origin of thermal drift in $LiNbO_3$ is mainly the pyroelectrically induced charges under normal operating temperatures [33,34]. Such drift can be significantly reduced by formation of semiconducting layers under the planar electrodes, which screen out the induced charges [35].

Despite the various fabrication approaches to improve DC and thermal drift, the drift can still exist in practical applications. They are deteriorating factors that affect second-order linearity performance, which a system designer must always pay attention to.

With the DC bias stabilization or DC-free bias as a prerequisite, let us now consider the linearization techniques for external modulation systems. There are two main categories of linearization techniques: electrical predistortion and optical linearization techniques. These techniques are reviewed in the following two sections.

7.4.1 Electrical Predistortion Technique

An electrical predistortion circuit is placed between the external modulator and the input signals, as shown in Fig. 7.14.

The transfer function of the predistortion circuit is expected to be

$$V'_{RF} = a_1 V_{RF} + a_3 V_{RF}^3 + a_5 V_{RF}^5 + \ldots, \tag{7.24}$$

whose third-order nonlinearity is used to cancel that of the sinusoidal transfer function as follows (let $G \cdot \pi / V_\pi = \chi'$, where G is the amplifier gain in Fig. 7.14):

$$\sin(\chi' V'_{RF}) = \sin(\chi' a_1 V_{RF} + \chi' a_3 V_{RF}^3 + \chi' a_5 V_{RF}^5 + \ldots). \tag{7.25}$$

Now according to Eq. (7.18), we have

$$\sin(\chi' V'_{RF}) = \chi' a_1 V_{RF} + (\chi' a_3 - (\chi')^3 a_1^3/6)\, V_{RF}^3 + \text{higher-order terms.} \tag{7.26}$$

Therefore, for a predistortion circuit with a fixed third-order coefficient, we should carefully control the amplifier gain G to satisfy $\chi' = (6a_3 / a_1^3)^{1/2}$, or $G = (V_\pi / \pi) \cdot (6a_3 / a_1^3)^{1/2}$, so that ideally the resultant third-order nonlinear distortion is completely canceled.

The preceding derivation assumes that the predistortion circuit and the modulator both exhibit static nonlinearities. A more general analysis can take the frequency-dependent characteristics of both the predistortion circuit and the modu-

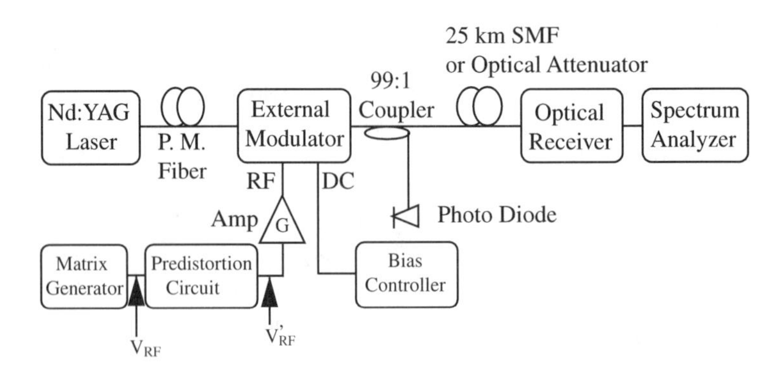

Figure 7.14: A complete experimental setup of a 1.3-μm external modulation system with bias control stabilization and a predistortion circuit.

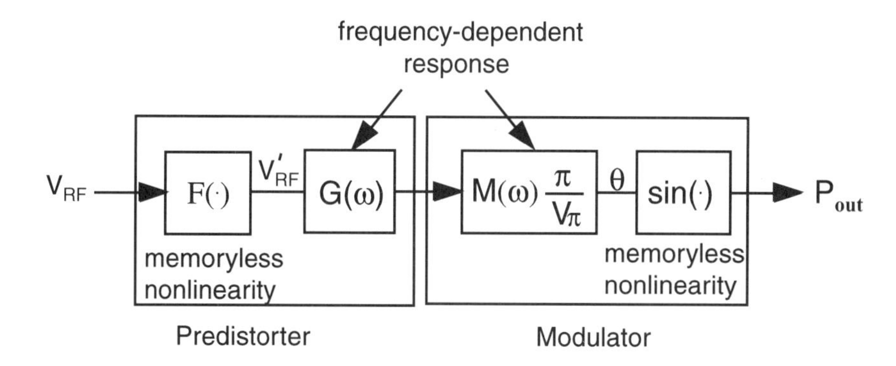

Figure 7.15: Block diagrams of a predistorter and an MZI modulator which include the frequency-dependent responses $G(\omega)$ and $M(\omega)$. After Ref. [14] (© 1993 IEEE).

lator into account, as shown in Fig. 7.15, by separating each nonlinearity into a memoryless (static) nonlinearity and a frequency-dependent transfer function [14]. Now Eq. (7.25) can be rewritten as

$$P_{out} = \sin(k(\omega) \cdot F(V_{RF})), \tag{7.27}$$

where

$$k(\omega) = G(\omega)M(\omega)\frac{\pi}{V_{\pi}},$$

and $G(\omega)$ and $M(\omega)$ are the frequency-dependent responses of the predistorter and the modulator, respectively. If we assume that the input signal consists of three tones, then from Eq. (7.24), V'_{RF} in Fig. 7.15 is given by

$$\begin{aligned} V'_{RF} &= F(V_{RF}) \\ &= a_1(e^{j\omega_1 t} + e^{j\omega_2 t} + e^{-j\omega_3 t} + c.c.) + a_3(\sum_{l \neq m \neq n}\sum\sum [e^{j(\omega_l + \omega_m - \omega_n)} + c.c.]) \end{aligned} \tag{7.28}$$

where c.c. stands for "complex conjugate," and we have considered the triple-beat $\omega_{\ell} + \omega_m - \omega_n$ $(l,m,n \in (1,2,3))$ as the dominant third-order distortion. From Eq. (7.27) and using the approximation $\sin\theta \approx \theta - \frac{1}{6}\theta^3$, we can obtain the triple beat at a particular frequency $\omega_1 + \omega_2 - \omega_3$ at the modulator output as

$$\begin{aligned} & a_3 \cdot k(\omega_1 + \omega_2 - \omega_3) \cdot [e^{j(\omega_1 + \omega_2 - \omega_3)t} + c.c.] \\ & -\frac{1}{6} a_1^3 \cdot k(\omega_1)k(\omega_2)k(-\omega_3)[e^{j(\omega_1 + \omega_2 - \omega_3)t} + c.c.], \end{aligned} \tag{7.29}$$

where the first term is generated by the predistorter and the second term is generated by the modulator. To let these two terms cancel each other, we must have

$$a_3 \cdot k(\omega_1 + \omega_2 - \omega_3) - \frac{1}{6} a_1^3 \cdot k(\omega_1) k(\omega_2) k(-\omega_3) = 0 . \tag{7.30}$$

If we let $k(\omega) = |k(\omega)| \cdot e^{j\phi(\omega)}$, Eq. (7.30) becomes

$$\begin{aligned} &a_3 \left| k(\omega_1 + \omega_2 - \omega_3) \right| e^{j\Phi(\omega_1 + \omega_2 - \omega_3)} \\ &- \frac{1}{6} a_1^3 \left| k(\omega_1) \right| \cdot \left| k(\omega_2) \right| \cdot \left| k(-\omega_3) \right| \cdot e^{j[\Phi(\omega_1) + \Phi(\omega_2) + \Phi(-\omega_3)]} = 0 . \end{aligned} \tag{7.31}$$

A necessary condition for this equation to hold is that

$$\Phi(\omega) = \tau_g \cdot \omega , \tag{7.32}$$

where τ_g is a constant group delay. Consequently, Eq. (7.31) becomes

$$a_3 \left| k(\omega_1 + \omega_2 - \omega_3) \right| - \frac{1}{6} a_1^3 \left| k(\omega_1) \right| \cdot \left| k(\omega_2) \right| \cdot \left| k(-\omega_3) \right| = 0 . \tag{7.33}$$

A sufficient condition for this equation to hold is

$$k(\omega) = \text{constant} = K, \tag{7.34}$$

so that as long as K satisfies

$$a_3 - \frac{1}{6} a_1^3 K^2 = 0 , \tag{7.35}$$

we can have the third-order intermodulation products generated from the predistorter completely cancel those generated from the modulator. From conditions (7.32) and (7.34), we see that broadband cancellation of third-order intermodulation products requires the constancy of the amplitude frequency response and group delay over the entire frequency range of interest. Condition (7.35) again indicates that the gain of the amplifier G, which is included in K, must be adjusted appropriately.

The next question we have to ask ourselves is how stringent the requirement is on constant amplitude and group delay as a function of frequency. To answer this question, we can treat Eq. (7.31) as the result of balancing two phasors. If the two phasors cannot cancel each other [i.e., conditions (7.32) and (7.34) cannot be met], there will be residual third-order intermodulation products (IMPs) which can be expressed as

$$IMP_{res} = 20 \log \left| 1 - \frac{\frac{1}{6} a_1^3 \left| k(\omega_1) \right| \cdot \left| k(\omega_2) \right| \cdot \left| k(-\omega_3) \right| \cdot e^{j[\Phi(\omega_1) + \Phi(\omega_2) + \Phi(-\omega_3)]}}{a_3 \left| k(\omega_1 + \omega_2 - \omega_3) \right| e^{j\Phi(\omega_1 + \omega_2 - \omega_3)}} \right| (\text{dBc}). \tag{7.36}$$

Let us first assume that $|k(\omega)|$ varies between $K(1 + \Delta)$ and $K(1 - \Delta)$, where K satisfies Eq. (7.35), and $\phi(\omega)$ varies between $(\tau_g \cdot \omega) + \Delta'$ and $(\tau_g \cdot \omega) - \Delta'$, where Δ and Δ'/ω represent the deviations of amplitude and group delay from their constant values, respectively. Then Eq. (7.36) can give us a worst-case IMP_{res} as [using Eq. (7.35)]

$$IMP_{res} = 20\log\left|1 - \frac{(1+\Delta)^3 \cdot e^{j(-3\Delta')}}{(1-\Delta)e^{j\Delta'}}\right| \quad (dBc)\,. \tag{7.37}$$

A plot of IMP_{res} for different values of amplitude ripples $(20\log(1 + \Delta)/(1 - \Delta))$ and phase ripple $(57.3° \times 2\Delta')$ is given in Fig. 7.16. We can see that the requirement for frequency response flattness and phase linearity (i.e., constant group delay) to obtain large third-order IMP suppression is very stringent (see Problem 7.4 for a more relaxed condition). For example, to obtain −12 dB suppression of IMP, the frequency response flattness must be bounded by ±0.5 dB and deviation from linear phase be at most 3.6°. Factors affecting the amplitude and group-delay flattness include (1) parasitics in the predistorter and modulator and (2) modulator velocity mismatch between the optical and electrical fields. To cope with 1, a broadband impedance-matching circuit is required between the predistorter, amplifier, and modulator. To cope with 2, a modulator with traveling-wave electrodes is preferred.

7.4.1.1 Predistortion Circuit An example of a predistortion circuit which can generate adjustable third-order distortion is shown in Fig. 7.17.

The circuit essentially takes advantage of a parallel combination of two diodes with reverse directions, and more (balanced) diodes can be used to further suppress the amplitude of third-order distortion. The DC current bias in Fig. 7.17a charges the two capacitors to sufficiently large voltages such that the equivalent circuit of

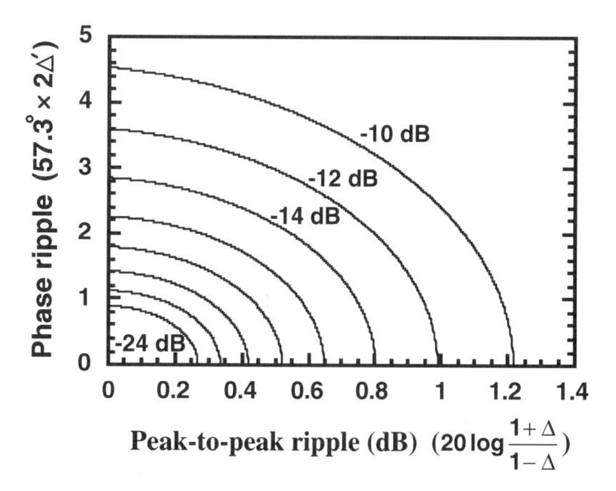

Figure 7.16: Suppression of the third-order IMP as a function of peak-to-peak ripple and phase ripple in the overall frequency response.

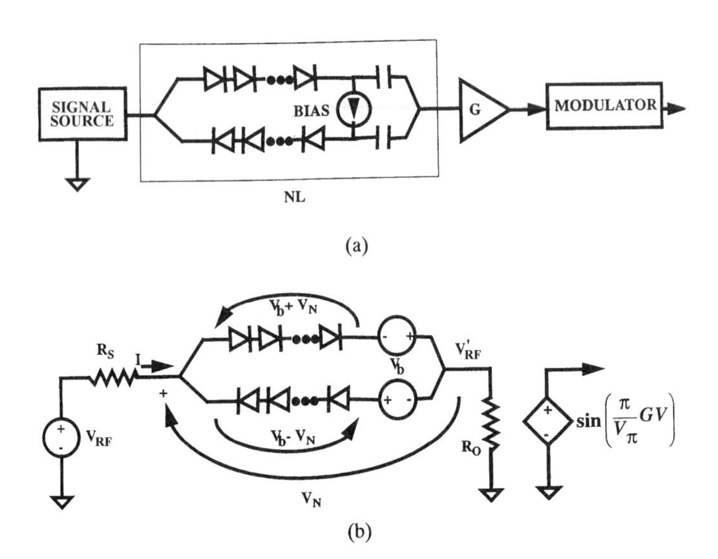

Figure 7.17: A typical predistortion circuit for $LiNbO_3$ amplitude modulators (a), and its equivalent circuit (b). After Refs. [4, 14] (© 1990, 1993 IEEE).

Fig. 7.17a can be expressed as in Fig. 7.17b, where the current flow through the output resistor R_0 is given by

$$\begin{aligned} I &= I_S \exp((V_b + V_N)/NV_T) - I_S \exp((V_b - V_N)/NV_T) \\ &= 2I_S \exp(\frac{V_b}{NV_T}) \sinh\frac{V_N}{NV_T}; \end{aligned} \tag{7.38}$$

where I_s is the reverse saturation current, V_T is the volt-equivalent of temperature given by $T/11{,}600$ (T is in kelvins), N is the number of diodes in each branch, and V_N and V_b are both defined in Fig. 7.17. Using Eq. (7.38), we can expand V_N in terms of I as

$$V_N = r_1 I + r_3 I^3 + \dots, \tag{7.39}$$

where

$$r_1 = \frac{NV_T}{2I_B}, \quad r_3 = -\frac{NV_T}{48I_B^3} \qquad \text{with} \quad I_B = I_S \exp(\frac{V_b}{NV_T}). \tag{7.40}$$

In order to obtain a final relation between the input V_{RF} and output V'_{RF}, we bring V_{RF} in by using

$$V_{RF} = R_s I + (r_1 I + r_3 I^3) + R_0 I = (R_s + R_0 + r_1)I + r_3 I^3, \tag{7.41}$$

from which we have the approximation

$$I \approx (R_s + R_0 + r_1)^{-1} V_{RF} + (-r_3 (R_s + R_0 + r_1)^{-4}) V_{RF}^3 . \quad (7.42)$$

Therefore,

$$V'_{RF} = I R_0 = a_1 V_{RF} + a_3 V_{RF}{}^3, \quad (7.43)$$

where $a_1 = R_0 (R_s + R_0 + r_1)^{-1}$ and $a_3 = -R_0 r_3 (R_s + R_0 + r_1)^{-4}$. By taking a close look at the coefficient a_3, we find that only r_1 and r_3 can be adjusted; all other parameters are fixed values. But from Eq. (7.40), r_1 and r_3 can be adjusted only by tuning the bias current in Fig. 7.17a, or equivalently, by tuning the bias voltage V_b. Therefore, in using the predistortion circuit given in Fig. 7.17, we have all together two parameters to control, to minimize CTB: the amplifier gain G [Eq. (7.35)] and the predistorter bias current (or voltage).

7.4.2 Optical Dual Parallel Linearization Technique

The optical dual parallel linearization technique was first proposed by Korotky and de Ridder [36] and actually implemented by Brooks *et al.* [37]. The basic configuration is shown in Fig. 7.18, where two MZI modulators are combined in parallel.

The top and bottom modulators, or the "primary" and "secondary" modulators, respectively, are biased at two 180°-out-of-phase quadrature points so that ideally their output optical power are given by Eq. (7.17) with "+" and "−" signs in the bracketed term, respectively. However, practically, the static phase shift ϕ of the primary MZI may be different from that of the secondary MZI. Therefore, a phase modulator at the output of the secondary MZI modulator is used to maintain the required 180° phase shift between the two MZ modulators.

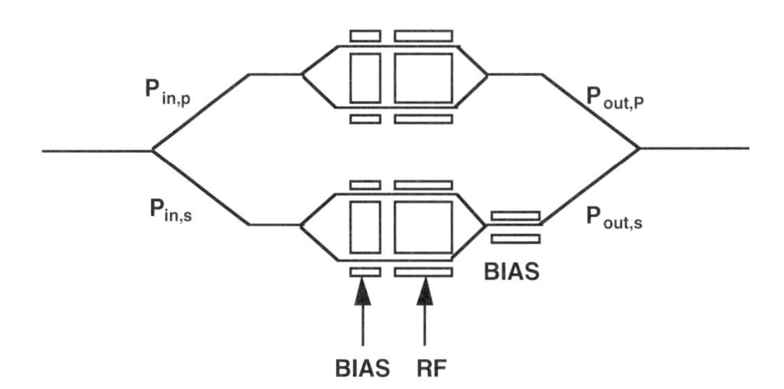

Figure 7.18: Basic configuration of a modulator using the dual parallel linearization technique. After Ref. [37] (© 1993 IEEE).

In Fig. 7.18, the input optical power and the driving RF per are both split unevenly to the two MZ modulators so that the primary modulator obtains higher optical power and the secondary modulator has a higher modulation index and greater distortion. When the ideal phase relationship between the primary and the secondary MZIs is maintained, the output optical power of the two MZ modulators is given by

$$P_{out,p} \cong (P_{in,p} L_a / 2) \cdot [1 - (\frac{\pi V}{V_\pi}) + \frac{(\pi V / V_\pi)^3}{6}] \tag{7.44}$$

and

$$P_{out,s} \cong (P_{in,s} L_a / 2) \cdot [1 + (\frac{c\pi V}{V_\pi}) - \frac{(c\pi V / V_\pi)^3}{6}], \tag{7.45}$$

where $c > 1$ and $P_{out,p}$, $P_{out,s}$, $P_{in,p}$, and $P_{in,s}$ are the output and input optical power of the primary and secondary MZI modulators, respectively. The two output signals are then combined to give

$$P_{total} \cong (L_a / 2) \cdot [P_{in,p} + P_{in,s} - (\frac{\pi V}{V_\pi})(P_{in,p} - kP_{in,s}) + \frac{(\pi V / V_\pi)^3}{6}(P_{in,p} - c^3 P_{in,s})]. \tag{7.46}$$

We can see that the third-order distortion disappears when

$$P_{in,p} = c^3 P_{in,s}. \tag{7.47}$$

Therefore, the optimum ratio between the two input optical power is determined by the RF voltage ratio c in the two modulators. The cancellation of the third-order nonlinear distortion is accompanied by a slight penalty in optical modulation index. It can be seen from Eqs. (7.46) and (7.47) that the signal amplitude and the DC optical power are given by

$$A = \frac{L_a \pi |V(t)| P_{in,p}}{2V_\pi}(1 - \frac{1}{c^2}) \tag{7.48}$$

and

$$P_{DC} = \frac{L_a P_{in,p}}{2}(1 + \frac{1}{c^3}). \tag{7.49}$$

Therefore, the OMI after linearization is decreased by a factor given by

$$[(1 - \frac{1}{c^2})/(1 + \frac{1}{c^3})].$$

For example, when $c = 3$, the OMI would be slightly decreased to about 86% of the OMI of a single modulator [37].

This technique, although interesting, is not as practical as the predistortion technique mentioned in the last section. The main reason is because the tolerance on the accuracy of the phase modulator voltage (which is required to achieve the phase relationship between the two modulators) is too tight to be reached realistically [37] (see Problem 7.3). In addition, to avoid laser phase noise being converted to intensity noise through the multipath-interferometer effect (see Section 9.3.1), that is, through the upper- and lower-arm signal (with a relative delay) beating at the photodetector, the phase modulator should also include electrodes for applying RF modulation to dither the laser phase. Even though two incoherent optical sources can be used to replace the phase modulator, the cost of the optical sources also forbid practical application of this technique.

7.4.3 Optical Dual Cascade Linearization Technique

Instead of dual parallel MZI modulators, another linearization technique uses dual series phase and DCMs, as shown in Fig. 7.19 [38,39]. This technique has been practically implemented, and products based on it are commercially available.

To analyze the linearity performance of this technique, we again let the electrical fields in the upper and lower branches after the first Y-splitter and at the output of the two-stage modulator be given by

$$\begin{pmatrix} A_{in,1} \\ A_{in,2} \end{pmatrix} \text{ and } \begin{pmatrix} A_{out,1} \\ A_{out,2} \end{pmatrix},$$

respectively, and they are now related by

$$\begin{pmatrix} \boldsymbol{A}_{out,1} \\ \boldsymbol{A}_{out,2} \end{pmatrix} = \overline{\boldsymbol{M}}_{DCM2}\overline{\boldsymbol{M}}_{PM2}\overline{\boldsymbol{M}}_{DCM1}\overline{\boldsymbol{M}}_{PM1}\begin{pmatrix} \boldsymbol{A}_{in,1} \\ \boldsymbol{A}_{in,2} \end{pmatrix}, \tag{7.50}$$

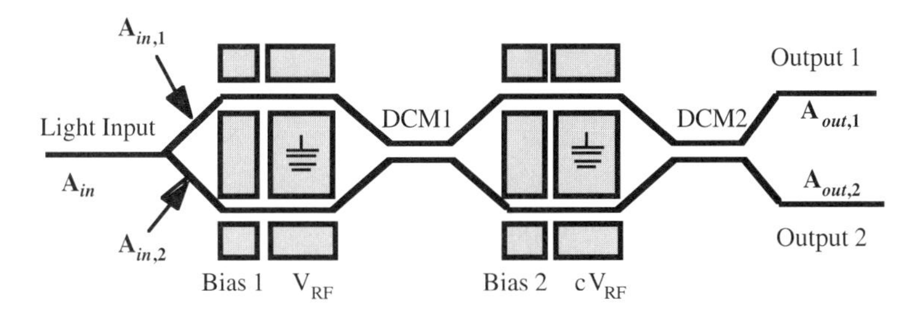

Figure 7.19: Two-stage optically linearized modulator. The RF drives to the first and second stages are in proportion c. The biases are used to maintain the modulator operating at quadrature. DCM1 and DCM2 represent the two directional coupler modulators, respectively. After Ref. [39].

where $\overline{M}_{PM1}$ is the matrix transfer function of the first phase modulator (after the Y-splitter in the first stage) given by

$$\overline{M}_{PM1} = \begin{pmatrix} \exp(i\pi V_{RF}/2V_\pi) & 0 \\ 0 & \exp(-i\pi V_{RF}/2V_\pi) \end{pmatrix}. \tag{7.51}$$

$\overline{M}_{PM2}$ is the matrix transfer function of the second phase modulator in the second stage, given by

$$\overline{M}_{PM2} = \begin{pmatrix} \exp(i\pi cV_{RF}/2V_\pi) & 0 \\ 0 & \exp(-i\pi cV_{RF}/2V_\pi) \end{pmatrix}, \tag{7.52}$$

and $\overline{\boldsymbol{M}}_{\boldsymbol{DCMn}}$ ($n = 1$ or 2) is the transfer matrix of the first or second directional couplers, which can be expressed as [recall Eqs. (7.7) and (7.12)]

$$\overline{\boldsymbol{M}}_{\boldsymbol{DCMn}} = \begin{pmatrix} \cos\gamma_n & -\boldsymbol{i}\sin\gamma_n \\ -\boldsymbol{i}\sin\gamma_n & \cos\gamma_n \end{pmatrix}, \quad n = 1,2, \tag{7.53}$$

where γ_n is equal to κL given in Eq. (7.7). After the receiving photodetector, the intensity transfer function for the upper-branch output can be written as

$$f(V_{RF}) = \left|\frac{A_{out,1}}{A_{in}}\right|^2 = \frac{1}{2}\left\{\begin{array}{l} 1-\sin 2\gamma_1 \cos 2\gamma_2 \sin\psi' \\ +\sin 2\gamma_2\left[-\cos 2\gamma_1 \sin\psi' \cos\psi'' - \cos\psi' \sin\psi''\right] \end{array}\right\}, \tag{7.54}$$

where $\psi' = (\pi V_{RF}/V_\pi) + \phi_1$ and $\psi'' = (\pi c V_{RF}/V_\pi) + \phi_2$. Note that $f(V_{RF})$ must be linearized by adjusting the bias voltages (bias 1 and bias 2 in Fig. 7.19) so that both the offset angles ϕ_1 and ϕ_2 are zero to ensure that even-order contributions vanish. Practically, this means that the bias voltages need to be continuously adjusted by servo control for satisfactory operation over temperature. For the ease of implementation, γ_1 and γ_2 are made equal and are not tuned after fabrication, and optimum values in the range of 58–65° were chosen to optimize the RF sensitivity (i.e., $df/d\Psi'$ is maximum) [39]. The only parameter which can be tuned is then the driving RF power ratio c. Now with $\gamma = \gamma_1 = \gamma_2$, the intensity transfer function becomes

$$f(V_{RF}) = \frac{1}{2}\{1-(1/2)\sin 4\gamma \sin\psi'(1+\cos\psi'') - \sin 2\gamma \cos\psi' \sin\psi''\}. \tag{7.55}$$

Based on this equation, Fig. 7.20 (with $\gamma = 63°$, $\phi_1 = 0$, and $\phi_2 = 0$) shows plots of the transfer function for various values of c. With $c = 0$, the transfer function is purely sinusoidal and we have no linearization. $c = 0.5$ corresponds to the optimum condition of $d^3f/d\Psi'^3 = 0$, where all third-order nonlinear distortions vanish. In practical CATV systems, a slightly larger value of c is optimal. This is due to the contribution of higher-order nonlinear distortions (fifth-order, seventh-order, etc.).

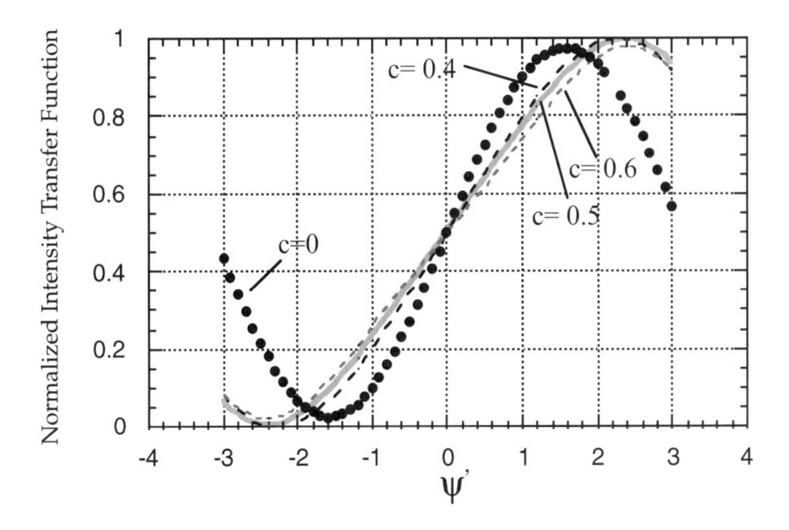

Figure 7.20: Normalized transfer function versus driving voltage parameter $\Psi' (= \pi V_{RF}/V_{\pi})$ for dual series linearization technique with different values of c. After Ref. [39].

7.4.4 Optical Feedforward Linearization Technique

The optical feedforward linearization technique was used in LEDs, laser diodes, and discrete MZI modulators [14,40], as illustrated in Fig. 7.21. The basic principle is very simple. An error voltage equal to the difference between the original modulating signal and the main modulator output signal is used to drive a feedforward modulator. Ideally, the error voltage contains only the nonlinear distortions generated in the main modulator. The output from the feedforward modulator is then combined with the main modulator output to cancel the related nonlinear distortions. The phase modulator before the main modulator is used to decorrelate the phase relationship between the two arms so the phase-to-intensity noise due to multipath-interferometric effects (see Section 9.3.1) will not occur. This technique has also been implemented in commercial products.

7.5 High-Power Optical Sources for External Modulation Systems

Since the total insertion loss of a $LiNbO_3$ modulator is rather high, one must use a high-power CW optical source so that the power budget of an external modulation system can compete with that of a direct-modulation system using high-power DFB lasers. In 1.55-μm systems, erbium-doped fiber amplifiers can be used at the output ports of MZI or BBI modulators to help boost the output power. Many of today's

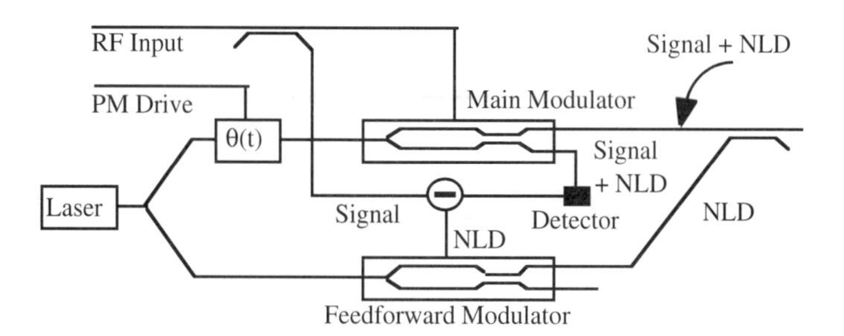

Figure 7.21: Lightwave CATV transmitter based on feedforward linearized external modulator. After Ref. [40].

commercial EDFAs can have output power > 20 dBm. Therefore, a high-power 1.55-μm DFB laser with a typical power of 20 to 30 mW is quite sufficient to serve as the 1.55-μm CW optical source, because the power loss due to a modulator can easily be compensated by a high-output-power EDFA. It can, however, be argued that the introduction of an EDFA degrades the system noise performance, and sometimes it is preferable to use high-power 1.55-μm DFB lasers with power > 50 mW without using EDFAs (assuming no photorefractive effects).

On the other hand, in a 1.3-μm multichannel AM-VsB system, unless practical 1.3-μm optical amplifiers are such as well-engineered PDFFAs are available one has to resort to a high-power solid-state Nd:YAG laser to provide up to 200 mW output power into a polarization-maintaining single-mode fiber [14].

Figure 7.22a shows the construction of a 1.3-μm diode-pumped Nd:YAG laser [14]. In response to the 0.5 to 1-W 0.8-μm AlGaAs laser array (emitted from a broad active area of about 200 μm $\times$ 1 μm), the Nd:YAG emits light at 1319 nm if a wavelength-selective 1.3-μm cavity can be carefully designed (see Fig. 7.22).

The Nd:YAG laser, whose energy diagram is shown in Fig. 7.22b, provides a spectrally narrow (< 0.5 nm wide), stable multimode emission spectrum at 1319 nm. The linewidth of each individual longitudinal mode is < ~20 KHz. Its RIN has been measured to be lower than -165 dB/Hz, with the peak of the relaxation oscillations typically occurring around 100–200 kHz (Fig. 7.23, see p. 246). This relaxation peak can be up-converted to each CATV channel because of second-order beating effects (this was described in Section 4.9.2), and therefore must be carefully eliminated using a noise-suppression feedback loop [14].

Typical output power from solid-state Nd:YAG lasers ranges from 50 mW (17 dBm) to 200 mW (23 dBm). Assuming the external modulator has a total loss of 8 dB (including the 3-dB loss due to DC bias at quadrature point, the 3-dB loss at an output Y-junction or a directional coupler, and waveguide–fiber coupling losses), the maximum output power from each SMF-arm of a MZI or BBI modulator can reach $23 - 8 = 15$ dBm.

High-power (>20 mW), low-intensity-noise (< -155 dB/Hz) 1550-nm DFB has been made by increasing the cavity length (~500 to 1000 μm) while decreasing

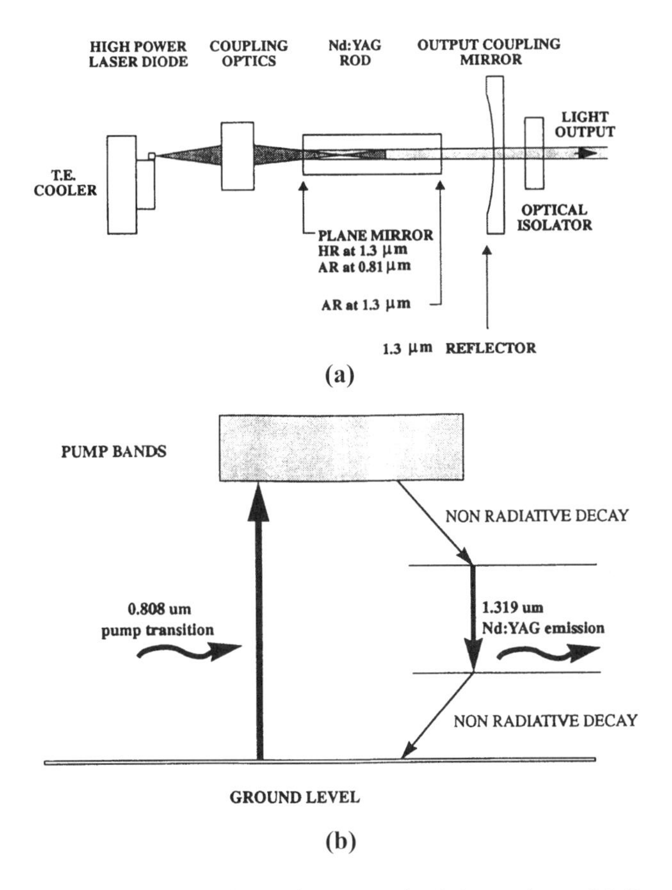

Figure 7.22: (a) Construction of a diode-pumped solid-state laser. (b) Simplified energy-level diagram of Nd:YAG. After Ref. [14] (© 1993 IEEE).

the distributed feedback coupling coefficient κ so that the coupling strength κL can be kept around 1, and by employing low-reflectivity/high-reflectivity coatings on the front/rear facets. Record CW optical power coupled into a single-mode fiber is as high as 108 mW [6].

High-power EDFAs which are needed to boost output power are discussed in Chapter 6.

7.6 Stimulated Brillouin Scattering

Although the optical power from an external-modulator-based transmitter can be increased almost indefinitely by using a high-power booster EDFA (see Section 6.1.8.2), the maximum optical power which can be launched into a single-mode fiber is limited by stimulated Brillouin scattering (SBS). SBS causes reflection of

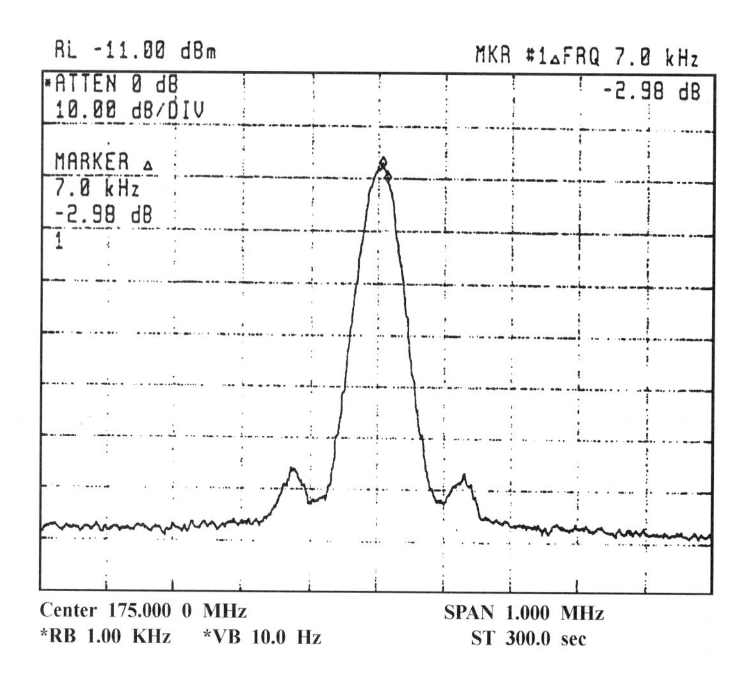

Figure 7.23: Measured linewidth of a 1319-nm Nd:YAG laser. The linewidth is 14 kHz in this case. Also shown in the relaxation oscillation peak occurring at about 130 kHz from the center.

input optical waves by the refractive-index grating which is formed by the acoustic (sound) waves traveling at a velocity of ≈6 km/s in the fiber [41]. Considering this "effective" fiber grating formed by the acoustic waves, we can see that the Bragg condition (see Section 3.3) is met when the grating period $\Lambda = \lambda/2n = \sim 1.55\text{-}\mu\text{m}/(2 \times 1.45) = 0.535\ \mu\text{m}$ for 1.55-μm incident lights. When the injected light is strong, the index grating is stimulatory enhanced by the input optical wave and causes more reflections. Therefore, we will have directly reflected light at a frequency lower than the 1.55-μm incident light by the acoustic phonon frequency of ~11 GHz (= 6 km/s/0.535 μm), which is termed the Brillouin–Stokes (frequency) shift, in order to conserve photon energy and momentum. In other words, the scattering of a photon from an acoustic phonon results in a backward-propagating lower-energy photon. Therefore, optical fiber loss is incurred if SBS occurs. But the scattering cross-sections are sufficiently small that power loss is negligible at low power levels. At high optical power levels, however, the nonlinear phenomenon of SBS can lead to significant fiber loss because the intensity of the backward-scattered light grows exponentially once the incident light power exceeds a threshold level.

Conventional SBS threshold power P_{th} is defined as the input power at which the backward Stokes power equals the pump power at the fiber input. For continuous-wave light, P_{th} is given approximately as [42–44]

$$P_{th,CW} = 21\frac{\zeta A_{eff}}{g_B L_{eff}} \cdot \frac{\Delta v_B \otimes \Delta v_L}{\Delta v_{\text{int}}}, \tag{7.56}$$

where ⊗ denotes convolution of the laser linewidth $\Delta\nu_L$ and the Brillouin bandwidth $\Delta\nu_B$; for Gaussian profiles, $\Delta\nu_L \otimes \Delta\nu_B = (\Delta\nu_L^2 + \Delta\nu_B^2)^{1/2}$, and for Lorentzian profiles, $\Delta\nu_L \otimes \Delta\nu_B = \Delta\nu_L + \Delta\nu_B$. $\zeta = 2$ for random polarization state, A_{eff} is the effective core area, g_B is the peak Brillouin gain coefficient (= 4.6×10^{-11} m/W) [45], and L_{eff} represents the effective interaction length, given by

$$L_{eff} = [\, 1 - exp(-\alpha L) \,] \,/\, \alpha, \tag{7.57}$$

where α is the fiber attenuation coefficient and L the fiber length; $\Delta\nu_L$ is the laser linewidth; $\Delta\nu_{int}$ is the intrinsic Brillouin gain bandwidth (≈ 35 MHz for silica-based fiber at 1.55 μm [46]., which is the local bandwidth of Brillouin gain. $\Delta\nu_B$ is the Brillouin bandwidth of the fiber, which expresses the bandwidth of the Brillouin gain accumulated along the fiber length. The $\Delta\nu_B$ values of conventional fiber with a uniform Brillouin frequency shift coincide with the intrinsic Brillouin gain bandwidth $\Delta\nu_{int}$. On the other hand, for a fiber with a nonuniform Brillouin frequency shift along its length, $\Delta\nu_B$ is larger than $\Delta\nu_{int}$. Equation (7.56) suggests that the threshold power can be increased by broadening the Brillouin bandwidth $\Delta\nu_B$ [44] of the fiber or the laser linewidth $\Delta\nu_L$. If we recall the popular structure of commercially available amplitude modulators given in Fig. 7.10c, the very first phase modulator integrated near its input port can be used to broaden $\Delta\nu_L$ if an out-of-band signal is applied to its electrodes [47].

From Eq. (7.56), when the laser linewidth is much smaller an the spontaneous Brillouin bandwidth, a simplified expression for the threshold power of CW light is given by

$$P_{th,CW} \approx 21\frac{\zeta A_{eff}}{g_B L_{eff}}. \tag{7.58}$$

Figure 7.24 shows the theoretical SBS threshold for narrow-linewidth CW light, using Eq. (7.58) as a function of fiber length for several fiber losses, where $\zeta = 2$ and effective core diameter $A_{eff} = 104\ \mu m^2$ were used in the calculation [43]. It can be seen that the SBS threshold decreases as the fiber length increases up to about 50 km and then asymptotically reaches a constant value. This is because L_{eff} becomes $1/\alpha$ for long fibers. For a fiber with a 0.2-dB/km loss at 1.55 μm and a length greater than 50 km, the SBS threshold will be as small as 4 mW. Also, the threshold at 1.3 μm is 8 mW for a long single-mode fiber with 0.4-dB/km loss. As we learned earlier, the output optical power in the two single-mode fiber arms can reach 10 to 15 mW for a 1.3 μm external modulator and can reach > 100 mW for amplified external modulators; hence, SBS can indeed occur for relatively long fiber links. More related system issues will be discussed in Chapter 9.

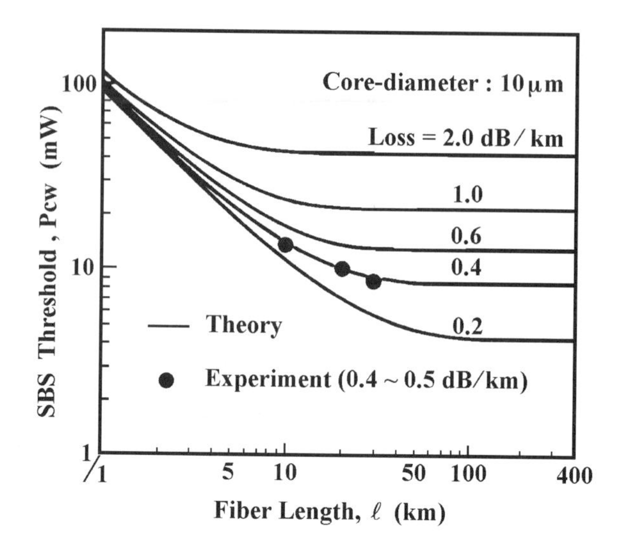

Figure 7.24: Calculated and measured SBS threshold as a function of transmission fiber length when $\Delta\nu_L << \Delta\nu_B$. After Ref. [43] (© 1988 IEEE).

Problems

7.1 Based on Eq. (7.6), derive Eqs. (7.8) and (7.9).

7.2 Derive Eq. (7.55). Based on Eq. (7.55) and Eqs. (7.19)–(7.21), derive the worst-case *C*/CTB for 42 and 80 channels, respectively, and compare your calculated results with Fig. 7.12. At $c = 0.5$, how significantly is *C*/CTB improved?

7.3 In Fig. 7.18, assume the transfer functions of the primary and secondary modulators are

$$P_{out,p} = \frac{c^3 P_{in,s} L_a}{2}\left[1 - \sin\left(\frac{c\pi V}{V_\pi} + \phi_1\right)\right]$$

and

$$P_{out,s} = \frac{P_{in,s} L_a}{2}\left[1 - \sin\left(\frac{c\pi V}{V_\pi} + \phi_2\right)\right],$$

where $\phi_1 \neq \phi_2$. Following the procedure given in Eqs. (7.46) and (7.47), estimate how critical it is to have $\phi_1 = \phi_2$ (controlled by the phase modulator at the output of the secondary modulator). What is the tolerance range to deviate away from this ideal condition? Your result should be able to explain why the dual parallel linearization technique is difficult to implement practically.

7.4 In obtaining Fig. 7.16, we used the worst-case condition given in Eq. (7.37). However, in practical implementations, we generally do not find the requirements on amplitude and phase ripples as strict as those shown in Fig. 7.16. Try modifying Eq. (7.37) so that the amplitude and phase ripple requirements can be lowered.

7.5 Double-check the results given in Table 7.1.

References

[1] C. Y. Kuo, "High-performance optically amplified 1550 nm lightwave AM-VSB CATV transport system," Tech. Digest, Conf. Optical Fiber Commun. paper WN2, 1996.

[2] H. Dai, S. Ovadia, and C. Lin, "Hybrid AM-VSB/M-QAM multichannel video transmission over 120 km of standard single-mode fiber with cascaded erbium-doped fiber amplifiers," IEEE Photon. Technol. Lett., vol.8, pp.1713–1715, December 1996.

[3] L. G. Kazovsky and D. A. Atlas, "Miniature Nd:YAG lasers: Noise and modulation characteristics," J. Lightwave Technol., vol.8, pp.294–301, 1990.

[4] R. B. Childs and V. A. O'Byrne, "Multichannel AM video transmission using a high-power Nd:YAG laser and linearized external modulator," IEEE J. Selected Areas Commun. vol.8, pp.1369–1376, 1990.

[5] T. R. Chen, J. Ungar, J. Iannelli, S. Oh, H. Luong, and N. Bar-Chaim, "High power operation of InGaAsP/InP multiquantum well DFB lasers at 1.55 μm wavelength," Electron. Lett., vol. 32, p.898, 1996.

[6] T. R. Chen, J. Iannelli, V. Leyva, M. Newkirk, H. Luong, and N. Bar-Chaim, "High power 1550 nm DFB lasers for long distance video transmission," Conf. Proc., IEEE LEOS'97, paper MI3, vol. 1, pp.69–70, November 1997.

[7] R. C. Alferness, "Titanium-diffused lithium niobate waveguide devices," in *Guided-Wave Optoelectronics*, 2nd Edition, Chapter 4, Springer-Verlag, Berlin, 1990.

[8] S. K. Korotky and R. C. Alferness, "Waveguide electrooptic devices for optical fiber communication," in *Optical Fiber Telecommunications II*, Eds., S. E. Miller and I. P. Kaminow, Chapter 11, Academic Press, Orlando, Florida, 1988.

[9] K. M. Kissa, P. G. Suchoski, and D. K. Lewis, "Accelerated aging of annealed proton-exchanged waveguides," J. Lightwave Technol., vol.13, pp.1521–1529, July 1995.

[10] M. A. Powell and A. O. Donnell, "What integrated optics is really used for," Optics Photonics News, pp.23–29, September 1997.

[11] R. Syms and J. Cozens, *Optical Guided Waves and Devices*, McGraw-Hill, New York, 1993.

[12] T. R. Halemane and S. K. Korotky, "Distortion characteristics of optical directional coupler modulators," IEEE Trans. Microwave Theory Tech., vol.38, pp.669–673, May 1990.

[13] S. Thaniyavarn, "Modified 1×2 directional coupler waveguide modulator," Electron. Lett., vol.22, pp.941–942, 1986.

[14] M. Nazarathy, J. Berger, A. J. Ley, I. M. Levi, and Yishai Kagan, "Progress in externally modulated AM CATV transmission systems," J. Lightwave Technol., vol. 11, pp.82–105, January 1993.

[15] A. M. Yurek, P. G. Suchoski, S. W. Merrit, and F. J. Leonberger, "Commercial $LiNbO_3$ integrated optic devices," Optics Photonics News, pp.26–30, June 1995.

[16] W. B. Bridges and J. H. Schaffner, "Distortion in linearized electro-optic modulators," IEEE Trans. Microwave Theory Tech., vol. 43, pp.2184–2197, September 1995.

[17] M. M. Howerton, C. H. Bulmer, and W. K. Burns, "Effect of intrinsic phase mismatch on linear modulator performance of the 1×2 directional coupler and Mach-Zehnder interferometer," J. Lightwave Technol., vol.8, pp.1177–1185, August 1990.

[18] R. L. Homan, J. R. Busch, and C. M. Verber, "Z-axis propagation and the avoidance of optical damage in lithium niobate waveguides," Tech. Digest, IGWO'84, paper WC2, 1984.

[19] G. T. Harvey, G. Astfalk, A. Y. Feldblum, and B. Kassahun, "The photorefractive effect in titanium indiffused lithium niobate optical directional couplers at 1.3 μm," J. Quantum Electron., QE-22, pp.939–946, 1986.

[20] G. T. Harvey, "The photorefractive effect in directional coupler and Mach-Zehnder $LiNbO_3$ optical modulators at a wavelength of 1.3 μm," J. Lightwave Technol., vol., p.872, 1988.

[21] T. Fujiwara, S. Sato, and H. Mori, "Wavelength dependence of photorefractive effect in Ti-indiffused $LiNbO_3$ waveguides," Apply. Phys. Lett., vol.54, pp.975–977, 1989.

[22] F. Koyama and K. Iga, "Frequency chirping in external modulators," J. Lightwave Technol., vol.6, pp.87–93, January 1988.

[23] G. E. Betts, F. J. O'Donnell, and K. G. Ray, "Effect of annealing on photorefractive damage in Titanium-indiffused $LiNbO_3$ modulators," IEEE Photo. Technol. Lett., vol.6, pp.211–213, February 1994.

[24] J. F. Lam and G. L. Tangonan, "Optical modulation system with enhanced linearization properties," IEEE Photon. Technol. Lett, vol.3, pp.1102–1104, December 1991.

[25] M. L. Farwell, Z. Q. Lin, E. Wooten, and W. S. C. Chang, "An electro-optic intensity modulator with improved linearity," IEEE Photon. Technol. Lett., vol.3, pp.792–795, September 1991.

[26] J. H. Schaffner, J. F. Lam, C. J. Gaeta, G. L. Tangonan, R. L. Joyce, M. L. Farwell, and W. S. C. Chang, "Spur free dynamic range measurements of a fiber optic link with traveling-wave linearized directional coupler modulators," IEEE Photon. Technol. Lett., vol.6, pp.273–275, February 1994.

[27] G. E. Bodeep and T. E. Darcie, "Semiconductor lasers versus external modulators: A comparison of nonlinear distortion for lightwave subcarrier CATV applications," IEEE Photon. Technol. Lett., vol.1, pp.401–403, November 1989.

[28] S. K. Korotky and J. J. Veselka, "RC circuit analysis of long-term Ti:$LiNbO_3$ bias stability," OSA Tech. Digest, Integrated Photon. Res., vol. 3, pp.187–189, 1994.

[29] A. R. Beasumont, B. E. Daymond-John, and R. C. Booth, "Effect of ambient water vapor on stability of lithium niobate electro-optic waveguide devices," Electron. Lett., vol.22, pp.262–263, 1986.

[30] Y. Kaneyama, T. Funaki, S. Yamanaka, H. Kawaji, K. Tomotsune, and A. Kawatani, "Reduction of DC drift in $LiNbO_3$ optical waveguide devices by ion exchange," Conf. Proc., LEOS'94, vol.2, pp.280–281.

[31] M. Seino, T. Nakazawa, Y. Kubota, M. Doi, T. Yamane, and H. Hakogi, "A low DC-drift Ti:$LiNbO_3$ modulator assured over 15 years," Tech. Digest, Conf. on Optical Fiber Commun. postdeadline paper PD3, 1992.

[32] R. W. Ade, et al., "Bias-free lithium niobate modulators for high speed digital applications," Proc. Opto'95, Paris, France, March 1995.

[33] C. H. Bulmer, W. K. Burns, and S. C. Hiser, "Pyroelectric effects in $LiNbO_3$ channel-waveguide devices," Apply. Phys. Lett., vol.48, pp.1036–1038, 1986.

[34] H. Nagata, K. Kiuchi, and T. Saito, "Studies of thermal drift as a source of output instabilities in Ti:$LiNbO_3$ optical modulators," J. Appl. Phys., vol.75, pp.4762–4764, 1994.

[35] H. Nagata and K. Kiuchi, "Temperature dependence of dc drift of Ti:$LiNbO_3$ optical modulators with sputter deposited SiO2 buffer layer," J. Appl. Phys., vol.73, pp.4162–4164, 1993.

[36] S. K. Korotky and R. M. de Ridder, "Dual parallel modulation schemes for low-distortion analog optical transmission," IEEE J. Selected Areas Commun., vol.8, pp.1377–1381, September 1990.

[37] J. L. Brooks, G. S. Mauer, and R. A. Becker, "Implementation and evaluation of a dual parallel linearization system for AM-VSB video transmission," IEEE J. Lightwave Technol., vol.11, pp.34–41, January 1993.

[38] H. Skeie and R. V. Johnson, "Linearization of electro-optic modulators by a cascade coupling of phase modulating electrodes," Proc. SPIE, vol.1538, pp.153–164, 1991.

[39] H. Skeie, "An optically linearized modulator for CATV applications," Proc. SPIE, vol.2291, pp.227–238, 1994.

[40] J. D. Farina, B. R. Higgins, and J. P. Farina, "New linearization technique for analog fiber-optic links," Tech. Digest, Optical Fiber Commun. Conf. (OFC), paper ThR6, San Jose, California, 1996.

[41] G. P. Agrawal, *Nonlinear Fiber Optics*, Academic Press, New York, 1995.

[42] R. G. Smith, "Optical power handling capacity of low loss optical fibers as determined by stimulated Raman and Brillouin scattering," Appl. Opt., vol.11, pp.2489–2494, 1972.

[43] Y. Aoki, K. Tajima, and I. Mito, "Input power limits of single-mode optical fibers due to stimulated Brillouin scattering in optical communication systems," J. Lightwave Technol., vol.6, pp.710–719, May 1988.

[44] K. Shiraki, M. Ohashi, and M. Tateda, "Performance of strain-free stimulated Brillouin scattering suppression fiber," J. Lightwave Technol., vol.14, pp.549–553, April 1996.

[45] R. H. Stolen, "Nonlinear properties of optical fibers," in *Optical Fiber Telecommunications*., Eds., S. E. Miller and A. G. Chynoweth, Chapter 5, 125–150. Academic Press, New York, 1979.

[46] T. O. Tsun, A. Wada, T. Sakai, and R. Yamauchi, "Novel method using white spectral probe signals to measure Brillouin gain spectra of pure silica core," Electron Lett., vol.28, pp.247–249, 1992.

[47] F. W. Willems, W. Muys, and J. S. Leong, "Simultaneous suppression of stimulated Brillouin scattering and interferometric noise in externally modulated lightwave AM-SCM systems," IEEE Photon. Technol. Lett., vol.12, pp.1476–1478, December 1994.

Chapter 8

RF Modem Design for HFC Systems

A radio frequency (RF) modulator/demodulator (modem) for transporting digital data/voice/video signals is the key electrical component in an HFC system. It is important to understand the design principles of an RF modem in order to have full control of transmission characteristics in an HFC system. The basic block diagram of an RF modem is shown in Fig. 8.1. The transmitter part of an RF modem consists of a data scrambler, a Reed–Solomon (RS) forward-error-correction encoder, a convolutional interleaver, a byte to m-tuple converter, a differential encoder, an M-ary quadrature (QAM) or vestigial sideband (VSB) modulator, and an RF upconverter. The receiver part of an RF modem consists of an RF tuner (tunable down-converter), a QAM or VSB demodulator with differential decoder, an m-tuple to byte converter, a convolutional de-interleaver, a RS decoder, and a descrambler. In this chapter, we will constantly refer to what have been standardized in DAVIC [1] (Digital Audio Video Council, for predominantly digital video broadcasting

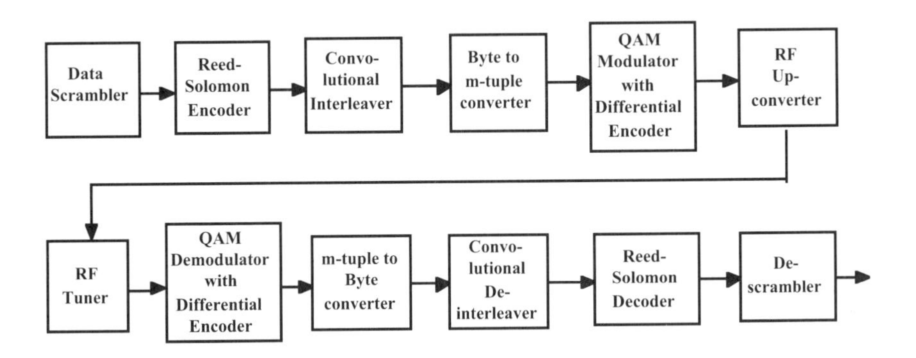

Figure 8.1: Basic building blocks in an HFC RF modem transmitter and receiver. After Ref. [2].

(DVB)), ITU J.83 Annex A [2] (based on European DVB), ITU J.83 Annex B [2] (based on General Instrument's DigiCipher II), IEEE 802.14 [3] (based on J.83 Annex A and B), MCNS DOCSIS [4] (based on J.83 Annex B), and ITU J.83 Annex D [2] (based on 16-VSB transmission for digital multiprogrammed television distribution by cable). In addition, we will review the basic principles of adaptive equalizers, carrier recovery, and timing recovery design for QAM and VSB modems.

8.1 Scrambler/Descrambler

The first block in a QAM or VSB modem transmitter is a data scrambler. A scrambler is used to achieve DC balance and eliminate long sequences of zeros to ensure accurate timing recovery in the receiver. There are two forms of scrambling—self-synchronization and frame synchronization. Both types of scramblers use maximal-length-shift register sequences or pseudo-random sequences, and these sequences are generated by the pseudo-random binary sequence (PRBS) generator shown in Fig. 8.2a. Note that the output of the PRBS generator $P(x)$ can be expressed by a *primitive*[1] polynomial with degree m given by

$$P(x) = 1 + \sum_{i=1}^{m} h_i x^i, \qquad h_i = \begin{cases} 1 & i = 0, m \\ 0,1 & i = 1,2,\ldots,m-1, \end{cases} \tag{8.1}$$

where x represents a bit delay (see Fig. 8.2a). A PRBS generator is made up of a feedback shift register with m stages and modulo–2 adders. The choice of m determines the period of the sequence, which is equal at most to $2^m - 1$ (the "all zero" state must be excluded, as it rules out any further changes in the registers' states).

Let us first consider the case of self-synchronization. As shown in Fig. 8.2b, the input data to the scrambler is expressed as a polynomial $I(x)$; the scrambled data is represented by $S(x)$. Note that $S(x)$ is fed back to the input of the PRBS generator. Therefore,

$$S(x) = I(x) + S(x) \sum_{i=1}^{m} h_i x^i . \tag{8.2}$$

Consequently, we have

$$S(x) = \frac{I(x)}{1 - \sum_{i=1}^{m} h_i x^i} . \tag{8.3}$$

[1]A *primitive* polynomial of degree m is an irreducible polynomial which is a divisor of $x^n + 1$, where $n = 2^m - 1$, and is not a divisor of $x^n + 1$ for $n < 2^m - 1$, while an irreducible polynomial is not divisible by any polynomial of degree less than m but greater than zero (see Appendix A).

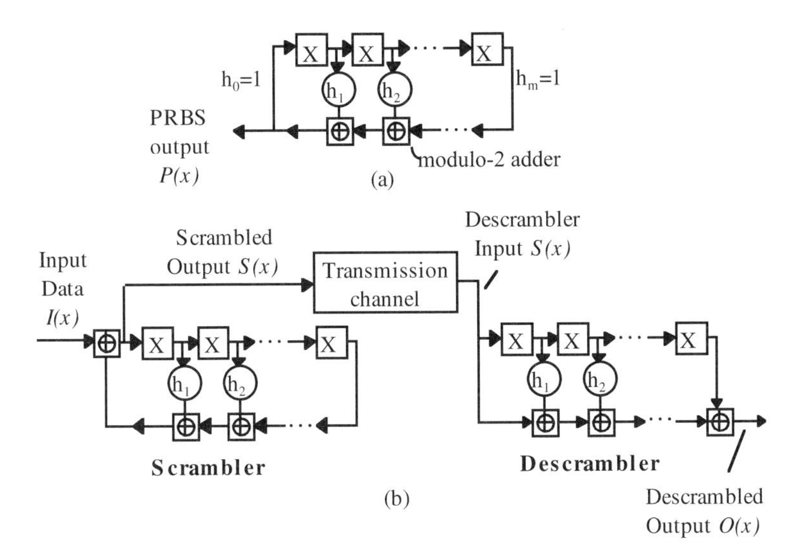

Figure 8.2: (a) A pseudo-random bit sequence (PRBS) generator. (b) A self-synchronized scrambler and descrambler. The summation is modulo-two.

In the absence of transmission errors, *S(x)* becomes the input of the descrambler, which uses an identical PRBS generator as that in the scrambler. We then have the output of the descrambler as

$$O(x) = S(x)[1 + \sum_{i=1}^{m} h_i x^i] . \tag{8.4}$$

Now from Eqs. (8.3) and (8.4), we have

$$O(x) = \frac{I(x)}{[1 - \sum_{i=1}^{m} h_i x^i]}[1 + \sum_{i=1}^{m} h_i x^i] . \tag{8.5}$$

Since $A + B = A - B$ in a binary system, we therefore prove that $O(x) = I(x)$. The problem with self-synchronization is error propagation. If the received data before the descrambler input is $S(x) + e(x)$, where $e(x)$ represents the error polynomial caused by the transmission channel, then the output of the descrambler becomes $I(x) + e(x)P(x)$. If $e(x) = 1$ (single error) and $P(x) = 1 + x^{14} + x^{15}$, then there are three errors in the descrambler output. The more nonzero taps ($h_i = 1$) there are in the shift register, the more errors there exist. Therefore, to simplify implementation and minimize error propagation, one usually designs a scrambler governed by a primitive polynomial with a minimum number of nonzero taps, that is, three.

To avoid the error propagation problem, the self-synchronization scheme can be changed to non-self-synchronization which is accomplished by a frame synchronization mechanism, as illustrated in Fig. 8.3a. The scrambler performs a modulo-2 summation of the user's bit stream $I(x)$ with the output of a PRBS generator $P(x)$ to generate the scrambled output $S(x)$. The descrambler contains an identical PRBS generator to recover the original user's bit streams. This recovery follows from the relation $S(x) \oplus P(x) = I(x) \oplus P(x) \oplus P(x) = I(x)$ since $P(x) \oplus P(x) = 0$. Figure 8.3b shows the construction of an ITU J.83 Annex A and DAVIC scrambler/descrambler which uses a PRBS generator with $1 + x^{14} + x^{15}$. Loading of the sequence "100101010000000" into the PRBS registers shall be initiated at the start of every eight transport packets.

Note that a scrambler can disperse the energy in the frequency domain by decreasing the spacing between neighboring spectral components to $1/[(2^m - 1)T_s]$, that

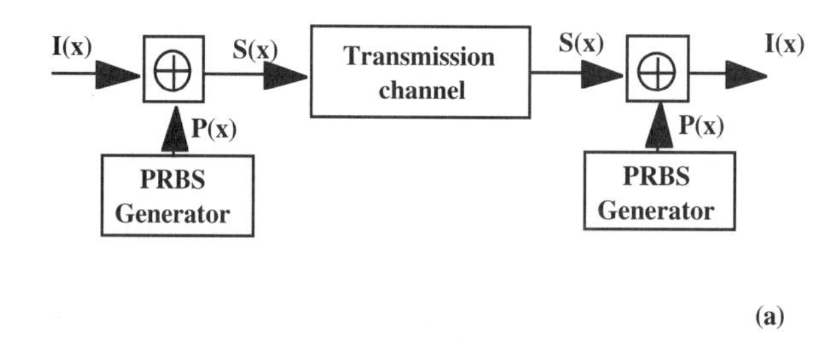

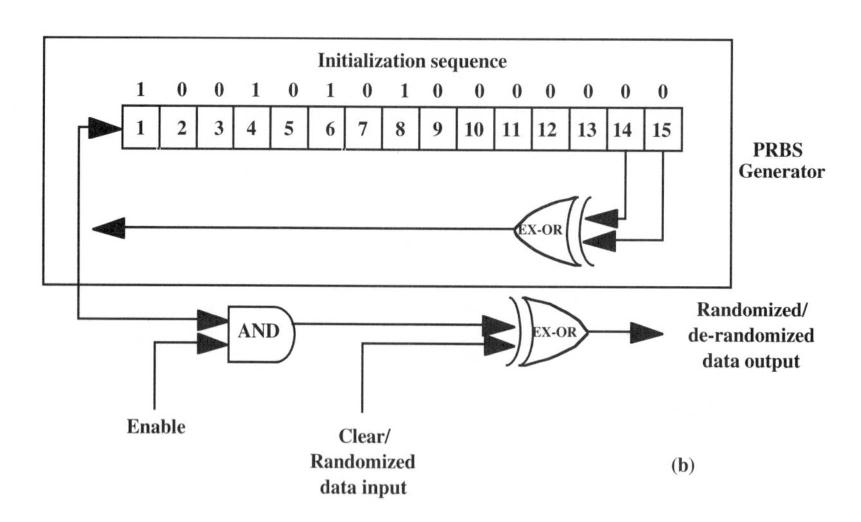

Figure 8.3: (a) A non-self-synchronized scrambler/descrambler. (b) The construction of a non-self-sychronized scrambler/descrambler in ITU J.83 Annex A and DAVIC.

is, the period of the scrambled output is increased to $(2^m - 1)T_s$ [5], where T_s is the symbol period.

8.2 Reed–Solomon Codecs

One special class of linear block codes[2] is the *nonbinary* RS codes which are particularly useful for multilevel signals such as M-QAM and N-VSB signals. A nonbinary block codes is well matched to an M-ary modulation technique for transmitting 2^m possible symbols ($M = 2^m$).

RS codes are constructed in a number system called Galois field or GF(2^m). In systematic RS coding, $n - k$ parity check symbols are appended at the end of each block of k information symbols. This brings the total size of the RS coded block to be n, and the code is expressed by a pair of numbers (n,k).

An advantage of nonbinary codes can be illustrated by the following example: Consider a binary (255, 239) code and a nonbinary (255, 239) RS code with $m = 8$, that is, each symbol is composed of 8 bits. The former has 2^{239} message vectors out of 2^{255} code vectors, while the latter has $(2^8)^{239} = 2^{1912}$ message vectors out of $(2^8)^{255} = 2^{2040}$ code vectors. Therefore, the latter has a smaller fraction of possible words become codewords, and can thus offer a much larger minimum distance d_{min},[3] and consequently is more immune to noise and interference. Note that RS codes achieve the largest possible code minimum distance for any linear code with the same encoder input and output block lengths. A t-symbol-error-correcting RS code with an alphabet of 2^m symbols can be described by the following parameters [6]:

$$\begin{aligned} n &= 2^m - 1 \\ k &= 2^m - 1 - 2t \\ d_{\min} &= n - k + 1 \\ t &= \frac{d_{\min} - 1}{2}. \end{aligned} \tag{8.6}$$

Note that RS codes are particularly useful for burst-error correction, and relative long codes can be implemented with an efficient hard-decision decoding algorithm [7].

For DAVIC and ITU J.83 Annex A standards, the RS code is constructed over GF(2^8) and is given by (204,188). Each code symbol is represented by a byte, or equivalently 8 bits. RS (204,188) is a *shortened* RS code which is implemented by appending 51 bytes of zero before the information bytes at the input of a systematic RS (255,239) encoder; after the coding procedure these bytes are discarded. The

---

[2] See Appendix B for definitions and explanations.

[3] A measure of the difference between any two codewords in an (n,k) block code is the number of corresponding (nonbinary) symbols in which they differ. This measure is called the *Hamming distance*. The smallest value of the Hamming distance is denoted as d_{min}.

reason for shortening the code is to match the MPEG-2 transport packet format of 188 bytes, as shown in Fig. 8.4a, or the ATM transport packets, as shown in Fig. 8.4b.

In ITU J.83 Annex A/D and DAVIC, the primitive polynomial over GF(2) that generates the elements of the particular extended field GF(2^8) is (field generator polynomial)

$$P(X) = X^8 + X^4 + X^3 + X^2 + 1 \tag{8.7}$$

(for a simpler example of how a primitive polynomial *P(X)* generates the elements of GF(2^4), see pp. 32–33 of Ref. [8]), and the RS code generator polynomial is

$$\begin{aligned} g(X) &= \prod_{i=0}^{2t-1} (X - \alpha^i) = \prod_{i=0}^{15} (X - \alpha^i) \\ &= g_0 + g_1 X + g_2 X^2 + \ldots + g_{15} X^{15} \quad , \end{aligned} \tag{8.8}$$

where α is the root of *P(X),* that is, $P(\alpha) = 0$. Note that *g(X)* has coefficients from GF(2^8). A systematic form (the message bits are shifted into the rightmost k stages of a codeword register, and then the parity digits are appended by placing them in the lefmost $n - k$ stages) of RS codeword *C(X)* can be obtained from manipulating the message codeword *m(X)* (188 bytes of MPEG-II packet) in the following way:

$$C(X) = m(X) \cdot X^{16} + [m(X) \cdot X^{16} \bmod \quad g(X)] \, . \tag{8.9}$$

If

$$m(X) = m_0 + m_1 X + \ldots + m_{187} X^{187}, \tag{8.10}$$

where the coefficients are also from GF(2^8), then

$$m(X) \cdot X^{16} = m_0 X^{16} + m_1 X^{17} + \ldots + m_{187} X^{203} \tag{8.11}$$

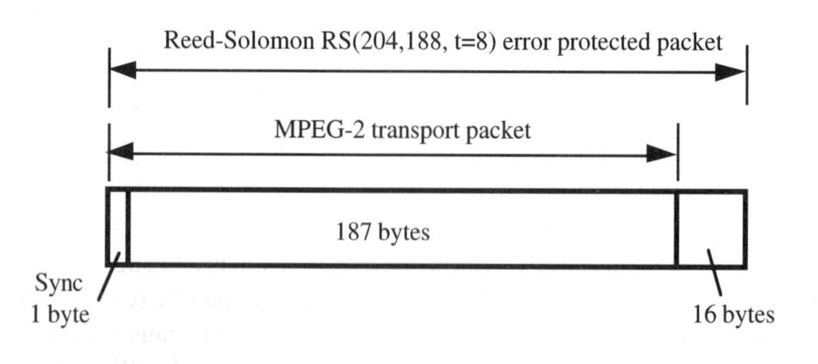

(a)

continued

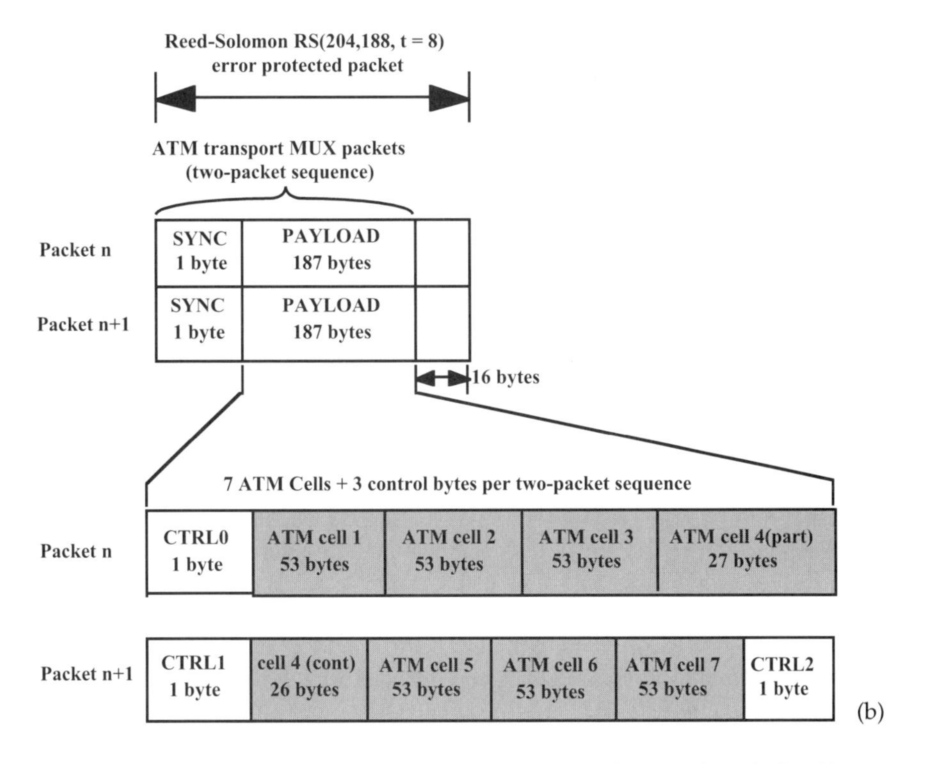

Figure 8.4: Reed–Solomon RS(204,188) error protected packet which includes (a) an MPEG–2 transport packet, and (b) ATM transport packets. After Ref. [1].

and we see that the message bits are right-shifted 16 positions. The second term on the right-hand side of Eq. (8.9) is a residual polynomial of degree 15 because the divisor $g(X)$ is of degree 16. Hence,

$$[m(X) \cdot X^{16} \quad \text{mod} \quad g(X)] = r_0 + r_1 X + \ldots + r_{15} X^{15}. \tag{8.12}$$

Consequently,

$$C(X) = r_0 + r_1 X + \ldots + r_{15} X^{15} + m_0 X^{16} + m_1 X^{17} + \ldots + m_{187} X^{203}. \tag{8.13}$$

8.3 Interleaver/De-interleaver

In a burst-error situation, the occurrence of a bit in error means that the next several bits are also likely to be in error. The large number of suddenly, sequentially occurring errors in a block may be more than an RS codec can handle (e.g., t = 8 or 8 bytes of errors can be corrected in RS(204,188)). An interleaver causes the burst errors to be spread out in time so that the errors can be handled by the RS decoder as if they were random errors. There are three types of interleavers: block, convolutional, and helical [9]. Since DAVIC, ITU J.83, and IEEE 802.14 all use convolutional interleavers, we shall concentrate only on the operation principle of convolutional interleavers.

A convolutional interleaver shuffles the coded RS symbols over a span of interleaving depth I. The span required is determined by the burst duration. A typical convolutional interleaver/de-interleaver is depicted in Fig. 8.5. Both the interleaver and the de-interleaver have I branches. The input and output commutator switches are assumed to operate synchronously. All switches change position after each RS symbol time so that successive encoder outputs enter different rows of the interleaver memory. Each interleaver and de-interleaver row is a first-in first-out (FIFO) shift register memory with depth $x \cdot J$ RS symbols (where $J = n/I$, n = RS code protected frame length, and $x = 0,1,2,\ldots,I-1$), as shown within the rectangles of Fig. 8.5. The first encoded output enters the top interleaver row, is transmitted over the channel immediately, and enters the de-interleaver memory of $(I-1)J$ symbols. The second encoded output enters the second row of the interleaver and is delayed J symbol times before transmission. Thus adjacent encoder outputs are transmitted J symbol times apart and are not affected by the same channel error burst. Upon reception, the second encoder symbol is delayed by an additional $(I-2)J$ RS symbol time for a total delay of $(I-1)J$ RS symbols. Observe that all symbols have the same delay after passing through both the interleaver and de-interleaver, so that the decoder input symbols at point B are in the same order as the encoder output symbols at point A (see Fig. 8.5). The end-to-end delay (latency) of the convolutional interleaver/de-interleaver is $IJ(I-1)$ RS symbols, and the memory required at both ends of the channel is $I[J(I-1)/2]$ RS symbols. The maximum noise burst length that can be tolerated by an RS codec is $(t \cdot y)IT_b$, where t is the number of correctable erroneous y-bit symbols per block [see Eq. (8.6)] and T_b is the bit period. For example, in ITU J.83 Annex A, the interleaver parameters are $I = 12$ and $J = 17$, and $t = 8$, $y = 8$, $T_b \approx 1/(5.274 \times 6)$ Mbps ≈ 0.0316 μs for a 64-QAM signal; therefore, the maximum noise burst length that can be tolerated by an RS(204,188) code is $8 \cdot 8 \cdot 12 \cdot 0.0316 \approx 24$ μs. For ITU J.83 Annex B, the above formulae for calculating latency and tolerable burst length must be modified (see Problem 8.5). This is much

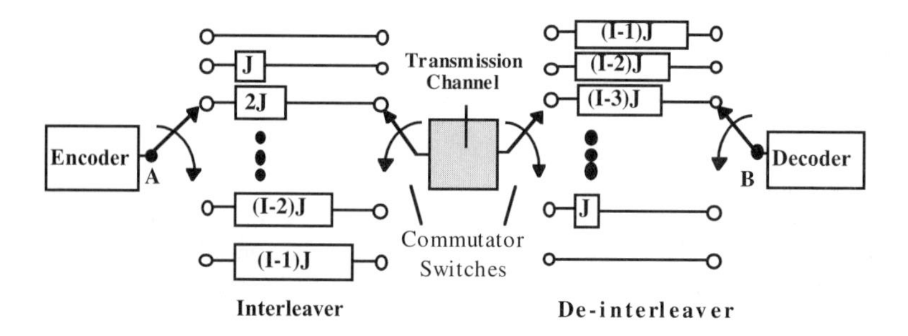

Figure 8.5: A convolutional interleaver and de-interleaver.

longer than the typical HFC system upstream bursty impulse of less than 5 μs (see Fig. 2.12 in Chapter 2). However, a much larger interleaver depth I may be required to cope with strong laser clipping-induced impulsive noise (see Sec. 9.2).

After convolutional interleaving, an exact mapping of bytes into symbols is performed, as shown in Fig. 8.1. The mapping relies on the use of byte boundaries in the modulation system. For the case of 2^m – QAM modulation, the process maps k bytes into n symbols, such that $8k = n \times m$.

8.4 M-QAM Modem Transmitter Design and General SER Performance

8.4.1 Baseband/IF Building Blocks for Analog and Digital Implementations

Quadrature amplitude modulation (QAM), a bandwidth-efficient modulation, was selected as the standard format for HFC transmissions by ITU, DAVIC, MCNS, and IEEE 802.14, primarily because it is the most common modulation format in use today for data transmission over bandwidth constrained channels such as twisted-pair telephone wires and digital microwave. The bandwidth efficiency of an M-QAM signal can be seen from the following example: To transmit a 30 Mb/s 64-QAM signal, only 30/ $\log_2 64$ or 5 MHz (= symbol rate) of bandwidth is required theoretically. In other words, the theoretical spectral efficiency is $\log_2 64$ or 6 bits/sHz. In practice, when considering the excess bandwidth factor (Section 8.4.3), the 64-QAM spectral efficiency is limited to about 5 bits/sHz.

A conventional QAM modulator is shown in Fig. 8.6a. The baseband non-return-to-zero (NRZ) data is first demultiplexed into two arms (in-phase and quadrature arms, or simply I and Q arms) by a serial to parallel (S/P) interface, which is followed by a Gray-code mapper, a differential encoder, a digital-to-analog converter (DAC), an amplitude equalizer (x/sin x) to equalize the sin x/x shape of the NRZ signal, and a square-root raised cosine (SRRC) shaping filter in each arm. Note that there is a corresponding SRRC filter in the receiver so that the total filter transfer function is raised cosine.[4] I and Q arms are named because the carriers in each arm (cos $\omega_c t$ and sin $\omega_c t$) are in quadrature with each other. Since the input data bits in an M-QAM signal are grouped into $\log_2 M$-bit symbols, half of $\log_2 M$ bits in each arm are converted to 2^K different amplitudes by the DAC, where $K = (\log_2 M)/2$. For example, in a 16-QAM signal, each symbol represents 4 bits, with 2 bits

---

[4]It is well known that in order to minimize the bit-error rate in the presence of white Gaussian noise in the communication channel, the optimum partitioning of the transmit and receive filters is the so-called "matched" filter pair whereby the transmit filter frequency response $H_T(\omega)$ and the receiver filter frequency response $H_R(\omega)$ satisfy $H_T(\omega) = H_R{}^*(\omega)$, *where* * denotes complex conjugate. Therefore, the transmitter and receiver low-pass filters are complex conjugates of one another and whose cascade is a Nyquist filter. It is for this reason that the transmitter and receiver baseband filters are chosen to be *SRRC* filters.

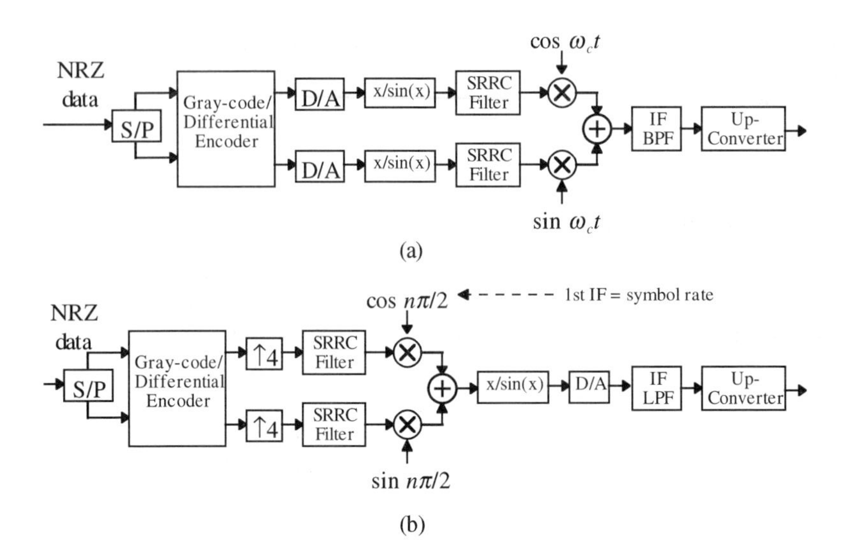

Figure 8.6: Block diagrams of (a) a conventional QAM modulator, and (b) a digitally implemented QAM modulator.

in each arm converted by the DAC to form 2^2 or 4 amplitude levels. With 4 amplitude levels in both the I and Q dimension, we then have $4 \times 4 = 16$ points on the I–Q diagram, as shown in Fig. 8.12a, which is called the constellation diagram.

The four-level pulse-amplitude-modulated pulse-amplitude-modulated (PAM) signals in each arm are then up-converted by the two quadrature (first IF) carriers, and then recombine and pass through an IF bandpass filter. Therefore, the mathematical expression of an M-QAM signal is given by

$$m(t) = I(t)\cos\omega_c t + Q(t)\sin\omega_c t\,, \tag{8.15}$$

where $I(t)$ and $Q(t)$ are the two 2^K-level pulse amplitude modulated (PAM) signals. It is apparent from Eq. (8.15) that an M-QAM signal has a form of the double-sideband–suppressed-carrier (DSB-SC) modulation. Note that the Gray-code mapper maps each group of $\log_2 M$ bits to represent one of the M symbols on the constellation diagram by following a certain rule, and the differential encoder is used to resolve the 90° phase ambiguity (see Section 8.4.2 for detailed discussions).

The IF portion of an M-ary QAM transmitter can also be implemented digitally (excluding the IF bandpass filter and the RF up-converter), as shown in Fig. 8.6b. Selecting a filter with an oversampling factor of 4 (i.e., 4 sampled points per symbol) and forcing the IF center frequency to equal the symbol rate result in oscillator samples of $\cos(n\pi/2)$ and $\sin(n\pi/2)$ which have trivial values of 1,0, –1,0,..., (with every 4 points forming a symbol cycle) thus eliminates the need for high-speed digital multipliers and adders to implement the mixing functions in both the modulator and demodulator [10]. The selections also cause half of the cosine and sine oscillator samples to be zero; therefore the complexity of the following Nyquist finite-impulse-response (FIR) filter is reduced by 50%. Note a single high-resolution DAC (e.g., 12 bits) in Fig. 8.6b can be used to replace two low-resolution DACs in Fig. 8.6a (e.g., 6 bits).

8.4.2 Differential Encoder/Decoder

We note that the M-QAM constellation is 90° symmetric. Consequently, there is an inherent 90°-phase ambiguity in the QAM demodulator, that is, the carrier phase detector in the QAM receiver can lock the carrier every 90° phase error. Differential encoding which encodes the next symbol based on the input symbol and the previous encoded symbol is one of the techniques to avoid phase ambiguity. A differential encoder that was defined in the ITU J.83 Annex A encoding scheme is given in Table 8.1 and schematically shown in Fig. 8.7. As an example, if the previous encoded symbol is 00 with an initial phase ϕ_0, then the input sequence

01 10 00 11

will be encoded as

01 00 00 11

and the corresponding transmitted phase will be

$$\phi_0 + \pi/2 \quad \phi_0 + 3\pi/2 \quad \phi_0 + 3\pi/2 \quad \phi_0 + \pi.$$

Table 8.1: Differential encoding rule for ITU J.83 ANNEX A.

I_n	Q_n	C_{n-1}	D_{n-1}	**Phase change from previous QPSK symbol (in radians)**	C_n	D_n
0	0	0	0	0	0	0
0	0	0	1	0	0	1
0	0	1	1	0	1	1
0	0	1	0	0	1	0
0	1	0	0	$\pi/2$	0	1
0	1	0	1	$\pi/2$	1	1
0	1	1	1	$\pi/2$	1	0
0	1	1	0	$\pi/2$,	0	0
1	1	0	0	$-\pi/2$	1	1
1	1	0	1	$-\pi/2$	1	0
1	1	1	1	$-\pi/2$	0	0
1	1	1	0	$-\pi/2$	0	1
1	0	0	0	π	1	0
1	0	0	1	π	0	0
1	0	1	1	π	0	1
1	0	1	0	π	1	1

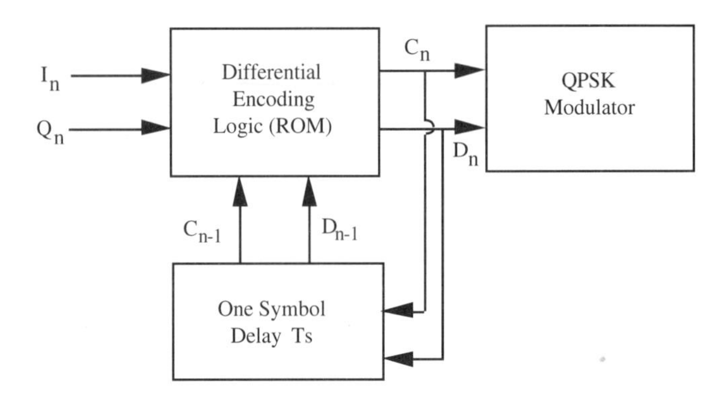

Figure 8.7: Block diagram of a QPSK differential encoder.

In the demodulator, the operation of the differential decoder is the reverse of Fig. 8.7, and is shown in Fig. 8.8. Using the relative phase decoding, we can obtain the original codes without phase ambiguity.

Note that in this scheme, since the decision is made on the basis of two consecutive symbols, there is a tendency for symbol error to occur in pairs. Therefore, the average probability of symbol error for differential encoding QPSK using *coherent* detection is about twice that for (uncoded) coherent QPSK. On the other hand, differential encoding QPSK signals with *noncoherent* detection requires about 2.3 dB more in E_b/N_o to achieve the same bit error rate performance as coherent QPSK [11].

To extend the differential encoding scheme of QPSK to M-QAM, say 64-QAM, the signal constellation with differential encoding is shown in Fig. 8.9. Note that the last four bits (LSBs), e.g., "1101" in Fig. 8.9, are rotationally invariant, that is, the LSBs remain the same no matter how the $n\pi/2$ phase error has been introduced. The

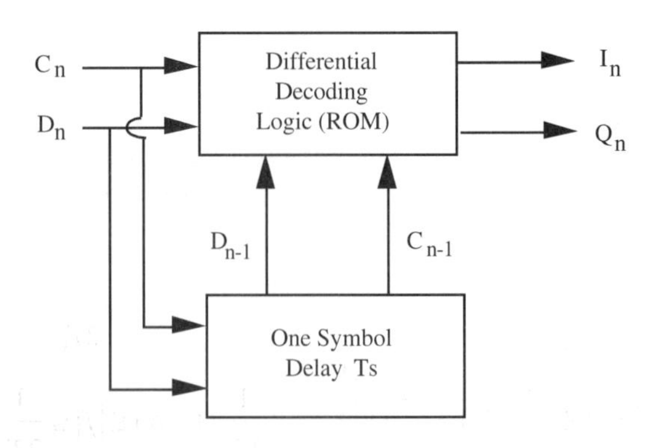

Figure 8.8: Block diagram of a QPSK differential decoder.

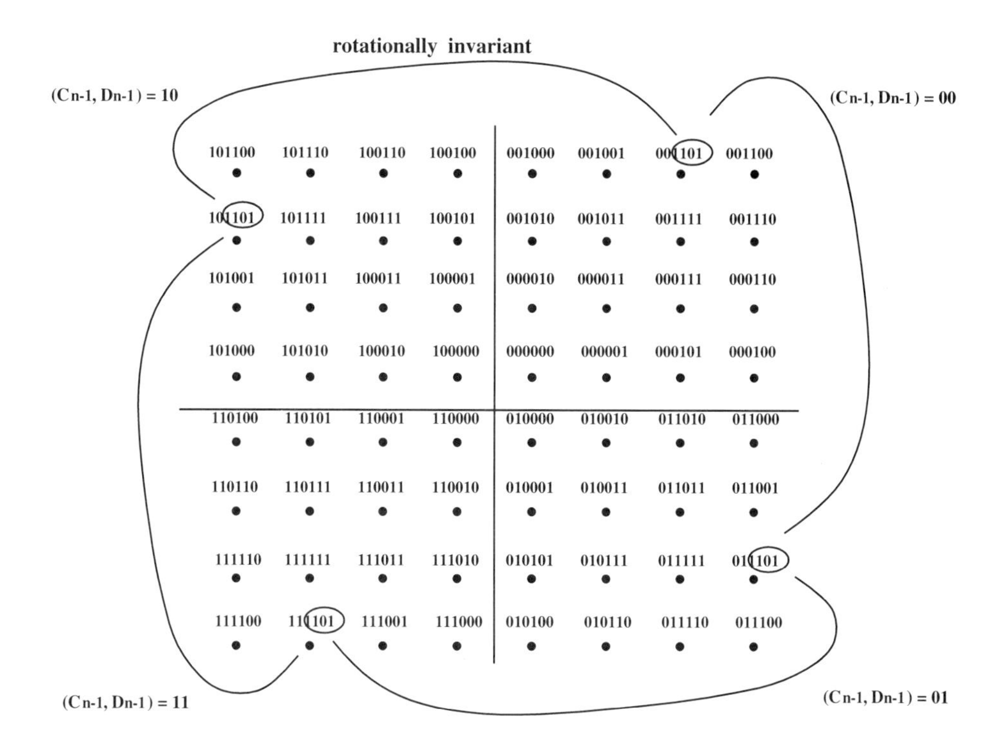

Figure 8.9: 64-QAM constellations for ITU J.83 Annex A.

rules of differential encoding for the first two bits (MSBs) are the same as those of QPSK. Thus, the phase ambiguity problem of 64-QAM can be reduced and simplified to that of QPSK. The block diagram of a 64-QAM differential encoder is shown in Fig. 8.10.

8.4.3 Pulse Shaping Filter

The most commonly used pulse shape in digital communication systems has the Nyquist raised-cosine ISI(intersymbol-interference)-free characteristic whose frequency domain function is given by (see Fig. 8.11a) [12,13]

$$R(f)=\begin{cases} T_S & 0\le |f|\le \dfrac{1}{2T_S}(1-\alpha) \\ \dfrac{T_S}{2}\cdot\left\{1-\sin\left[\dfrac{\pi T_S}{\alpha}\left(|f|-\dfrac{1}{2T_S}\right)\right]\right\}, & \dfrac{1}{2T_S}(1-\alpha)\le |f|\le \dfrac{1}{2T_S}(1+\alpha) \end{cases}, \tag{8.16}$$

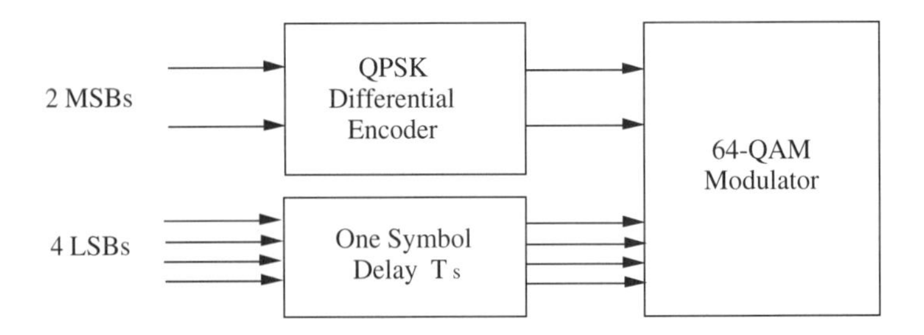

Figure 8.10: Block diagram of a 64-QAM differential encoder. (MSBs = most significant bits; LSBs = least significant bits.)

where T_s is the symbol rate, and α is defined as the percentage of excess bandwidth (relative to the 3-dB bandwidth) that the signal occupies. The corresponding time response is given by (see Fig. 8.11b)

$$r(t) = \frac{\sin\frac{\pi t}{T_s}}{\frac{\pi t}{T_s}} \frac{\cos\left(\frac{\pi \alpha t}{T_s}\right)}{1-\left(\frac{2\alpha t}{T_s}\right)^2} \tag{8.17}$$

From Fig. 8.11b, it can be observed that the pulse goes through zero at every sampling point except for the point at time zero, and therefore we have no ISI (i.e., the oscillating tails of the previous or the following pulses will not interfere with the current pulse). We also notice from Figs. 8.11a and b that the smaller α is, the higher the amplitudes of the pulse tails are. Although the excess bandwidth is small when α decreases, the complexity of implementing a digital filter increases. In addition, longer tails in the filter impulse response would require more equalizer taps in the receiver to combat multipath-induced intersymbol interference (ISI). Typical α values chosen in various standards, as will be shown in Table 8.2, range from 0.13 to 0.18 for 64-QAM signals.

The impulse response of the SRRC filter in Fig. 8.6 is given by [40]

$$r'(t) = \frac{1}{\sqrt{T_s}} \frac{\sin[\frac{\pi t}{T_s}(1-\alpha)] + \frac{4\alpha t}{T_s}\cos[\frac{\pi t}{T_s}(1+\alpha)]}{\frac{\pi t}{T_s}[1-(\frac{4\alpha t}{T_s})^2]}. \tag{8.18}$$

Note that Eq. (8.16) is used to describe the frequency response of a Nyquist channel for impulse transmission. In actual transmission systems, the input data is usually a random, binary non-return-to-zero (NRZ) sequence which has a sin x/x spectrum. Therefore, it is necessary to have an amplitude equalizer which has an amplitude response of $x/\sin x$ to compensate for the NRZ spectrum (the location of the equalizer was previously shown in Fig. 8.6).

ITU J.83 defines that the complete baseband Nyquist filter must have its frequency response of out-of-band $f \geq (1 + \alpha)(1/(2T_s))$ be lower than that of in-band

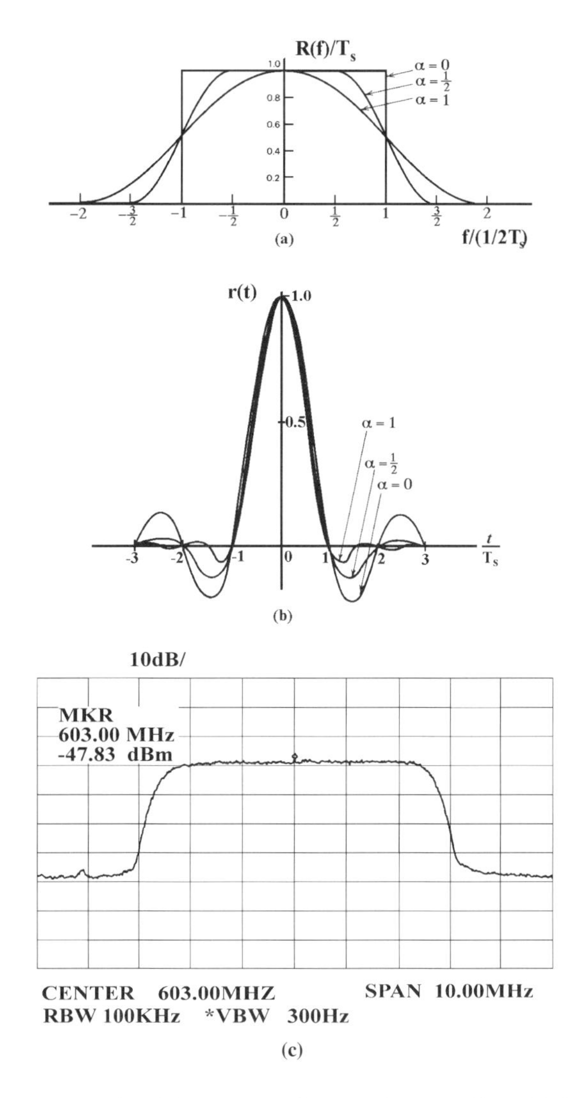

Figure 8.11: Responses for different α's of a raised cosine filter: (a) frequency-domain response, (b) time-domain response (after Ref. [13]); and (c) a measured upconverted 64-QAM RF spectrum.

$0 \le f \le (1-\alpha)(1/(2T_s))$ by 43 dB, and the in-band ripple must be lower than 0.4 dB [2]. It should be noted, however, that for the passband at the transmitter output, the attenuation at the band edge is not as deep as 43 dB, but is ~25 dB, as shown in Fig. 8.11c. This is because the SRRC filter, rather than the raised cosine filter, is implemented in the transmitter. The required filter band-edge rejection which is necessary to avoid adjacent channel interference can still be achieved by using a high-Q, sharp SAW filter as an IF BPF.

8.4.4 Diagnosis via Constellation Diagrams

In a two-dimensional constellation diagram, the x- and y-coordinates represent the amplitudes of the I- and Q-arms, respectively. For example, in a 16-QAM signal, its amplitude levels at the transmitter end are ±1, ±3 V for both I- and Q-arms. Therefore, under normal conditions, we should see 16 points, whose coordinates are (–1,1), (–3,1), (1,1), etc., as shown in Fig. 8.12a. However, this constellation diagram can change because of transmitter and receiver impairments, and because of transmission channel noise and distortions. We can diagnose system impairments by analyzing an "abnormal" constellation diagram, and then fix the problem accordingly. A number of examples are given as follows [14].

8.4.4.1 Transmitter Impairments Figure 8.12b is due to misadjusted modulator levels. We can see that while the levels of I signal states are produced correctly, the levels of Q signal states are not.

Figures 8.12c and d are due to the nonlinear gain and phase characteristics of transmitter power amplifiers. In Fig. 8.12c, the outer four corner points represent the high signal levels, and so gain compression shows up as a crushing of the outer states. In Fig. 8.12d, due to an overcompensating predistorter, all constellation points except the four corner points are compressed.

Figure 8.12e is due to the phase noise of up-converter or down-converter local oscillators. This phenomenon can be illustrated by the following simple derivations. Assuming that the phase noise caused by the upconversion process is $\phi(t)$, and the receiver has a zero-phase noise local oscillator, the received x- and y-values of a constellation point are given by

$$\begin{aligned} x(t,\phi) &= LPF\{[I(t)\cos(\omega_c t+\phi(t))+Q(t)\sin(\omega_c t+\phi(t))]\bullet\cos\omega_c t\} \\ &= I(t)\cos\phi(t)+Q(t)\sin\phi(t) \\ y(t,\phi) &= LPF\{[I(t)\cos(\omega_c t+\phi(t))+Q(t)\sin(\omega_c t+\phi(t))]\bullet\sin\omega_c t\} \\ &= -I(t)\sin\phi(t)+Q(t)\cos\phi(t), \end{aligned} \tag{8.19}$$

where LPF stands for low-pass-filter operations. From those two equations, we can see that the radius of each constellation point is

$$R(t)=\sqrt{x^2+y^2}=\sqrt{I(t)^2+Q(t)^2} \tag{8.20}$$

and is independent of the phase noise, whereas the phase of each constellation point is

$$\theta(t)=\tan^{-1}\frac{y}{x}=\tan^{-1}\sqrt{\frac{-I(t)\sin\phi(t)+Q(t)\cos\phi(t)}{I(t)\cos\phi(t)+Q(t)\sin\phi(t)}} \tag{8.21}$$

and is dependent on the time-varying phase noise. Therefore, we obtain the rotated traces in Fig. 8.12e.

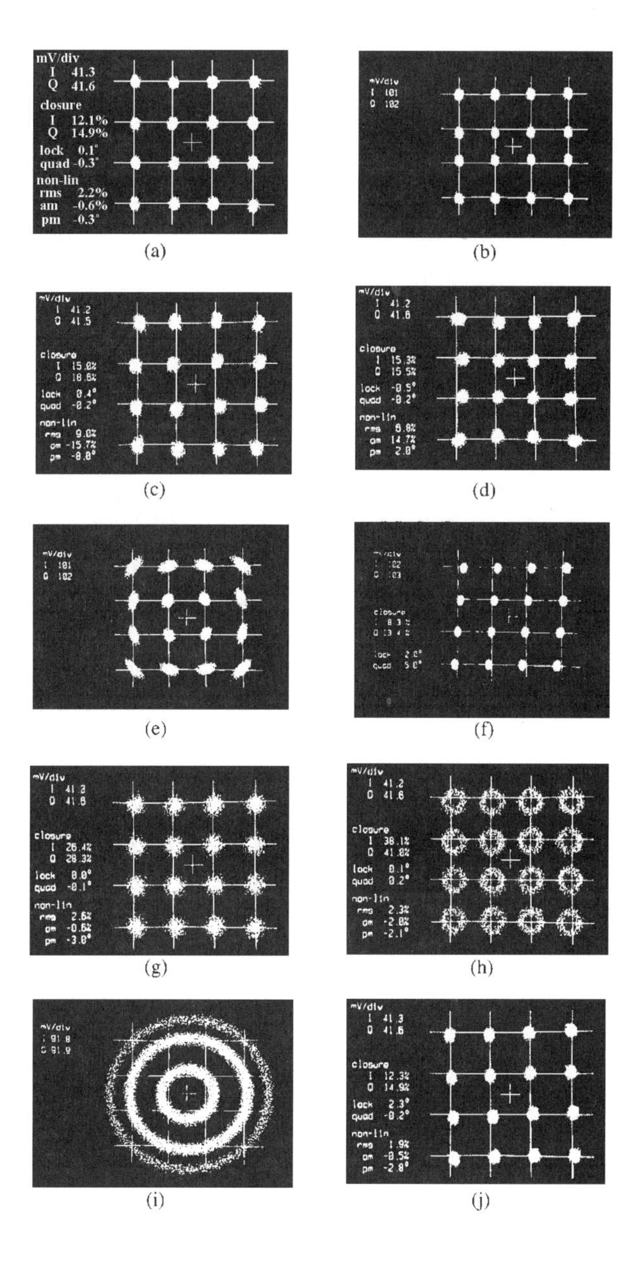

Figure 8.12: Constellation pattern analysis. (a) Normal, (b) misadjusted modulator levels, (c) nonlinear distortions due to overdriven amplifier, (d) same as in (c) but with underdriven amplifier and an overcompensating predistorter, (e) phase noise, (f) quadrature angle error, (g) low C/N, (h) interfering tone, (i) carrier recovery loop out-of-lock, (j) lock angle error. After Ref. [14].

Figure 8.12f is due to the quadrature angle error, that is, the deviation from 90° between the I and Q signals. This can also be easily verified by following the same procedure in obtaining Fig. 8.12e (see Problem 8.2).

8.4.4.2 Transmission Channel Impairments Figure 8.12g is due to higher than normal level of noise or ISI. It is a normal constellation pattern, but the clusters are larger.

Figure 8.12h is due to an interfering tone of the form $\xi \cdot \cos[(\omega_c + \omega_1)t]$. This interfering tone, when received together with a normal input signal [Eq. (8.15)] and down-converted by a local oscillator of the form $\cos(\omega_c t)$, the resultant I and Q signals are

$$\begin{aligned} x(t) &= I(t) + \xi \cdot \cos(\omega_1 t) \\ y(t) &= Q(t) + \xi \cdot \sin(\omega_1 t), \end{aligned} \tag{8.22}$$

and consequently,

$$[x(t) - I(t)]^2 + [y(t) - Q(t)]^2 = \xi^2, \tag{8.23}$$

which means that the loci are circles centered around normal states, with the size of the circles indicating the strength of the interfering tone.

8.4.4.3 Receiver Impairments Figure 8.12i is due to the carrier recovery loop out-of-lock. The "bullseye" pattern occurs because there is no carrier phase reference. This phenomenon is an exaggeration of what was caused by small phase noise shown in Fig. 8.12e. TCM modulator is located after the RS encoder (Fig. 8.1).

Figure 8.12j is due to carrier phase-tracking misadjustment. The rotation of the constellation diagram is similar to Figure 8.12e and i, except that it is due to a fixed, non-time-varying phase-tracking error.

8.4.5 Trellis-Coded Modulation

ITU J.83 Annex B uses a combined QAM modulation and trellis coding, called trellis-coded modulation (TCM), that achieves coding gain without any bandwidth expansion [6,15,16]. TCM modulator is located after the RS encoder (Fig. 8.1).

Conventionally, the required trellis coding redundancy is taken care by an increased signal alphabet (in QAM) instead of bandwidth. As an example, given a channel with a bandwidth limitation, one can first determine the symbol rate that must be transmitted, then determine the size 2^m of the alphabet that would be required (without coding) to transmit the source bits at the desired bit rate. One can then subsequently double the size of the alphabet to 2^{m+1} and introduce a convolutional coder that produces one extra bit. Recently, however, to improve the poor code rate of a conventional trellis encoder, punctured convolutional (PC) encoders [17] which can flexibly adjust the coding rate have been used to achieve the required trellis coding redundancy. An example is shown in Fig. 8.13 for a PC encoder with an improved code rate of 4/5 which is constructed from a convolutional code with a worse code rate of 1/2 [18]. Note that a zero in the puncturing table represents the

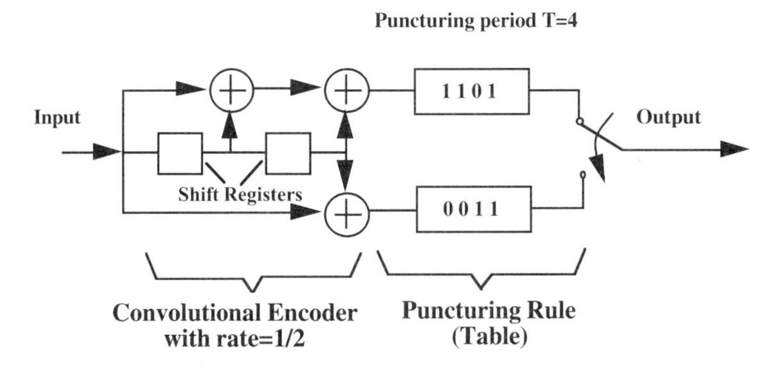

Figure 8.13 A rate 4/5 PC encoder which is constructed from a convolutional encoder with a rate of 1/2. After Ref. [18] (© 1995 IEEE).

deletion of a symbol and a one represents the transmission of a symbol. We can see that when there are 4 input information bits, 5 coded bits (out of 8 bits in the puncturing table) are transmitted and hence the effective coding rate becomes 4/5.

A slightly more complicated combination of a 1/2 binary convolutional coder and a 4/5 puncture matrix can found in Ref. [2] for ITU J.83 Annex B.

Figure 8.14 shows the structure of a 64-QAM trellis coded modulator using the coding method defined by ITU J.83 Annex B. The target of this modulator is to achieve a code rate of 14/15 using 64-QAM modulation. In ITU J.83 Annex B, each RS symbol is composed of 7 bits. Therefore, for every 4 RS symbols (28 bits), shown as the input signals, the resultant code should have 30 bits or 5 64-QAM symbols. Annex B achieves this highly efficient 14/15 code rate by letting 20 out of the 28 bits remain uncoded (as shown by the Is and Qs in the upper four branches), while having the other 8 bits (as shown by the Is and Qs in the lower two branches) pass through two 4/5 PC encoders, as shown in Fig. 8.14. In other words, for every two I or Q RS symbols consisting of 14 bits, 4 bits are coded by the PC encoder to produce 5 coded bits. Ten bits are not coded, therefore yielding a total output of 15 bits. Thus, the overall trellis-coded modulation yields a 14/15 rate.

8.4.6 Symbol-Error Rate and Bit-Error Rate of M-QAM Signals with AWGN Noise—without and with FEC Coding

We start with one dimensional M-ary PAM signals with values [19]

$$s_m = d \cdot A_m, \qquad m = 1,2,\ldots,M, \tag{8.24}$$

where d^2 is the energy of the basic signal pulse $r(t)$ in Eq. (8.17). The amplitude values may be expressed as

$$A_m = (2m - 1 - M), \qquad m = 1,2,\ldots,M. \tag{8.25}$$

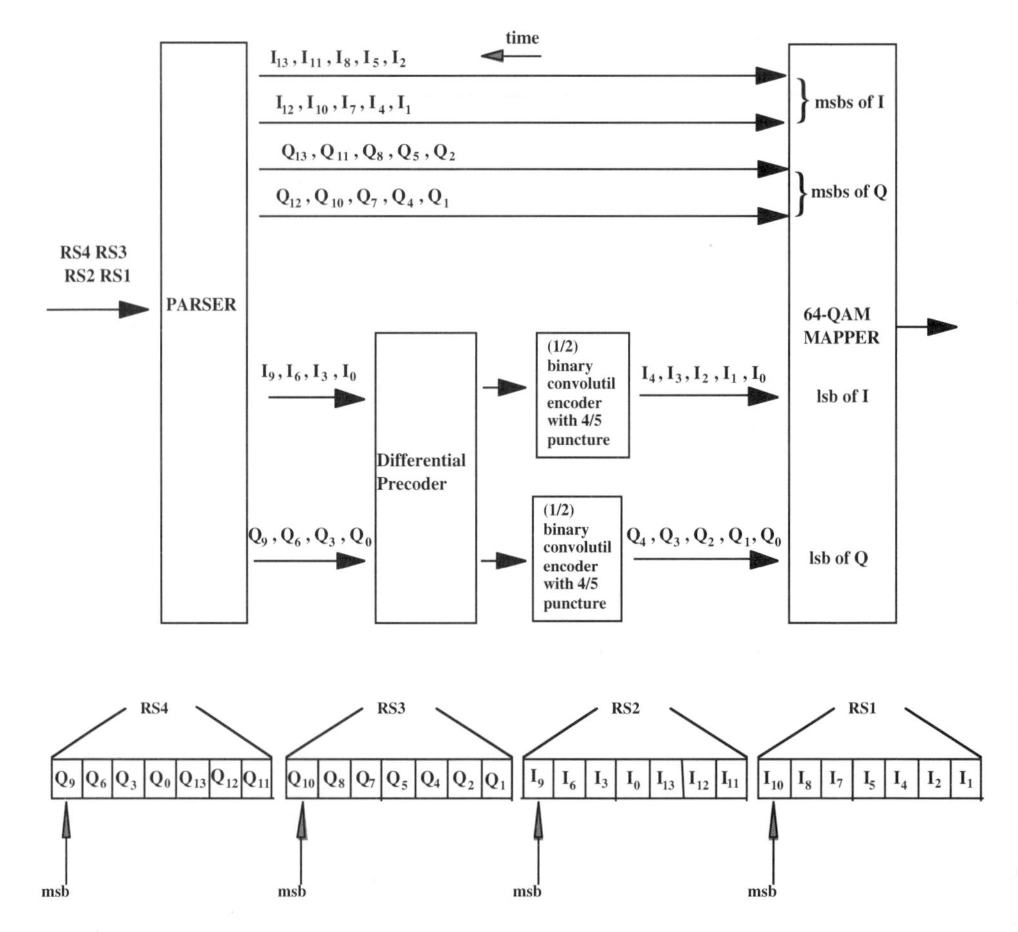

Figure 8.14: 64-QAM trellis-coded-modulator for ITU J.83 Annex B. Note that the neighboring bits in the upper four branches (e.g., I_2 and I_5) are taken from every other bits of the original RS symbols (e.g., RS1). After ref. [2].

For example, $A_m = -3, -1, +1, +3$ for 4-level PAM. Therefore, the distance between adjacent signal points is $2d$. The average symbol energy is given by

$$E_s = \frac{1}{M}\sum_{m=1}^{M}(d \cdot A_m)^2 = \left(\frac{M^2-1}{3}\right)d^2 . \tag{8.26}$$

Assuming that all amplitude levels are equally likely a priori, and the added white Gaussian noise due to the transmission channel has zero mean and variance, $\sigma_n^2 = N_o/2$, we then have the average probability of a symbol error rate (SER) given by (with $Q(d/\sigma_n) \ll 1$)

$$P_{se} \approx \frac{2(M-2)+2}{M} Q(\frac{d}{\sigma_n}), \tag{8.27}$$

where the error rates due to the middle $M - 2$ levels (error occurs in two directions) and the two outside levels (error occurs in one direction only) are included.

$$Q(\frac{d}{\sigma_n}) = \int_{d/\sigma_n}^{\infty} p_n(x)dx = \frac{1}{\sqrt{2\pi}} \int_{d/\sigma_n}^{\infty} e^{-x^2/2} dx \tag{8.28}$$

is the well-known Q-function, and $p_n(x)$ is the probability density function of a Gaussian random variable. Substituting (8.26) into (8.27) yields

$$P_{se} = 2\left(1 - \frac{1}{M}\right) Q\left(\sqrt{\frac{6E_s}{(M^2-1)N_o}}\right). \tag{8.29}$$

For Gray coding, each symbol has $\log_2 M$; bits and the relation between symbol and bit energy is given by $E_s = E_b \cdot \log_2 M$; therefore, we can express the bit error rate (BER) in terms of E_b/N_o as

$$P_{be} = \frac{2}{\log_2 M}\left(1 - \frac{1}{M}\right) Q\left(\sqrt{\frac{6\log_2 M}{(M^2-1)} \cdot \frac{E_b}{N_o}}\right), \tag{8.30}$$

Now let us extend the one-dimensional M-ary PAM signal symbol error rate to that of a two-dimensional M-ary QAM signal with rectangular signal constellation. The symbol error probability is given by

$$P_{se,QAM} = 1 - (1 - P_{\sqrt{M}})^2 , \tag{8.31}$$

where $P_{\sqrt{M}}$ is the symbol probability of error of a $\sqrt{M}$-ary PAM with one-half the average power in each quadrature signal of the equivalent QAM system, that is,

$$P_{\sqrt{M}} = 2\left(1 - \frac{1}{\sqrt{M}}\right) Q\left(\sqrt{\frac{3E_s}{(M-1)N_o}}\right). \tag{8.32}$$

When $P_{\sqrt{M}} << 1$, Eq. (8.31) can be approximated by

$$P_{se,QAM} \approx 2P_{\sqrt{M}} = 4\left(1 - \frac{1}{\sqrt{M}}\right) Q\left(\sqrt{\frac{3E_s}{(M-1)N_o}}\right). \tag{8.33}$$

Alternatively, Eq. (8.33) can be obtained by noting that the average symbol energy is

$$E_s = \frac{d^2}{M} \sum_{n=1}^{\sqrt{M}} \sum_{m=1}^{\sqrt{M}} \left(A_m^2 + A_n^2\right) = \left(\frac{2(M-1)}{3}\right) d^2 \tag{8.34}$$

and the fact that SER can be obtained from (with $Q\left(\frac{d}{\sigma_n}\right) << 1$)

$$\begin{aligned} P_{se,QAM} &\approx \frac{4}{M} \bullet 2Q(\frac{d}{\sigma_n}) + \frac{(\sqrt{M}-2)4}{M} \bullet 3Q(\frac{d}{\sigma_n}) + \frac{(\sqrt{M}-2)^2}{M} \bullet 4Q(\frac{d}{\sigma_n}) \\ &= 4(1 - \frac{1}{\sqrt{M}}) \bullet Q(\frac{d}{\sigma_n}) , \end{aligned} \tag{8.35}$$

where the first term in Eq. (8.35) accounts for the SER of the constellation points at the four corners (error occurs in two directions), the second term accounts for the SER of the constellation points at the four edges excluding the four corner points (error occurs in three directions), and the third term accounts for the SER of all other constellation points (error occurs in four directions). Substituting $\sigma_n = \sqrt{N_o/2}$ and $d = \sqrt{3E_s/2(M-1)}$ [from Eq. (8.34)] into Eq. (8.35), we can again obtain Eq. (8.33). For Gray coding, the BER of an uncoded M-ary QAM signal can be obtained as

$$P_{be,QAM} \approx \frac{4}{\log_2 M}\left(1-\frac{1}{\sqrt{M}}\right)Q\left(\sqrt{\frac{3\log_2\sqrt{M}}{(M-1)}\cdot\frac{E_b}{N_o}}\right). \tag{8.36}$$

The RS symbol error probability, $P_{se,\,QAM\,+\,RS}$, can be written in terms of the channel symbol error probability, $P_{se,\,QAM}$, as [20]

$$P_{se,QAM+RS} \approx \frac{1}{n}\sum_{i=t+1}^{n} i\binom{n}{i}P_{se,RS}^{i}\,(1-P_{se,RS})^{n-i}, \tag{8.37}$$

where, $(1 - P_{se,QAM})^a = (1 - P_{se,RS})^b$ and $a = \log_2(n)$, $b = \log_2(M)$ for an n-symbol RS block code which is transmitted in M-symbol QAM format.

The BER as a function of E_s/N_o for a (204,188) RS code is shown in Fig. 8.15. With this code, it can be seen that at a BER of 10^{-8}, the required E_s/N_o of the original uncoded 64-QAM signal can be decreased by about 4.6 dB. Although not shown explicitly in the figure, it is known that this type of code can improve the uncode 64-QAM BER performance from 10^{-4} to a range of 10^{-10} to 10^{-11}. For 30 Mb/sec 64-QAM, this ensures "quasi-error-free" (QEF) operation with approximately one uncorrected error event per transmission hour [2].

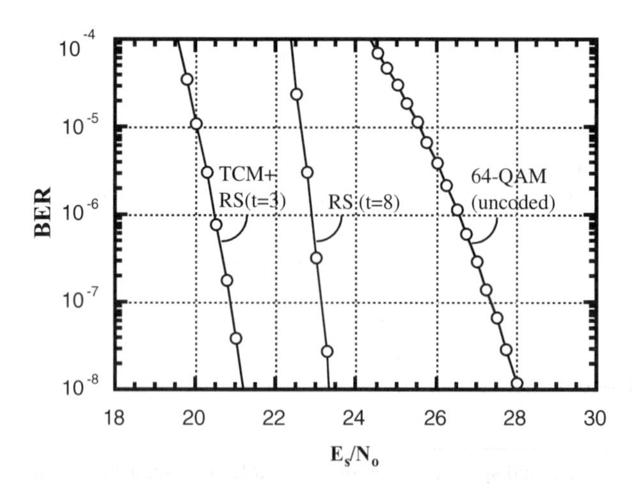

Figure 8.15: Theoretical BER versus E_s/N_o for uncoded 64-QAM, 64-QAM with RS(204,188) (ITU J.83 Annex A), and with TCM + RS(128,122) (ITU J.83 Annex B).

When comparing the performance of TCM 64-QAM signals with that of the uncoded 64-QAM signals, we usually calculate the minimum Euclidean distances[5] between the corresponding signal points of two distinct code sequences. For soft decision decoding, the larger the Euclidean distance, the lower the probability of error. For the TCM 64-QAM signals, the minimum squared Euclidean distance for uncoded and coded bits is $12d^2$; whereas for the uncoded 64-QAM signals, it is $4d^2$. Therefore, the coding gain is 3 times or 4.77 dB [17,21]. The calculated results of TCM + RS(128,122) for ITU J.83 Annex B is shown in Fig. 8.15. It can be seen that the concatenated TCM and RS(t = 3) code has the lowest requirement on E_s/N_o (when Gaussian noise dominated), such as, $E_s/N_o = 21.2$ dB at BER = 10^{-8}.

8.5 Summary of QAM and VSB Modem Parameters in Various Standards

Having gone through Sections 8.1–8.4, we now summarize in Table 8.2 the QAM and VSB modem parameters (VSB modem design will be discussed in Section 8.8) for various important standards that are applicable to HFC systems.

8.6 RF Up-converter and Tuner

The RF up-converter in Fig. 8.1 converts the modulator output's second IF frequency[6] (cable IF is 43.75 MHz in North America and ~36 MHz in Europe; satellite IF is ~600 MHz in North America and ~480 MHz in Europe) into the typical CATV band, that is, 50 to 750 MHz (MCNS compliant systems are from 88 to 860 MHz). Conversely, the RF tuner in Fig 8.1 down-converts one of the CATV channels to the IF frequency. The down-converted IF signal is subsequently sent to the M-QAM demodulator.

The basic structure in a typical receiver tuner is shown in Fig. 8.16. The input signal is assumed to range from 50 to 750 MHz. The first down-converter, which can be a monolithically integrated circuit, consists of a preamplifier, a mixer, a local oscillator (LO#1), and a postamplifier. LO#1 is a phase-locked tunable frequency synthesizer which can be tuned from 750 to 1450 MHz so that the down-converted channel always falls into a fixed bandpass filter centered at 700 MHz. The band-passed signal is then sent to a second down-converter which is similar to the first, except that the second local oscillator (LO#2) is a fixed-frequency synthesizer at 656.25 MHz. The output of this second down-converter is centered around the first IF or 43.75 MHz. Note that this is not the last stage of the down-conversion process — the final down-converter further down-converts the second IF to the first IF signal,

[5]The distance in signal space between the allowable code sequences, that is, the energy required to alter one code sequence to another, is termed the squared Euclidean distance.

[6]Recall from Fig. 8.6 that the first IF is equal to the symbol frequency for digital implementation.

Table 8.2: Summary of QAM and VSB Modem Parameters in Various Standards

Parameter	**DAVIC**	**ITU J.83 Annex A (IEEE 802.14)**	**ITU J.83 Annex B (IEEE 802.14) (MCNS DOCSIS)**	**ITU J.83 Annex D (U.S. ATSC cable-mode 16-VSB)**
Scrambler (generating polynomial)	$1 + X^{14} + X^{15}$	$1 + X^{14} + X^{15}$	$\alpha^3 + X + X^3$ (after the convolutional interleaver)	$1 + X + X^3 + X^6 + X^7 + X^{11} + X^{12} + X^{13} + X^{16}$
Reed-Solomon code	GF(256) (8-bit symbols) $(n,k) = (204,188)$ Code generator polynomial: $g(x) = (x + \alpha^0)(x + \alpha^1)\ldots(x + \alpha^{15})$ Primitive generator polynomial: $p(x) = x^8 + x^4 + x^3 + x^2 + 1$	GF(256) (8-bit symbols) $(n,k) = (204,188)$ Code generator polynomial: $g(x) = (x + \alpha^0)(x + \alpha^1)\ldots(x + \alpha^{15})$ Primitive generator polynomial: $p(x) = x^8 + x^4 + x^3 + x^2 + 1$	GF(128) concatenated RS(128,122) and convolutional coding (7-bit symbols) Code generator polynomial: $g(x) = (x + \alpha^1)(x + \alpha^1)\ldots(x + \alpha^5)$ Primitive generator polynomial: $p(x) = x^7 + x^3 + 1$	GF(256) $(n,k) = (207, 187)$
Convolutional interleaver	(16/64 QAM) $I = 12, J = 17$ (256 QAM) $I = 204, J = 1$	$I = 12, J = 17$	(64 QAM) $I = 128, J = 1$ (64/256 QAM) variable $I = 128, 64, 32, 16, 8$; variable $J = 1,2,4,8,16$ (see standards for details)	$I = 52, J = 4$
Symbol rate	5.274 Mbaud (for 6 MHz) 6.887 Mbaud (for 8 MHz)	same as DAVIC	5.056944 Mbaud (64 QAM) 5.360537 or 5.19 Mband (256 QAM)	10.76 Mbaud
Pulse shaping (root-raised cosine filter)	$\alpha = 0.13$ (for 6 MHz) $\alpha = 0.15$ (for 8 MHz)	$\alpha = 0.15$ (for 64 QAM)	$\alpha = 0.18$ (for 64 QAM) $\alpha = 0.12$(for 256 QAM)	$\alpha = 0.115$
Constellation	16/64-QAM	16,32,64-QAM	64/256-QAM	16-VSB
TCM	No	No	Yes	No

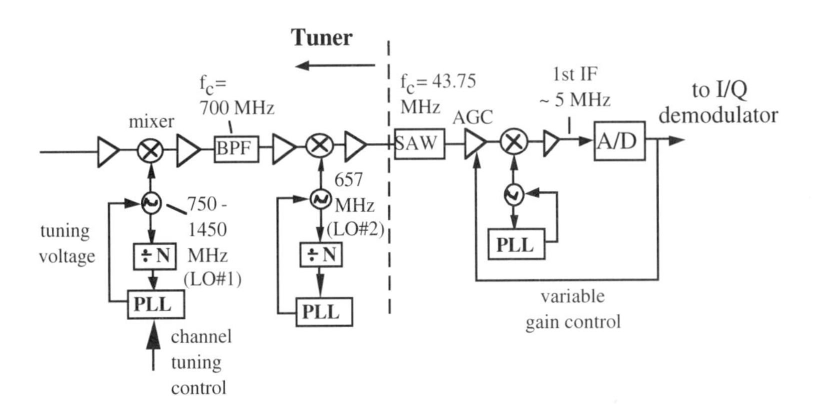

Figure 8.16: Receiver RF-front-end tuner, a last-stage down-converter, and an A/D sampled at four times of the first IF frequency (i.e., sampling frequency ~ 20 MHz).

which is equal to the symbol rate (~5 MHz for North American systems and ~7 MHz for European systems). The first IF is then sent to an analog-to-digital converter (A/D) with a sampling clock which is equal to four times of the symbol rate f_s for QAM signals and is equal to two times of the symbol rate for VSB signals (see Section 8.8).

It should be noted that the output of the first mixer may consist of nonlinear distortion products owing to the fact that the input broadband signal is composed of multiple channels. In addition, the bandpass filter centered at 700 MHz (mainly used for image suppression and LO radiation reduction) should be carefully designed so as not to cause additional linear distortions such as amplitude or group delay tilt.

The SAW filter has a very sharp bandpass shape so it significantly suppresses adjacent-channel signals. Following the SAW filter is an automatic gain control (AGC) amplifier whose gain is controlled by a feedback signal from the output of the A/D.

The tuner for a cable modem or digital set-top box will require better phase noise performance than standard analog NTSC tuners.[7] We see that there are three LOs in the down-conversion process, which in combination with the three LOs in the up-conversion process result in a total of six LOs. The sum of the six LOs' phase noise could cause significant rotation or smear of the recovered constellations (see Figs. 8.12e and j). Generally speaking, the higher the frequency of an LO, the larger the phase noise. Therefore, we usually pay more attention to LO#1 than to the lower-frequency LOs. Today's commercially available LO#1 typically exhibits a phase noise characteristic of -85 – -90 dBc at 10-kHz offset from the carrier frequency. This is a much better phase-noise performance than that of a typical analog set-top box tuner's LO#1—with about –75 dBc at 10-kHz offset.

[7]The phase noise threshold of the 16-VSB modem has been measured by CableLabs at –82 dBc at 20 kHz offset from carrier.

8.7 M-QAM Receiver Design

Following the A/D in Fig. 8.16 is the fully digital part of a receiver. A typical design of this part is illustrated in Fig. 8.17 [22]. Immediately after the A/D is a digital I/Q demodulator which is basically a digital mixer that produces I- and Q-arm baseband signals. These signals are then fed to a pair of SRRC Nyquist filters which are the matched filters of the transmitter SRRC filters. After the Nyquist filters is the heart of the entire receiver, which consists of essentially three parts: equalizer, timing recovery, and carrier recovery.

The interdependence among equalization, timing recovery, and carrier recovery is a key design issue. Generally speaking, in an M-QAM system, equalization and carrier recovery are jointly conducted in order to minimize the initial acquisition time [23], while timing recovery can be dependent on [24–26] or independent of [27] an equalizer. Figure 8.17 is a state-of-the-art architecture with an equalizer-independent timing recovery, and with a joint blind equalization and carrier recovery. It can be seen that timing recovery is implemented by directly utilizing the output of Nyquist filters.

In the following, we will review the state-of-the-art adaptive equalization, carrier recovery, and timing recovery techniques, respectively, and discuss their interdependence.

8.7.1 Adaptive Equalizer Fundamentals and Design

In a communication receiver, in addition to bandpass filters and square-root raised-cosine filters which are used to decrease the effects of additive noise without caus-

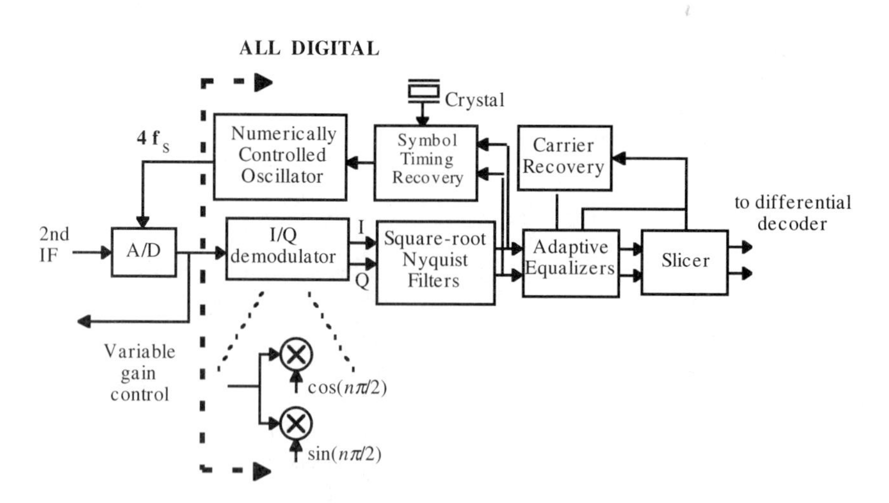

Figure 8.17: All-digital part of an M-QAM receiver which is located after an A/D. After Ref. [22].

ing additional ISIs, a filter which can undo the channel linear distortions or ISI (caused by cable reflections) is also needed. This type of ISI-reducing filter is called an equalizer. Ideal equalization requires the construction of a filter having a transfer function inverse to that of the transmission channel.

The simplest equalizer is the linear transversal equalizer (LTE), which is constructed using a tapped delay line and a series of tap weights, as shown in Fig. 8.18.

The LTE impulse response and its corresponding Fourier transform are given by

$$\varepsilon_k(t) = \sum_{n=0}^{N} W_{nk}\delta(t - nT) \tag{8.38}$$

and

$$E_{kT}(f) = \sum_{n=0}^{N} W_{nk}e^{-jn\omega T}\,, \tag{8.39}$$

respectively. In these two equations, k denotes the kth sampling time. On the other hand, the received signal $x(t)$, shown in Fig. 8.18, is composed of the superposition of the impulse responses of the channel to each transmitted symbol (we temporarily neglect all noise terms), and we assume that it has been sampled by an ADC at $t_o + kT$ and is given by

$$x(t_o + kT) = \sum_m a_m h(t_o + kT - mT)\,, \tag{8.40}$$

where $h(t)$ is the pulse response of the transmission channel. Note that only the term $a_k h(t_o)$ (with $m = k$) is the desired output and all other terms represent the ISI

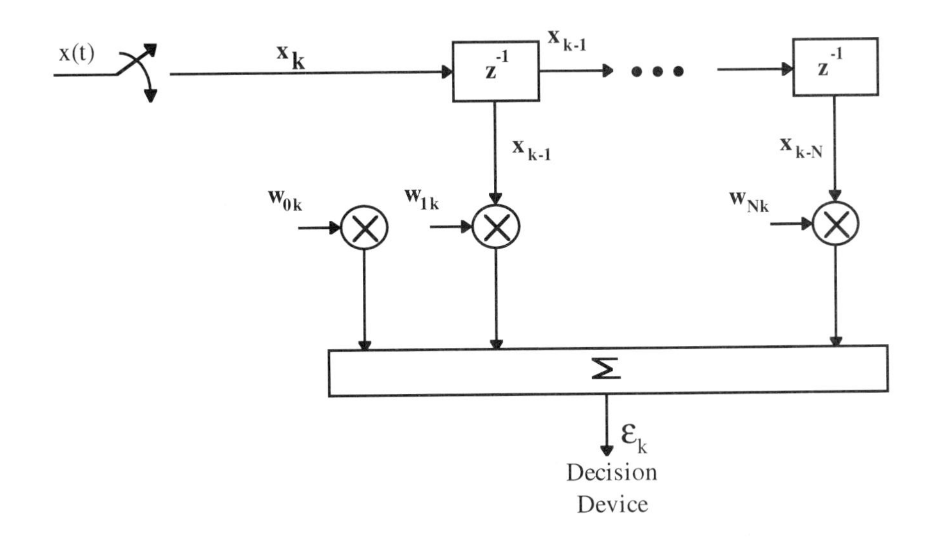

Figure 8.18: A linear transversal equalizer.

contributions (with $m \neq k$). From Eq. (8.40), the folded (aliased) spectrum of the sampled $h(t_o + t)$ is given by

$$H_{samp}(f) = \sum_l H(f - \frac{l}{T}) \exp(j2\pi \pm_o (f - \frac{l}{T}) t_o), \tag{8.41}$$

where $H(f)$ is the Fourier transform of $h(t)$ i.e., from $h(t + t_o) \Leftrightarrow H(f)e^{j2\pi f t_o}$, we obtain $h(t + t_o) \otimes \sum \delta(t - lT) \Leftrightarrow \sum H(f - \frac{l}{T})e^{j2\pi t_o (f - \frac{l}{T})}$, where "$\Leftrightarrow$" represents the Fourier transform pair, and "$\otimes$" represents time-domain convolution. The process of obtaining $H_{samp}(f)$ is illustrated in Fig. 8.19a.

In an ideal case, the combined frequency response of the sampled spectrum and the equalizer should be a constant, that is,

$$E_{KT}(f) H_{samp}(f) = const. \tag{8.42}$$

In a practical condition, however, the sampler jitter t_o in Eq. (8.41) strongly influences the amplitude and delay characteristics in the spectral overlap region of the sampled equalizer input. In particular, if a dip or a null is created at the bandedges ($|f| = 1/2T$) in Fig. 8.19a because of the variation of t_o from sample to sample, then all the equalizer can do is to synthesize a large gain in the affected region. Consequently, significant noise enhancement in this region can occur. This phenomenon naturally leads one to increase the sampling frequency and use a *fractionally spaced* equalizer (FSE) [28] with the tap spaced $\boldsymbol{T'} < T$, as shown in Figs. 8.19b and 8.20. The input signal is also sampled every $\boldsymbol{T'}$ period.

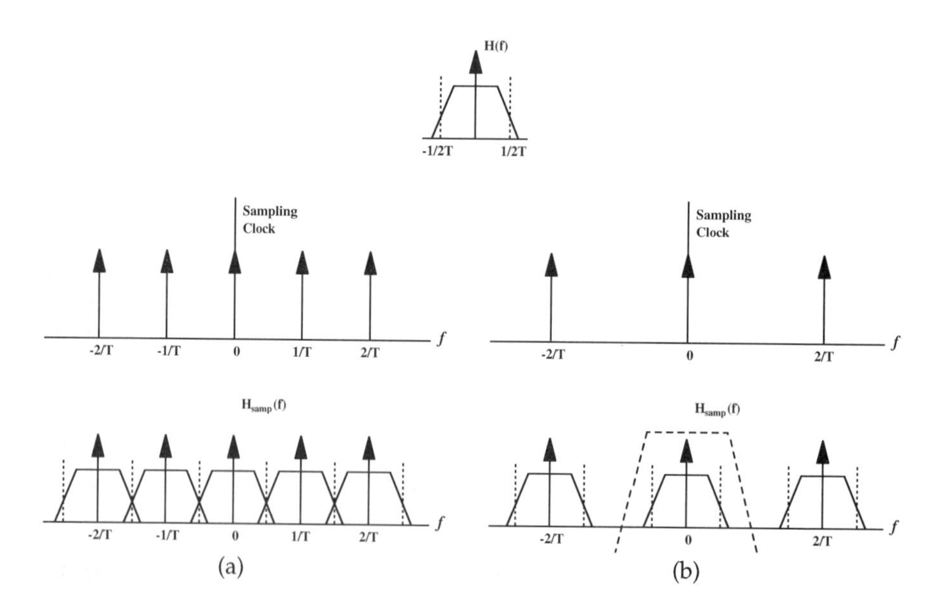

Figure 8.19: Folded spectra of a sampled signal with a sampling frequency f_s equal to (a) $1/T$ and (b) $2/T$.

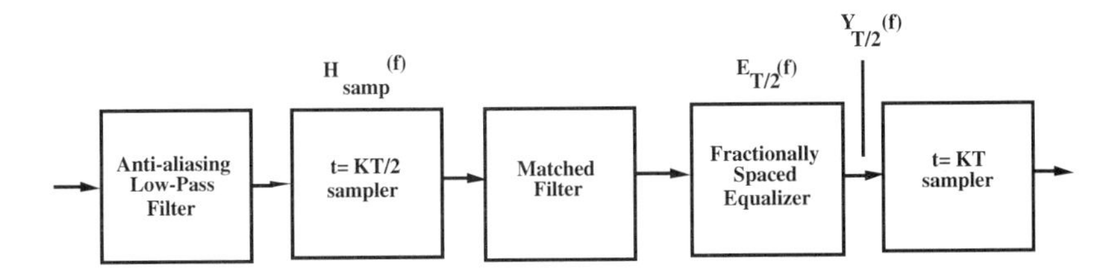

Figure 8.20: Implementation of a fractionally spaced equalizer. (After Ref. [40], with kind permission from Kluwer Academic Publishers.)

The equalized spectrum $Y_{T'}(f)$, just prior to the output sampler, is periodic (with a frequency period $1/T'$) and is given by

$$Y_{T'}(f) = E_{KT'}(f) \sum_{l} H\left(f - \frac{l}{T'}\right) \exp\left(j2\pi\left(f - \frac{l}{T'}\right)t_o\right). \quad (8.43)$$

In most practical implementations, $\boldsymbol{T'} = T/2$ (see Fig. 8.20 for the location of $Y_{T/2}(f)$). In this case, the resultant folded spectrum of the sampled $h(t_o + t)$ is shown in Fig. 8.19b, and we can use a matched filter, as shown in Fig. 8.20, to keep only the $l = 0$ term to give

$$Y_{T/2}(f) = E_{KT/2}(f) H(f) \exp(j2\pi f t_o). \quad (8.44)$$

The important aspect of Eq. (8.44) is that $E_{KT/2}(f)$ acts on $H(f)\exp(jft_o)$ before spectrum folding is performed (by the t = KT sampler in Fig. 8.20), as opposed to the case in Eq. (8.42) where $E_{KT}(f)$ acts on a spectrum that is already folded.

The last step in using an FSE, as shown in Fig. 8.20, is to sample the equalizer output at a rate $1/T$ to give a periodic output spectrum,

$$Y_T(f) = \sum_{l} Y_{T/2}\left(f - \frac{l}{T}\right) = \sum_{l} E_{KT/2}\left(f - \frac{l}{T}\right) H\left(f - \frac{l}{T}\right) \exp\left[j2\pi\left(f - \frac{l}{T}\right)t_o\right]. \quad (8.45)$$

Notice that Eq. (8.45) differs from Eq. (8.42) in that it is the sum of equalized aliased components rather than an equalization of an already-formed sum of aliased components. It is also clear from Eq. (8.44) that $E_{KT/2}(f)$ can compensate for any timing jitter due to $\exp(j2\pi f t_o)$ and avoid the extreme sensitivity to timing phase associated with the T-spaced equalizer. Therefore, FSEs have been used in almost all of today's QAM or VSB modems.

Having explained the operation principle of an FSE, let us rewrite a digital equalizer output in a simple form as

$$y_k = \sum_{n=0}^{N} w_{nk} x_{k-n}, \quad (8.46)$$

where again k denotes the kth sampling time. The LTE weights, w_{nk} $(n = 0, 1, ..., N)$ can be chosen to force the samples of the combined channel and equalizer impulse to zero at all but one of the N T'-spaced instants in the span of the equalizer. Such an equalizer is called a zero-forcing (ZF) equalizer [11]. A simple one-tap ZF equalizer shown in Fig. 8.21 shows how zero-forcing works. It can be seen that the one-tap equalizer reduces ISI to 1% from the original 10%. A finite-length ZF can provide a transfer function which approximates the inverse of the channel frequency response. However, its drawback is that when a deep null occurs in the channel frequency response because of a large unattenuated reflection, the LTE-ZF provides a large gain to compensate for the null. This large gain also enhances any noise that is present in the channel which can significantly reduce the SNR in the receiver.

To overcome the noise enhancement problem of the LTE, one can use an *adaptive* equalizer using the well-known least-mean-squared (LMS) algorithm. The LMS algorithm ajusts the equalizer weighting coefficients to minimize the mean-squared error, which includes the squares of all the ISI terms and the noise power at the output of the equalizer. It will be seen in the following explanation of LMS operation principle that the LMS algorithm does not provide an exact solution to the problem of minimizing the mean-squared error, but rather only approximates the solution. This approximation is the price paid for not requiring that the channel be known or

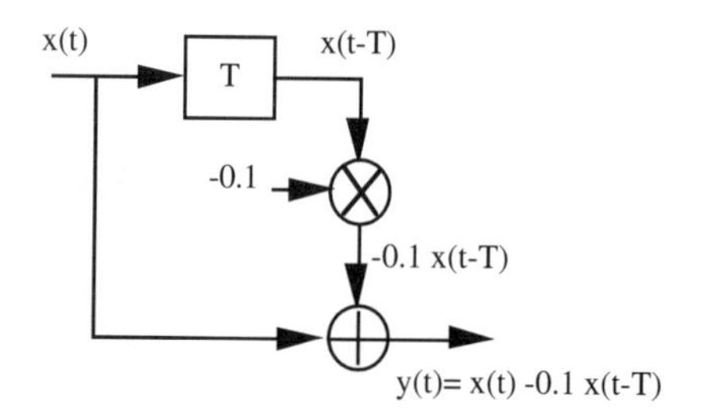

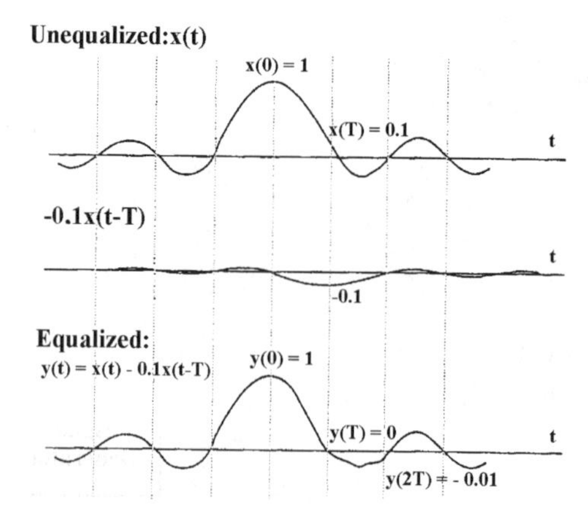

Figure 8.21: A simple example showing how one-tap zero-forcing equalizer works. (Courtesy of Dr. K. T. Wu)

stationary. The operation principle of the LMS algorithm is explained as follows: the *adaptation* of the equalizer is driven by an error signal so that the equalizer coefficients can be adjusted (during each symbol interval or a fraction of symbol interval) to more accurately represent the data symbols at the slicer input. The question is, however, where does the error signal come from? As shown in Fig. 8.22, before regular data transmission begins, the equalizer must identify the unknown channel parameters as well as to track their time variations. Initial acquisition of the equalizer can be accomplished by having the transmitter send a training signal sequence known to the receiver, which regenerates the same sequence locally to be used to form the prediction error for coefficient updating.

According to Fig. 8.22 and Eq. (8.46), we have

$$e_k = d_k - \sum_{n=0}^{N} w_{nk} x_{k-n} . \tag{8.47}$$

The LMS algorithm specifies a steepest descendant type of adaptation of the weighting coefficients with respect to the square error as [29]

$$w_{n(k+1)} = w_{nk} - \frac{\mu}{2} \frac{\partial e_k^2}{\partial w_{nk}} \qquad (n = 0, 1, ..., N), \tag{8.48}$$

where μ is the adaptation step size that regulates the speed and stability of adaptation, and the division by 2 is included to avoid a factor of 2 in the subsequent adaptation algorithm. Note that in Eq. (8.48), although we take gradient with respect to

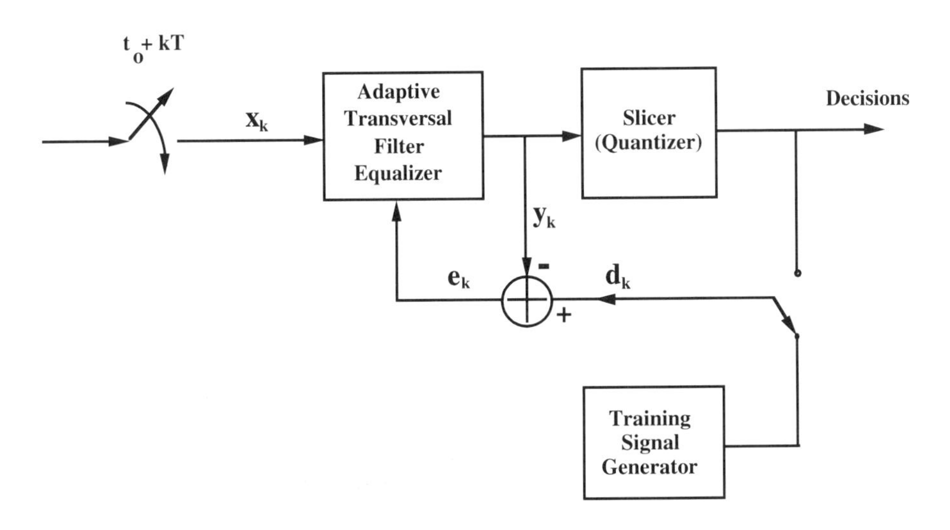

Figure 8.22: Adaptive transversal equalizer with a training signal generator for initial acquistion.

the noisy squared error instead of mean squared error $\overline{e_k^2}$, the noise is attenuated with time by the adaptive process, which acts as a low-pass filter in this respect. From Eqs. (8.47) and (8.48), we can obtain

$$w_{n(k+1)} = w_{nk} + \mu e_k x_{k-n} \qquad (n = 0, 1, ..., N). \qquad (8.49a)$$

For a complex input x_k such as a QAM-signal, Eq. (8.49a) can be modified as

$$w_{n(k+1)} = w_{nk} + \mu e_k x^*_{k-n} \qquad (n = 0, 1, ..., N) \qquad (8.49b)$$

where * stands for complex conjugate. Equation (8.49a) or (8.49b) is the so-called LMS algorithm. We can see that the LMS algorithm can be implemented in a practical system without squaring, averaging, or differentiation and is elegant in its simplicity and efficiency.

After the initialization process is completed, the switch in Fig. 8.22 flips to the output of the slicer, and the adaptation of the equalizer becomes *decision-directed* in the steady state. This means that the receiver decisions are used to generate the error signal, and the LMS algorithm can still be used.

Both linear feedforward equalizers (FFEs) and nonlinear decision-feedback equalizers (DFEs) take advantage of this decision-directed LMS operation mode. The locations of an FFE and a DFE in a digitally implemented QAM receiver are shown in Fig. 8.23 [30]. The detailed block diagrams of a digital FFE and DFE using LMS algorithm are shown in Figs. 8.24 and 8.25 [31], respectively (readers should check the matching of Eq. (8.49) in both diagrams).

As a side point, we note that there is no carrier recovery loop in Fig. 8.23. This is because the first IF center frequency is set to be equal to the symbol rate. Once the symbol clock has been recovered, the carrier is obtained for free.

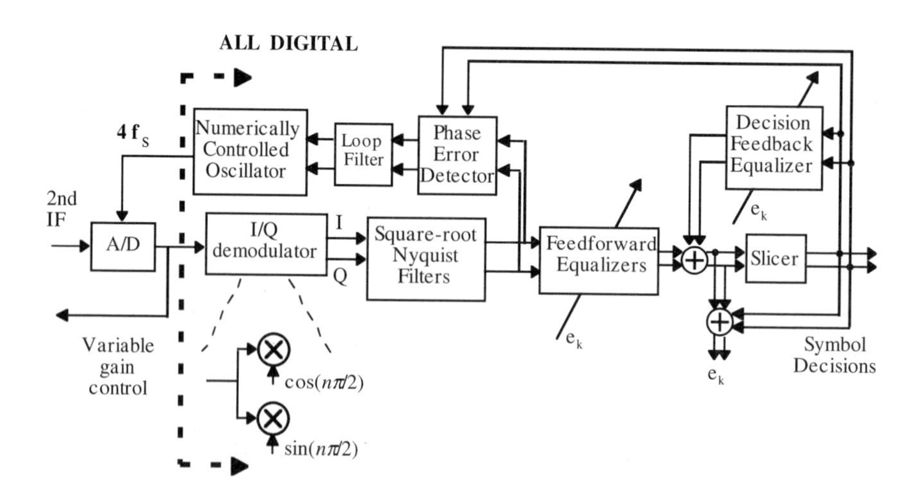

Figure 8.23: Block diagram of a QAM receiver with both a feedforward equalizer and a decision-feedback equalizer. After Ref. [30] (© 1992 IEEE).

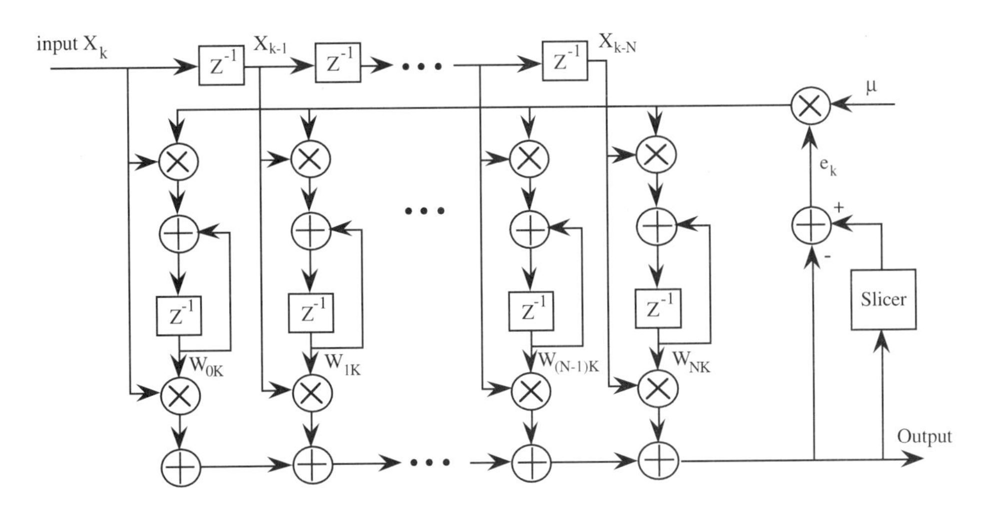

Figure 8.24: A digital implementation of a feedforward equalizer (FFE). After Ref. [31] (© 1996 IEEE).

Note that because of the presence of a DFE, there is more freedom in choosing the coefficients of an FFE, because the FFE need not approximate the inverse of the channel response and so avoids excessive noise enhancement and sensitivity to sampler phase. In most of today's HFC QAM modems, nonlinear DFEs are used in various combinations with linear FFEs [22,30].

If we take a closer look of Eq. (8.49), we can see that the coefficient updates depend on the step size μ and a *known* error signal e_k. As for step size, the larger it

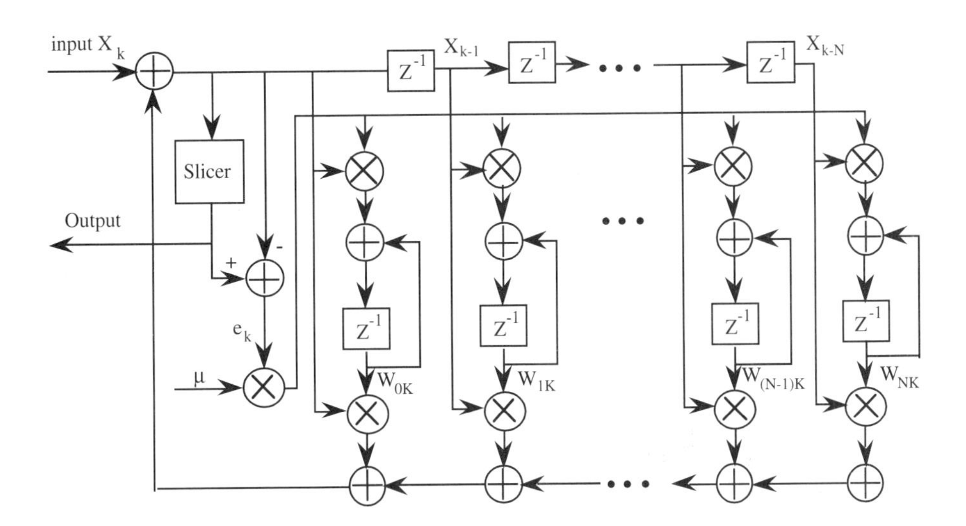

Figure 8.25: A digital implementation of a decision-feedback equalizer (DFE). After Ref. [31] (© 1996 IEEE).

is, the faster is the equalizer tracking capability. However, a compromise must be made between fast tracking and the excess MSE of the equalizer. In practice, the value of the step size is selected for fast convergence during the training period and then reduced for fine tuning during steady-state operation. Concerning the known error signal, especially during initial acquisition, we note that the Grand-Alliance ATSC standard [32] indeed provides this kind of training signal sequence (Section 8.8) and the equalizer in a VSB receiver takes advantage of the frame syncs to update its filter coefficients. However, such a training session cannot be implemented in typical HFC or DVB systems, and *blind* equalizers [33–37] must be used (at least for initial acquisition). A blind equalizer achieves parameter adaptation based on its observation of the channel output and its prior knowledge of some statistical properties of the input data sequences. Adaptation algorithms for parameter updating can be classified as modified error signal algorithms (MESAs) since they replace the prediction error in the traditional LMS algorithm with a modified (locally generated) error signal (MES) not involving the reference signal. The rudest blind equalizer is by replacing d_k in Eq. (8.47) by $Q(y_k)$, where Q is the operator of the slicer. But this kind of blind equalizer has very poor convergence behavior under high ISI. Consequently, better algorithms were developed [23,38,39]. The most commonly used MES for QAM signals is based on the constant modulus algorithm (CMA) given by

$$e'_k = \left(|y_k|^2 - \frac{E(|y_k|^4)}{E(|y_k|^2)} \right) y_k \quad , \tag{8.50}$$

and the resulting blind equalization algorithm can be written as

$$w_{n(k+1)} = w_{nk} + \mu e'_k x_{k-n} \qquad (n = 0,1,...,N) \tag{8.51a}$$

or

$$\mathbf{w}_{n(k+1)} = \mathbf{w}_{nk} + \mu \mathbf{e}'_k \mathbf{x}^*_{k-n} \qquad (n = 0,1,...,N), \tag{8.51b}$$

which is the same as the LMS algorithm, Eqs. (8.49), except that e_k has been replaced by e'_k. Blind acquisition has several performance advantages over systems with training sequences. First, the acquisition time is much faster: The equalizers are trained on all the incoming data, whereas training sequences occur very infrequently, and therefore the equalizers are updated with a low duty cycle. Fast acquisition time is an important feature in digital CATV systems, since reacquisition is performed on every channel change. Second, there is no wasted bandwidth for training sequences in blind equalization.

8.7.2 Carrier Recovery

We have illustrated in Figs. 8.12e, i and j, that if the received carrier is out-of-lock or has a fixed phase-tracking error, the detected signal point positions will spin or rotate a fixed angle about the origin of the constellation diagram. The rate of spinning of the detected signal point positions is dependent on the carrier phase jitter and frequency offset, and the angle of rotation is dependent on the level of the tracking error. Therefore, a robust carrier recovery circuit is very important in satisfactory system performance.

In the extreme case, as was shown in Fig. 8.22, when the IF frequency is set equal to the symbol rate, carrier recovery can be completely eliminated. However, many other receiver designs would still require a stand-alone or equalizer-dependent carrier recovery circuit.

Stand-alone carrier recovery for M-QAM signals can be classified into two major categories: (1) power-of-*N* [40] and Costas loops [41,42], and (2) decision-directed loops [41–43]. The operating principles of each category are given next.

Both the power-of-*N* loop and the Costas loop use a nonlinear circuit to regenerate the carrier frequency component out of double-sideband suppressed-carrier (DSB-SC) signals such as BPSK or M-QAM signals. A simplified block diagram for a power-of-*N* loop is shown in Fig. 8.26. A basic Costas loop for BPSK signals is shown in Fig. 8.27, where we can see that the VCO is driven by a voltage proportional to sin 2θ, where θ is the phase difference between the received BPSK carrier and the VCO. We can see that the power-of-*N* loop operates in the passband with a bandpass filter (BPF), whereas the Costas loop operates in the baseband with a low-pass loop filter to ease the implementation complexity. Furthermore, it can be verified that power-of-*N* and Costas loops have the same performance theoretically.

For an M-QAM signal, a fourth-order Costas loop [41,44] can be used to replace the conventional Costas loop. Figure 8.28 shows a typical fourth-order Costas loop. It can be derived that (see Problem 8.4) the mean of the phase detector output ($P(t)$ in Fig. 8.28) is given by

$$P(t) = E\{[I(t)]^2 [Q(t)]^2\} \sin 4\theta \,, \tag{8.52}$$

where $E\{\cdot\}$ is the expected value, $I(t)$ and $Q(t)$ were defined in Eq. (8.15), and θ is the phase offset between the local oscillator and the carrier of the received signal. We see that $P(t)$ is directly related to the phase offset and it drives the VCO to match the phase of the received signal carrier.

In case of QPSK modems, a polarity-type Costas loop [42] or a fourth-order Costas loop have been practically used.

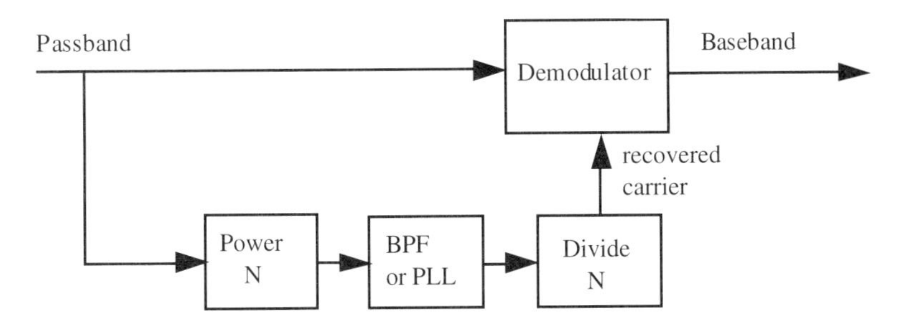

Figure 8.26: Block diagram of power-of-*N* loop.

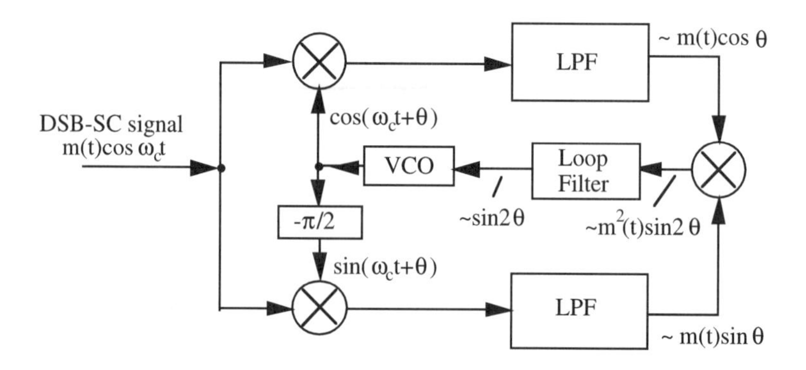

Figure 8.27: Block diagram of a Costas loop.

The second type of carrier recovery circuit is the decision-directed loop [41,43]. A general block diagram of a popular decision-directed loop carrier recovery is shown in Fig. 8.29. "Decision-directed" means that the receiver decisions ($\hat{A}_k$) are used to generate the phase error signal (through the phase discriminator Im[$q_k \hat{A}_k^*$] (see Problem 8.5) to adjust the incorrect rotation of the constellation points (by using exp($-j\hat{\phi}_k$) from the complex VCO). Practical examples of this type of carrier recovery loop are discussed in [45–47].

In today's M-QAM receiver modem where digital circuit implementation is predominant, not only is the I/Q demodulator (immediately following the A/D) implemented digitally, but also carrier recovery (*after* a blind equalizer) is digitally implemented [40]. A general block diagram showing a joint adaptive equalization and carrier recovery is given in Fig. 8.30.

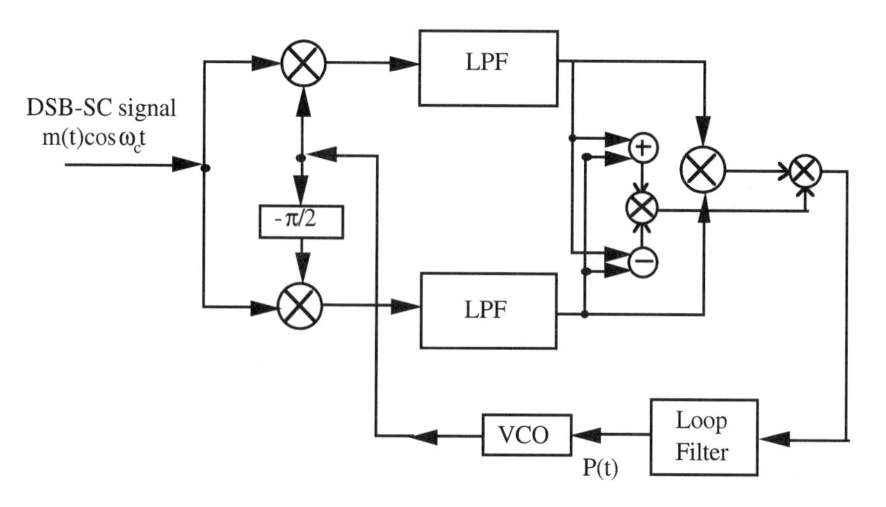

Figure 8.28: Block diagram of a fourth-order Costas loop for M-QAM carrier recovery.

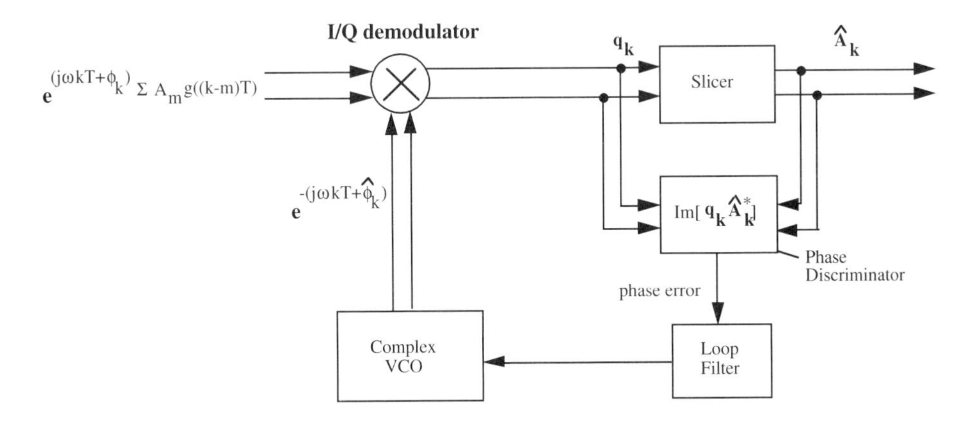

Figure 8.29: Decision-directed carrier recovery loop. (After Ref. [40], with kind permission from Kluwer Academic Publishers.)

We can see that the I/Q demodulator uses free-running oscillators before equalization, and leaves the carrier phase error to be corrected after the equalization by using a complex derotator which provides an angle rotation of $\exp[-j\phi_k]$. The adaptive equalizer is supposed to generate a good estimate of the *rotated* symbols, which will later be derotated by the carrier recovery process; that is why there is a complex rotator $\exp[j\phi_k]$ between the error generator and the equalizer (this kind of equalizer is sometimes termed a "passband equalizer"; see pp. 546–548 of Ref. [40]).

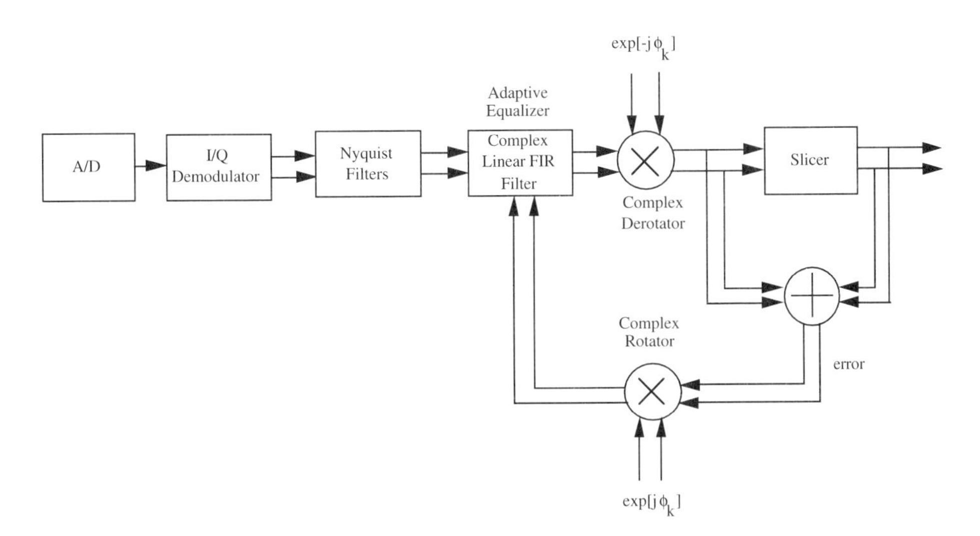

Figure 8.30: A digital receiver with a complex derotator located after the adaptive equalizer. After Ref. [31] (© 1996 IEEE).

The operation sequence is as follows: by using the CMA algorithm [Eq. (8.50)] with a step size which leads to fastest convergence, the adaptive equalizer approaches convergence while the carrier recovery loop is off. At convergence, the equalizer performance is not fully satisfactory because of the large residual variance of the phase and amplitude error signal. Next, the carrier recovery loop is turned on to rotate the signal constellation into place and track any frequency offset and low-frequency phase jitter.

Note that the carrier recovery loop is mainly used to remove the major burden of phase tracking from the slowly adapting equalizer, and therefore its time constant is much shorter than that required for the equalizer coefficient update.

8.7.3 Timing Recovery

Basic timing recovery schemes can be dependent on or independent of the adaptive equalization and carrier recovery. Generally speaking, the long delays in digital equalizers preclude the use of decision-directed timing recovery algorithms [48] because of the difficulty in avoiding jitter in adjustment loops with long delay, and because of the interaction with carrier tracking and equalizer updating loops. On the other hand, we can use a timing recovery scheme which is independent of equalization and carrier recovery, provided the timing acquistion can be done in a period that is short compared to the equalizer convergence time. In addition, we have noticed earlier in deriving the $T/2$-spaced equalizer, the spectrum folding effect is avoided and the mean square error at equalizer output becomes practically independent of the clock phase. Therefore, many state-of-the-art M-QAM modem designs use timing recovery schemes which are independent of the equalizer [22,27]. For M-QAM signals, this kind of timing recovery scheme can generally be classified into two categories [41,49], the spectral-line method, and the feedback loop method.

The spectral-line method uses zero crossings of the timing tone and a phase-locked loop, as illustrated in Fig. 8.31. The timing tone detector is commonly implemented by using a prefilter centered at $1/(2T)$ (this is where the useful clock energy resides in an NRZ signal) and an envelope detector or a nonlinear device (such as a rectifier, squaring, or fourth-power circuit).

The feedback loop method uses a feedback loop to reiteratively adjust the sampling phase (through a VCO) to minimize the timing error, as illustrated in Fig. 8.32.

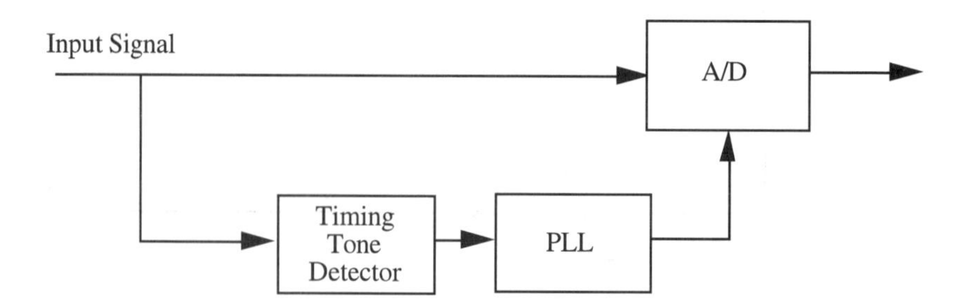

Figure 8.31: The spectral-line timing recovery technique.

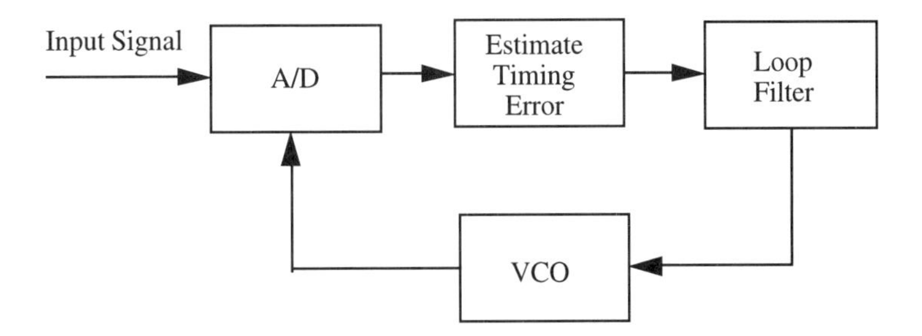

Figure 8.32: A feedback-loop timing recovery technique.

The MMSE (minimum mean-square error) technique using a stochastic gradient algorithm is a typical example [40].

Three examples are given next to illustrate the preceding two timing recovery techniques for M-QAM signals. The first is a passband envelope-derived technique, called band-edge timing recovery (BETR) [50], as shown in Fig. 8.33.

The timing tone at f_{sym} is extracted by using two bandpass filters to get the frequency components at $f_c + (f_{sym}/2)$ and $f_c - (f_{sym}/2)$, respectively, a multiplier to multiply the filtered outputs, and finally a third bandpass filter centered at f_{sym}. Assuming all filters have bandwidths << $1/2T_s$, where T_s is the symbol rate, we can see that the extraction of the timing tone can be approximated by

$$BPF\,@\,f_{sym}\left[\;E_1(t)\cos\{2\pi[f_c+(f_{sym}/2)]t+\phi(t)\}\times E_2(t)\cos\{2\pi[f_c-(f_{sym}/2)]t+\phi(t)\}\;\right]$$

$$= E_3(t)\cos\{2\pi f_{sym}t+\phi(t)\} \tag{8.53}$$

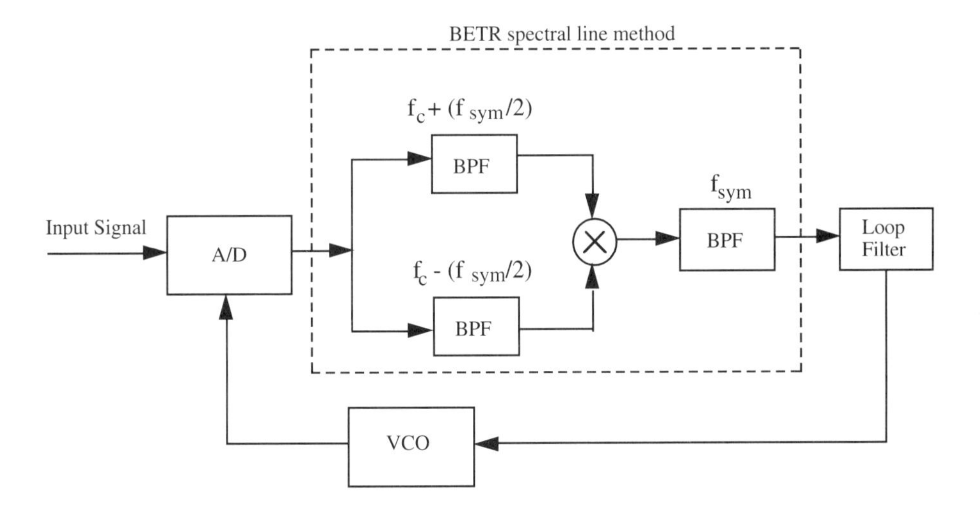

Figure 8.33 Block diagram of the BETR spectral line timing recovery system which can be implemented digitally.

where BPF@f_{sym} represents the bandpass filter operation (with the center frequency of the BPF at f_{sym}), and $E_x(t)cos[2\pi f_x t + \phi(t)]$ ($x = 1,2,3$) represents a narrowband random process with an envelope $E_x(t)$ and a phase function $\phi(t)$. The residual noise around f_{sym} can be minimized through a carefully designed PLL. Note, however, that when the bandpass filters are implemented digitally, many more multipliers are needed than appear in the figure [50] and a large integrated-circuit chip area may be required.

A second example is the Costas loop technique. The main goal of this technique is again to extract the useful clock energy for an NRZ signal which resides around $1/2T$, while avoid operating at high frequencies. An architecture based on dual Costas loops is shown in Fig. 8.34.

Its operating principle can be observed directly from the mathematical expressions given in the figure. In those expressions, ω_c and ω_s represent the carrier and symbol frequencies, respectively; θ_c and θ_s represent the carrier phase error due to the demodulator and the timing phase error due to the NCO, respectively. In addition, since we can treat the prefiltered narrowband signal as a new double-sideband suppressed-carrier signal centered at $f_s/2$, at the prefilter output, we can replace the original $I(t)$ and $Q(t)$ by the general DSB-SC expressions $I'(t)\cos(\omega_s t/2)$ and $Q'(t)\cos(\omega_s t/2)$, respectively. Note that the final voltage which was used to control the NCO depends only on the timing phase error θ_s and is independent of the carrier phase error θ_c. The major disadvantage of this architecture is that it requires six multipliers (which does not include those that may need to implement the prefilters), which could potentially occupy a significant IC chip area.

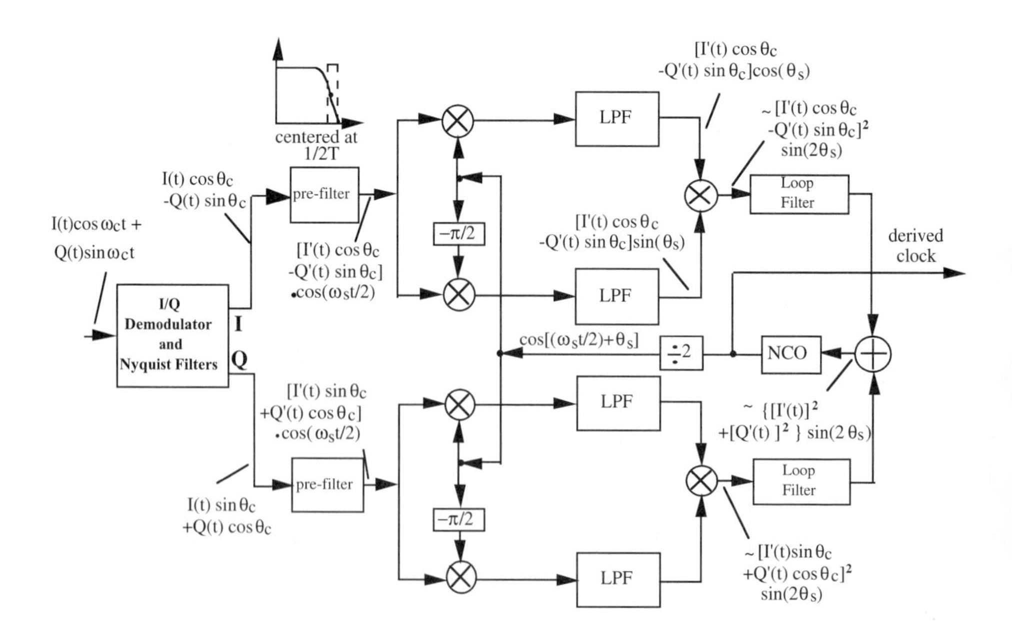

Figure 8.34: Using Costas loops for QAM signal timing recovery.

The third method which is also frequently used is the early–late timing recovery technique, as illustrated in Fig. 8.35. This method belongs to the feedback loop method and is an approximation to the MMSE method. The basic assumptions made by the early–late method are that the peaks in the incoming waveform are at the correct sampling points and that the waveforms around the peak are symmetrical. As shown in Fig. 8.35, the incoming M-QAM signals are first squared to make all peaks positive, and are then sampled at two different times, one advanced by δ and the other delayed by δ with respect to the time when a *predicted* peak value may occur. If the predicted sampling instant is correct, then the values of the early samples and the late samples should be equivalent. If the sampled values are not equivalent, the difference voltage is then used to tune a VCO to the correct sampling phase.

The disadvantage of the early–late scheme is that not all QAM sequences result in time-domain waveform peaks occurring every sampling period [51]. A modified early–late scheme via the decision-directed timing recovery algorithm, however, has been successfully used in QAM modem design [48,52].

8.8 N-VSB Modem Design

N-ary vestigial sideband (N-VSB) modulation was initially selected by the digital High Definition Television (HDTV) Grand Alliance, a consortium formed by a group of six companies and one university, as the digital HDTV transmission standard for the United States. Later this standard changed its name to ATSC (Advanced Television System Committee). In 1994, two VSB modes were approved by the

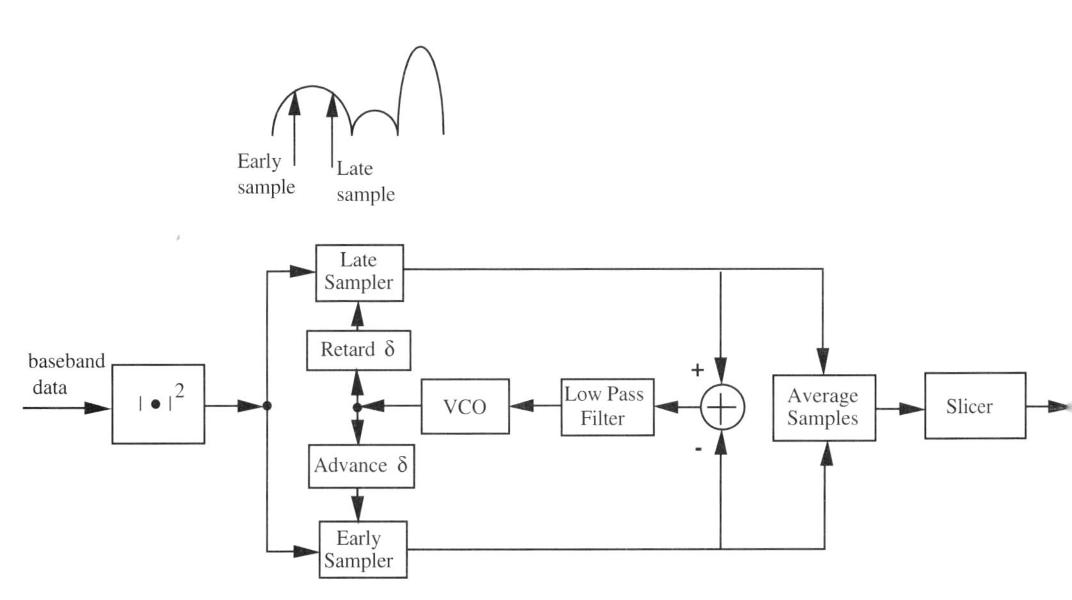

Figure 8.35: Block diagram of the early–late timing recovery scheme. After Ref. [51].

technical subgroup of the United States FCC Advisory Committee on Advanced Television Systems for field testings (1) the trellis-coded 8-VSB terrestrial mode, and (2) the 16-VSB cable mode. In December 1996, FCC adopted the major elements of the ATSC digital television standard for the nation's next generation of broadcast television. However, it should be noted that VSB was not selected as the modulation standard for digital cable or satellite video transmission by ITU-T, DAVIC, and IEEE 802.14. Therefore, in the future, it is possible that a digital television may have to be equipped with both M-QAM and N-VSB modems.

8.8.1 N-VSB Modulation Principle

The mathematical expression of an N-VSB signal is given by [11]

$$s(t) = g(t)\cos\omega_c t \pm \hat{g}(t)\sin\omega_c t\ , \tag{8.54}$$

where

$$\hat{g}(t) = \frac{1}{2\pi}\int_{-\infty_c}^{\infty_c} G(\omega)\hat{W}(\omega)\exp(i\omega t)d\omega \tag{8.55}$$

and

$$\hat{W}(\omega) = \begin{cases} -2iV_0(\omega), & |\omega| < \pi\ B_s\alpha \\ -i\,\mathrm{sgn}\,\omega, & |\omega| > \pi\ B_s\alpha \end{cases}, \tag{8.56}$$

where B_s is the Nyquist bandwidth or half of the symbol rate, α is the filter roll-off factor, sgn is the sign function, $G(\omega)$ is the Fourier transform of $g(t)$, and V_o is an odd function of frequency as shown in Fig. 8.36. In ATSC VSB systems, $B_s = 5.38$ MHz, $\alpha = 0.115$ (see Table 8.2), therefore, $B_S\ \alpha/2 \approx 310$ KHz in Fig. 8.36. Equation (8.54) implies that the quadrature component $\hat{g}(t)$ is essentially a Hilbert transform of the in-phase component $g(t)$, except in the vestigial band $|\omega| < \pi B_s\alpha$. This fact should be compared to the case of a QAM signal [Eq. (8.15)], where the in-phase and

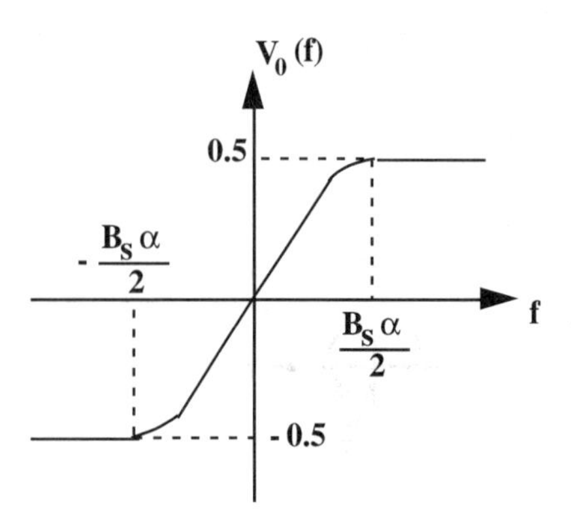

Figure 8.36: Frequency characteristics of $V_0(f)$.

quadrature-phase components are independent. Therefore, a VSB signal is basically a one-dimensional multilevel signal, as opposed to a QAM signal, which is a two-dimensional multilevel signal. Later we will see that an M-VSB achieves the same spectral efficiency as an M^2-QAM signal by using the same transmission bandwidth.

Conventionally, an N-VSB signal can be generated using the Weaver method [53], as shown in Fig. 8.37. According to the figure, we have

$$
\begin{aligned}
&C(t) = r(t)\exp[j\pi B_s t] \quad \Leftrightarrow \quad C(f) = R(f - \frac{B_s}{2}) \\
&y(t) = C(t) \otimes h(t) \quad \Leftrightarrow \quad Y(f) = R(f - \frac{B_s}{2}) \cdot H(f) \\
&y^*(t) = C^*(t) \otimes h(t) \quad \Leftrightarrow \quad Y^*(f) = R(f + \frac{B_s}{2}) \cdot H(f) \\
&s(t) = 2y_R(t)\cos\omega_c t + 2y_I(t)\sin\omega_c t \\
&\quad\;\; = y(t)\exp[j\omega_c t] + y^*(t)\exp[-j\omega_c t] \\
&\Leftrightarrow \quad S(f) = Y(f - f_c) + Y^*(f + f_c) \\
&\qquad\quad = R(f - f_c - \frac{B_s}{2}) \cdot H(f - f_c) + R(f + f_c + \frac{B_s}{2}) \cdot H(f + f_c),
\end{aligned}
\tag{8.57}
$$

where * and $\otimes$ represent complex conjugate and convolution, respectively, $\Leftrightarrow$ stands for Fourier transform pair; and $y_R(t)$ and $y_T(t)$ are the real and imaginary parts of $y(t)$, respectively. The resultant spectrum $S(f)$ is also shown in the shaded area in Fig. 8.37.

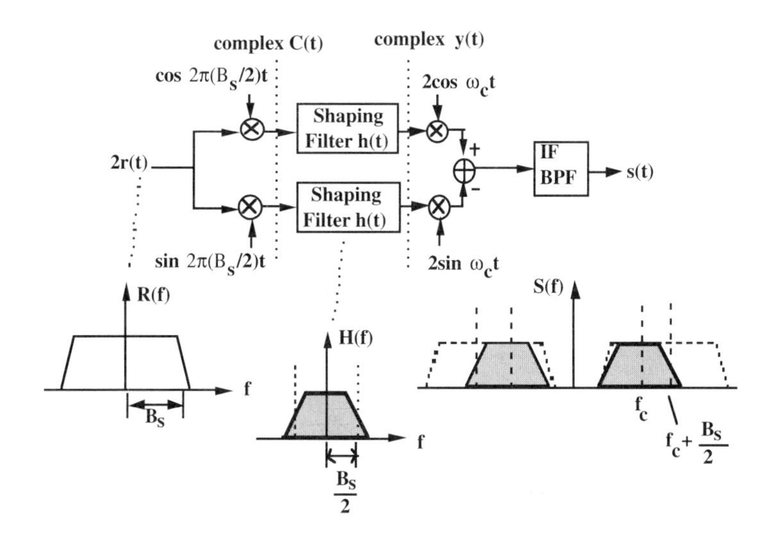

Figure 8.37: The Weaver method of generating a VSB RF signal. In ATSC VSB systems, B_S = 5.38 MHz.

To digitally implement the VSB transmitter based on Weaver's method, we can further simplify the transmitter as shown in Fig. 8.38 [54].

In this case, if we let the Hilbert-transform-pair filters in Fig. 8.38 be

$$\begin{aligned} h'(t) = h(t)\cos[\pi B_s t] \quad &\Leftrightarrow \quad H'(f) = \frac{1}{2}H(f+\frac{B_S}{2}) + \frac{1}{2}H(f-\frac{B_S}{2}) \\ \hat{h}'(t) = h(t)\sin[\pi B_s t] \quad &\Leftrightarrow \quad \hat{H}'(f) = \frac{j}{2}H(f+\frac{B_S}{2}) - \frac{j}{2}H(f-\frac{B_S}{2}). \end{aligned} \tag{8.58}$$

We can have $H'(f)$ and $\hat{H}'(f)$ of the two digital filters shown in Fig. 8.38. From that figure, we also have

$$s(t) = [r(t)\otimes h'(t)]\cdot 2\cos[\omega_c t] \mp [r(t)\otimes \hat{h}'(t)]\cdot 2\sin[\omega_c t], \tag{8.59}$$

where the minus sign is for upper sideband (USB) and the plus sign is for lower sideband (LSB) VSB signals. The Fourier transform of the output $s(t)$ can be written as

$$\begin{aligned} S(f) &= R(f-f_c)H'(f-f_c) + R(f+f_c)H'(f+f_c) \\ &\quad + jR(f-f_c)\hat{H}'(f-f_c) - jR(f+f_c)\hat{H}'(f+f_c) \\ &= R(f+f_c)H(f+f_c+\frac{B_S}{2}) + R(f-f_c)H(f-f_c-\frac{B_S}{2}) \quad (USB) \\ \textit{or} &\quad R(f+f_c)H(f+f_c-\frac{B_S}{2}) + R(f-f_c)H(f-f_c+\frac{B_S}{2}) \quad (LSB). \end{aligned} \tag{8.60}$$

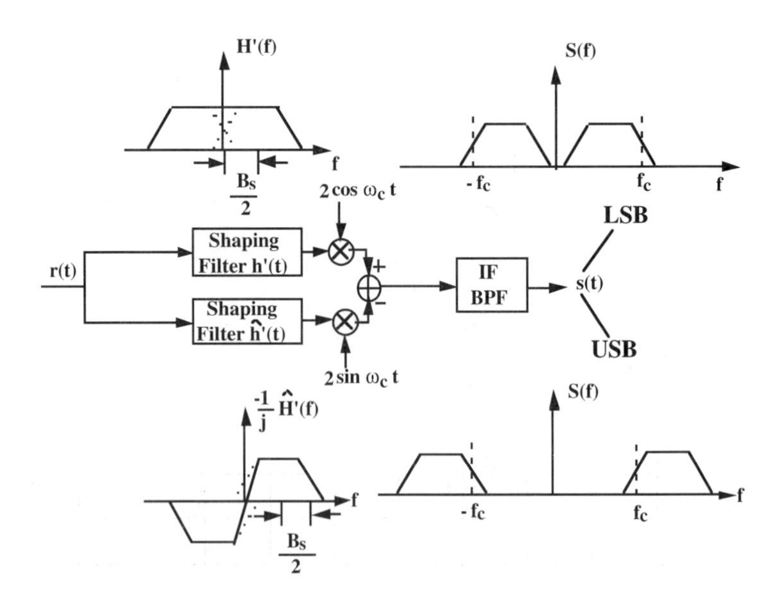

Figure 8.38: Digital implementation of VSB transmitter and the spectra of filters and VSB signals. In ATSC VSB systems, B_S = 5.38 MHz.

The resultant spectra $S(f)$ of the USB and LSB VSB signals are shown in Fig. 8.38, and we can see that the LSB part is similar to that in Fig. 8.37 except for a center frequency offset of $B_s/_2$. The occupied spectrum of a VSB signal is half of that of an M-QAM signal (the DSB-SC spectrum shown by the dashed lines in $S(f)$ of Fig. 8.37). Therefore, for the same transmission bandwidth, say 6 MHz, a 2^{2m}–QAM signal can achieve a theoretical data rate of 2m × 6 Mb/s, with each orthogonal dimension carrying 6m Mb/s; while a 2^m–VSB signal can achieve the same data of 2m × 6 Mb/s because it requires only 3 MHz bandwidth to transport 6m Mb/s.

We also note from Eq. (8.59) that if a DC offset is added to the baseband input signal $r(t)$, a delta function at $f = \pm f_c$, or the so-called *pilot tone,* can be formed. The pilot tone is used to aid carrier recovery in the ATSC standard VSB receivers (Sections 8.8.2 and 8.8.3). Whether the pilot tone is located at the right- or left-hand-side vestigial sideband depends on whether the signal is an LSB or USB VSB signal, and whether it is an IF or an up-converted RF signal (a right-side pilot tone switches to left-side after up-conversion, and vice versa). A typical RF spectrum (after up-conversion) showing the location of the pilot tone of a 16-VSB signal is shown in Fig. 8.39 [55].

The exact location of the pilot tone can be examined from the resolved spectrum shown in Fig. 8.40, where we can see that the pilot tone is inserted 310 kHz from the band edge. Consequently, we know that the excess bandwidth factor α in ATSC standard is equal to (310 kHz)/(2.69 MHz) ≈ 11.5%.

8.8.2 ATSC Standard N-VSB Transmitter

A block diagram of the United States ATSC standard *cable-mode* N-VSB transmitter is shown in Fig. 8.41 (the *terrestrial-mode* transmitter is described in [32]). It is similar to an M-QAM transmitter, except that the VSB transmission system uses three supplementary signals for synchronization: (1) a small pilot tone is inserted

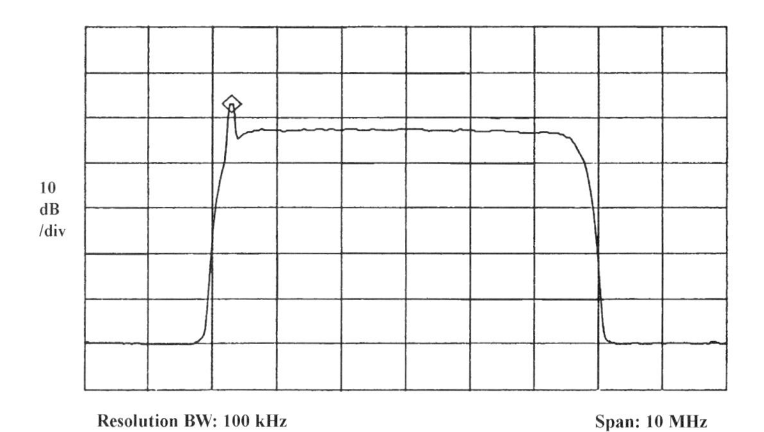

Figure 8.39: An upconverted RF spectrum of a 16-VSB (USB) signal. After Ref. [55].

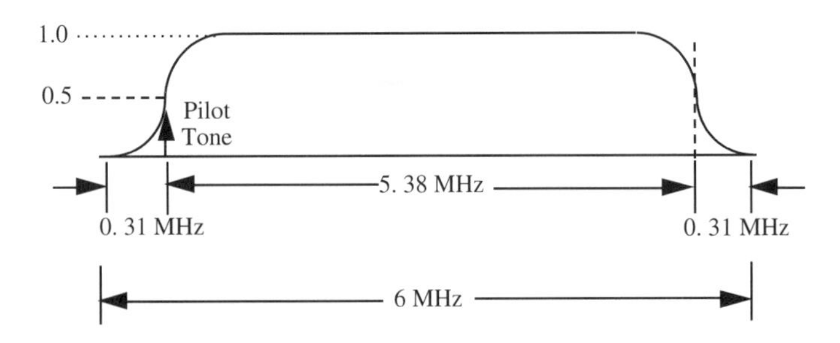

Figure 8.40: A resolved spectrum of an N-VSB signal showing the exact location of the inserted pilot tone. After Refs. [32,56, and 57] (© 1995 IEEE).

for carrier acquistion, (2) a data segment sync is added for synchronizing the data clock in both frequency and phase, and (3) a data field sync is added for data framing and equalizer training [32,56,57].

As mentioned previously, a pilot tone is inserted by adding a small DC level (V_{DC} = 2.5 V) to every symbol of the digital baseband data plus sync signal $r(t)$ (see Fig. 8.38), that is, $V_{DC} + r(t)$. We can easily calculate that the power difference before and after inserting the pilot tone is (assuming the amplitude levels are ±1, ±3, ..., ±15)

$$\Delta P = 10 \cdot \log\left[\frac{\sum_{n=-7}^{8}(2n-1+2.5)^2}{\sum_{n=-7}^{8}(2n-1)^2}\right] \approx 0.3 \text{ dB}, \tag{8.61}$$

which is considered an insignificant amount of power penalty for the transmission system.

A *binary* 4-symbol data segment sync is inserted into the 16-level digital data stream at the beginning of each data segment, as can be seen in Figs. 8.42 and 8.43.

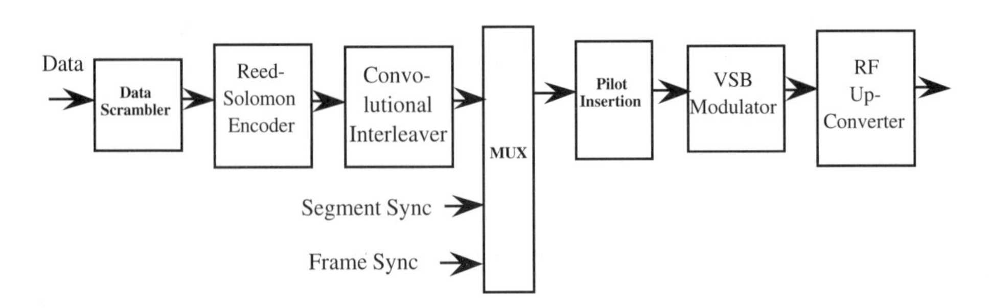

Figure 8.41: Cable-mode 16-VSB transmitter block diagram. After Ref. [32].

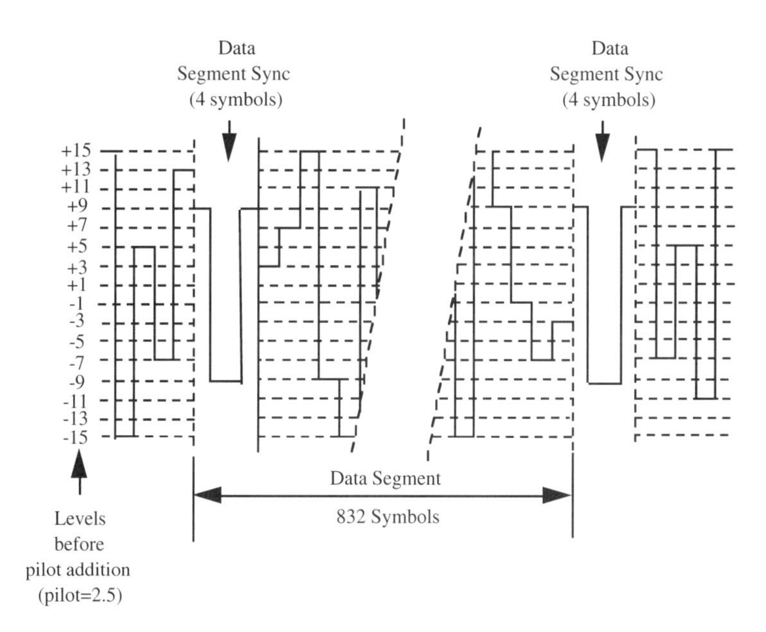

Figure 8.42: Two-level, 4-symbol cable-mode VSB data segment syncs are inserted in the 16-level random data every 832 symbols or 77.3 μs. After Ref. [32].

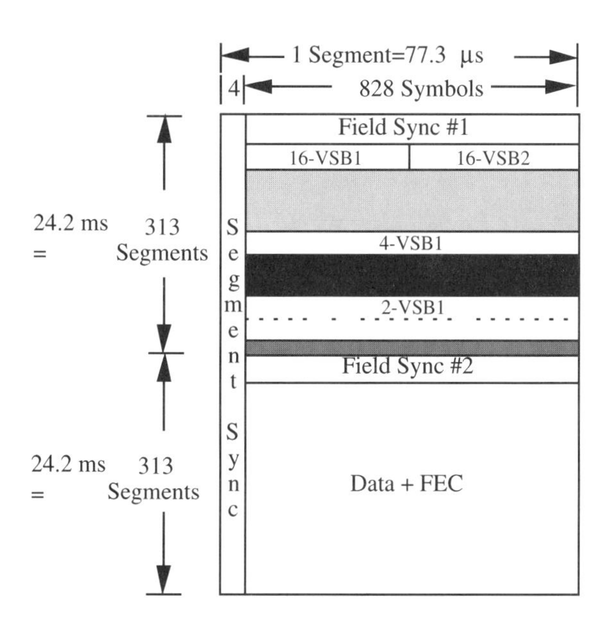

Figure 8.43: VSB data frame and its accommodations of variable data rate. After Ref. [55].

A complete segment consists of 832 symbols: 4 sync symbols (= 1 byte, since there are 2 bits per symbol), and 828 data plus parity symbols (= 828 × 4/8 = 414 bytes = 207-byte/block × 2 RS-coded blocks, and there are 4 bits per symbol). Consequently, there are two 16-VSB blocks in each segment, as can be seen in Fig. 8.43. Similarly, it can be derived that each segment can accommodate (832 symbols/segment)/((208 bytes/block × 8 bits/byte)/(3 bits/symbol) = 1.5 8-VSB blocks/segment, or one 4-VSB block/segment, or 0.5 2-VSB block/segment, and the latter two are illustrated in Fig. 8.43.

The same segment sync pattern occurs regularly at 77.3-μs intervals, the only signal repeating at this rate. The symbol rate is thus 832/(77.3 μs) = 10.76 Msymbol/s, now we can see why the data segment sync is used to regenerate a properly phased 10.76-MHz symbol clock.

Each data field starts with a data field sync, as shown in Fig. 8.43. The field sync signal is made up of one 511-symbol and three 63-symbol binary pseudorandom sequences and has the same amplitude as the data segment sync. The three 63-symbol PRBS sequences in the field sync alternate polarity from one field to the next, providing field identification information to the receiver. Therefore, there are two data field syncs and their follow-up data that make up one data frame which lasts 48.4 ms.

8.8.3 ATSC Standard N-VSB Receiver

A simplified block diagram of an N-VSB receiver is shown in Fig. 8.44. We can immediately see that the analog front end includes not only the tuner and the SAW filter, but also a carrier recovery circuit based on a frequency- and phase-locked loop (FPLL) [58]. This arrangement is quite different from that in QAM receivers (see Figs 8.17 and 8.23)-a VSB demodulator acquires the carrier pilot via a relatively

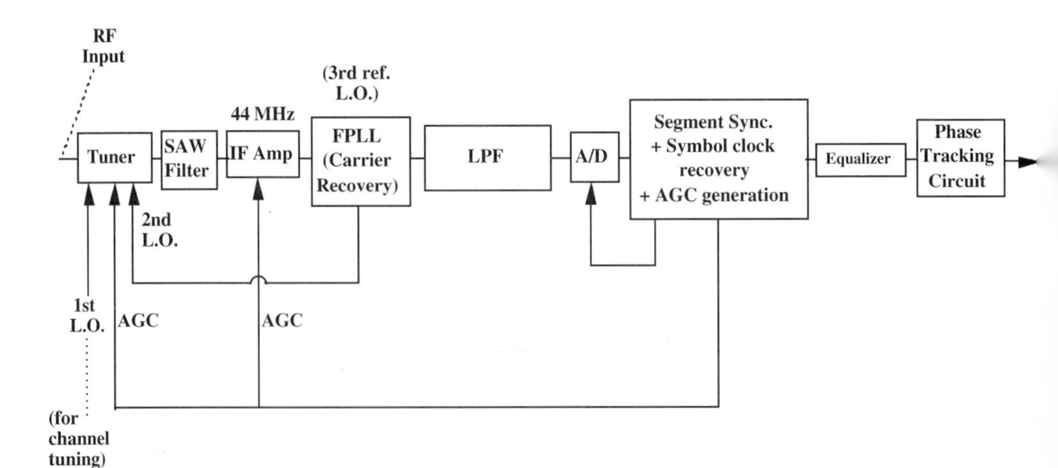

Figure 8.44 A simplified block diagram of an N-VSB receiver.

narrow tracking loop, prior to any equalization, while a QAM demodulator can digitally perform blind equalization in the presence of carrier offset.

8.8.3.1 FPLL in VSB Receiver A detailed block diagram of FPLL is shown in Fig. 8.45. The operation principle of FPLL can be explained as follows: as shown in Fig. 8.45a, the control for the second LO (the third LO is a fixed reference oscillator) is from the FPLL which contains both a frequency loop and a phase-locked loop in one circuit. The frequency loop provides a wide frequency pull-in range of ±100 KHz while the phase-locked loop has a narrow bandwidth less than 2 KHz. Initially, assuming the frequency and phase differences between the incoming pilot tone ($\cos\omega_0 t$) and the third LO ($\cos(\omega_0 t + \Delta\omega t + \theta)$) are $\Delta\omega$ and θ, respectively. The I- (after the automatic frequency control (AFC) low-pass-filter (LPF)) and Q- signals after the down-conversion process are given by $\cos(\Delta\omega t + \theta + \phi(\Delta\omega))$ and $\sin(\Delta\omega t + \theta)$ (we drop the high frequency component at $2\omega_0$ because it will later be filtered out by the automatic phase control (APC) LPF), respectively, where $\phi(\Delta\omega)$ is the phase characteristic of the AFC LPF as a function of the frequency offset $\Delta\omega$, shown in Fig. 8.45b. We can see that $\phi(\Delta\omega)$ approaches $\pi/2$ and $-\pi/2$ for $\Delta\omega > 0$ and $\Delta\omega < 0$, respectively. Therefore, the limiting amplifier output is in-phase or 180° out-of-phase with the *Q*-signal, depending on if $\Delta\omega < 0$ or $\Delta\omega > 0$, as shown in Fig. 8.45c. Consequently, the average DC level after the APC LPF is either positive or negative, depending on whether $\Delta\omega < 0$ or $\Delta\omega > 0$. The VCO frequency will then be pulled in to minimize $\Delta\omega$.

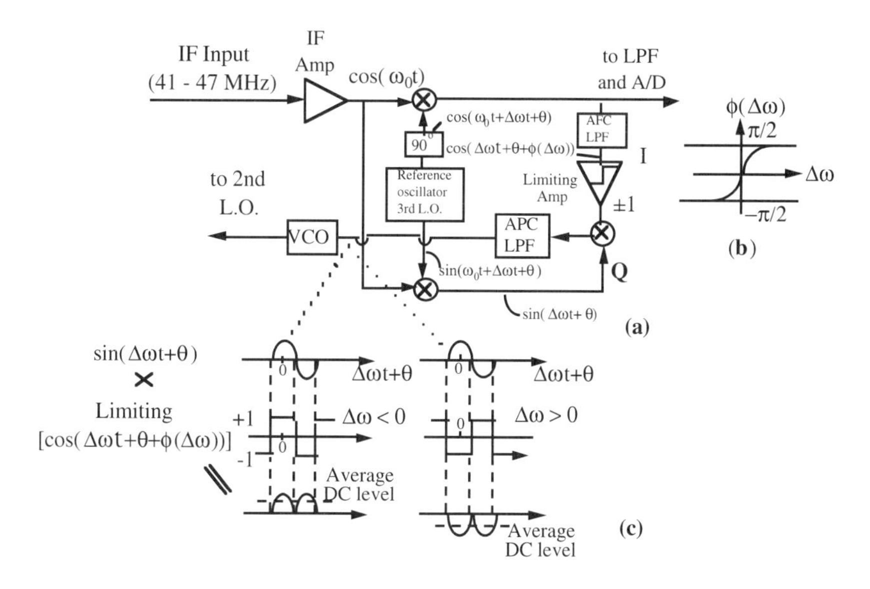

Figure 8.45: FPLL operating at IF: (a) the FPLL block diagram, (b) Phase-frequency characteristics of the AFC low-pass-filter, and (c) the resultant signal after the APC low-pass-filter which has a positive ($\Delta\omega < 0$) or negative ($\Delta\omega > 0$) average DC level.

When $\Delta\omega$ approaches zero, the I-signal (cos θ) has a 90° phase shift with respect to the Q-signal (sin θ), and the DC level after the APC LPF approaches zero; only the phase-locked loop is active. The output signal after the APC filter is then proportional to sin θ, which is the typical PLL characteristics, with the exception that it is biphase stable.

8.8.3.2 Phase Tracking Loop As shown in Fig. 8.44, there is a phase tracking loop after the equalizer to further track out phase noise which has not been removed by the IF FPLL operating on the pilot tone. Because the system is already frequency-locked to the pilot tone by the IF FPLL, the phase tracking loop bandwidth is maximized for phase tracking by using a first-order loop. Higher order loops, which are needed for frequency tracking, do not perform phase tracking as well as first-order loops. Since the phase tracker is operating on the 10.76 Msymbol/s data, the bandwidth of the phase tracking loop is fairly large, approximately 60 kHz. The basic function of the phase tracker is to use the equalized I-arm data to generate its corresponding Q-arm data [remember they are related by the approximate Hilbert transform, Eq. (8.55)] through an FIR filter. Then by using the generated I/Q constellations, the phase tracker can decide the amount of "derotation" that is needed (a normal 16-VSB constellation should contain 16 horizontal lines, but when phase noise occurs, the entire constellation rotates around the origin, making the symbols with large Q-values rotate toward the slice levels more than those symbols with smaller Q-values).

8.8.3.3 Segment Sync and Symbol Clock Recovery As shown in Fig. 8.44, following the carrier recovery circuit and a low-pass filter is an A/D converter, and the segment sync and clock recovery circuit. Unlike an M-QAM receiver, the N-VSB receiver uses a *data-aided* synchronization method, where the "aiding" data is the periodic segment sync inserted at the N-VSB transmitter site.

In the block diagram shown in Fig. 8.46, the correlation filter uses a fixed two-level sync segment (i.e., (+9, –9, –9, +9) as was shown in Fig. 8.42), which is the same sync segment that was periodically inserted in the data sequence at the transmitter, as a local reference. Repeated correlations at a data segment rate are integrated and amplified by the segment integrator. The repetitive segment sync is detected, while the random data integrates to zero. The point at which maximum correlation occurs is selected to be the optimum sampling point.

The confidence counter is used to reduce the decision error probability of the correlation filter. When the segment sync is first detected, the correlation filter triggers the confidence counter to start counting symbols. After the counted number reaches 832 (a segment length), the counter rechecks whether the correlation filter also finds the next segment sync word. If the answer is yes, the sync word detection is considered successful with a high confidence level.

Precise phasing is obtained by sampling the zero crossing of the quadrature filter output, which corresponds to the center of the sync waveform (note that the quadrature filter is a 90°-phase shifter). Once a predefined level of confidence is

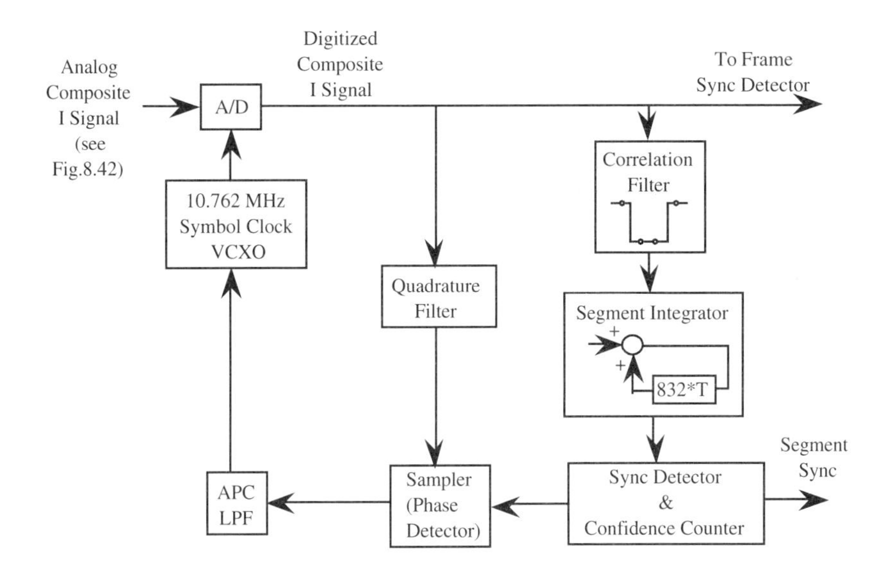

Figure 8.46: Segment sync and clock recovery block diagram. After Ref. [57] (© 1996 IEEE).

reached that the segment sync has been found (using the confidence counter), subsequent receiver loops are enabled.

As an example, one can calculate the amplitude difference between the two sampled middle points, as shown in Fig. 8.47. If the difference approaches zero, there is negligible sampling phase error. If the difference is much greater than zero, one can estimate the phase error according to the filtered waveforms (e.g., raised cosine) and the amplitude difference Δv. One can also adjust the sampling phase by examining how the middle two sampled values differ from the reference values of (–9, –9). If the sampled values are, say, (–11, –7), as shown in Fig. 8.47, we know that there is a phase lag in the sampling instant; if the sampled values are (–7, –11), we know there is a phase lead in the sampling instant. The sampling phase can then be adjusted accordingly.

In summary, the scheme uses a correlation filter and a confidence counter to find the location of segment syncs, and the phase detector is used to adjust the phase error of the extracted segment syncs. The scheme has been adopted by ATSC and ITU J.83 Transmission Standards and works reliably at low SNR (~ 0 dB) and in the presence of heavy interference. However, the periodically increased inserted sync segments apparently increase the transmission overhead.

8.8.3.4 Equalizer The VSB receiver equalizer has a 64-tap feedforward transversal filter followed by a 192-tap decision feedback filter, and operates at the 10.76-MHz symbol rate (T- instead of $T/2$-spaced). The equalizer first uses the data field sync in each frame as the training signal to open the eye diagram. The data field

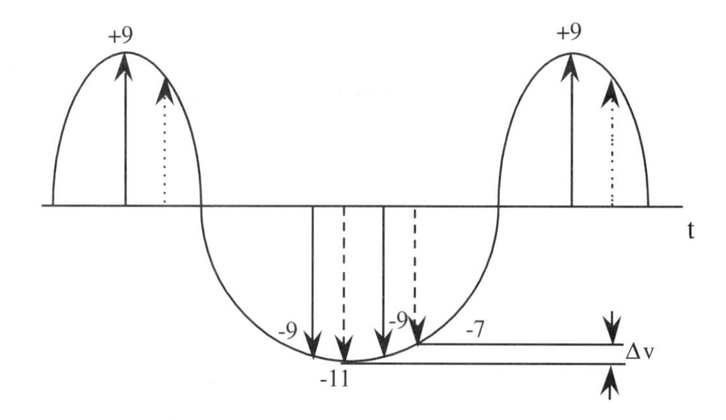

Figure 8.47: Methods for deriving phase error information.

sync consists of pseudonoise (PN) sequences without coding and interleaving in order to provide direct measurement and compensation for channel linear distortion. After adaptation with the training signal, the eye has opened, and the equalizer may be switched to a decision-directed mode of operation. Decision-directed blind equalization techniques become necessary when the channel impairments vary more rapidly than the transmission of the training signals (recur with a period of 24.2 ms). Recall that the decision-directed mode uses the symbol values at the output of the decision device instead of the training signal.

Because the VSB is a one-dimensional signal, techniques such as the CMA algorithm [Eq. (8.50)] or the RCA algorithm [59] that have been extensively used for QAM systems are not directly applicable. Instead, a technique called the modified reduced constellation algorithm (MRCA) has been developed for VSB systems [32].

Table 8.3 summarizes the key parameters of a cable-mode VSB modem. Additional parameters of this kind of modem can also be found in Table 8.2, ITU J.83 Annex D. Table 8.4 compares the modem design and performance of QAM and VSB modems.

8.9 Upstream Physical-Media-Dependent Sublayer Parameters

In this Section, we will use the MCNS standard [4] to illustrate the important upstream physical-media-dependent (PMD) sublayer parameters. The upstream PMD sublayer uses a FDMA/TDMA burst modulation format, which provides five symbol rates (160, 320, 640, 1280, and 2560 ksymbol/s) and two modulation formats (QPSK and 16 QAM). The modulation includes pulse shaping for spectral efficiency ($\alpha = 0.25$), is carrier-frequency agile, and has selectable output power level (+ 8 dBmV to +58 dBmV for QPSK and to +55 dBmV for 16 QAM). The PMD sublayer format includes a variable-length modulated burst (with 0 to

Table 8.3: Parameters for cable-mode VSB transmission.

Parameter	Cable-mode specifications
Channel bandwidth	6 MHz
Excess bandwidth (α)	11.5%
Symbol rate	10.76 Msymbol/s
Bits per symbol	4
Trellis FEC	None
Reed–Solomon FEC	$t = 10$ (207,187)
Segment length	832 symbols
Segment sync	4 symbols per segment
Frame sync	1 per 313 segments
Payload data rate	38.57 Mbps
Pilot-tone power contribution	0.3 dB
C/N threshold	28.3 dB

Table 8.4: Comparison of QAM and VSB modem characteristics.

Parameters	16-VSB	64/256-QAM[a]
Symbol rate (Msps)	10.76	5.0 to 5.27 (6 MHz bandwidth)
Receiver structure (after IF BPF)	Analog Front-end demodulator	All digital
Carrier recovery	Aided by pilot-tone and FPLL (both independent of equalizer) and phase-tracking circuit	data-directed, blind, interacts with equalizer
Symbol timing recovery	Aided by training signal (data segment sync) and independent of equalizer	Nonlinear processing (independent of equalizer), or data-directed (dependent on equalizer)
Equalizer	Aided by training sequence (data field sync) and blind equalization using MRCA algorithm	Blind equalization using CMA or RCA algorithms

[a] Theoretically, the capacity of a 16-VSB modem is equivalent to that of a 256-QAM modem.

255 minislots[8]) with precise timing beginning at boundaries spaced at integer multiples of 6.25 μs apart (which is 16 symbols at the highest data rate of 2560 ksymbol/s).

The PMD sublayer can support a near-continuous mode of transmission, wherein ramp-down of one burst may overlap the ramp-up of the following burst,

[8]See Chapter 11 for definition and detailed descriptions.

so that the transmitted envelope is never zero. However, the system timing of the TDMA transmission from the various cable modems must ensure that the center of the last symbol of one burst and the center of the first symbol of the preamble of an immediately following burst are separated by at least the duration of five symbols (see illustration in Fig. 11.8). The guard time must be greater than or equal to the duration of five symbols plus the maximum timing error. The maximum guard time is 255 symbols.

There can be no FEC or a RS FEC with $t = 1$ to 10. The RS primitive polynomial is $P(x) = x^8 + x^4 + x^3 + x^2 + 1$. A 15-bit seed scrambler must also be provided whose primitive polynomial is $x^{15} + x^{14} + 1$. The preamble length must be programmable from 0 to 1024 bits.

Readers can see that many upstream parameters are very flexible (e.g., variable data rate, variable FEC strength, QPSK or 16-QAM, carrier frequency agility, etc.). This flexibility enables a system designer to optimize upstream system performance under different ingress and impulsive noise conditions.

Problems

8.1. Assume a shift register with four bits as shown in Fig. P8.1. Let the initial state of the register be 1 0 0 0. Show that the register state can have $2^4 - 1$ patterns.

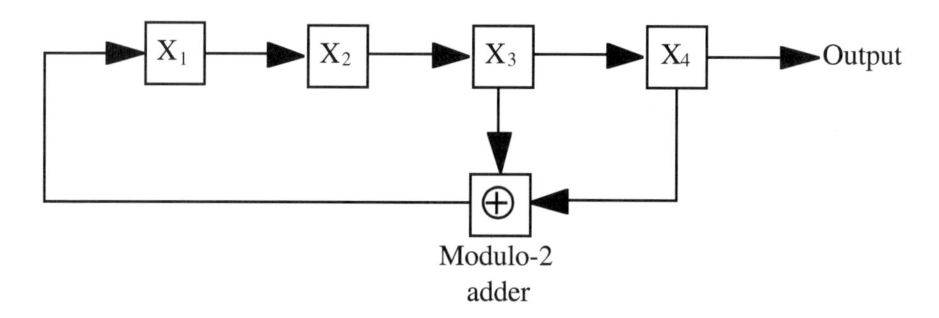

Figure P8.1 A linear feedback shift register.

8.2. Prove that when there is quadature angle error (Fig. 8.12f), the constellation diagram looks like a diamond shape as a whole.

8.3. Derive the operation principle of a power-of-2 loop that can be used to regenerate a carrier.

8.4. Derive the operation principle of a fourth-order Costas loop whose end result is given by Eq. (8.52). Explain why the conventional Costas loop cannot be used for M-QAM signals.

8.5. The 64-QAM interleaver characteristics of ITU J.83 Annex B is shown in Table P8.1.

Table P8.1

I (Number of Taps)	J (Increment)	Burst Protection	Latency
8	16	5.9 μs	0.22 ms
16	8	12 μs	0.48 ms
32	4	24 μs	0.98 ms
64	2	47 μs	2.0 ms
128	1	95 μs	4.0 ms

Consider the 64-QAM trellis-coded-modulator given in Fig.8.14 and the formulae given in Section 8.3, calculate the burst protection time length and the latency, see if you can obtain the same values as those given in Table P8.1. Discuss what the tradeoffs in system performance are between burst protection and latency.

8.6. In an M-QAM demodulator shown in Fig. P8.2, prove that $CNR_i + 0.63dB = SNR_o$.

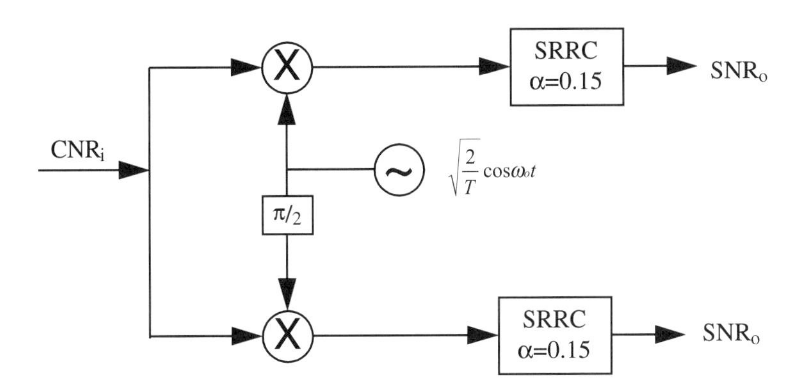

Fig. P8.2 CNR_1 and SNR_1 are measured at the input and output of a QAM demodulator, respectively.

References

[1] Digital Audio-Visual Council, DAVIC 1.1 Specifications (http:/www.davic.org/DOWNI.htm).

[2] ITU-T Recommendation J.83, "Digital multi-programme systems for television sound and data services for cable distribution," October 1995/1997.

[3] IEEE 802.14 Physical Layer Specifications, v1.0, (http:/www.walkingdog.com/catv/index.html).

[4] MCNS Specification, "Data-over-cable interface specifications, radio-frequency interface specification," SP-RFII02–9701008, (http:/www.cablemodem.com).

[5] K. Feher and Engineers of HP Ltd., *Telecommunications Measurements, Analysis and Instrumentation*, Prentice-Hall, Englewood Cliffs, New Jersey, 1987.

[6] B. Sklar, *Digital Communications*, Prentice-Hall, Englewood Cliffs, New Jersey, 1988.

[7] W. W. Peterson and E. J. Weldon, Jr., *Error-Correcting Codes*, 2nd Edition, MIT Press, Cambridge, Massachusetts, 1972.

[8] S. Lin and D. J. Catello, *Error-Correction Coding for Digital Communications*, Prentice-Hall, Englewood Cliffs, New Jersey, 1983.

[9] M. K. Simon, S. M. Hinedi, and W. C. Lindsey, *Digital Communication Techniques*, Prentice-Hall, Englewood Cliffs, New Jersey, 1995.

[10] B. C. Wong and H. Samueli, "A 200 MHz all-digital QAM modulator and demodulator in 1.2 μm CMOS for digital radio applications," IEEE J. Solid-State Circuits, vol.26, pp.1970–1979, December 1991.

[11] R. W. Lucky, J. Salz, and E. J. Weldon, Jr., *Principles of Data Communications*, McGraw-Hill, New York, 1968.

[12] K. Feher, *Digital Communications: Microwave Applications*, Prentice-Hall, Englewood Cliffs, New Jersey, 1981.

[13] S. Haykin, *Communication Systems*, Wiley, New York, 1994.

[14] HP 3709 B Constellation Analyzer Training Manual, Hewlett-Packard Co.

[15] G. Ungerboeck, "Trellis-coded modulation with redundant signal sets, Part I; introduction," IEEE Commun. Magazine, vol.25, pp.5–11, February 1987.

[16] G. Ungerboeck, "Trellis-coded modulation with redundant signal sets, Part II; state of the art," IEEE Commun. Magazine, vol.25, pp.12–21, February 1987.

[17] H. K. Wolf and E. Zehavi, "P^2 codes: Pragmatic trellis codes utilizing punctured convolutional codes," IEEE Commun. Magazine, vol. 33, pp.94–99, February 1995.

[18] Y. Bian and J. O'Reilly, "An accelerated search algorithm for punctured convolutional codes," Tech Digest, IEEE GLOBECOM '95, vol.3, pp.2262–2266, 1995.

[19] J. G. Proakis and M. Salehi, *Communication Systems Engineering*, Prentice-Hall, Englewood Cliffs, New Jersey, 1994.

[20] J. P. Odenwalder, *Error Control Coding Handbook*, Linkabit Corp., San Diego, July 15, 1976.

[21] Y. Yasuda, K. Kashiki, and Y. Hirata, "High-rate punctured convolutional codes for soft decision Viterbi decoding," IEEE Trans. Commun., vol.32, pp.315–319, March 1984.

[22] P. McGoldrick, "Demodulator with FEC decoder has on-chip timing, recovery," Electron. Design, pp.55–64, September 3, 1996.

[23] D. N. Godard, "Self-recovering equalizer and carrier tracking in two dimensional data communication system," IEEE Trans. Commun., vol.28, pp.1867–1875, November 1980.

[24] G. Ungerboeck, "Fractional tap-spacing equalizer and consequences for clock recovery in data modems," IEEE Trans. Commun., vol.24, pp.856–864, August 1976.

[25] R. D. Gitlin and H. C. Meadors, Jr., "Center-tap tracking algorithms for timing recovery," AT&T Tech. J., vol.66, pp.63–78, November/December 1987.

[26] R. L. Cupo and C. W. Farrow, "Equalizer-based timing recovery," U.S. Patent 4, 815, 103, March 21, 1989.

[27] N. K. Jablon, "Joint blind equalization, carrier recovery, and timing recovery for high-order QAM signal constellations," IEEE Trans. Signal Processing, vol.40, pp.1383–1398, June 1992.

[28] R. D. Gitlin and S. B. Weinstein, "Fractionally-spaced equalization: An improved digital transversal equalizer," Bell Syst. Tech. J. vol.60, pp.275–296, February 1981.

[29] B. Widrow and S. D. Stearns, *Adaptive Signal Processing*, Chapter 6, Prentice-Hall, Englewood Cliffs, New Jersey, 1985.

[30] R. B. Joshi, B. Daneshrad, and H. Samueli, "A VLSI architecture for a single 5-Mbaud QAM receiver," Tech. Digest, GLOBECOM, pp.1265–1268, 1992.

[31] A. J. King, "Digital demodulation, equalization, and error correction," ISSCC Short Course, February 1996.

[32] *Guide to the Use of Digital Television Standard for HDTV Transmission,* U.S. Advanced Television Systems Committee, April 1995.

[33] A. Benveniste and M. Goursat, "Blind equalizers," IEEE Trans. Commun., vol.32, pp.871–882, August 1984.

[34] S. U. H. Qureshi, Adaptive Equalization, in *Advanced Digital Communications*, Ed. K. Feher, Chapter 12, Prentice-Hall, Englewood Cliffs, New Jersey, 1987.

[35] G. Picchi and G. Prati, "Blind equalization and carrier recovery using a "stop-and-go" decision-directed algorithm," IEEE Trans. Commun., vol.35, pp.877–887, September 1987.

[36] S. Bellini, "Blind equalization," Alta Frequenza, LVII.7, pp.445–450, September 1988.

[37] Z. Ding, C. R. Johnson, Jr., and R. Kennedy, "Global convergence issues with linear blind adaptive equalizers," in *Blind Deconvolution*, Ed., S. Haykin, Prentice-Hall, Englewood Cliffs, New Jersey, 1994.

[38] Y. Sato, "A method of self-recovering equalization for multilevel amplitude-modulation systems," IEEE Trans. Commun., vol.23, pp.679–682, June 1975.

[39] G. Picchi and G. Prati, "Blind equalization and carrier recovery using a "stop-and-go" decision-directed algorithm," IEEE Trans. Commun., vol.35, pp.877–887, September 1987.

[40] E. A. Lee and D. G. Messerschmitt, *Digital Communication*, 2nd Edition, Kluwer Academic Publishers, Dordrecht, The Netherlands, 1994.

[41] W. C. Linsey, *Synchronization Systems in Communication*, Prentice-Hall, Englewood Cliffs, New Jersey, 1972.

[42] J. A. C. Bingham, *The Theory and Practice of Modem Design*, Wiley, New York, 1988.

[43] A. Leclert and P. Vandamme, "Universal carrier recovery loop for QASK and PSK signal sets," IEEE Trans. Commun., vol.31, pp.130–136, January, 1983.

[44] A. J. Rustako, "Using times-four carrier recovery in M-QAM digital radio receivers," IEEE J. Selected Areas Commun. vol.SAC–5, April 1987.

[45] K. T. Wu and J. D. McNicol, "Carrier lock detector for a QAM system," U.S. Patent 4,987,375, January 1991.

[46] M. J. Head, "QAM demodulator carrier recovery loop using N-M LSB's," U.S. Patent 4,571,550, February 1986.

[47] P. P. Giusto, "Circuit recovering the carrier of a signal amplitude and phase modulated by digital signals," U.S. Patent 4,538,111, August 1985.

[48] K. H. Muller and M. Muller, "Timing recovery in digital synchronous data receivers," IEEE Trans. Commun., vol.24, pp.516–521, May 1976.

[49] L. E. Franks, "Carrier and bit synchronization in data communication—a tutorial review," IEEE Trans. Commun., vol.28, pp.1107–1120, August 1980.

[50] D. N. Godard, "Passband timing recovery in all-digital modem receiver," IEEE Trans. Commun., vol.26, pp.517–523, May 1978.

[51] W. Webband and L. Hanzo, *Modern Quadrature Amplitude Modulation*, Pentech Press (London) and IEEE Press (New York), 1994.

[52] B. Daneshard and H. Samueli, "A carrier and timing recovery technique for QAM transmission on digital subscriber loops," Tech. Digest, Int. Conf. Commun., pp.1804–1808, June 1993.

[53] N. Boutin, "Complex signals: Part IV," RF Design, pp.65–75, May 1990.

[54] S. C. Yin, C. C. Su, M. T. Shiue, L. Y. Huang, C. K. Wang, S. J. Jou, and W. I. Way, "A new VSB modulator technique and shaping filter design," Tech. Digest, vol. 4, pp. 312–315, Int. Symp. Circuits Systems, 1996.

[55] R. B. Lee, "High data rate VSB modem for cable applications including HDTV: Description and performance," Tech. Digest, NCTA, pp.274–282, New Orleans, Louisiana, May 1994.

[56] W. Bretl, G. Sgrignoli, and P. Snopko, "VSB modem subsystem design for Grand Alliance digital television receivers," IEEE Trans. Consumer Electron. vol.41, pp.773–786, August 1995.

[57] G. Sgrignoli, W. Bretl, and R. Citta, "VSB modulation used for terrestrial and cable broadcasts," IEEE Trans. Consumer Electron. vol.41, pp.367–382, August 1995.

[58] R. Citta, "Frequency and phase lock loop," IEEE Trans. Consumer Electron. vol.CE-23, pp.358–365, August 1979.

[59] D. N. Godard and P. E. Thirion, "Method and device for training an adaptive equalizer by means of an unknown data signal in a QAM transmission system," U.S. Patent 4,227,152, October 1980.

PART III

Systems and Protocols

Chapter 9

Subcarrier Multiplexed Lightwave System Design Considerations

With the device and component background provided in Chapters 3 to 8, we are now ready to discuss the overall system design issues of subcarrier multiplexed (SCM) lightwave systems for broadband HFC access networks.

9.1 Overall System Considerations

9.1.1 Selection of 1.3 versus 1.55-μm Sources, Direct versus External Modulation Systems

For downstream transmissions, there are obviously four different SCM systems to choose from: 1.3-μm direct modulation systems, 1.3-μm external modulation systems, 1.55-μm direct modulation systems, and 1.55-μm external modulation systems. The technical and economic design considerations for these four systems are very distinct and sometimes complicated, and require in-depth understanding of device/component physics and overall system-integration-related issues. The purpose of this chapter is to provide readers with the general system design concepts and the necessary analytical tools, so that they can optimize their component selections and overall system design, under various real-world operating conditions.

Before starting with the detailed discussions, we first summarize in Table 9.1 the major advantages and disadvantages of the four different systems just mentioned. Several basic *facts* have been used in this summary: (1) Almost all existing optical fibers that have been deployed today are the "conventional" 1.3-μm zero-dispersion single-mode optical fibers; (2) if the wavelength of an optical source is 1.55 μm, then a high optical fiber dispersion of 17–20 ps/nm/km may be incurred;

and (3) despite the fiber dispersion penalties, operating at 1.55 μm can benefit from the lower fiber loss (~0.22 dB/km as opposed to ~0.35 dB/km at 1.31 μm) and the power boosting of erbium-doped fiber amplifiers. However, extra bending loss penalty may be incurred when transmitting 1.55-μm signals in conventional single-mode fibers (see Chapter 3).

Most of the descriptions in Table 9.1 have been mentioned previously in Chapters 3 to 7; however, additional system design issues which will be covered later in this chapter are (a) fundamental channel capacity of a laser or a Mach–Zehnder modulator due to clipping effects; (b) multiple-reflection-induced interferometric noise and nonlinear distortions; (c) various system penalties due to laser frequency chirping and optical fiber dispersion; and (d) WDM and high input optical power related issues. Readers who come back to Table 9.1 after reading through this chapter, which covers (a)–(d), should be able to comprehend the table's contents more thoroughly.

9.1.2 High-Power versus Low-Power Sources, Broadcasting versus Narrowcasting

Although an optical transmitter with high optical power is attractive for overcoming splitting and transmission losses, low-power optical transmitters such as a directly modulated DFB laser with <4 mW output power also have their niches in providing *narrowcasting* services [1a]. This advantage is illustrated in Figs. 9.1a to c. Figure 9.1a is a typical arrangement for broadcasting multiple channels to multiple serving areas by using a single high-power optical transmitter (e.g., 20 mW). Assuming that 200-MHz bandwidth (in the spectrum between 550 and 750 MHz) is used for downstream two-way interactive services (the spectrum between 5 and 42 MHz is for upstream transmission), we note that in this arrangement not only the high laser output power, but also the 200-MHz bandwidth is shared by the four optical nodes (or serving areas). In other words, each serving area can only use $(750 - 550)/4 = 50$ MHz of downstream bandwidth. This situation can be improved when four low-power sources (e.g., 4 mW) are used instead; each optical node can then enjoy the full $(750 - 550) = 200$ MHz of bandwidth. This *narrowcasting* example tells us that a high optical power transmitter is not the only solution in two-way HFC networks where there are multiple and intensive interactive services.

Figure 9.1c shows another narrowcasting application by using low-power transmitters. We can see that midway in a long link of the broadcasting system, a low-power optical transmitter is used to locally inject narrowcasting services such as telephony or Internet provision.

Figure 9.1d shows how dense wavelength-division-multiplexing (DWDM) technologies can be used to save the number of optical fibers significantly, while in the same time achieve simultaneous broadcasting and narrowcasting [1b]. All λ_i's ($i = 0,1,...,8$) can be allocated in 1.55μm band so that EDFAs (also shown in the figure) can be used.

Another consideration in taking full advantage of transmitter power is to ask how much optical power can actually be launched into a certain distance of optical fiber. For example, a 100-mW 1.55-μm external-modulator-based transmitter with-

Table 9.1: Overall comparison of different optical fiber CATV downlinks (all using conventional 1.3-μm zero-dispersion single-mode optical fibers).

	Direct modulation systems	**External modulation systems**
1.3-μm (loss = 0.35–0.4 dB/km, fiber dispersion < 2.8 ps/nm/km)	(+) Low cost, high reliability (+) Wide range of output power, typically 4 to 20 mW, can be used for both broadcasting and narrowcasting. (+) Higher SBS threshold power. (−) Limited transmission distance and power budget due to higher attenuation loss and the lack of reliable and cost-effective 1.3-μm optical amplifiers.	Using Nd:YAG solid-state laser as source: (+) Low- intensity noise (< −165 dB/Hz) and narrow linewidth (<10 kHz), insensitive to multiple-reflection-induced interferometric effects. (+) Dual output power with 30 to 40 mW each, suitable to upgrade system power budget for older networks with poor fiber installation conditions (e.g., large bending loss, poor connectors/splices). (−) Higher cost, limited transmission distance and power budget (including limited SBS threshold power).
1.55-μm (loss = 0.20–0.3 dB/km, fiber dispersion ≈ 17 ps/nm/km, severe bending loss)	(+) Low cost, yet with upgradable power budget when using EDFAs. (+) High SBS threshold. (−) Requires optical fiber dispersion compensation techniques, or electrical predistortion techniques, or both. (−) Has many frequency-chirp-related system problems (e.g., EDFA gain tilt, various fiber dispersion penalties).	Using high-power (20- to 60-mW) MQW DFB laser (RIN <−160 dB/Hz) as source: (+) High output power (100 to 500 mW) when using EDFAs, small fiber-dispersion penalties; thus, long-distance transmission feasible. (−) Requires suppression of SBS-induced and multiple-reflection-induced intensity noise and NLD penalties. (−) Higher cost.

Note: External modulator exhibits static nonlinearities and can be easily compensated by building a general electrical predistortion circuits; directly modulated DFB lasers usually exhibit dynamic clipping-induced nonlinear distortions and are device-dependent.

out stimulated Brillouin scattering suppression can only launch about 4 mW into a >10-km SMF link (see Section 7.6). To use the full power of this transmitter, one can either apply an SBS suppression technique (Section 9.6.1) to increase the launching power to >40 mW, or one can split the transmitter optical power several ways before launching each part toward different serving areas. More details about utilizing the high transmitter output power will be discussed in Section 9.6.

9.1.3 Modulation Formats, AM versus QAM

Despite the fact that voice telecommunications have been transported and switched in digital form for a number of years, the analog AM-VSB video modulation format remains in use today. There are several tens of millions of analog NTSC or PAL tele-

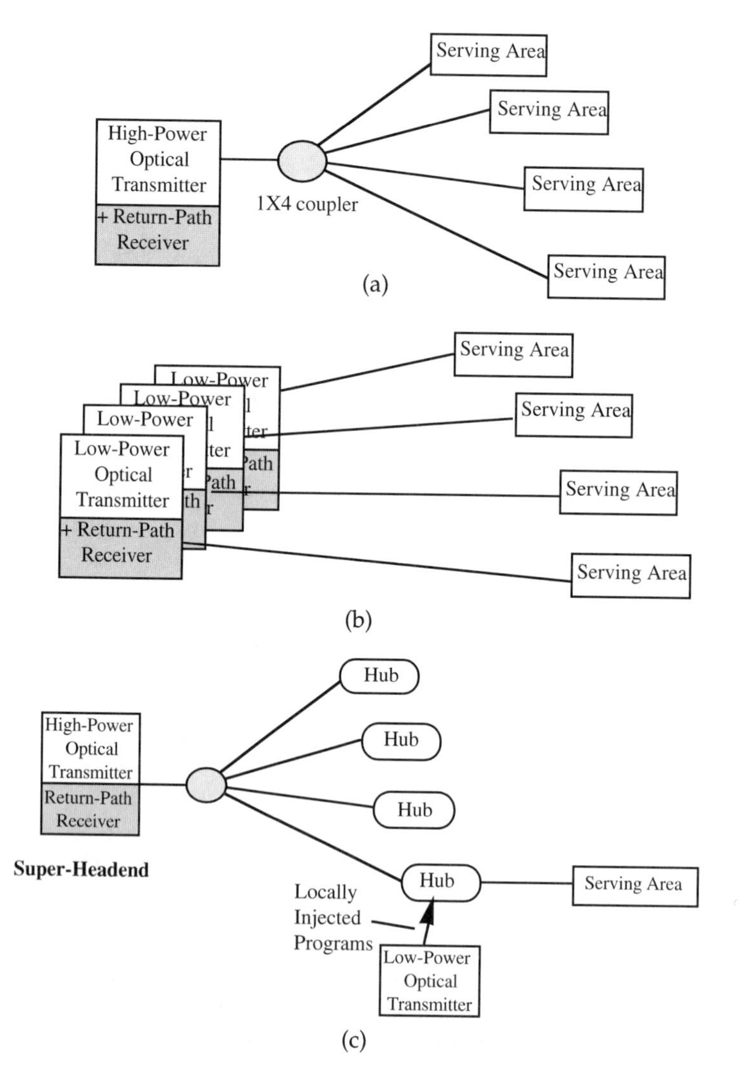

Figure 9.1: Transmitter arrangements for broadcasting (a) and narrowcasting (b) and (c). *(continued)*

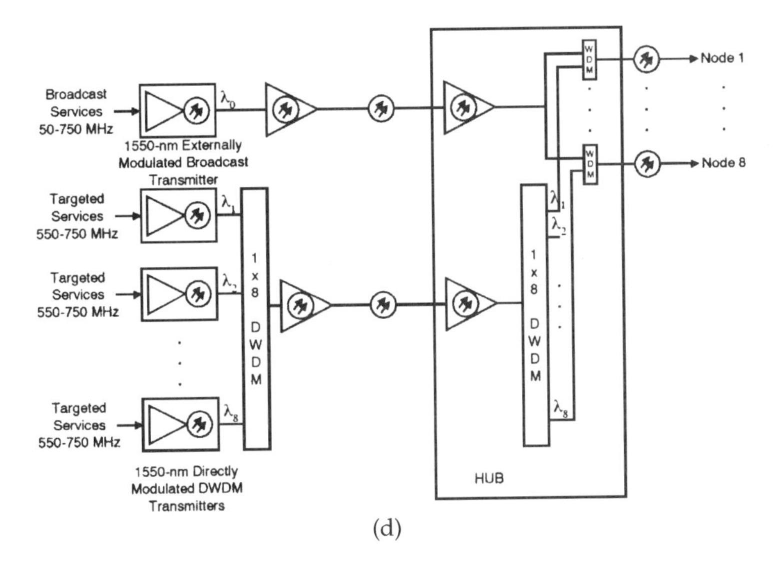

Figure 9.1: (d) 1.55 μm DWDM CATV optical transport network. After Ref. [1b].

vision receivers in the world today, so the analog AM modulation format will still be used for a number of years. However, recently we have seen QAM-modulated digital video signals (mainly MPEG-2) being broadcast to millions of subscribers via direct digital satellite, MMDS, and LMDS services (see Chapter 1). In addition, the U.S. FCC has set 1998 as the time for the network TV stations to broadcast digital VSB video or HDTV signals. All this implies that digital set-top boxes, cable modems, and even digital-video-compatible TV sets are becoming more and more affordable to end users. Therefore, we can clearly see the evolutionary path in the modulation signal formats for broadband signal transportation: An SCM system will still transport just multiple channels of AM-VSB signals for a few years; then it will have to transport hybrid, coexisting AM-VSB and QAM channels (see Fig. 2.5); and finally, all analog AM-VSB signals will be replaced by digitally modulated QAM signals (with the possibility of mixing some digital VSB signals). Therefore, in Section 9.2, we will discuss SCM transmitter design considerations for hybrid AM and QAM signals, and for all QAM signals.

9.1.4 Return-Path Transmissions

For return-path transmissions, uncooled, high-temperature MQW lasers (see Section 4.7) are probably the most cost-effective and practical candidates. Optical isolator(s) in the laser may or may not be required, depending on the CNR requirement of the modulating signals. For QPSK or 16-QAM modulation formats (used by MCNS/IEEE 802.14/DAVIC standards), the required CNR generally is less than 20 dB. However, since system designers always want to leave a sufficient link margin, the CNR requirement can be as high as 40 dB over temperature extremes.

The return-path signals can be transported via a different optical fiber from that of the downstream signals, in which case at least two optical fibers are required for each optical node. On the other hand, the return-path signals can also be transported on the same optical fiber as that of the downstream signals, in which case WDM technologies can be used. For example, 1.55 μm downstream signals and 1.3 μm upstream signals can use the same optical fiber. With the recent advance of WDM technologies, this latter approach can be a cost-effective method in saving significant numbers of optical fibers.

In the rest of this chapter, we will discuss clipping-induced nonlinear distortions in Section 9.2, laser frequency chirp-induced impairments in Section 9.3, system impairments due to laser frequency chirping and optical fiber dispersion effects in Section 9.4, WDM transmission of SCM video channels in Section 9.5, the effects of high input optical power in an optical link in Section 9.6.

9.2 Clipping-Induced Nonlinear Distortions

Clipping-induced nonlinear distortions (NLDs) set the fundamental limitation on the channel count and power budget in a lightwave CATV system. Two basic clipping phenomena are illustrated in Fig. 9.2.

When transporting multiple (e.g., >40) channels of AM-VSB video signals, the composite input signal can only be weakly clipped, as shown in Fig. 9.2a, so that every video channel can achieve high CNR and low CTB/CSO values. On the other hand, when transporting multiple channels of QAM signals, a laser diode can be strongly clipped, as shown in Fig. 9.2b, because the required CNR per QAM channel is much less stringent than those (CNR/CTB/CSO) required by AM-VSB sig-

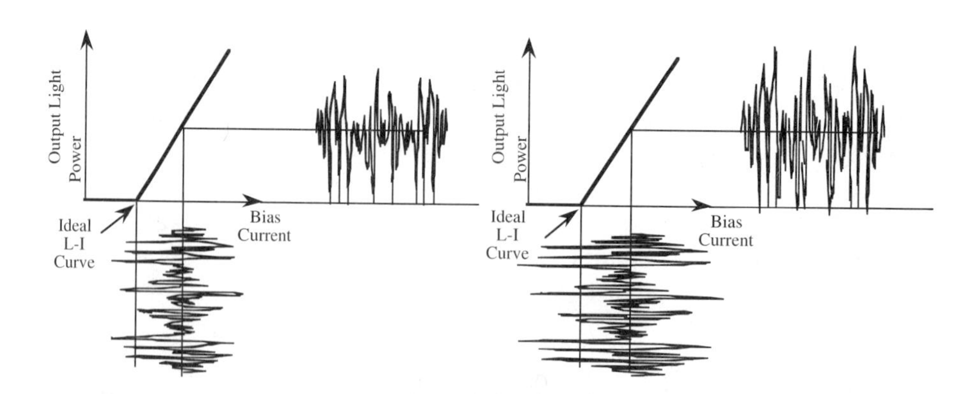

Figure 9.2: (a) Weakly clipped modulation waveforms. (b) Strongly clipped modulation waveforms. In (a), one must also pay attention to the smooth knee of the non-ideal *L–I* curve, and the "dynamic clipping" which causes turn-on delay and additional nonlinear distortions.

nals. As was mentioned in Chapter 8, the required CNR for a 64-QAM signal to achieve a BER of 10^{-9} is only 30 dB. Both types of clipping phenomena will be discussed in the following sections.

9.2.1 Static Clipping and Ultimate AM-VSB Channel Capacity

Consider an SCM system with multiple subcarriers such that the received optical power can be written as

$$P(t) = P_o\left[1 + \sum_{i=1}^{N} m_i \cos(\omega_i t + \phi_i)\right], \tag{9.1}$$

where P_o is the average received optical power, N is the number of channels, and m_i is the modulation index of the ith channel. At the output of the photodetector, the received signal can be written as

$$\begin{aligned} I(t) &= \mathcal{R} \bullet P(t) \\ &= I_p + \sum_{i=1}^{N} (\mathcal{R} p_o) m_i \cos(\omega_i t + \phi_i), \end{aligned} \tag{9.2}$$

where $\mathcal{R} = e\eta/h\nu$ is the responsivity of the photodiode, and I_p is the DC photocurrent.

Let us now consider the probability density function (PDF) of the resultant photocurrent amplitude $I(t)$. We assume each ϕ_i in Eq. (9.2) is random and uniformly distributed in the interval $(-\pi, \pi)$, and let $m_i = m$, $\mathcal{R} p_o \cdot m = A$. If there is only a single subcarrier channel, then $I(t) = I_p + A \cos(\omega t + \Phi)$, and the PDF of $I(t)$ is given by[1]

$$f_I(I) = \frac{1}{\pi\sqrt{A^2 - (I - I_p)^2}} \tag{9.3}$$

and is plotted in Fig. 9.3a. As the number of channels increases, the PDF approaches a normal distribution, as shown in Fig. 9.3b–f. For $N = 5$ or more, the PDF converges to the same PDF. Therefore, when N is large, $I(t)$ can be modeled as a Gaussian random process with a mean value of I_p and a variance of $\sigma^2 = NA^2/2$ [2], which can be obtained from $E((I(t) - I_p)^2,)$ where $E(\bullet)$ stands for the expected value.

Since $A = \mathcal{R} p_o m = I_p \cdot m$, we have $\sigma_s^2 = N \cdot m^2 I_p^2/2$. As long as $m\sqrt{N}$ is held constant, σ_s will be held constant. We therefore define a normalized total RMS modulation index μ as

$$\mu = \frac{\sigma_s}{I_p} = m \cdot \sqrt{\frac{N}{2}}. \tag{9.4}$$

[1] See, for example, A. Papoulis, *Probability and Statistics*, Chapter 4, pp. 120–121, Prentice-Hall, Englewood Cliffs, New Jersey, 1990.

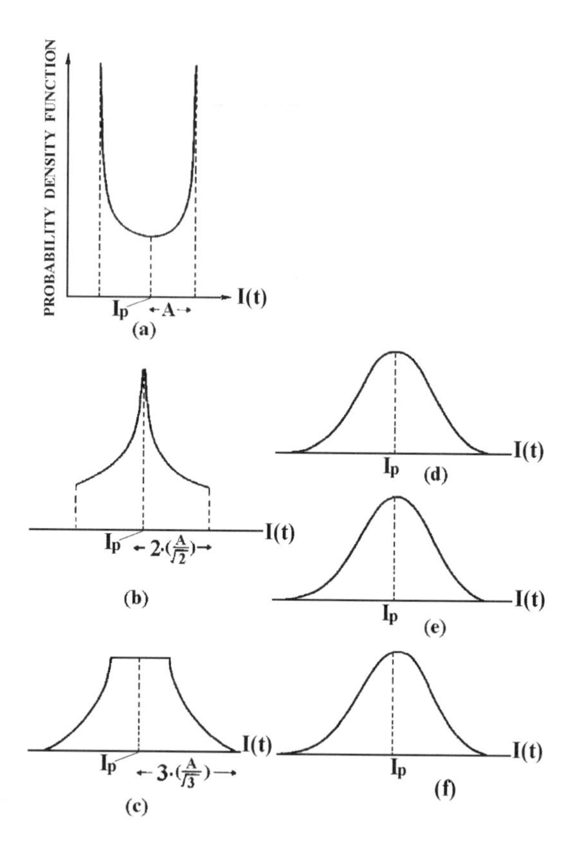

Figure 9.3: Amplitude probability density function for a total channel number equal to (a) 1, (b) 2, (c) 3, (d) 5, (e) 10, and (f) 30. The width of each figure is the rms current amplitude of multiplexed SCM signals superposed on the bias current. After Ref. [3] (© 1990 IEEE).

This is an important parameter which will later be used constantly in our nonlinear distortion analyses.

In an ideal case where the *L–I* curve is perfectly linear for bias current above threshold, as was shown in Fig. 9.2, the total variance of the clipped portion of $I(t)$ is given by

$$\sigma_d^2 = \overline{I_{NLD}^2} - \overline{I_{NLD}}^2 \tag{9.5}$$

where the mean and mean square of I_{NLD} are given by [6]

$$\overline{I_{NLD}} = \frac{1}{\sqrt{2\pi}\sigma_s}\int_{-\infty}^{0} I \bullet \exp\left[-\frac{(I-I_p)^2}{2\sigma_s^2}\right] dI$$

$$= \frac{I_p}{\sqrt{2\pi}}\left[\sqrt{\frac{\pi}{2}}erfc\left(1/\sqrt{2}\mu\right) - \mu\exp(-1/2\mu^2)\right] \tag{9.6}$$

and

$$\overline{I_{NLD}^2} = \frac{1}{\sqrt{2\pi}\ \sigma_S} \int_{\infty}^{0} I^2 \cdot \exp\left[-\frac{(I - I_p)^2}{2\sigma_s^2} \right] dI$$

$$= \frac{I_p^2}{2} \left[(1+\mu^2) erfc\left(1 \Big/ \sqrt{2}\mu \right) - \sqrt{\frac{2}{\pi}} \mu \exp(-1 \big/ 2\mu^2) \right], \quad (9.7)$$

respectively. On the other hand, the carrier power per channel is given by $\frac{1}{2} m^2 I_p^2$ (see Section 5.4) and is equivalent to $\frac{1}{N}\ \mu^2 I_p^2$ according to Eq. (9.4). Therefore, we can obtain the carrier-to-nonlinear distortion ratio (C/NLD) per channel as

$$C/NLD = \frac{\frac{1}{N}\mu^2 I_p^2}{\Gamma \cdot \frac{1}{N} \cdot \sigma_d^2}, \quad (9.8)$$

where Γ accounts for the percentage of the distortion power which is distributed inside the CATV band, and the factor $1/N$ in the denominator is based on the assumption that the Γ factor of total nonlinear distortion power is uniformly distributed over the N channels. Note that the value of Γ varies for different frequency plans and for different values of the total RMS modulation index μ [4,5]. For conventional 50–450 or 50–550 MHZ frequency plans, and $\mu < 0.3$, a good approximation for Γ is about 0.5 [5]. A detailed discussion of estimating Γ is given in Section 9.2.5. Now, according to Eqs. (9.5)–(9.8) and assuming a small μ, we can obtain a final result of C/NLD as a function of μ:

$$C\!\!\Big/_{NLD} = \frac{1}{\Gamma}\sqrt{\frac{\pi}{2}}\mu^{-3}(1+6\mu^2)\cdot \exp\left(\frac{1}{2\mu^2}\right). \quad (9.9)$$

If the effect of NLD can be treated as a noise term just like shot and RIN noise, we can write [recall Eq. (5.29)]

$$CNR = \frac{\left(\frac{1}{N}\mu^2 I_p^2\right)}{\left(2qI_p \cdot \Delta f + RIN \cdot I_p^2 \cdot \Delta f + \frac{\Gamma I_p^2}{N}\sqrt{\frac{2}{\pi}}\frac{\mu^5}{(1+6\mu^2)} \cdot \exp\left(\frac{-1}{2\mu^2}\right)\right)}, \quad (9.10)$$

where the receiver thermal noise has been neglected to obtain the fundamental CNR, considering NLDs, shot noise, and RIN noise. Equation (9.10) in turn can be rewritten for the number of channels N as

$$N = \left(\frac{\mu^2}{CNR} - \sqrt{\frac{2}{\pi}}\Gamma\frac{\mu^5}{1+6\mu^2}\exp\left(\frac{-1}{2\mu^2}\right)\right) \Bigg/ \left(\frac{2q\Delta f}{\mathcal{R} \cdot P_0} + RIN \cdot \Delta f\right). \quad (9.11)$$

When CNR, RIN, Γ, and Δf are given, we can find an optimum μ which gives a maximum channel number N per milliwatt of received optical power P_o. The result is shown in Fig. 9.4. We can see that for a typical DFB laser with RIN = -155 dB/Hz, the ultimate TV-channel number is over 150 when CNR = 50 dB. This number may be overestimated for a practical laser, because the effects of dynamic laser clipping were not included, as will be explained later in Section 9.2.4. Note that in Fig. 9.4, the dominant noise terms are the shot and RIN noise, and C/NLD > 65, 62, and 55 dB, for C/(shot noise + RIN noise + NLDs) > 55, 50, and 45 dB, respectively. Figure 9.4 only provides a rough estimation of fundamental channel number per milliwatt of received laser power; several modifications are needed as follows.

The first modification to the preceding analysis is that the NLD within a channel is actually the sum of all discrete intermodulation products (CSO, CTB, etc.), and the results of NLD models can only be applied in a qualitative sense to CSO and CTB measurements, because the relative contributions of CSO and CTB change with channel frequency. The second modification is due to the fact that dynamic clipping, that is, a frequency-dependent clipping phenomenon at weak clipping, must be considered, as will be discussed in Section 9.2.4.

The lack of theoretical connection between the integrated NLD and the resolved (CSO, CTB) descriptions motivated the development of the "effective transfer function model" [6]. The "effective transfer function" is the effective L–I curve seen by a subset of k carriers (e.g., $k = 3$ if we are considering triple beats) when all other $N - k$ channels are treated as a background noise process with Gaussian probability

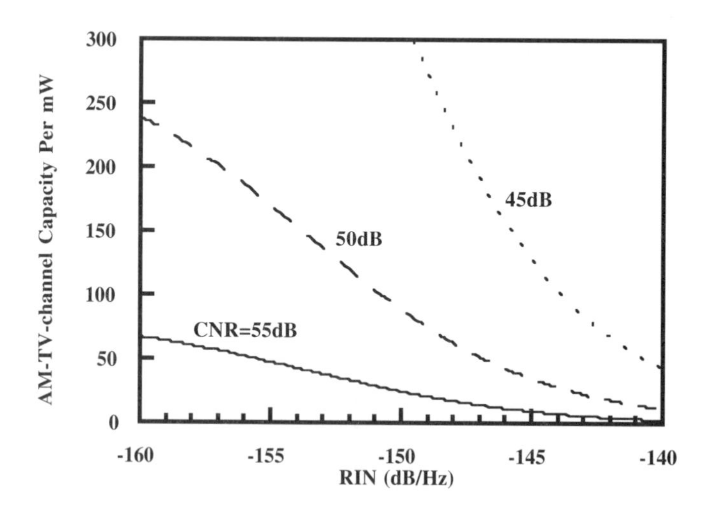

Figure 9.4 Ultimate AM-VSB channel capacity of a static clipping-limited laser diode as a function of RIN and CNR, where CNR is the power ratio of carrier and (shot noise + RIN noise + NLDs). The calculations are based on Eq. (9.11) with an optimum RMS modulation index μ. For all three cases of CNR, the dominant noise terms are the shot and RIN noise.

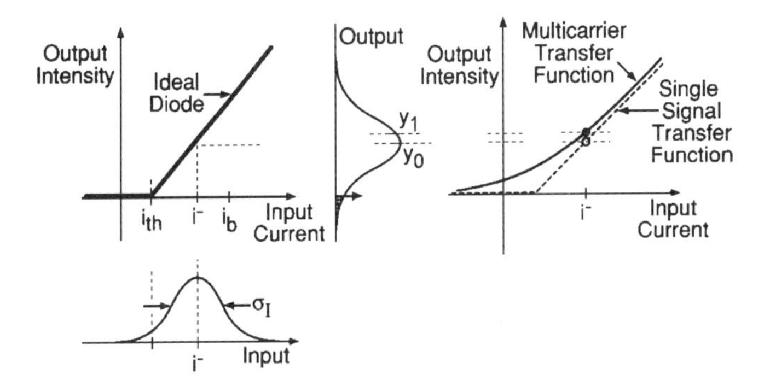

Figure 9.5: Effective transfer function of an ideal laser diode. Input signal plus Gaussian noise from other carriers (left) is clipped, resulting in expected output (y_1) greater than the output if there were no clipping (y_0). The set of all such inputs results in effective transfer curve (solid curve, right). After Ref. [6] (© 1993 IEEE).

density function. As shown in Fig. 9.5, the subset of k carriers see an expected output value (y_1) under clipping to be greater than the expected value when there is no clipping (y_0). This is because the clipping-induced NLDs due to the background Gaussian random noise yield an additional expected value and therefore the original *L–I* curve (dashed curve) is lifted up slightly (solid curve). The difference between y_1 and y_0 increases when the clipped portion of the output Gaussian probability density function increases.

A Taylor series expansion of a normalized "effective transfer function" (or "effective *L–I* curve") about the operating point is given as

$$f(a) = E_x(g(a+x)) = \sum \chi^{(n)} a^n(t), \tag{9.12}$$

where the normalized $a(t) = \sum_{i=1}^{k} m \cos(\omega_i t + \phi_i)$ is formed by the k carriers, $g(x)$ is the ideal transfer function (or ideal static *L–I* curve), and $x(t)$ is the Gaussian random process formed by the $N - k$ carriers. $\chi^{(1)}$ accounts for the compression of the fundamental carriers because of the nonlinear transfer function. Note that taking away two or three carriers from $x(t)$ (which consists of >40 channels typically) for CSO or CTB considerations does not affect the statistical property of $x(t)$. To obtain the values of the coefficients $\chi^{(n)}$, we first observe that [7]

$$f(a) = \frac{1}{\mu\sqrt{2\pi}} \int_{-\infty}^{\infty} g(x) \exp\left[-\frac{(x-a)^2}{2\mu^2}\right] dx. \tag{9.13}$$

Using the generating function of Hermite polynomials [8] given by

$$\exp\left(-\frac{a^2}{2} + ax\right) = \sum_{n=0}^{\infty} \frac{a^n}{n!} H_n(x) \tag{9.14}$$

and exchanging the order of summation and integration, we have

$$f(a)=\sum_{n=0}^{\infty}\frac{a^n}{\mu^{n+1}n!\sqrt{2\pi}}\int_{-\infty}^{\infty}g(x)\exp\left(-\frac{x^2}{2\mu^2}\right)H_n(\frac{x}{\mu})dx, \tag{9.15}$$

which implies that

$$\chi^{(n)}=\frac{1}{\mu^{n+1}n!\sqrt{2\pi}}\int_{-\infty}^{\infty}g(x)\exp\left(-\frac{x^2}{2\mu^2}\right)H_n(\frac{x}{\mu})dx. \tag{9.16}$$

The intermodulation power at a given frequency ν is computed by counting the number of ways the k carriers can be combined to form an intermodulation product (IMP) at ν, and finally by squaring this amplitude. Thus, the IMP power at frequency ν is given by

$$P_\nu^{(n)}=\frac{1}{2}\left[\frac{\chi^{(n)}m^n}{2^{n-1}}\right]^2\cdot N_\nu^{(n)}, \tag{9.17}$$

where $N_\nu^{(n)}$ stands for the product count for the nth-order terms that contribute power at frequency ν.

Since we know that the normalized carrier power (per channel) is given by $P_c=\mu^2[\chi^{(1)}]^2/N$, we can then obtain the carrier to IMP power ratio as [6]

$$C/IMP_\nu^{(n)}=P_c/P_\nu^{(n)}=(4N)^{n-1}\pi(\chi^{(1)}n!)^2\exp(1/\mu^2)\Big/H_{n-2}^2(1/\sqrt{2}\mu)N_\nu^{(n)}. \tag{9.18}$$

According to Eq. (9.18), CSO (n = even) and CTB (n = odd) can be obtained separately. In addition, C/NLD can be obtained by summing over all n with IMPs on the desired channel. The results for CSO/CTB are shown in Fig. 9.6. The obvious discrepancy in the region of weak clipping where $\mu \leq 0.25$ is possibly due to the fact that dynamic clipping-induced NLDs (Section 9.2.4) could not have been neglected.

9.2.2 Static Clipping of an Ideal External Modulator and a Linearized Modulator

Static clipping phenomena also occur in an external modulator (see Chp. 7) when it is overmodulated beyond the peak and zero light intensity. This may be unavoidable if one wishes to use a higher OMI per channel (to obtain higher CNR or optical power budget). In this case, a predistortion circuit (predistorter) which has a transfer function different from the conventional arcsine (see Section 7.4.1) may be used [9]. As shown in Fig. 9.7, with increasing order of arcsine predistortion, the central region of the curve becomes increasingly linear. However, at the same time, the "knees" formed at the extrema become increasingly sharp. An overmodulating input

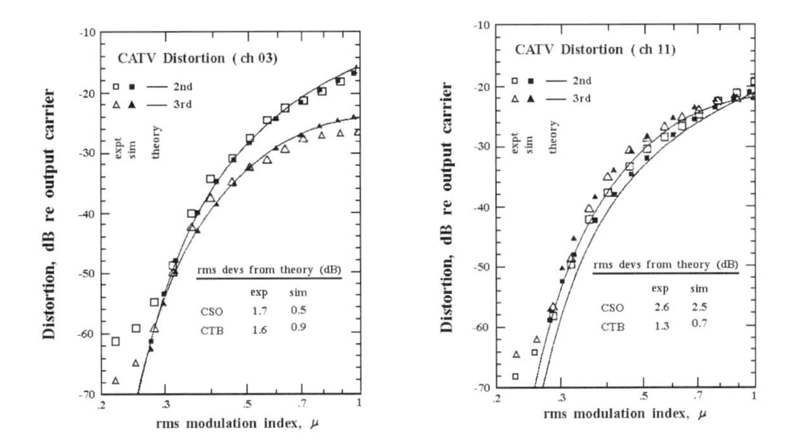

Figure 9.6: CSO (squares and solid curve), CTB (triangles and dotted curve), for CATV channels 3 and 11. Simulations are solid figures, experiments are open figures. After Ref. [6] (© 1993 IEEE).

$a(t)$ will occasionally drive the modulator beyond these knees into clipping. Above some level of overmodulation, the modulator arcsine-predistorted to infinite order produces more distortion than the unpredistorted modulator [10].

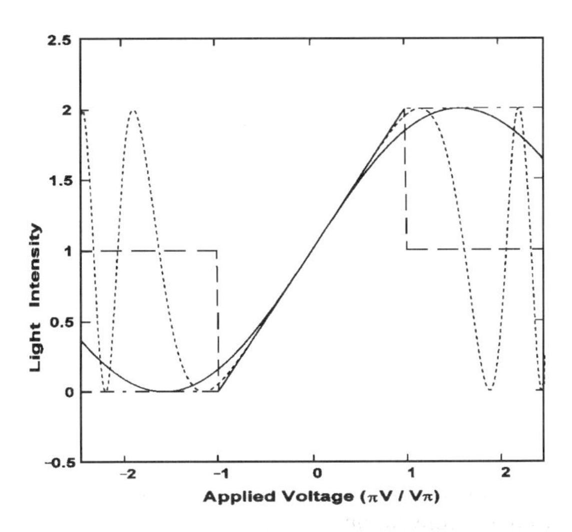

Figure 9.7: Transfer function of conventional MZI modulator (—), third- and fifth-order predistortion-linearized modulator (---), perfectly predistortion-linearized modulator (– – –), and a limiter (- - -). After Ref. [10] (© 1996 IEEE).

To see how we can obtain an optimum predistorter transfer function, we assume that the predistorter transfer function and the combined predistorter/modulator transfer function are given by

$$g(a) = a + b_3 a^3 + b_5 a^5 + b_7 a^7 + ... \tag{9.19}$$

$$p(g(a)) = \sum_{n=1,3,5,...}^{\infty} c_n a^n , \tag{9.20}$$

respectively, where again the normalized $a(t) = \sum_{i=1}^{k} m \cos(\omega_i t + \phi_i)$. Following Eq. (9.16), we can have

$$\chi^{(n)} = \frac{1}{\mu^{n+1} n! \sqrt{2\pi}} \int_{-\infty}^{\infty} p(g(x)) \exp\left(-\frac{x^2}{2\mu^2}\right) H_n\left(\frac{x}{\mu}\right) dx \tag{9.21}$$

Carrying out the integration by using the relationship

$$\frac{d}{dx}\left(e^{-x^2/2\mu^2} H_n\left(\frac{x}{\mu}\right)\right) = -\frac{1}{\mu} e^{-x^2/2\mu^2} H_{n+1}\left(\frac{x}{\mu}\right), \tag{9.22}$$

we can finally obtain [9]

$$\chi^{(n)} = c_n + \frac{1}{n!} \sum_{l=1}^{\infty} \frac{(2l+n)!}{2^l l!} \mu^{2l} c_{2l+n} . \tag{9.23}$$

Therefore, we can see that the optimal nth-order predistorter is to null $\chi^{(i)}$ for $1 < i \leq n$. On the other hand, arcsine predistortion nulls c_i for $1 < i \leq n$, but increases $|c_i|$ for $i > n$. From (9.23) we can see that these higher-order terms of c_i contribute to $\chi^{(i)}$ for $i \leq n$ and hence to nth- and lower-order distortion.

It can be shown that the CTB of an ideal MZI modulator is 6 dB higher than the ideal laser under static clipping-dominated condition [10]. We can thus obtain a similar prediction but lower ultimate TV channel capacity per milliwatt received, as was calculated for an ideal laser diode in Fig. 9.4. Since only static nonlinearities should be considered in $LiNb0_3$-based MZI modulators, the calculated ultimate TV channel capacity will *not* be reduced because of *dynamic* clipping (Section 9.2.4), as in a laser diode.

9.2.3 Effect of Static Clipping on Hybrid AM-VSB/QAM SCM Systems

9.2.3.1 Impulsive Nature of Clipping Noise Clipping impulses can be modeled as the difference between the clipped (output) and the unclipped (input) Gaussian process, as shown in Fig. 9.8. The fact that this clipping impulse exists when a laser diode is modulated by multiple channels of AM carriers can be easily measured in an empty channel by using a spectrum analyzer with zero frequency span, as illustrated in Fig. 9.9. We can see that when spectrum analyzer was set with a zero fre-

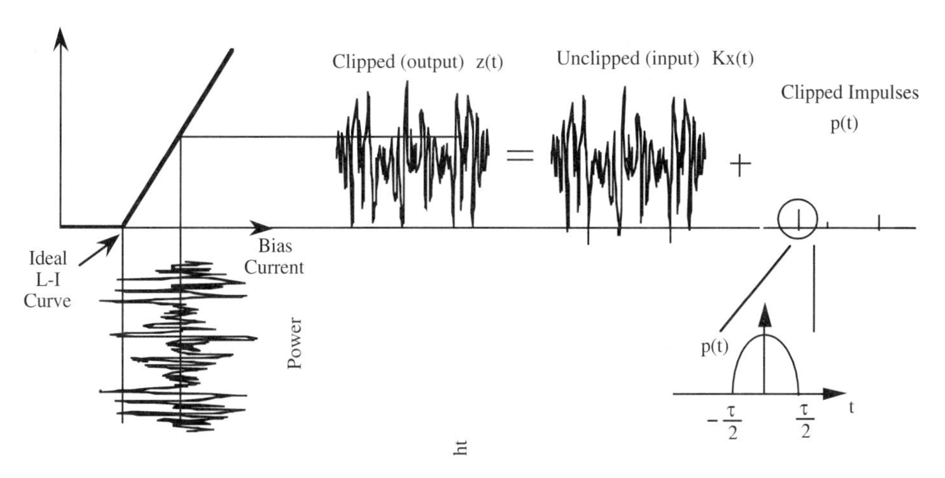

Figure 9.8: The clipped impulses are the difference between the clipped and unclipped signals. The clipped impulses can be assumed to be parabolic with a Rayleigh distributed pulse duration τ.

quency span, it measures the time-domain clipped pulses in a time duration equal to the sweep time across the screen (10 ms in Fig. 9.9). The clipped pulses in Fig. 9.9 occur in a 10-MHz band (equal to the resolution bandwidth of the spectrum analyzer) centered at 601.1 MHz.

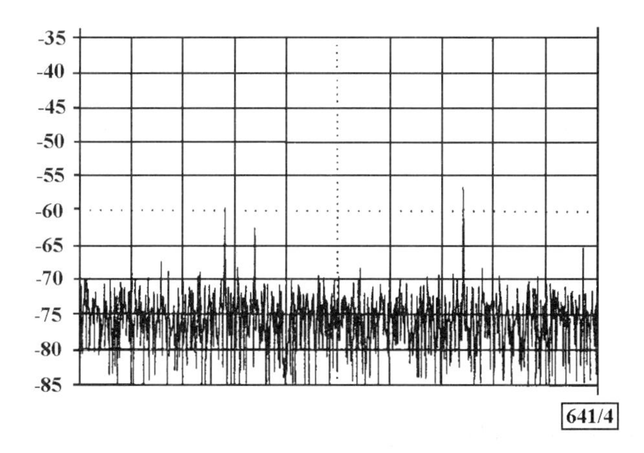

Figure 9.9: Measured time domain laser clipping noise (due to 42 channels of AM-VSB signal) which occurs in an empty channel centered at 601.1 MHz. The spectrum analyzer parameters are span = 0 Hz, resolution bandwidth = 10 MHz, video bandwidth = 10 MHz, and sweep time = 10 ms. After Ref. [11].

It was observed that the time interval between clipping events has a high probability of being 167 ns, 1.5 μs, or 4 μs in an NTSC system with unmodulated carriers [12]. The 4-μs interval is due to the common divisor (1 / 4 μs = 0.25 MHz) of all carriers. The 167-ns interval is attributed to the 1/167 ns = 6-MHz spacing between NTSC channel carriers. The 1.5-μs interval exists because the two 6-MHz-spacing nonconformant carriers, channels 5 and 6, are periodic with a 1.5-μs period. These results have potential impact on forward-error-correcting coded (FECed) QAM channels where the FEC code is used to ameliorate the degrading error rate due to clipping.

Assuming the Gaussian process formed by the multiple AM signals has a flat spectrum over the band of $[f_a, f_b]$ and the rate of clipping is low, Mazo [5] approximated the clipping impulses by a Poisson sequence of random parabolic pulses with a Rayleigh distributed pulse width τ whose probability density function is given by

$$f_\tau(\tau) = \frac{\pi}{2}\frac{\tau}{\bar{\tau}^2}\exp\left[-\frac{\pi}{4}\left(\frac{\tau}{\bar{\tau}}\right)^2\right], \qquad \tau \geq 0\ , \tag{9.24}$$

where $\bar{\tau}$ is the average time duration of the clipping pulses. The parabolic arc shown in Fig. 9.8 can be expressed by the equation

$$\begin{aligned} \boldsymbol{p(t)} &= \frac{1}{2}\boldsymbol{g''t^2} - \frac{1}{8}\boldsymbol{g''\tau^2} && |\boldsymbol{t}| \leq \frac{\tau}{2} \\ &= 0 && \textbf{\textit{otherwise}}\ , \end{aligned} \tag{9.25}$$

where g'' is given by [5]

$$g'' \approx I_p (2\pi)^2 \frac{f_b^3 - f_a^3}{3(f_b - f_a)} \tag{9.26}$$

and is derived from the second derivative of the autocorrelation function of the Gaussian random process. I_p is the average received photo-current corresponding to the laser operating point. Mazo used the previous assumptions of a Poisson sequence of clipped pulses, Rayleigh distributed pulse duration, and parabolic shape of pulses to obtain the power spectral density $S(f)$ of the pulse train,

$$z(t) = \sum_{k=-\infty}^{\infty} p_k\,(t - t_k), \tag{9.27}$$

where $p_k(t)$ is the parabolic arc described earlier. From the resultant $S(f)$, one can estimate the factor Γ, which represents the percentage of the distortion power distributed inside the CATV band, in Eq. (9.8).

From Eq. (9.25), we can obtain the pulse area as $A(\tau) = \int_{-\tau/2}^{\tau/2} p(t)dt = -g''\tau^3/12$, and its probability density function as

$$f(A) = f(\tau)/\left|\frac{dA}{d\tau}\right| = \frac{2\pi}{\bar{\tau}^2 (g'')^{2/3}(12A)^{1/3}}\exp\left[-\frac{\pi}{4\bar{\tau}^2}\left(\frac{12A}{g''}\right)^{2/3}\right],\ A > 0\ . \tag{9.28}$$

The expected rate of clipping (upward crossing of the zero level in Fig. 9.8) λ was derived to be [5]

$$\lambda = \left[\frac{f_b^3 - f_a^3}{3(f_b - f_a)} \right]^{1/2} \exp\left(-\frac{1}{2\mu^2} \right). \tag{9.29}$$

Since we know the clipping probability is given by $\lambda \bar{\tau}$, we can have [recall Eq. (9.6)]

$$\begin{aligned} \lambda \bar{\tau} &= \frac{1}{\sqrt{2\pi}\sigma_s} \int_{-\infty}^{0} \exp\left[-\frac{(I - I_p)^2}{2\sigma_s^2} \right] dI \\ &= \frac{1}{2} erfc\left(\frac{1}{\mu\sqrt{2}}\right) \end{aligned} \tag{9.30}$$

Therefore, for a given operating frequency range $[f_a, f_b]$, total RMS modulation index μ (μ <<1 so that the assumptions of Poisson process and parabolic pulses are valid), we can estimate the average pulse duration. For example, in the case of a 78-channel AM video system with the signal frequency spectrum from f_a = 55.25 MHz to f_b = 547.25 MHz, and a normalized total RMS modulation index μ = 0.25, the average pulse duration is about 285 ps.

We note that for typical values of μ, the average pulse duration is about three orders of magnitude smaller than the reciprocal of a typical 6-MHz CATV channel bandwidth (1/(6 MHz) ≈ 167 ns) where an M-QAM signal is accommodated. Therefore, to each M-QAM channel, the clipping pulse train can be approximated by [13]

$$w(t) \approx \sum_{k=-\infty}^{\infty} A_k \delta(t - t_k), \tag{9.31}$$

where $\delta(\bullet)$ is the impulse function, t_k are Poisson-distributed event times, and A_k are random variables representing the parabola area whose PDF was given in Eq. (9.28). Both Eqs. (9.28) and (9.31) will be used in the analysis in the following section.

9.2.3.2 BER of a QAM Channel due to Clipping-Induced Impulse Noise To smoothly handle the evolutionary stages described in Section 9.1.3, it is most economical and convenient to use a *single* DFB laser transmitter to achieve hybrid downstream transmission of multiple *existing* analog AM-VSB channels and *new* M-QAM digital channels, with the frequency bands of M-QAM signals on top of the conventional AM-VSB signal bands (see Fig. 2.5). However, it has been shown in a number of reports [11–16] that the clipping noise generated by the laser diode threshold results in a severe bit-error rate performance degradation in the transmission of digital signals.

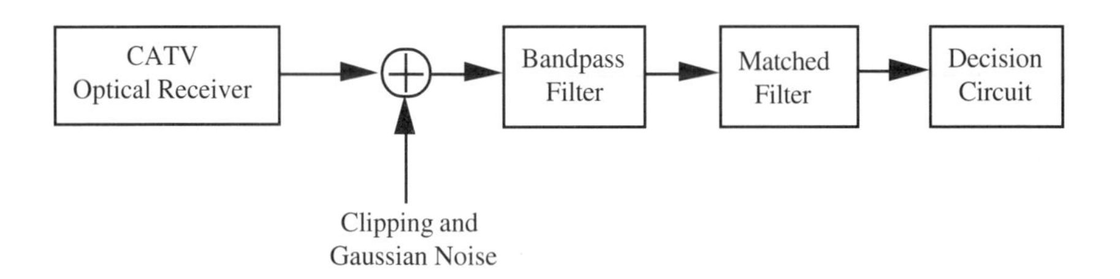

Figure 9.10: Analytical noise model for the receiver part for receiving an M-QAM signal (the optical receiver simultaneously receives an M-QAM signal and multichannel AM-VSB signals). After Ref. [13].

We can use the analytical noise model shown in Fig. 9.10 to analyze the effect of clipping noise on an M-QAM channel [13]. Our first goal is to find the PDF of the combined impulsive and Gaussian noise after a matched filter in an M-QAM receiver. Once the PDF $f_x(x)$ of the combined noise is known, we can obtain the BER of an uncoded M-QAM signal from (see Eq. (8.36))

$$BER \approx \frac{4}{\log_2 M}(1-\frac{1}{\sqrt{M}})\int_{d_{\min}}^{\infty} f_X(x)dx\,, \tag{9.32}$$

where d_{min} is half of the minimum distance between adjacent constellation points given by (see Eq. (8.34))

$$d_{\min} = \sqrt{\frac{3P_{av}T_s}{2(M-1)}}\,. \tag{9.33}$$

$P_{av} = (m_q I_o)^2$ is the average received signal power in each M-QAM channel, m_q is the rms OMI per M-QAM channel, I_p is the received photocurrent, and T_s is the symbol duration. Since the impulsive noise and the Gaussian noise processes are independent, the PDF $f_x(x)$ of the combined noise is just the convolution of the PDFs of the two noise processes, that is,

$$f_X(x) = f_G(x) * f_S(x), \tag{9.34}$$

and $f_G(x)$ is the conventional PDF of a Gaussian noise process whose variance includes the single-sided PSD of receiver thermal noise, shot noise, and system RIN noise (see Section 5.4) and is given by

$$\sigma_G^2 = [\boldsymbol{i}_{th}^2 + 2\boldsymbol{q}\boldsymbol{I}_p + \boldsymbol{RIN}\cdot \boldsymbol{I}_p^2]\Delta f\,. \tag{9.35}$$

While $f_S(x)$ in Eq. (9.34) is the PDF of the impulsive clipping noise process to be determined. If we further assume the bandpass filter is ideal and contributes little to the overall filtering characteristics, and the matched filter has an impulse response given by $h(t) = \sqrt{2/T_s} \cos\omega_o t$ for $0 \le t \le T_s$ (where ω_o is angular frequency of the QAM carrier), then the problem simply becomes finding the PDF of the pulse train at the output of the matched filter. The output pulse train $S(t)$ is given by [using Eq. (9.31)]

$$S(t) = w(t) * h(t) = \sum_{i=-\infty}^{\infty} A_k h(t - t_k). \tag{9.36}$$

The PDF of $S(t)$ can be obtained from the conventional shot noise process [8,17] PDF as

$$f_s(x) = \frac{1}{2\pi} \int_{-\infty}^{\infty} \exp(-j\omega x) \exp\left\{\lambda \int_{-\infty}^{\infty} f(A) \int_0^T [\exp[j\omega A h(t) - 1] dt dA\right\} d\omega, \tag{9.37}$$

where $f(A)$ is given by Eq. (9.28). Using Eqs. (9.34)–(9.37), $f_X(x)$ can be obtained and numerically calculated. Finally, from Eq. (9.32), the BER in an M-QAM signal can be obtained. A typical result is shown in Fig. 9.11. We can clearly see that as the modulation index of the AM channel increases, that is, $m_a \ge 5\%$, there is an error

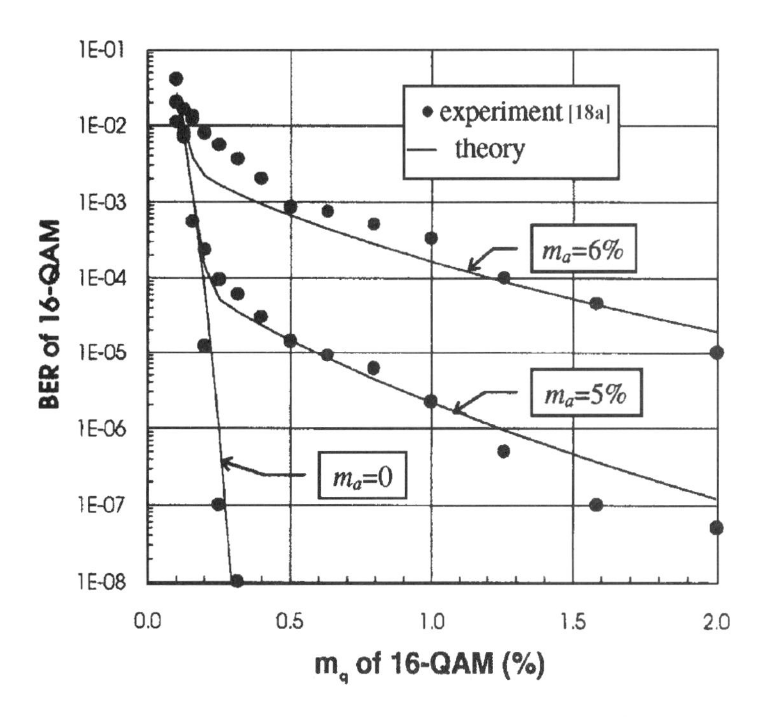

Figure 9.11: BER of a 16-QAM signal co-transmitted with 42 AM channels. m_a and m_q are the AM- and QAM-channel modulation indices, respectively. After Ref. [13] (© 1996 IEEE).

floor which cannot be reduced even when the modulation index of the QAM channel increases beyond 2%.

Besides the analytical approach just presented, an alternative method estimating the BER in an M-QAM channel is by directly counting the clipping impulse rate (number of impulses per second) using a spectrum analyzer with a zero frequency span [14] or a digital sampling scope with a histogram display feature [16]. This can be done because after the matched filter in Fig. 9.10, each clipping pulse, whose original average duration is about a few hundred pico-seconds, exhibits a pulse duration comparable to the symbol duration of ~1/(5 Mbps) or ~200 ns, as illustrated in Fig. 9.12. Furthermore, if the time interval between clipping events has a high probability of being 167 ns, 1.5 μs, or 4 μs in an NTSC system with unmodulated carriers (see Section 9.2.3.1), then each clipping impulse in average can cause one symbol error [14.23]. The symbol error rate can thus be estimated by using the ratio of the counted clipping impulses per second and the total symbol rate.

How can this clipping-induced BER be reduced? The most obvious solution is to use Reed–Solomon forward-error-correction codecs (see Section 8.2). However, it was observed that (204,188) RS FEC codecs cannot reduce the BER of a 64-QAM channel below 10^{-8} when the 64-QAM channel (centered at 735 MHz) is combined with 80 channels of NTSC unmodulated carriers with OMI/channel = 4% [19a]. This is mostly due to the bursty CSO (which is composed of many bursty spikes, each occurring every 167 ns, with a burst length of ~0.2 μs), whose total burst length is much longer than the 8 symbol-correction capability of an (204,188) RS

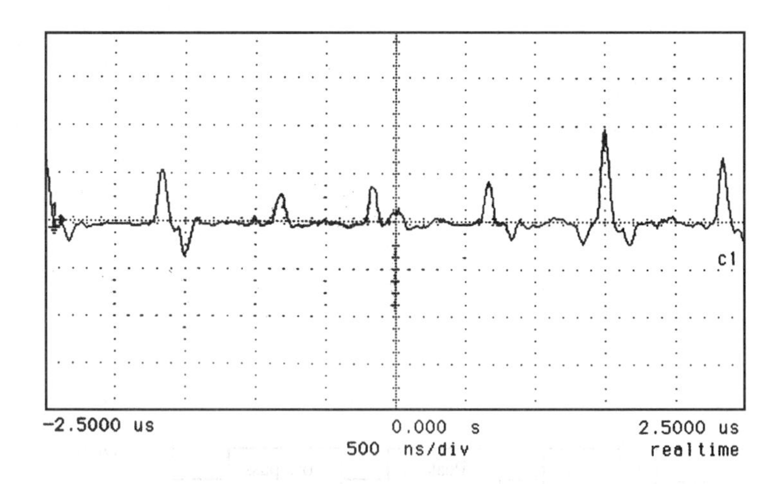

Figure 9.12: Measured (from a real-time sampling scope) time-domain laser clipping pulses. Laser operating condition: 75 AM-VSB channels, OMI/channel = 0.04, $\mu = 0.245$

codec (about 25 μs). An interleaver with a long interleaving depth I (recall that the burst length that can be handled by an interleaver is $I \cdot 8 \cdot t \cdot T_b$ where $t = 8$ for the (204,188) RS code and T_b is the bit rate) can probably be used to reduce the BER [18b], depending on whether the CSO burst length is longer than $I \cdot 8 \cdot t \cdot T_b$.

Besides the traditional modem techniques which use FEC codecs and interleavers, another "clipping-reduction" technique [19a], shown in Fig. 9.13, uses a fast threshold detector and a peak detector. Whenever a clipping event is detected, the laser bias current is momentarily increased so that the clipping effect can be greatly reduced. A recently proposed pre-coding technique [19b] can also be of practical use.

9.2.4 Static *L–I* Curve with a Smoothed Knee, and Dynamically Clipped Laser with Turn-on Delay

As can be observed in Fig. 9.6, measured CTBs and CSOs (Chapters 3 and 11) in the region of $\mu \leq \sim 0.25$ differ from the theoretical and simulation results significantly. This large discrepancy has two possible explanations. The first is that the ideal *L–I* curve, which goes abruptly to zero at threshold, may have to be replaced by a static *L–I* characteristic which goes gradually to zero at the threshold [20,21]; the second is that the ideal *L–I* curve with static characteristics should actually have a dynamic characteristic with memory [22-26].

The first explanation can be verified by tuning the parameters in the single-mode rate equations to smooth the *L–I* curve knee, and observing that the nonlinear distortions in the case of weak clipping are indeed increased when compared with the results obtained from an ideal *L–I* curve [21]. For example, a lower optical gain and higher optical loss can let a laser transit from its spontaneous emission regime into the stimulated emission regime more slowly. On the other hand, a lower electron lifetime can increase the slope of the *L–I* curve in the spontaneous emission region, and a higher β can increase the coupling of the spontaneous emission to the lasing mode, both resulting in a smoother *L–I* characteristic. However, even though the NLDs due to a smoother *L–I* knee can indeed be increased to fit the experimental results, the tuned parameters do not always have reasonable values [21].

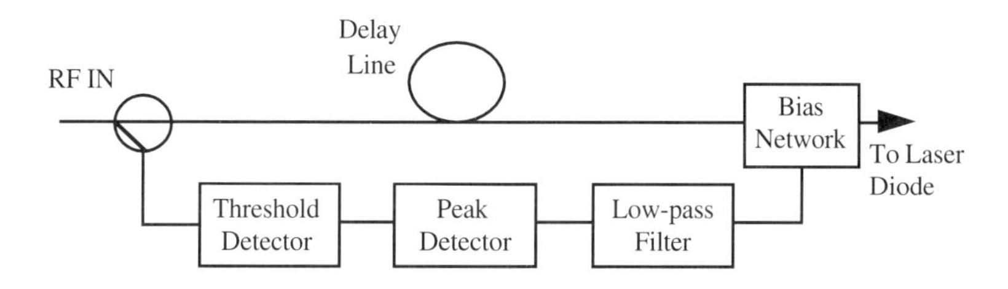

Figure 9.13: A clipping reduction circuit. After Ref. [19a].

The other explanation, which has more experimental support and has recently aroused significant interest, is called "dynamic clipping" [22-26]. Dynamic clipping causes nonlinearities that are different from those caused by laser relaxation oscillation (Section 4.10.1). As shown by the experimental result given in Fig. 9.14, we see that when a 151.25-MHz modulating tone with a peak OMI = 100% is applied to a CATV-quality DFB laser diode, the output light exhibits a clear turn-on delay, in addition to some oscillations which correspond to the laser resonance frequency. The effect of this turn-on delay due to overmodulation of a laser diode on SCM systems is to generate frequency-dependent NLDs, with higher NLDs at higher frequencies [22, 25, 26], or to generate a frequency-dependent QAM channel bit-error rate in hybrid AM/QAM transmission systems, as illustrated in Fig. 9.15 [23].

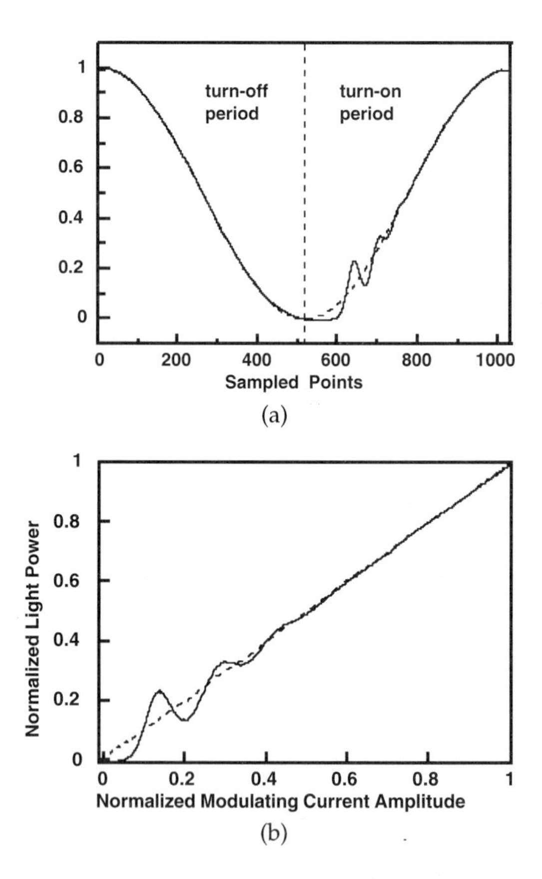

Figure 9.14: (a) One cycle of a modulating current of 151.25 MHZ with OMI = 100% (dashed line) and the corresponding dynamically-clipped sinusoidal light output (solid line). (b) The corresponding normalized light power versus modulating current, where the solid and dotted lines were converted from the turn-on and turn-off periods, respectively.

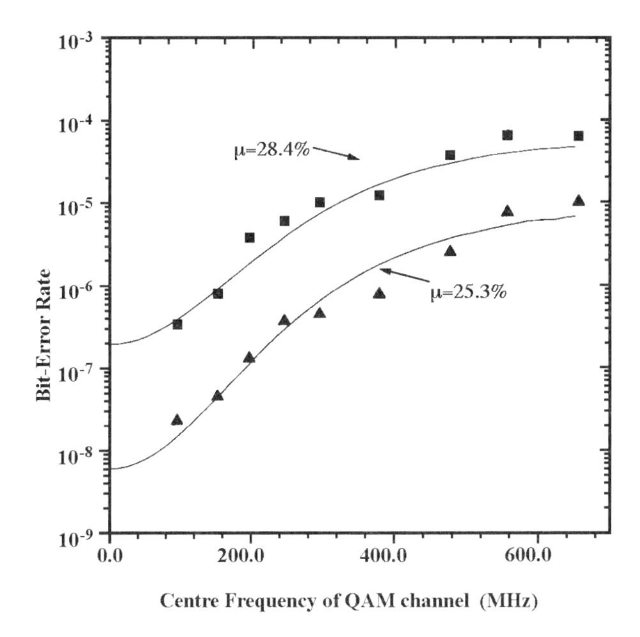

Figure 9.15: Frequency-dependent NLDs which affect the QAM BER when the center frequency of the QAM channel is changed. The laser diode was modulated with 63 analog carriers and a 64-QAM signal 10 dB below the analog carriers. Measurements were performed at total RMS modulation depths of 28.4% (squares) and 25.3% (triangles). The theoretical curves assumed a turn-on delay of 0.4 ns. After Ref. [23] (© 1996 IEEE).

The dynamic response and NLDs owing to a single sinusoidal modulating current [22, 25] or multicarriers [26] can be obtained by numerically solving the time-dependent rate equations in the time domain.

This dynamic nonlinear effect caused by overmodulating a laser is very important in obtaining the ultimate channel capacity of a laser diode, and in analyzing the hybrid AM/QAM system performance. Until recently, research attention has been focused on static clipping, whose dominant NLDs in the operating OMI range are of the seventh to ninth order (see Fig. 9.20) [4]. However, measurement and analysis have shown that the dominant NLDs due to a weakly clipped laser with a turn-on delay should be second and third order [27]. Therefore, even though a number of papers [28–30] have been devoted to designing *pre-clipping* techniques to compensate the statically clipped NLDs, the ultimate laser capacity can be reached only if dynamic clipping effects are remedied (i.e., by using a predistortion circuit which generates the right order of NLDs). In addition, since one can rarely find a laser diode with an ideal static limiter L–I characteristic, typical calculated results based on static L–I modeling predict better C/NLD performance than can be experimentally measured; this is especially true in the weak clipping regime (see Fig. 9.6).

9.2.5 Strong Static Clipping and the Ultimate Channel Capacity of QAM Signals

As mentioned in Section 9.1.3, in the long run, all AM-VSB analog video channels will be replaced by QAM digital video channels. This is expected for both upstream and downstream transmissions, and therefore it is important for us to estimate the ultimate QAM channel capacity that a laser diode can transport.

The QAM channel capacity of a laser diode can be estimated from the C/NLD of the worst-case channel. The combination of multiple channels of QAM signals can be treated as a Gaussian random process, just like the combination of multiple channels of AM-VSB signals. Consequently, Eq. (9.8), which was derived for multiple CW tones, should also be applicable to the case of multiple QAM channels. However, there are two reasons why we may not directly use the results in Section 9.2.1 directly. First, it is not so obvious how the factor Γ (which accounts for the percentage of the distortion power falling inside the CATV band) differs from the case of multiple CW tones. This fact is illustrated in Fig. 9.16, where we can see that the third-order IMPs of two QAM channels spread over six times the channel bandwidth, while the IMPs of two tones concentrate on $2f_2 - f_1$ and $2f_1 - f_2$, respectively. Therefore, the distribution of NLDs among channels may not be the same in those two cases. Another major difference between multiple QAM and multiple AM channels is that, in the case of multichannel QAM signals, we are considering *strong* laser clipping as opposed to *weak* laser clipping (both cases were illustrated in Fig. 9.2) because the SNR requirement on M-QAM signals is much lower than that on

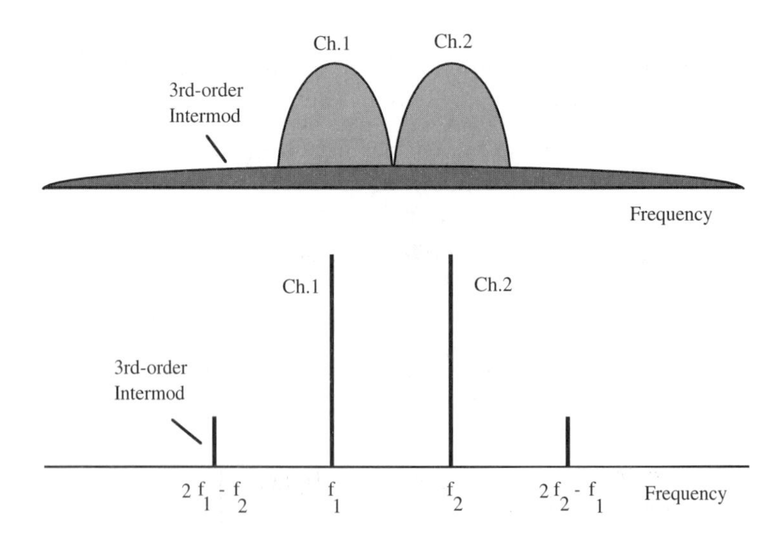

Figure 9.16: Comparison of the spectral distribution of the third-order intermodulation products for two QAM channels and two tones, respectively.

AM-VSB signals. Under a strong clipping condition, whether there are more (or fewer) NLDs spread outside the useful CATV band than in the weak clipping case is something that needs careful examination.

Based on these two considerations, therefore, it is worthwhile to analytically examine the depence of Γ on the number of QAM channels N and the total RMS modulation index μ (from ~20% to 100%).

Spectral analysis has been used to resolve this issue [4]. The spectral analysis is based on modeling of the composite input signal as a Gaussian random process, and using the series expansion of the output autocorrelation function [5,7,31,32] which results from laser clipping to obtain the corresponding output power spectral density [33–35]. The general spectral analysis presented next can be applied not only to downstream, but also to upstream multichannel QAM transmission [4].

The second-order density of a zero-mean Gaussian process can be expanded as a power series of normalized autocorrelation function $\rho(\tau)$ as [32, 35]

$$f(x_1,x_2)=\frac{1}{2\pi\sigma^2}\exp\left[-\frac{x_1^2+x_2^2}{2\sigma^2}\right]\sum_{n=0}^{\infty}\frac{\rho^n(\tau)}{n!}H_n\left(\frac{x_1}{\sigma}\right)H_n\left(\frac{x_2}{\sigma}\right), \tag{9.38}$$

where $x_1 = x(t_1)$, $x_2 = x(t_2)$, $\tau = t_2 - t_1$, and σ^2 is the variance of the Gaussian process. $R_x(\tau) = \sigma^2 \cdot \rho(\tau)$ is the autocorrelation function of the input signal; $H_n(x)$ is the nth-order Hermite function, which is defined as

$$H_n(x)=(-1)^n e^{\frac{x^2}{2}}\frac{d^n}{dx^n}e^{-\frac{x^2}{2}}. \tag{9.39}$$

Assume the transfer function of the lightwave transmission system is $y = g(x)$. Then the output autocorrelation function is given by

$$\begin{aligned} R_y(t_1,t_2) &= E[y(t_1)y(t_2)] \\ &= \int_{-\infty}^{\infty}\int_{-\infty}^{\infty} f(x_1,x_2)\,g(x_1)g(x_2)\,dx_1\,dx_2. \end{aligned} \tag{9.40}$$

By substituting Eq. (9.38) into Eq. (9.40), we obtain

$$\begin{aligned} R_y(t_1,t_2) &= \int_{-\infty}^{\infty}\int_{-\infty}^{\infty}\frac{1}{2\pi\sigma^2}\exp\left[-\frac{x_1^2+x_2^2}{2\sigma^2}\right]\sum_{n=0}^{\infty}\frac{\rho^n(\tau)}{n!}H_n\left(\frac{x_1}{\sigma}\right)H_n\left(\frac{x_2}{\sigma}\right) \\ &\quad\cdot g(x_1)g(x_2)\,dx_1\,dx_2 \\ &= \sum_{n=0}^{\infty}\frac{\rho^n(\tau)}{n!}\left[\frac{1}{\sqrt{2\pi}}\int_{-\infty}^{\infty}g(\sigma x)\,e^{-\frac{x^2}{2}}H_n(x)\,dx\right]^2 \\ &= \sum_{n=0}^{\infty}\frac{\rho^n(\tau)}{n!}b_n^2, \end{aligned} \tag{9.41}$$

where b_n is defined as

$$b_n \equiv \frac{1}{\sqrt{2\pi}} \int_{-\infty}^{\infty} g(\sigma x) e^{-\frac{x^2}{2}} H_n(x) dx, \qquad n \geq 0. \tag{9.42}$$

For a laser diode with a normalized *L–I* transfer function *g(x)* shown in Fig. 9.2, $\sigma = \mu = \sqrt{N} \cdot m_q$, and b_n can be evaluated as

$$b_n = \begin{cases} -\frac{1}{2}\mathrm{erfc}(\frac{1}{\sqrt{2}\sigma}) + \frac{\sigma}{\sqrt{2\pi}} e^{-\frac{1}{2\sigma^2}}, & n = 0 \\ \frac{\sigma}{2}\mathrm{erfc}(-\frac{1}{\sqrt{2}\sigma}), & n = 1 \\ \frac{\sigma}{\sqrt{2\pi}} H_{n-2}(-\frac{1}{\sigma}) e^{-\frac{1}{2\sigma^2}}, & n \geq 2 \end{cases} \tag{9.43}$$

Note that the RMS OMI m_q is used here instead of the peak OMI, because the peak amplitude of an M-QAM signal varies not only with the alphabet size *M*, but also with the characteristics of a pulse shaping filter.

The output power spectral density (PSD) $S_y(f)$ is the Fourier transform of the autocorrelation function $R_y(\tau)$,

$$\begin{aligned} S_y(f) &= \int_{-\infty}^{\infty} R_y(\tau)\, e^{-j2\pi f\tau}\, d\tau \\ &= \sum_{n=0}^{\infty} \frac{b_n^2}{n!} \int_{-\infty}^{\infty} \rho^n(\tau)\, e^{-j2\pi f\tau}\, d\tau \\ &= \sum_{n=0}^{\infty} \frac{b_n^2}{n!} \varphi^{(n)}(f) \end{aligned} \tag{9.44}$$

where $\varphi^{(1)}(f) = \int_{-\infty}^{\infty} \rho(t) e^{-j2\pi ft} dt$ is the normalized input PSD, and

$$\varphi^{(n)}(f) = \underbrace{\varphi^{(1)}(f) \otimes \varphi^{(1)}(f) \otimes \cdots\cdots \otimes \varphi^{(1)}(f)}_{n}$$

is the $n - 1$ times self-convolution of $\varphi^{(1)}(f)$ $(n \geq 2)$. Note that the first term b_0^2 $\varphi^{(0)}(f)$ in the summation of Eq. (9.44) represents the DC component in the spectrum. The second term $(n = 1)$ $b_1^2\, \varphi^{(1)}(f)$ is the signal component in the output spectrum with attenuation 20 $\log(b_1)$ dB compared to the input signal. The rest of the terms $(n \geq 2)$ represent the second- to higher-order nonlinear distortion powers due to the laser transfer function $g(x)$.

From Eq. (9.44), the factor Γ in which we are interested can be expressed by

$$\Gamma = \frac{\sum_{n=2}^{\infty} \frac{b_n^2}{n!} \int_{f_a}^{f_b} \varphi^{(n)}(f) df}{\sum_{n=2}^{\infty} \frac{b_n^2}{n!} \int_0^{\infty} \varphi^{(n)}(f) df}, \tag{9.45}$$

where b_n is a function of μ, $\varphi^{(n)}(f)$ is a function of N, and f_a and f_b are the minimum and maximum frequencies of the input signal, respectively. Figure 9.17 shows Γ as a function of N and μ. It can be seen that the value of Γ can be as low as 0.27, or as high as 0.64, depending on the values of N and μ. We note that as the number of channels increases, a higher fraction of the NLDs will fall into the composite signal band; also, for a given N, Γ increases faster for lower μ than for higher μ.

Now we can plug this accurately calculated Γ into Eq. (9.9) and obtain the results (shown by the solid line in Fig. 9.18) of SNLD for all channels as a function of μ. The results in Fig. 9.18 were obtained from a specific upstream transmission example: The input spectrum consists of 36 upstream M-QAM channels from 5 to 50 MHz with a channel spacing = 1.25 MHz and a symbol rate = 1 Msps. The results given by squares and circles will be explained later in the section.

A nice feature of the spectral analysis is that different orders of NLDs can be easily resolved, as can be observed from Eq. (9.44). This feature enables a system designer to know the dominant NLD orders for any given laser transfer function $g(x)$. Consequently, a proper linearization technique (such as a predistorter to generate the right order of nonlinearities) can be applied. For example, when RMS OMI $\mu = 0.4$, by using Eqs. (9.43) and (9.44) and the same 36 M-QAM upstream channels as given earlier, we can obtain the calculated clipped spectra shown in Fig. 9.19. Several interesting phenomena can be observed. (1) There are wiggles in the second- and third-order nonlinear distortion spectra, owing to the fact that there are filtered notches among input QAM channels. As for higher orders, the wiggles disappear because of more convolutional averages. (2) The highest frequency of the nth-order nonlinearity is n-fold of the maximum frequency (50 MHz) of the input spectrum. (3) The left-most channel (channel #1) always has the highest NLD power. (4) Lower-order NLD power is not always higher than that of higher orders; for example, when $\mu = 0.4$, the fourth-order NLD power is greater than that of the second order.

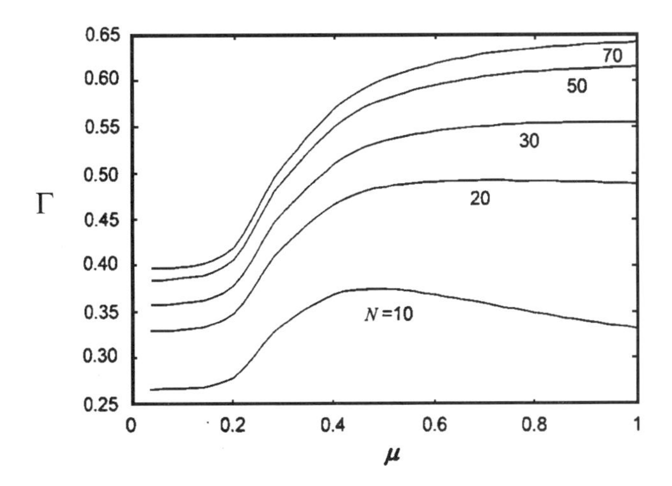

Figure 9.17: Γ as a function of μ with various values of N. After Ref. [4] (© 1997 IEEE).

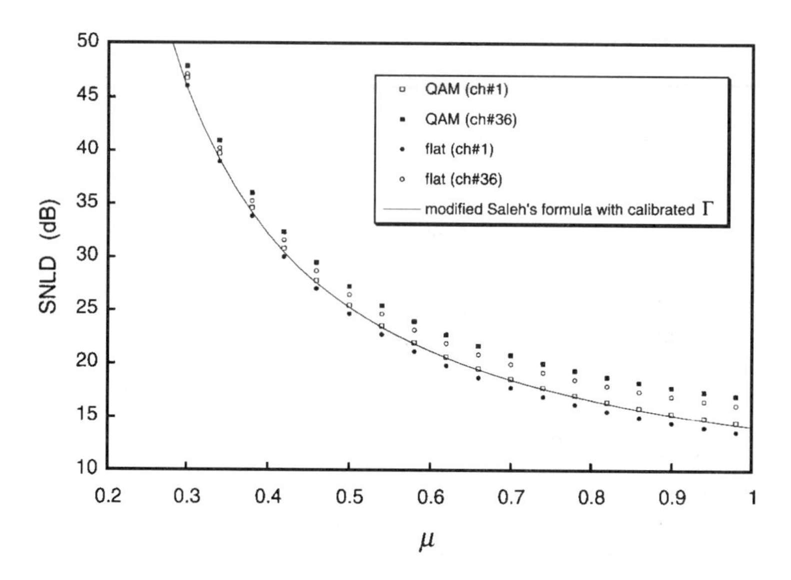

Figure 9.18: SNLD as a function of μ for $N = 36$. Squares and circles are calculated results obtained from 36 RRC filtered QAM input signals and an input signal with a flat spectrum. The solid line is calculated from Eq. (9.9). After Ref. [4] (© 1997 IEEE).

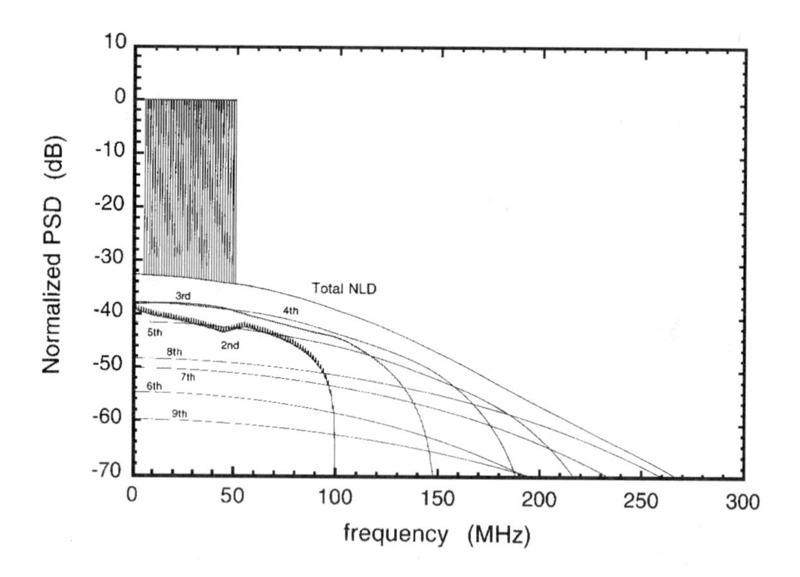

Figure 9.19: Calculated spectral distribution of clipping-induced second, third, ..., ninth-order NLDs generated from an ideal static L–I curve. Thirty-six channels of upstream M-QAM channels with RRC filtering were used. $\mu = 0.4$. After Ref. [4] (© 1997 IEEE).

In addition to the feature of resolving the order of NLDs, spectral analysis can easily let us obtain the signal-to-nonlinear distortions (SNLDs) of the ith channel centered at frequency f_i with a bandwidth $2\Delta f$. Once we know the normalized input signal PSD $\varphi^{(1)}$, by dividing the integrated signal PSD $(b_1^2\varphi^{(1)}(f))$ by the total integrated NLD PSD (from Eq. (9.44) with $n \geq 2$), we can have

$$\mathrm{SNLD}_i = \frac{b_1^2 \int_{f_i-\Delta f}^{f_i+\Delta f} |H(f)|^2 \, \varphi^{(1)}(f)\,df}{\sum_{n=2}^{\infty} \frac{b_n^2}{n!} \int_{f_i-\Delta f}^{f_i+\Delta f} |H(f)|^2 \varphi^{(n)}(f)\,df}, \tag{9.46}$$

where H(f) is the transfer function of the receiver filter. The signal to a specific nth-order NLD ratio for the ith channel, $\mathrm{SNLD}_i^{(n)}$, can be easily resolved from Eq. (9.46). The calculated $\mathrm{SNLD}_i^{(n)}$ $(n = 2,3,...,9)$ with an RRC receiver filter $(\alpha = 0.2)$, are plotted in Fig. 9.20, which shows that SNLD in channel #1 (the worst-case channel) varies with μ for different orders of NLDs $(N = 36)$. Note that there are spikes in each $\mathrm{SNLD}_i^{(n)}$ which represent singularities caused by $b_n = 0$ $(n = 2,3,...,9)$ for certain values of μ. It can be observed from Fig. 9.20 that, when the laser transfer function is a static limiter as shown in Fig. 9.2, the second and third NLDs are not dominant terms for $\mu \leq 0.45$. In fact, as μ decreases toward zero, the order of the dominant NLD increases. This apparently has important implications for the design of predistortion circuits or any other nonlinearity compensation method. It should be noted, however, that when the laser has a gradual turn-on knee of the *L–I* curve, or has significant dynamic clipping phenomena, the dominant terms can become the second- and third-order nonlinearities [27].

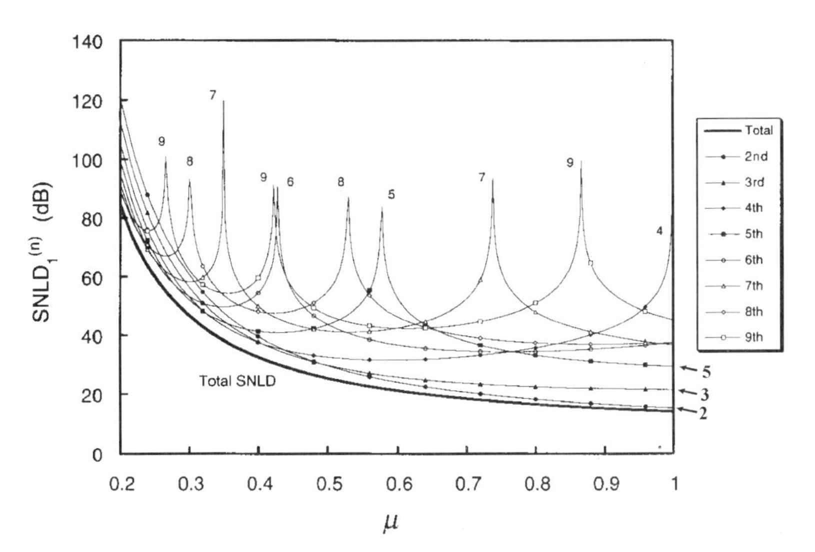

Figure 9.20: SNLD of channel #1 (due to various orders of NLDs) as a function of μ for $N = 36$. After Ref. [4] (© 1997 IEEE).

The SNLD obtained from spectral analysis can be compared with that based on Eq. (9.9). For the same 36 upstream M-QAM channels which were used in Fig. 9.18, we use Eq. (9.46) to study two cases. The first case assumes that the input spectrum is flat between f_a and f_b, and the second case assumes a more realistic input spectrum which is composed of 36 channels of root-raised cosine filtered M-QAM signals. The calculated results are shown by squares and circles in Fig. 9.18. We can see that they match fairly well with those obtained from Eq. (9.9) using calibrated Γ. The largest difference of about 3 dB occurs in channel #36, while the difference is negligible for the worst-case channel, channel #1.

Another interesting phenomenon which can be observed from Fig. 9.18 is that even when μ is as high as 0.8, at which very strong laser clipping occurs, it is still possible to maintain an SNLD > 16 dB. Therefore, we should be able to obtain error-free transmission for upstream QPSK signals provided the NLDs can be treated as Gaussian noise (so that we have SNR > 16 dB). This fact was confirmed in a corresponding experiment [4].

By using Eq. (9.46) we can calculate the ultimate channel number that an ideal static-limiter-like laser can transport with an assumed poor RIN noise level of -115 dB/Hz (assumed to be caused by optical reflections in an unisolated laser). The result is shown in Fig. 9.21 for upstream (QPSK or 16-QAM) channel number as a function of OMI per channel. For the worst-case channel to reach a bit-error rate of 10^{-9}, we set the minimum required SNR to be 16 and 24 dB for QPSK and 16-QAM signals, respectively. The useful OMI/channel values are bounded by two sides: the strong-clipping-induced NLDs on the right and the RIN noise on the left. We can see

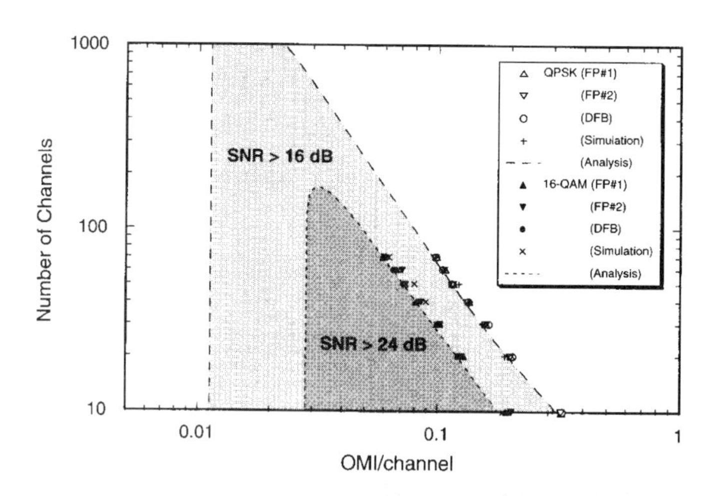

Figure 9.21: Number of 1 Msps QPSK or 16-QAM channels as a function of OMI/channel. Shaded regions represent where one can find the useful values of OMI/channel and the corresponding transportable number of channels. Left boundaries in both shaded areas are set by RIN=−115 dB/Hz. After Ref. [4] (© 1997 IEEE).

that a typical FP or DFB laser can transmit more than 1000 QPSK or 170 16-QAM channels even in the presence of strong optical reflections that caused a poor RIN of −115 dB/Hz. We note that the available data rate per 16-QAM channel is about two times that for a QPSK channel, yet the clipping-limited channel number of 16-QAM is only about 0.1 times that for QPSK modulation. Therefore, given sufficient laser bandwidth (<2 GHz) and using frequency-stacking techniques [4], the QPSK modulation format can transport more information in the upstream link.

The large capacity of an upstream laser can be severely affected by collision-based medium-access control (MAC) protocols (see Chapter 11). In the worst-case condition, when collisions occur in all but one channel, the ultimate QPSK channel capacity of an upstream laser can be dramatically reduced to 125 and 55, for 8 and 16 collisions/channel, respectively [4].

Following the same derivations just given, one can also estimate the ultimate capacity of a downstream laser diode (assuming its RIN= −135 dB/Hz). We found that as high as 600 and 128 channels of 5 Msps 64-QAM and 256-QAM signals (equivalent to 3600 and 1152 channels of 4 Mb/s MPEG-2 live video signals) can be transported, respectively, as shown in Fig. 9.22. If a laser with a good RIN value of -155 dB/Hz is used, the laser channel capacity is essentially limited only by the modulation bandwidth (BW) of a laser (e.g., BW = 88 + 6 · N (MHz) where 88 MHz is the MCNS DOCSIS-compliant starting frequency of low-band channel and N is the total number of QAM channels), and is not limited by the NLDs generated in the laser.

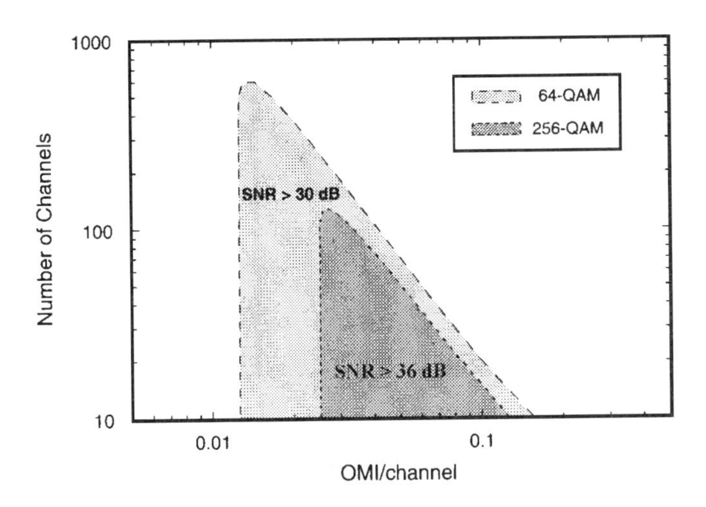

Figure 9.22: Number of 5 Msps 64/256-QAM channels as a function of OMI/channel for strongly clipped downstream laser diode. Shaded regions represent where one can find the useful values of OMI/channel and the corresponding transportable number of channels. Left boundaries in both shaded areas are set by RIN= −135dB/Hz. After Ref. [4] (© 1997 IEEE).

Note that the analytical method introduced in this section can be easily extended to external modulator-based transmitters, that is, we can simply replace the laser diode transfer function in Eqs. (9.40)–(9.42) by those transfer functions given in Chapter 7 (see problem 9.1).

9.3 Laser Chirp-Induced Impairments

9.3.1 Effects of Interferometric Phenomena

Interferometric phenomena occur in an optical fiber link, as illustrated in Fig. 9.23, because of two or more discrete index discontinuities, or continuous Rayleigh backscattering in a long length of optical fibers. Those discrete index discontinuities can be caused by imperfect optical fiber connectors, splices, or nonideal packaging inside the laser transmitter and photoreceiver. Rayleigh backscattering is caused by the intrinsic microreflection due to the inhomogeneities of the fiber refractive index along the transmission path.

The system impact of interferometric phenomena, when combined with laser frequency chirping, is twofold: Both the system CNR and the system linearity performance are affected. We separately discuss these two system issues in the following two sections.

9.3.1.1 Phase to Intensity Noise Conversion due to Multiple Discrete Reflections and Laser Chirping Let us consider the case of a laser operating under CW conditions, for which the emitted optical field is expressed as

$$E_i(t) = A \cdot \exp(i\Omega_o t + \phi(t)), \tag{9.47}$$

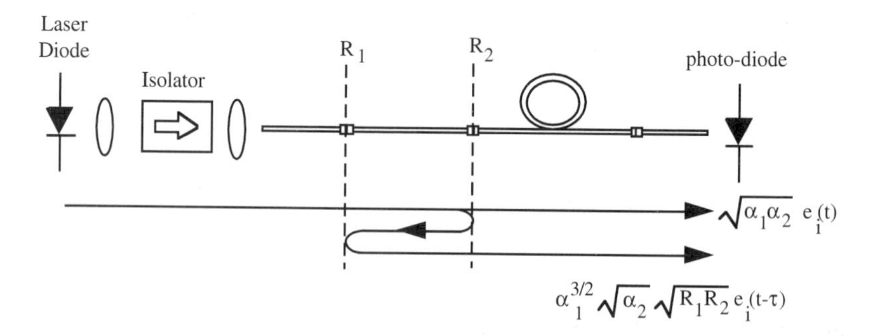

Figure 9.23: Illustration of an interferometric phenomenon caused by multiple reflections in a fiber link. The multiple reflections are due to the residual reflectivities of connectors in this case. After Ref. [36] (© 1989 IEEE).

where Ω_o is the laser center frequency and $\phi(t)$ represents the laser random phase noise. The light intensity due to the interferometer shown in Fig. 9.23 (which includes direct and doubly reflected light with the same polarization) is given by

$$\begin{aligned} I(t) &= \left|\sqrt{\alpha_1\alpha_2}\, e_i(t) + \alpha_1^{3/2}\sqrt{\alpha_2}\sqrt{R_1R_2}\, e_i(t-\tau)\right|^2 \\ &\approx \alpha'|A|^2\left[1+\rho(t,\tau)\right] \end{aligned}, \tag{9.48}$$

where R_1 and R_2 are the intensity reflection coefficients of the two connectors, respectively; α_1 is the optical attenuation due to the path between the two connectors; α_2 is the attenuation between the second connector and the optical receiver; τ is the relative delay between the direct and the doubly reflected light; $\alpha' = \alpha_1\alpha_2$; and $\rho(t, \tau)$ is given by

$$\rho(t,\tau) = 2R\cdot\cos(\Omega_o\tau + \phi(t) - \phi(t-\tau)), \tag{9.49}$$

where R is the effective reflection coefficient given by $R = \alpha_1\sqrt{R_1R_2}$. Note that in Eq. (9.48) the direct and doubly reflected light were assumed to have the same polarization so that the worst-case condition can be investigated.

The key now is to find the power spectral density of $\rho(t,\tau)$ from its autocorrelation function defined by

$$\begin{aligned} R_\rho(t,\tau) &= \langle \rho(t',\tau)\rho(t'+t,\tau)\rangle \\ &= 4R^2 \langle \cos[\Omega_o\tau + \Psi(t',\tau)]\cdot\cos[\Omega_o\tau + \Psi(t'+t,\tau)]\rangle, \end{aligned} \tag{9.50}$$

where $\langle\bullet\rangle$ represents an ensemble average. The random variable $\Psi(t,\tau) = \phi(t) - \phi(t-\tau)$ is normally distributed with a PDF given by [36]

$$P(\Psi) = \frac{1}{\sqrt{2\pi\sigma^2(\tau)}}\exp(-\Psi^2/2\sigma^2(\tau)), \tag{9.51}$$

where $\sigma^2(\tau) = 2\pi\cdot\Delta\upsilon\cdot\tau$ for a laser power spectral density function with a Lorenzian line shape. Equation (9.50) can be solved by considering two conditions shown in Figs. 9.24a and b, respectively.

Condition (a) indicates $|t| > \tau$, which implies that $\Psi(t', \tau)(= \phi(t') - \phi(t' - \tau))$ and $\Psi(t' + t, \tau)(= \phi(t'+t) - \phi(t'+t-\tau))$ are independent, and therefore

$$\begin{aligned} R(t,\tau) &= 4R^2\int_{-\infty}^{\infty}\int_{-\infty}^{\infty}\cos(\Omega_o\tau+\Psi_1)\cdot\cos(\Omega_o\tau+\Psi_2)\cdot P(\Psi_1)P(\Psi_2)d\Psi_1 d\Psi_2 \\ &= 2R^2 e^{-\sigma^2(\tau)}(1+\cos(2\Omega_o\tau)) \end{aligned}, \tag{9.52}$$

where we have used the integral formulas

$$\int_{-\infty}^{\infty}\cos\Psi_1\frac{1}{\sqrt{2\pi\sigma^2(\tau)}}e^{-\frac{\Psi_1^2}{2\sigma^2(\tau)}}\cdot d\Psi_1 = e^{-\frac{\sigma^2(\tau)}{2}} \tag{9.53}$$

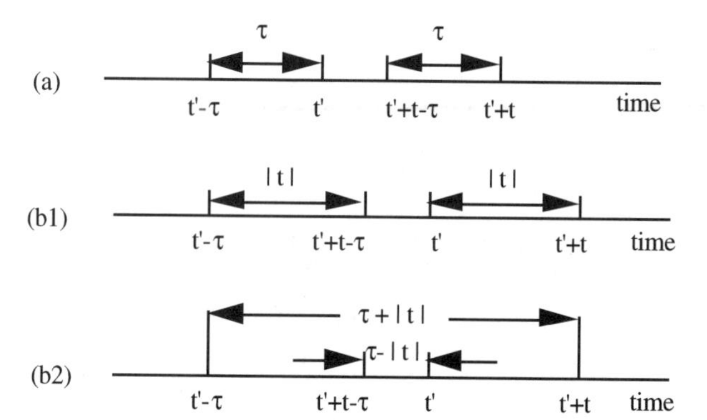

Figure 9.24: Relative time positions of the four time instants, (a) $|t| > \tau$, (b1) and (b2) $|t| < \tau$. Note that $t > 0$ in this figure.

and

$$\int_{-\infty}^{\infty} \sin\Psi_1 \frac{1}{\sqrt{2\pi\sigma^2(\tau)}} e^{-\frac{\Psi_1^2}{2\sigma^2(\tau)}} \cdot d\Psi_1 = 0 \quad . \tag{9.54}$$

Condition (b) indicates $|t| < \tau$ which implies that $\Psi(t', \tau)$ and $\Psi(t', + t, \tau)$ are dependent. In this case, we can rearrange the terms in Eq. (9.50) to obtain

$$R(t,\tau) = 2R^2\left\langle \cos\left(\Psi_a - \Psi_b\right) + \cos\left(2\Omega_o\tau + \Psi_c + \Psi_d\right)\right\rangle \tag{9.55}$$

where

$$\begin{aligned} \Psi_a &= \phi(t'+t-\tau) - \phi(t'-\tau), \qquad & \Psi_b &= \phi(t'+t) - \phi(t') \\ \Psi_c &= \phi(t') - \phi(t'+t-\tau), \qquad & \Psi_d &= \phi(t'+t) - \phi(t'-\tau)\,. \end{aligned} \tag{9.56}$$

According to Fig. 9.24b1, Ψ_a and Ψ_b are independent; therefore, the expected value of the first term in Eq. (9.55) can be obtained as

$$2R^2\left\langle \cos\left(\Psi_a - \Psi_b\right)\right\rangle = 2R^2 \exp[-\sigma^2(|t|)]\,. \tag{9.57}$$

However, according to Fig. 9.24b2, Ψ_c and Ψ_d are dependent, and the second term in Eq. (9.55) can be evaluated using the results of

$$\begin{aligned} \left\langle \cos\left(\Psi_c + \Psi_d\right)\right\rangle &= \int_{-\infty}^{\infty}\int_{-\infty}^{\infty} \cos\left(\Psi_c + \Psi_d\right) P(\Psi_d \mid \Psi_c) P(\Psi_c) d\Psi_c d\Psi_d \\ &= \exp\left[-2\sigma^2(\tau - |t|)\right]\exp\left[-\sigma^2(2|t|)/2\right] \end{aligned} \tag{9.58a}$$

and

$$\left\langle \sin\left(\Psi_c + \Psi_d\right)\right\rangle = 0\ , \tag{9.58b}$$

which in turn gives

$$2R^2\langle\cos(2\Omega_o\tau+\Psi_c+\Psi_d)\rangle$$
$$=2R^2\left[\cos(2\Omega_o\tau)\cdot\exp\left[-2\sigma^2(\tau-|t|)\right]\cdot\exp\left[-\sigma^2(2|t|)/2\right]\right]. \quad (9.59)$$

Combining the results of Eqs. (9.52), (9.57), and (9.59), and recalling the assumption $\sigma^2(\tau)=2\pi\cdot\Delta\nu\cdot\tau$, we have the final results for the autocorrelation function as

$$R(t,\tau)=2R^2\begin{cases}e^{-2\pi\cdot\Delta\nu\cdot\tau}\cdot\left[1+\cos(2\Omega_o\tau)\right], & |t|>\tau\\ e^{-2\pi\cdot\Delta\nu\cdot|t|}\cdot\left\{1+\cos(2\Omega_o\tau)e^{-4\pi\cdot\Delta\nu\cdot(\tau-|t|)}\right\}, & |t|<\tau.\end{cases} \quad (9.60)$$

After Fourier transforming Eq. (9.60) and subtracting the DC term, we obtain the *single-sided* noise power spectral density:

$$RIN(f)=4R^2[1+\cos(2\Omega_o\tau)]\left(-\frac{1}{\pi f}\sin 2\pi f\tau\right)e^{-2\pi\Delta\nu\tau}$$
$$+\frac{4R^2}{\pi}\left[\frac{\Delta\nu}{f^2+(\Delta\nu)^2}\right]\cdot\left\{\begin{aligned}&1-e^{-2\pi\Delta\nu\tau}\left[\cos(2\pi f\tau)-\frac{f}{\Delta\nu}\sin(2\pi f\tau)\right]\\&-\cos(2\Omega_o\tau)\left[e^{-4\pi\Delta\nu\tau}-e^{-2\pi\Delta\nu\tau}\left(\cos(2\pi f\tau)+\frac{f}{\Delta\nu}\sin(2\pi f\tau)\right)\right]\end{aligned}\right\}. \quad (9.61)$$

From Eq. (9.61) we can see that the maximum conversion of phase noise into intensity noise occurs when the direct and doubly reflected light interfere in quadrature, that is, $\Omega_0\tau=(n\pm 1/2)\pi$ or $\cos(2\Omega_0\tau)=-1$. For this case, we have

$$\boldsymbol{RIN_Q(f)}=\frac{\boldsymbol{4R^2}}{\pi}\left[\frac{\Delta\nu}{\boldsymbol{f}^2+(\Delta\nu)^2}\right]\cdot\left\{1+\boldsymbol{e}^{-4\pi\Delta\nu\tau}-\boldsymbol{2e}^{-2\pi\Delta\nu\tau}\cos(2\pi f\tau)\right\}. \quad (9.62)$$

In the limit of $2\pi\Delta\nu\tau \ll 1$, $\mathrm{RIN_Q}(f)$ approaches $16\pi R^2\,\Delta\nu\tau^2\,sinc^2(f\tau)$, that is, in the limit of small phase fluctuations, maximum RIN is proportional to $\Delta\nu$. In the limit of $2\pi\Delta\nu\tau \gg 1$, the interfering terms combine *incoherently*, and $\mathrm{RIN_Q}(f)$ approaches $(4R^2/\pi)\{\Delta\nu/[f^2+(\Delta\nu)^2]\}$. Therefore, for both limits, the interferometric RIN is negligible when the linewidth of the light source is extremely narrow (e.g., in an Nd:YAG laser) or broad (e.g., in an LED). It can be derived from Eq. (9.62) that when τ is fixed and $f \ll \Delta\nu$ (note that the lower the frequency is, the higher $\mathrm{RIN_Q}(f)$ is, as shown in Fig. 9.25), the worst $\mathrm{RIN_Q}(f)$ occurs at $\Delta\nu\cdot\tau\approx 0.2$. It is quite coincidental that for a typical DFB laser with a linewidth of 10–30 MHZ, and an optical fiber jumper with a typical length of 1–3 m, the product of $\Delta\nu\cdot\tau$ is very close to the worst condition of 0.2

A measured interferometric RIN for a DFB laser under CW and modulation conditions are shown in Fig. 9.25 [37]. It can be observed that in the CW condition, the RIN resembles the original Lorentzian lineshape of the source laser [this fact can also be observed from Eq. (9.62), where $RIN_Q(\Delta\nu) \approx 1/2 \cdot RIN_Q(0)$ with $2\pi\Delta\nu\tau >> 1$] and has a higher level at the frequency region close to DC. Therefore, under this operating condition (as is true in an external modulation system based on a high-power CW MQW-DFB laser), if there are bad connectors or splices in the CATV optical fiber link, the CNR of the VHF-band TV channels (e.g., ch #2, 3) can be seriously affected.

On the other hand, when the laser is modulated by multiple TV channels, the dynamic linewidth is broadened significantly, and the interferometric RIN noise is spread over a much wider frequency range. Consequently, a lower RIN results, as can be seen in Fig. 9.25. For the general case of interest, we have $2\pi\Delta\nu_{eff}\tau >> 1$, which gives an approximate expression of interferometric RIN, according to Eq. (9.62), as

$$RIN_{dynam} \approx 4R^2/(\pi \cdot \Delta\nu_{eff}), \qquad 2\pi\Delta\nu_{eff}\tau >> 1, \Delta\nu_{eff} >> f_{\max}, \quad (9.63)$$

where f_{max} is the maximum frequency component of the modulating signals. $\Delta\nu_{eff}$ is the effective linewidth of the Gaussian shape power spectral density under modulation (typically about 4–6 GHz for OMI/ch = 4–5% in a 40-channel system) and is given by

$$\Delta\nu_{eff} \approx \eta_{FM}(I_b - I_{th})\mu \ , \quad (9.64)$$

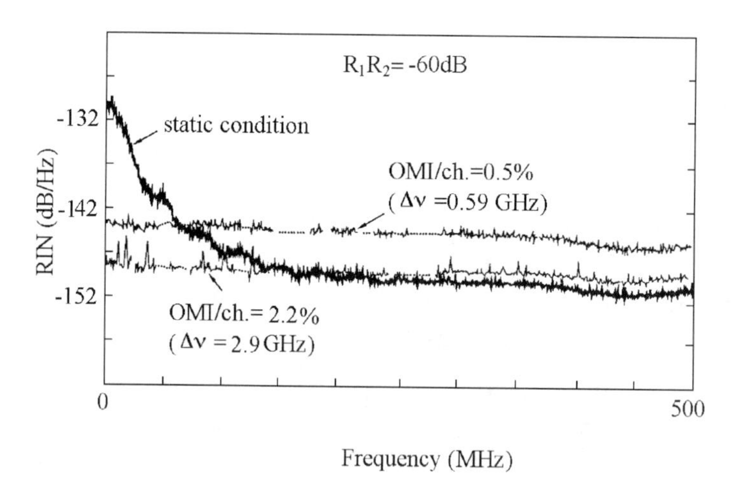

Figure 9.25: Measured interferometric RIN spectral distribution dependence on signal modulation index (the laser diode was modulated by 19-channel AM-VSB video signals). The dots in the dynamic RIN curves represent where the modulating signals were located. After Ref. [37] (© 1990 IEEE).

where η_{FM} is the light source chirping efficiency in MHz/mA, I_b and I_{th} are the bias and threshold currents, respectively, and μ is the RMS modulation index previously defined in Eq. (9.4).

When more than two reflection points exist in an optical fiber link, the interferometric RIN (in decibels) will increase according to [37]

$$RIN_N = RIN_2 + 10\log\left\{\frac{1}{\eta^2}\sum_{i=1}^{N-1}\sum_{j=i+1}^{N}\eta^{2(j-i)}\right\}$$
$$= RIN_2 + 10\log\left\{\frac{1}{1-\eta^2}\left[(N-1)+\frac{\eta^2\cdot\left(\eta^{2(N-1)}-1\right)}{1-\eta^2}\right]\right\}, \tag{9.65}$$

where RIN_N and RIN_2 are the interferometric RINs due to N and 2 connectors, respectively, and η is the loss factor of each connector. As η approaches unity, that is, each connector is lossless, RIN_N differs from RIN_2 by a factor $10\cdot\log(N(N-1)/2)$. Experimental results of N-cascaded connectors (with the distance between adjacent connectors equal to 2 m) are shown in Fig. 9.26 [38].

If there exists an optical amplifier between the two connectors shown in Fig. 9.23, the optical power of each double reflected light will be increased more than the direct light in the amplifier by a factor G^2, where G is the fiber-to-fiber gain of the optical amplifier [37]. Therefore, the square of the effective reflection coefficient and the resultant interferometric RIN will be increased by a factor G^2. It is thus a common practice to have built-in optical isolators and to use fusion splices instead of optical fiber connectors when building an erbium-doped fiber amplifier [38].

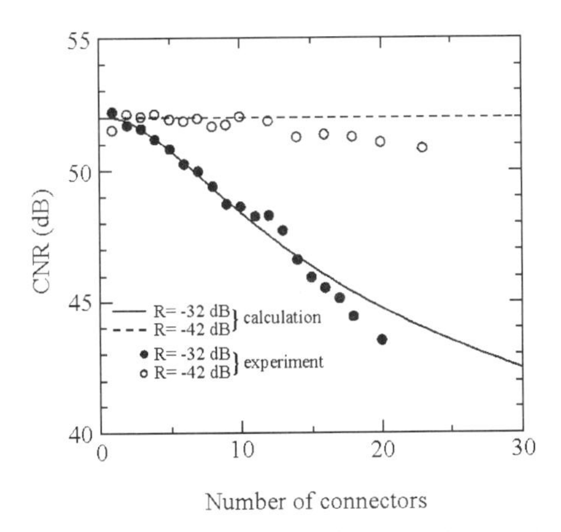

Figure 9.26: CNR versus number of cascaded connectors. Calculated results based on Eq. (9.65) are shown as solid and dotted lines. Experimental results are shown as open and solid circles. After Ref. [38] (© 1992 IEEE).

9.3.1.2 Phase to Intensity Noise Conversion due to Double Rayleigh Backscattering Multiple discrete reflections can be minimized by using obliquely polished connectors, fusion splices, and high return loss passive components. However, the interferometric RIN due to the interference between the main beam and the light which has been retroreflected twice by Rayleigh scattering imposes a fundamental limitation in long-distance optical fiber links [39,40], as illustrated in Fig. 9.27. Rayleigh scattering is due to the inhomogeneities of the fiber refractive index along the transmission path, and therefore is a fundamental property that cannot be neglected even in optical fibers with lengths as short as 10 km.

When considering the DRB light in Fig. 9.27, we can assume that the single mode fiber with a length L is made up of N ($\rightarrow\infty$) equal discrete microreflectors, whose reflected light is incoherent. Each microreflector has a differential fiber length of Δx, a loss of $\eta = exp(-\alpha\Delta x)$, and a reflectivity of $R = S\alpha_s\Delta x$, where α_s is the proportion of signal scattered per unit length, and S is the fraction of scattering that is captured by the fiber. From Eq. (9.65), we can use two "equivalent" reflectors, each with an equivalent reflectivity R_{eq}, to represent the effect of all DRB microreflectors:

$$R_{eq}^2 = R^2 \sum_{i=1}^{N-1} \sum_{j=i+1}^{N} \eta^{2(j-i)} \approx \frac{S^2\alpha_s^2}{4\alpha^2}\left(2\alpha L - 1 + e^{-2\alpha L}\right), \tag{9.66}$$

where we have used $L = N \cdot \Delta x$. Depending on whether the state of polarization of the DRB light is the same as that of the direct light, R_{eq} can be reduced accordingly [41]. We can get a feeling for the magnitude of the R_{eq} given in Eq. (9.66) as a function of L at 1310 and 1550 nm by using the parameters given in the caption of Fig. 9.28a. We can see from the figure that when the fiber length is as long as 50 km, the effective reflectivity caused by DRB is equivalent to that caused by a pair of poor connectors with about -29-dB reflectivities.

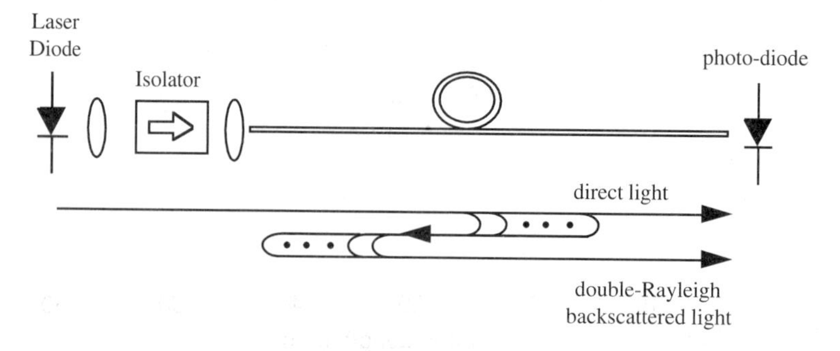

Figure 9.27: Interferometric RIN can occur due to the interference between the direct light and the double-Rayleigh-backscattered (DRB) light. When considering the DRB light, we can assume that the single mode fiber is made up of N ($\rightarrow\infty$) discrete reflectors.

By replacing R^2 with R_{eq}^2 in Eq. (9.62), we can obtain the power spectral density of interferometric noise due to the interference between direct light and DRB light. For the incoherent case of $2\pi\,\Delta\nu\tau >> 1$, the calculated results of *RIN(f)* for different lengths of fiber are plotted in Fig. 9.28b.

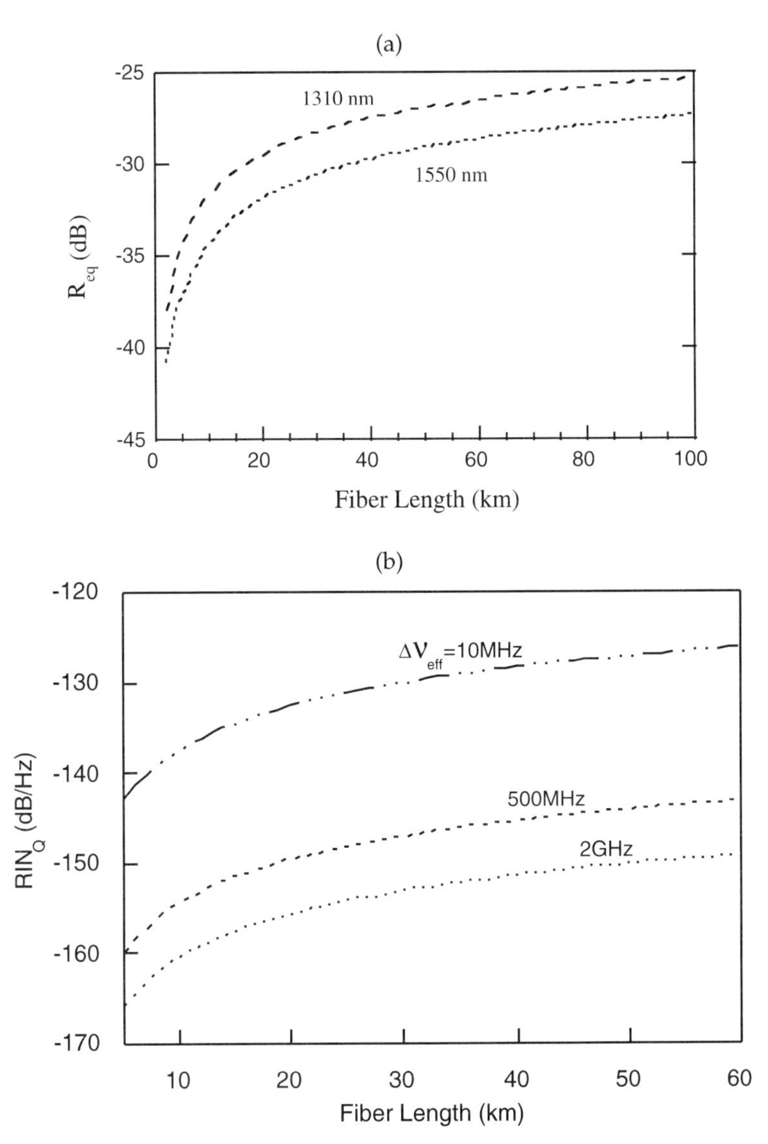

Figure 9.28: (a) Equivalent reflectivity R_{eq} due to DRB from optical fibers with different length. (S = 0.0015, α at 1310 (1550) nm = 0.35 (0.22) dB/km, α_s at 1310 (1550) nm = 0.35 (0.18) dB/km). (b) RIN_Q versus fiber length L for various effective laser linewidths $\Delta\nu_{eff}$.

When the light source is both intensity and *frequency* modulated because of multiple channels of subcarrier-multiplexed signals, its resultant power spectral density is Gaussian. Therefore, we can again use the results of Eq. (9.63) with R^2 replaced by R_{eq}^2, and use Eq. (9.66) to obtain

$$
\begin{aligned}
RIN_{dynam} &\approx 4R_{eq}^2/(\pi \cdot \Delta\nu_{eff}), \qquad\qquad 2\pi\Delta\nu_{eff}\tau >> 1 \\
&\approx \frac{4}{\pi}\left[\frac{S^2\alpha_s^2}{4\alpha^2}\left(2\alpha L - 1 + e^{-2\alpha L}\right)\right]\cdot\frac{1}{\eta_{FM}(I_b - I_{th})\mu}.
\end{aligned} \tag{9.67}
$$

This simple approximation turns out to be within 3 dB of those obtained from tedious derivations [Eq. (23) in Ref. 42].

However, if the light source is both amplitude and *phase* modulated, as in an external modulation system, the resultant spectral power density may not be Gaussian-like. In that case, an exact solution must be used [43].

Before we leave this section, we should mention that the interferometric RIN described in this section can also occur inside a short length of erbium-doped fiber amplifier, and could limit the maximum CNR achievable in an SCM system. Several papers reporting this phenomenon were cited in Section 6.1.9.2.

9.3.1.3 Nonlinear Distortions due to Interferometric Phenomena and Laser Chirping Interferometric phenomena due to discrete multiple reflections not only degrade the overall system RIN level, but also increase the resultant system nonlinear distortions. This problem can be analyzed as follows [44]. We start with a laser optical field with combined IM and FM (due to the modulation of N-SCM channels) as

$$E(t) = A\left(1 + m\sum_{i=1}^{N}\cos\omega_i t\right)^{1/2} \exp\left[j(\Omega_o t + \theta(t))\right], \tag{9.68}$$

with optical phase shift given by $\theta(t) = \sum_{i=1}^{N}(2\pi\Delta\nu_{eff,i}/\omega_i)\sin\omega_i t$. Following the derivations of Eqs. (9.48) and (9.49), we have the detected photocurrent as

$$
\begin{aligned}
I(t) &\approx \alpha'|A|^2\left[1 + (m\sum_{i=1}^{N}\cos\omega_i t) + \rho'(t,\tau)\right] \\
&= I_{DC}\left[1 + (m\sum_{i=1}^{N}\cos\omega_i t) + \rho'(t,\tau)\right],
\end{aligned} \tag{9.69}
$$

where

$$\rho'(t,\tau) = 2R\cdot\cos(\Omega_o\tau + \theta(t) - \theta(t-\tau)). \tag{9.70}$$

Equation (9.70) can be expanded by using the expression

$$e^{j(\theta(t)-\theta(t-\tau))} = \exp\left[j\sum_{i=1}^{N} b_i\cos\omega_i(t-\frac{\tau}{2})\right] = \prod_{i=1}^{N}\left[\sum_{k_i=-\infty}^{\infty} J_{k_i}(b_i)e^{jk_i[\varphi_i+\frac{\pi}{2}]}\right], \tag{9.71}$$

where J_k is the Bessel function of order k, $b_i = (4\pi\Delta\nu_{eff,i}/\omega_i)\sin(\omega_i\tau/2)$, and $\varphi_i = \omega_i[t - (\tau/2)]$. In the simplified case of two tones, $N = 2$, we have

$$e^{j(\theta(t)-\theta(t-\tau))} = \sum_{k_1=-\infty}^{\infty}\sum_{k_2=-\infty}^{\infty}[J_{k_1}(b_1)J_{k_2}(b_2)]e^{j\theta_o}e^{j(k_1\varphi_1+k_2\varphi_2)} \tag{9.72}$$

where $\theta_o = \frac{\pi}{2}\sum_{i=1}^{2}k_i$ .Therefore, the square of the intermodulation (IM) product photocurrent in the worst case is given by

$$I\,(f_{IM}) = 4I_{DC}R\cdot\cos(\Omega_o\tau+\theta_o)J_{k_1}(b_1)J_{k_2}(b_2), \tag{9.73}$$

where frequency of the IM is at $f_{IM} = k_1f_1 + k_2f_2$. The term $\cos(\Omega_0\tau + \Theta_0)$ is a rapidly changing function and equals unity in the worst case. Recall from Eq. (9.69) that the photocurrent of each video channel is $I_{DC}m\cos(\omega_1 t)$ (assuming $m >> J_0(b_1)J_1(b_2)$ and $m >> J_0(b_2)J_1(b_1)$), we thus have the (electrical) power ratio of the IM and a carrier as

$$IM/C = \frac{16R^2}{m^2}\left[J_{k_1}(b_1)J_{k_2}(b_2)\right]^2, \tag{9.74}$$

where again $R = \alpha_1R_1R_2$. For example, the second-order IM at $f_1 - f_2$ is

$$IM_2/C = \frac{16R^2}{m^2}\left[J_1(b_1)J_{-1}(b_2)\right]^2 \tag{9.75}$$

and the third-order IM at $2f_1 - f_2$ is

$$IM_3/C = \frac{16R^2}{m^2}\left[J_2(b_1)J_1(b_2)\right]^2. \tag{9.76}$$

It is clear that both the second- and third-order NLDs depend critically on parameters b_1 and b_2, respectively. However, b_1 and b_2 depend not only on the effective chirped linewidth $\Delta\nu_{eff,1}$ and $\Delta\nu_{eff,2}$, but also on the modulating frequencies ω_1 and ω_2, respectively. Therefore, the NLDs are modulating frequency-dependent, and product count (see Section 2.3.1) cannot be used to predict the NLDs of multiple channels.

CSO/CTB degradations because of interferometric-type interference can be observed in links with discrete reflectivities such as poor fiber connectors. Subtle linearity degradations can also be observed in a packaged laser transmitter with residual reflectivities inside the package. An example is given in Fig. 9.29. The poor reflectivities due to an optical isolator in a packaged laser transmitter can affect the laser transmitter's linearity performance when ambient temperature varies. Therefore, the alignment of lenses, optical isolator, and so on to reduce multiple reflections in a packaged laser is critical.

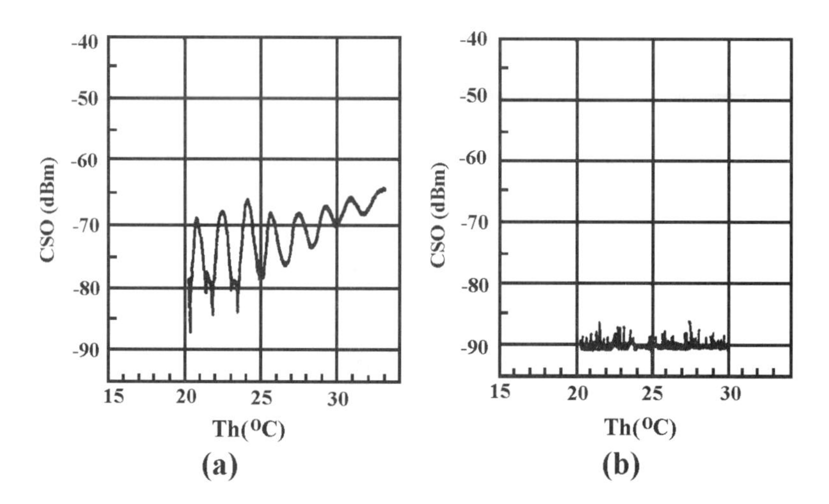

Figure 9.29: Submount temperature dependence of CSO characteristics of the DFB laser module under 42 channel modulation (a) with and (b) without interferometric interference. After Ref. [45] (© 1990 IEEE).

9.3.2 Effects of Polarization-Dependent Devices

Almost all optical isolators contain a birefringent crystal (see Fig. 3.25). The crystal is usually set at an angle of several degrees with respect to the input light to avoid optical reflection at the front facet of the crystal, as shown in Fig. 9.30. However, the refraction angles at the surface of the birefringent crystal, ϑ_A and ϑ_B, are different in the two input polarization states. This is because these refraction angles are determined by the vertical and horizontal (fast- and slow-axes) refractive indices of the birefringent crystal. The smaller refraction angle in Fig. 9.30a (input light is vertically polarized) enables both the first-order light and the doubly reflected light (other higher order-reflected lights are neglected) to launch into the output port fiber, whereas in Fig. 9.30b only the first-order light can be launched into the output port fiber. The two beams in Fig. 9.30a have a relative delay τ due to the roundtrip traveling time of the doubly reflected light through the finite thickness of the crystal (typically about 5 mm), and when they arrive at the square-law detector, interferometric-phenomena-induced nonlinear distortions could occur, as was explained in the previous section. Therefore, the nonlinear distortion (mainly CSO) generation mechanism is due to the combination of laser frequency chirping and input light polarization state-dependent multiple reflections. This type of polarization-dependent nonlinear distortions can be avoided if external modulation system is used, as can be seen in Fig. 9.31.

In an optical fiber transmission system where many EDFAs are concatenated, the same nonlinear distortions could be generated, owing to the fact almost all EDFAs contain one or two optical isolators. For a direct modulation system, de-

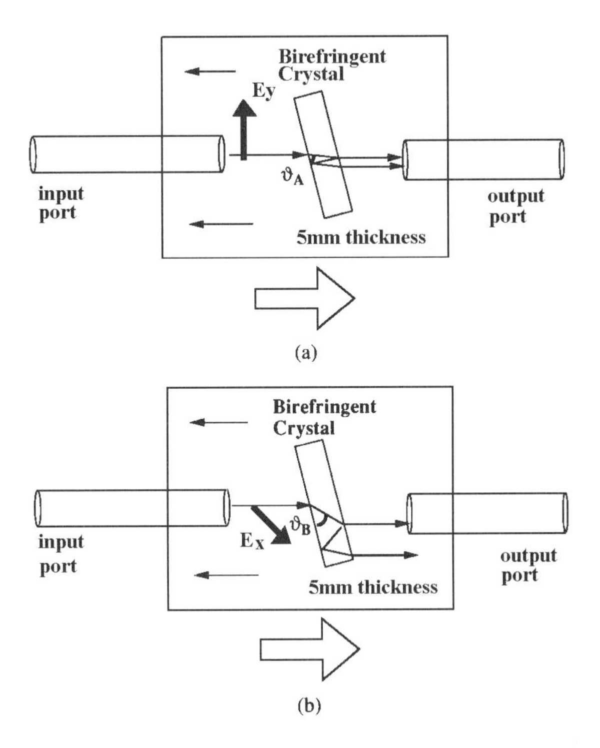

Figure 9.30: Mechanism of the polarization dependent distortion in an optical isolator. (a) For vertically polarized light, all multiple reflection light is injected into the output port, and (b) for horizontally polarized light, only the first-order light is injected into the output port. After Ref. [46] (© 1994 IEEE).

pending on the input polarization state, the CSO variations could be in the range of 3 to 15 dB per EDFA stage [46]. For $LiNbO_3$ external modulator-based systems, the CSO variation is within a few decibels for different input polarization states when the external modulator is operated under the *maximum extinction ratio condition.* However, input polarization-dependent CSO variations increase significantly (up to a maximum of about 7 dB per EDFA) if the external modulator is operated under the *minimum loss condition* [46].

The difference between the two external modulator operating conditions is that the preferred output polarization state can be maintained in the former case, whereas the output polarization angle rotates in the latter case. The rotation of the polarization states causes CSO variations in systems with optical isolators or EDFAs, because of the two varying conditions shown in Fig. 9.30.

It was also found that stimulated Brillouin scattering–induced CSOs could also depend on the input light polarization states of an optical fiber link (with the maximum and minimum CSO values differing by 5–7 dB). The reason is that the Brillouin gain is a factor of 2 higher for linearly polarized input light than for circularly

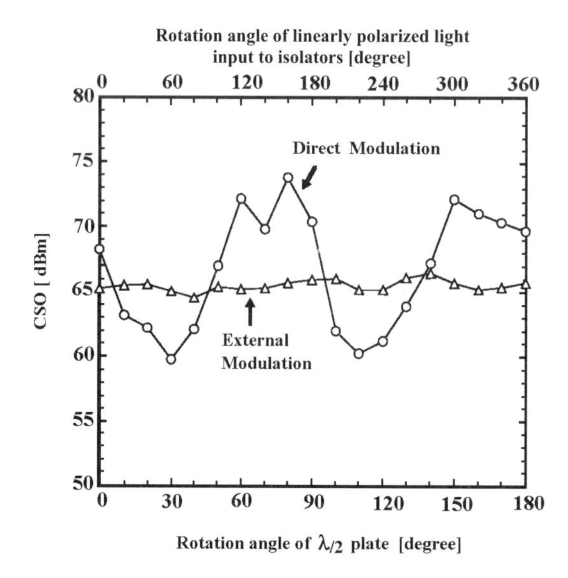

Figure 9.31: Measured CSO distortion values against the rotation angle of a $\lambda/2$ plate which is used to adjust the input light polarization state of an isolator. The circle plots are measured values under the direct modulation scheme. The triangles are measured values under the external modulation scheme. After Ref. [46] (© 1994 IEEE).

polarized input light [47]. More details of SBS-induced effects will be discussed in Section 9.6.1.

9.3.3 Effects of EDFA Gain Tilt

In a direct modulation multichannel AM-VSB system which uses an EDFA as the postamplifier for a laser transmitter, it was reported [48–50] that second-order nonlinear distortions can be caused by the interaction between the frequency chirping (FM) of the directly modulated DFB laser and the wavelength-dependent gain or gain tilt (AM) of the EDFA, mainly through the FM-to-AM conversion, as illustrated in Fig. 9.32.

Typical gain tilt values are within a range of ± 1 dB/nm [48,49], depending on the EDFA input power level and wavelength. It was found that the laser chirp-induced gain tilt values resemble the measured spontaneous emission gain tilt values, and that EDFAs whose spontaneous emission spectrum is flat over a wide wavelength range can minimize the CSO due to gain tilt and FM chirping more effectively [51].

Although this phenomenon can be explained using rigorous analysis [48], a simplified but useful analysis can be derived as follows [49]. We first assume that

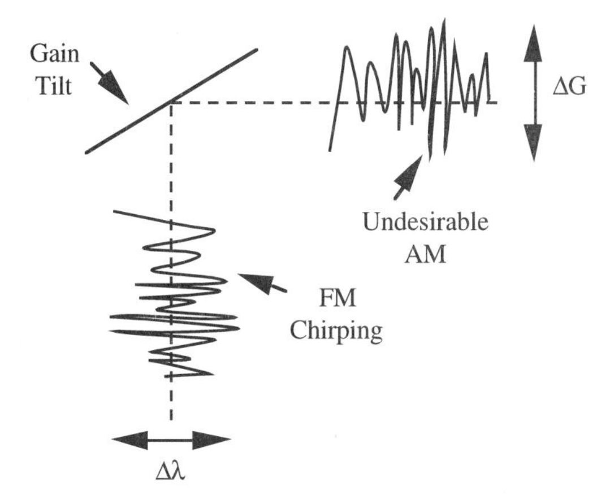

Figure 9.32: The interaction of laser FM chirping and EDFA gain tilt results in undesirable FM-to-AM conversion and causes significant CSOs (© 1996 IEEE).

the DFB laser is modulated by a single AM carrier signal, and that the EDFA input power and the optical wavelength are given by

$$P_{in}(t) = P_o[1 + m\sin(2\pi ft)] \tag{9.77}$$

$$\lambda(t) = \lambda_o - \Delta\lambda \cdot \sin(2\pi ft), \tag{9.78}$$

respectively, where P_0 is the average input optical power, m the optical modulation index per channel, f the modulation frequency, λ_0 the average signal wavelength, and $\Delta\lambda$ the peak wavelength deviation due to modulation. Note that we have assumed amplitude and wavelength modulations are in and 180° out of phase with current modulation. The input power- and wavelength-dependent chirping-induced varying gain of an EDFA can be expressed as

$$G(P_{in}(t), \lambda(t)) = G(P_o, \lambda_o) + \left[\frac{dG(P_0, \lambda)}{d\lambda}\right]_{\lambda_o} \cdot [\lambda(t) - \lambda_o], \tag{9.79}$$

where $dG(P_0,\lambda)/d\lambda$ at λ_0 is the gain tilt of the EDFA at λ_0. The EDFA output is given by

$$\begin{aligned} P_{out}(t) &= P_{in}(t) \cdot G(P_{in}(t), \lambda(t)) \\ &\approx G(P_o, \lambda_o)P_o + G(P_o, \lambda_o)P_o m\sin(2\pi ft) + \frac{1}{2}\left[\frac{dG(P_o, \lambda)}{d\lambda}\right]_{\lambda_o} \Delta\lambda \cdot mP_o\cos(4\pi ft), \end{aligned} \tag{9.80}$$

where negligible terms were dropped. From Eq. (9.80), the power ratio of second-order harmonic distortion and the fundamental signal (2HD/C) can be easily obtained. The CSO distortions can be obtained from the second-order harmonic using the product count concept introduced in Section 2.6, that is, by adding 10log N_{cso} to 2HD/C, and the resultant CSO can be expressed by

$$CSO = 20 \cdot \log \left[\sqrt{N_{cso}} \frac{\left[\frac{dG(P_o, \lambda)}{d\lambda} \right]_{\lambda_o} \Delta\lambda}{2G(P_o, \lambda_o)} \right]. \tag{9.81}$$

It should be noted that the final system CSOs are due to the combination of laser transmitter nonlinearity, EDFA gain tilt, and optical fiber dispersion. The improvement or degradation of final CSOs due to EDFA gain tilt depends on the phase relationship among those CSOs which were generated from different mechanisms [48,49].

The possible CSO degradation due to EDFA gain tilt and laser FM chirping can be avoided if one resorts to a lightwave transmitter based on external modulators such as $LiNbO_3$ Mach–Zehnder amplitude modulators. Other possible solutions include adding an electrical predistorter to compensate the CSO, or use a very low-chirp (e.g., $\eta_{FM} < 60$ MHz/mA) laser diode under direct modulation [52].

9.4 Optical Fiber Dispersion and Laser Chirp-Induced Nonlinear Distortions

As was explained in Section 4.8.3, when a laser diode is directly modulated by a single RF tone, its emitted optical field can be expressed as (similar to Eq. (9.68))

$$\begin{aligned} \mathrm{E}(t) &= \left[1 + m\cos\omega_m t\right]^{1/2} \exp\left[\Omega_o t + \beta \sin\omega_m t\right] \\ &= \sum_{n=0}^{\infty} a_n \exp\left[j(\Omega_o \pm n\omega_m)t\right], \end{aligned} \tag{9.82}$$

where Ω_0 is the laser center frequency, ω_m is the modulating signal frequency, m is the intensity modulation index, β is the frequency modulation index, and a_n are the Fourier series expansion coefficients. If there is no transmission dispersion, all sideband signals arrive at the receiver simultaneously; therefore, the original signal can be recovered upon square-law detection at the optical receiver. However, when there are chromatic fiber dispersions, each sideband signal will experience a different amount of phase shift, given by

$$\begin{aligned} \theta_n &= \beta \cdot z \\ &= \left[\beta(\Omega_o) + \beta_1(\Omega_o)(\Omega - \Omega_o) + \frac{\beta_2(\Omega_o)}{2}(\Omega - \Omega_o)^2 + \frac{\beta_3(\Omega_o)}{6}(\Omega - \Omega_o)^3 + \ldots \right] \cdot z, \end{aligned} \tag{9.83}$$

where $\Omega = \Omega_o \pm n\omega_m$, β is the propagation constant, the dotted βs are derivatives with respect to ω, and z is the distance traveled. The first term in Eq. (9.83) is a constant phase, the second term is a constant delay (because $d\theta/d\omega$ equals $\dot{\beta}(\Omega_o)z$ and is a constant), and the fourth- and higher-order terms dominate system performance at or very near the zero dispersion point, but can be neglected when the operating wavelength is not near the zero dispersion point. In the latter case, the only term that dominates is the third term. At a frequency of $\Omega_o \pm n\omega_m$, this term is given by [53,54]

$$\theta_n \approx \frac{\beta_2(\Omega_o)}{2}\left(n\omega_m\right)^2 z\,, \tag{9.84}$$

which is a phase proportional to the square of the harmonics of ω_m. We also know that chromatic dispersion is given by

$$D = \frac{1}{Z}\frac{d\tau}{d\lambda} = \frac{1}{Z}\frac{d}{d\lambda}\left(Z\frac{d\beta}{d\Omega}\right) \approx \frac{d}{d\lambda}\left(\beta_2(\Omega_o)(\Omega-\Omega_o)\right) = \left(-\frac{\Omega}{\lambda}\right)\beta_2(\Omega_o). \tag{9.85}$$

From Eqs. (9.84) and (9.85), we have

$$\theta_n \approx -\frac{Z\lambda_o}{2\Omega_o} D\cdot\left(n\omega_m\right)^2 \tag{9.86}$$

which implies that the fundamental subcarrier at ω_m and its own harmonics $n\omega_m$ ($n \geq 2$) generated from FM chirping arrive at the photodetector with different phases θ_n, that is, the optical field before the photodetector is given by

$$E'(t) = \sum_{n=0}^{\infty} a_n \exp\left[j\left(\Omega_o \pm n\omega_m\right)t + j\theta_n\right]. \tag{9.87}$$

Obviously, $E'(t)$ no longer resembles $E(t)$ in Eq. (9.82). After square-law detection, $|E'(t)|^2$ will consist of harmonic distortions, as was calculated by Meslener [53].

Instead of the frequency-domain analysis we have discussed so far, the nonlinear distortions due to optical fiber dispersions and laser frequency chirping can also be obtained from a time-domain analysis [55,56], which is less tedious than frequency-domain analysis. The derivation is as follows. When a laser is directly modulated by two tones, the launched laser power can be represented as

$$\boldsymbol{P_{in}(t) = P_0(1 + m\sum_{i=1}^{2}\sin(\omega_i t + \phi_i))}, \tag{9.88}$$

where P_0 is the average transmitted optical power, m is OMI per channel, and ω_i and ϕ_i ($i = 1,2$) are the modulation frequencies and phases of the two tones, respectively. The output optical power, after fiber transmission, can be expressed as

$$P_{out}(t) = \alpha \cdot P_{in}(t - \Delta\tau(I(t)))\cdot\gamma(t)\,, \tag{9.89}$$

where α is the optical fiber attenuation, $\Delta\tau(I(t))$ is the driving-current-dependent group delay variation given by

$$\Delta\tau(I(t)) = \Delta\lambda(t)\cdot D\cdot L = I(t)\frac{d\lambda}{d\nu}\frac{d\nu}{dI}\cdot D\cdot L = -\frac{\lambda^2}{c}\cdot D\cdot L\cdot\frac{d\nu}{dI}I(t), \quad (9.90)$$

where

$$I(t) = A[\sin(\omega_1 t + \phi_1 + \theta) + \sin(\omega_2 t + \phi_2 + \theta)], \quad (9.91)$$

and $\gamma(t)$ in Eq. (9.89) is the "deformation factor" (due to power level compression or expansion after fiber transmission) given by

$$\gamma(t) = \frac{d(t-\Delta\tau)}{dt}. \quad (9.92)$$

In Eq. (9.90), $d\nu/dI$ represents the frequency chirp or FM efficiency of the laser. Now by letting

$$P_{in}(t-\Delta\tau) = P_{in}(t) - \Delta\tau\frac{dP_{in}}{dt} \quad (9.93)$$

and substituting Eqs. (9.88) and (9.90)–(9.93) into Eq. (9.89), we can obtain the second harmonic distortions and second-order intermodulation products at the corresponding distortion frequencies as

$$\begin{aligned} 2HD/C &= \frac{1}{4}(2\omega_1)^2 z_\eta^2 \\ IM_2/C &= (\omega_1+\omega_2)^2 z_\eta^2, \end{aligned} \quad (9.94)$$

where $z_\eta = A(\lambda^2/c)DL(d\nu/dI)$. Therefore, we can see that the second-order nonlinear distortions are proportional to the square product of distortion frequency, modulation current, frequency chirp, and fiber dispersion. For the case of multichannel transmission, CSO distortions can be obtained by simply multiplying IM_2/C by the CSO product count, N_{CSO}, in the distortion frequency, that is,

$$CSO = N_{cso}\cdot(\omega_1+\omega_2)^2\cdot z_\eta^2. \quad (9.95)$$

Following a similar approach, one can also obtain the third-order nonlinear distortions [56]. However, it should be noted that for multichannel CATV transmissions over a relatively short distance, second-order nonlinear distortions dominate. Readers should think carefully why product count can be used here while cannot be used in cases such as Eqs. (9.75) and (9.76).

Generally speaking, if one uses a directly-modulated 1.55-μm DFB laser and standard single-mode fibers for transmission (with D = 17 ps/nm/km), the transmission distance is limited to only a few kilometers [57,58] if the CSOs at all channels are required to be less than the current CATV industry recommendation of about

−60 dB, as illustrated in Fig. 9.33. Therefore, to achieve a long-distance transmission at 1.55 μm, external-modulator-based optical transmitters should be used [59,60]. However, it is still possible to use chirped-fiber gratings [61,62] or dispersion compensation fibers [63] to compensate the optical fiber dispersion and consequently decrease the nonlinear distortions due to frequency chirping and fiber dispersion. Experimental demonstration was carried out for the case of using dispersion-compensation fibers, but not yet for chirped fiber gratings.

9.5 WDM Transmissions of SCM Video Channels

Wavelength-division muliplexing (WDM) has been used (1) to alleviate the channel loading on a single optical transmitter [64], (2) to provide a mixture of analog broadcast video and interactive digital services (e.g., targeted internet or telephony services [1b], broadband ISDN [65], or HDTV [66]), or (3) to greatly increase the information capacity of a transmission system [67,68].

An example of (1) by using a 1.3/1.55-μm WDM analog lightwave system which transports 112 channels (55.25–751.25 MHz) over 10 km of conventionally single-mode fiber is shown in Fig. 9.34 [64]. The lower 60 channels are transmitted by externally modulating a 1.55-μm DFB laser diode, while the upper 52 channels are transmitted by directly modulating a 1.3-μm DFB laser. The upper 52 channels are kept within one octave (442.25–751.25 MHz) so that a typical DFB laser can easily meet all the CNR/CTB requirements, because a DFB laser diode usually has very good third-order linearity performance (see Chapter 4). The directly modulated DFB laser is not used for the 1.55-μm system to avoid system penalties due to

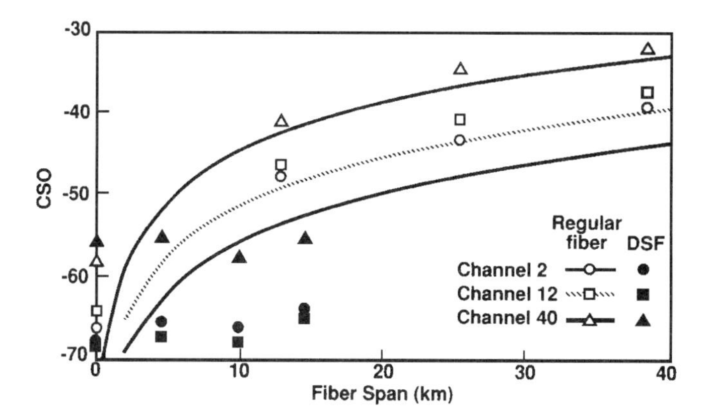

Figure 9.33: CSO dependence on conventional and dispersion-shifted fiber span for a 42-AM-channel link. The laser chirp is 720 MHz/mA. After Ref. [57] (© 1991 IEEE).

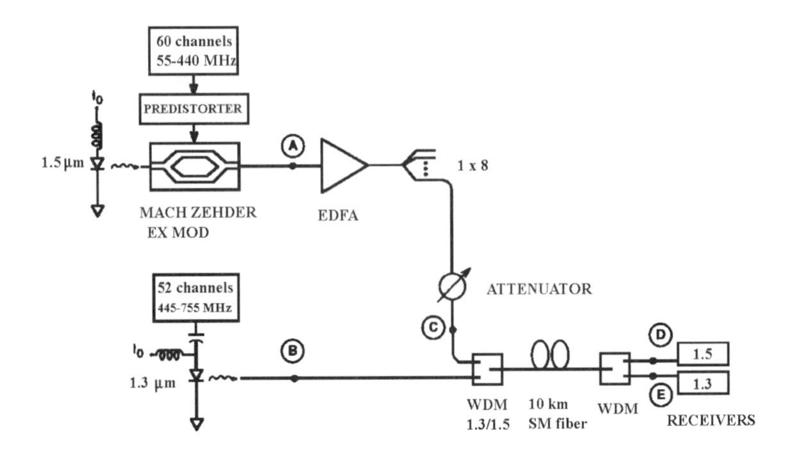

Figure 9.34: A 1.3/1.55-μm WDM analog lightwave system which transports 112 channels (55.25–751.25 MHz) over 10 km of conventionally single-mode fiber. After Ref. [64] (© 1992 IEEE).

(1) frequency chirping combined with EDFA gain tilt and (2) frequency chirping combined with fiber dispersion, as were explained in previous sections. An important point to note in this experiment is that there are two separate optical receivers to detect the two subbands. There are two reasons why a single optical receiver cannot be used: (1) The NLDs generated from one subband can interfere with another subband, and (2) if the maximum received optical power of the receiver must be maintained for satisfactory linearity performance, say at 0 dBm, then only -3 dBm received power can be allocated to each subband if a single receiver were used, which in turn could reduce the CNR per channel in each subband by 3 dB (if shot noise predominates) to 6 dB (if thermal noise predominates).

When the WDM transmission mixes broadcast analog AM-VSB video signals with digital services [65,66], several important system design concepts must be emphasized. The first is that the receiver sensitivities differ significantly for the two types of signals. For example, the minimum received optical power required are 0, -33, and -27 dBm for a practical well-designed analog CATV receiver, a 2.488 Gb/s receiver, and a 10 Gb/s receiver, respectively. Therefore, with a WDM demultiplexer (at the receiver site) which has imperfect isolation, the large optical power of the analog AM-VSB signals may interfere with the weak optical power of the digital signal; the small amount of interference from the digital signal could also interfere with the highly vulnerable AM-VSB signals. From the experimental results of Phillips *et al.* [64] and Muys *et al.* [65], it was observed that the demultiplexing WDM must keep the unwanted digital signal power on the analog receiver below about -40 to -45 dBm to maintain the analog signal CNR above ~47 dB. The actual value strongly depends on the spectral overlap of the two signals and is bit-rate dependent. On the other hand, the unwanted optical power of the SCM signal on the digital receiver should be about 3 to 10 dB below the digital receiver sensitivity and is also dependent on the spectral overlap between the two signals.

The second WDM system design issue is when there are in-line optical amplifiers present in the system. The minimum required input optical power to the amplifier ($P_{in,min}$) is again quite different among the digital and analog wavelengths. For example, $P_{in,min}$ for analog AM-VSB signals usually is greater than about 0 dBm [59,60], whereas for digital baseband signals such as 622 Mb/s, as long as $P_{in,min}$ is greater than about −40 dBm, the receiver power penalty can be kept below 1 dB [69]. This $P_{in,min}$ has to be increased to about −25 dBm for 10 Gb/s, but it is still much lower than the 0 dBm required by AM-VSB signals. This large difference in $P_{in,min}$ implies that AM-VSB signals are very "incompatible" with digital signals while using in-line optical amplifiers, and may cause significant waste of digital signal power budget or mutual deterioration of both signals. The latter is because when P_{in} of the digital signal wavelength is too large, it may reduce the gain of the AM-VSB signal wavelength [68]. From these discussions, it seems that the right time to wavelength-division-multiplex SCM and baseband digital signals in systems with in-line optical amplifiers is when all AM-VSB signals are replaced by, say, 64-QAM signals whose CNR requirement is only ~30 dB (without forward error correction). According to Eq. (6.38) which is repeated here for convenience,

$$\text{CNR}^{-1}(\text{due to sig-sp beat noise only}) = \frac{8n_{sp}h\nu}{m^2 c_1 P_{in}}(1-\frac{1}{G}) , \tag{9.96}$$

where we assumed signal-spontaneous beat noise predominates, $P_{in,min}$ for QAM signals can be as low as 0− (51 − 30) = −21 dBm if we assume the original AM-VSB CNR requirement is 51 dB and the 64-QAM CNR requirement is 30 dB. In addition, it is possible that the OMI per 64-QAM channel can be larger than that for AM-VSB channels for a laser operated in the strong-clipping region (Section 9.2.5), and therefore $P_{in,min}$ can be further reduced. Consequently, we can see that 64-QAM signals are more compatible with digital baseband transmission using WDM and optical amplifiers.

Another application of WDM transmission is to use more wavelengths to transport more SCM AM-VSB channels. However, when the launched optical power per wavelength is too large, significant crosstalks between wavelengths may occur [70]. For example, when two wavelengths in the 1.3-μm region with 42 AM-VSB channels/wavelength were combined and amplified by an PDFFA optical amplifier, severe crosstalk from shorter to longer wavelength (and vice versa) can be observed because of the high launched optical power per wavelength (as high as 14.4 dBm per wavelength in Ref. [70]). This is illustrated in Fig. 9.35, where we can see that when the AM-VSB channel at 211.25 MHZ is turned off or on in $\lambda 1$ (1292 nm), the crosstalk at 211.25 MHz in λ_2 (1307 nm) can be as low as ≤ -69.3 dBc or as high as −18.3 dBc, respectively. This crosstalk is mainly caused by stimulated Raman scattering and is discussed later in the next section.

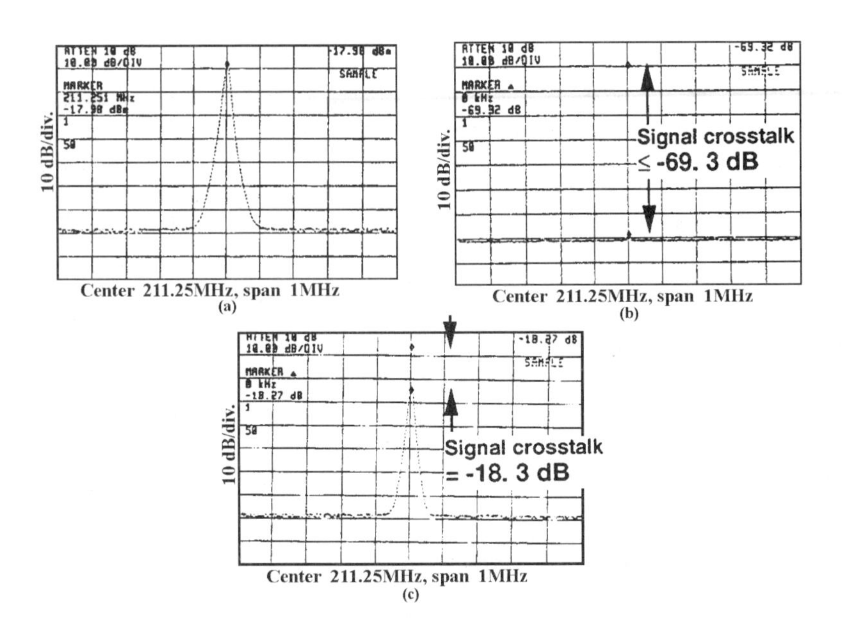

Figure 9.35: Demonstration of crosstalk between WDM-SCM channels. Electrical spectra of signal and crosstalk for an AM-VSB channel at 211.25 MHz of λ_2 (1307 nm). The transmission fiber is 20 km of conventional 1.3-μm SMF. Fiber input light power is 14.4 dBm per wavelength. (a) Signal, (b) crosstalk without 211.25 MHz AM-VSB channel in λ_1 (1292 nm), (c) crosstalk with 211.25 MHz AM-VSB signal present in λ_1. After Ref. [70].

9.6 Effects of High Input Optical Power in an Optical Link

For the purpose of achieving longer transmission distance and/or obtaining more split power branches, the development of high-power erbium-doped fiber amplifiers is an ongoing trend (see Section 6.1.8). Today, the amplified optical output power can be as high as +27 dBm for commercial products. However, the maximum injected power in an HFC system with moderate or long optical fiber span is limited by optical fiber nonlinearities. The important optical nonlinear phenomena in fibers fall into two main categories. One is stimulated scattering processes such as stimulated Brillouin scattering (SBS) [71–75] and stimulated Raman scattering (SRS) [76]. The other is the self-phase modulation (SPM) [58,77,78]. We shall start with the discussion of SBS effects.

9.6.1 Effects of Stimulated Brillouin Scattering

SBS is typically the first optical nonlinearity encountered is lightwave systems using external modulators and single-mode optical fibers. The origin of SBS and its fundamental limitation on the maximum power that can be launched into a long-link (>10 km) optical fiber were explained in Section 7.6. Basically, in silica fibers, SBS is a narrowband process with a bandwidth of about 30 to 200 MHz [79] and therefore can affect the maximum launched power in external modulation systems where most of the optical power is concentrated near the optical carrier.

As mentioned in Section 7.6, the SBS threshold power that can be launched into single-mode fibers is about 6 and 9 dBm at 1550 nm and 1310 nm, respectively. However, the threshold power, which is defined as the input power at which the backscattered power equals the pump power at the fiber input [see Eqs. (7.56) and (7.58)] [80], is not experimentally observable, because the backscattered power is always less than the injected power as a result of pump power depletion. Therefore, SBS threshold power usually depends on the specific system requirements that must be met. For example, SBS can be avoided in digital systems if the total fiber launched power is kept below about 10 dBm for bit rates above several hundred megabits for any modulation format (ASK, FSK, PSK) [81]. In SCM systems, however, the SBS threshold can be defined as the fiber injected power at which the received RIN noise is increased by a certain value (e.g., 0.2 dB in Ref. [73]), or the fiber injected power at which the received worst-case CSO value is increased by a certain value.

We will see in this section that with the help of phase dithering, SBS can be avoided in SCM lightwave systems if the total fiber launched power is kept below about 13–16 dBm. However, in the future, even higher launched optical power may be desirable for longer repeaterless transmission distance.

In this section, we will first examine the effects of SBS on SCM CNR and linearity performance, and then examine methods to increase SBS threshold power for SCM systems.

9.6.1.1 Effects of SBS on System CNR Performance The effects of SBS on SCM system CNR performance can be examined using experimental results obtained from a system with a 1550-nm light source and a $LiNb0_3$ Mach–Zehnder modulator [71]. In this system, 42 AM-VSB CATV channels with 4% OMI per channel were applied to the modulator, and the fiber injected power (for a 13 km dispersion-shifted fiber) at which the SBS scattered power starts to pick up (vaguely defined as the SBS threshold power), as shown in Fig. 9.36a [71], was about 6 dBm and is almost identical to that of the unmodulated case. It was observed that, as soon as the fiber injected power is increased beyond 6 dBm, a rapid degradation in CNR occurs, as illustrated in Fig. 9.36a. This is mainly due to the increased SBS-induced noise spectral density at the output of the receiver, as shown in Fig. 9.36b.

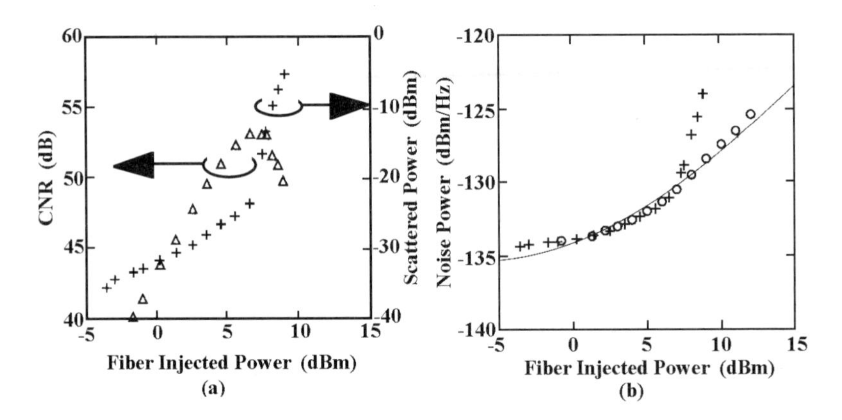

Figure 9.36: Demonstration of the SBS effect on CNR and received noise power. (a) CNR and backscattered power versus fiber injected power. (b) Transmitted noise power versus injected power with and without 13-km fiber. After Ref. [71] (© 1992 IEEE).

9.6.1.2 Effects of SBS on System Linearity Performance A simplified analysis of the effects of SBS on the linearity of an externally modulated CATV system can again begin with the modulated electrical field given as

$$\mathrm{E}(t) = \mathrm{E}_A(t)\cos(\Omega_o t) \tag{9.97}$$

where

$$\mathrm{E}_A(t) = \left(P_o(1 + \sum_{i=1}^{N} m_i \cos(\omega_i t + \theta_i)) \right)^{1/2} . \tag{9.98}$$

P_o is the average power of the modulated optical signal, and m_i, ω_i, and θ_i denote the OMI, RF frequencies, and initial phase of the ith carrier channel. It was observed that for a typical total rms modulation depth μ less than about 25%, >90% of the optical power is carried in the optical carrier $\cos(\Omega_0 t)$ [75,82]. Furthermore, we learned earlier (Section 7.6) that for silica fiber around 1550 nm the Brillouin bandwidth is about 35 MHz, which is less than the lowest CATV NTSC signal frequency of 54 MHz. Therefore, when the launched power is above the SBS threshold and SBS occurs, almost all the reflected light is the DC component. In this case, the envelope of the received electrical field can be modified from Eq. (9.98) to become (assuming all $m_i = m$)

$$\mathrm{E}_{SBS}(t) = \sqrt{P_o}\left(1 + m\sum_{i=1}^{N} \cos(\omega_i t + \theta_i) \right)^{1/2} - \eta_s \sqrt{P_o} , \tag{9.99}$$

where η is the backscattered coefficient of the DC electrical field, and the received photocurrent is proportional to

$$E_{SBS}^2(t) = P_o\left\{ 1 + m\sum_{i=1}^{N}\cos(\omega_i t + \theta_i) + \eta_s^2 - 2\eta_s\left[1 + m\sum_{i=1}^{N}\cos(\omega_i t + \theta_i)\right]^{1/2} \right\}$$

$$= P_o\left\{ \begin{array}{l} 1 + m\sum_{i=1}^{N}\cos(\omega_i t + \theta_i) + \eta_s^2 \\ -2\eta_s\left\{ \begin{array}{l} 1 + \frac{m}{2}\left[\sum_{i=1}^{N}\cos(\omega_i t + \theta_i)\right] - \frac{m^2}{8}\left[\sum_{i=1}^{N}\cos(\omega_i t + \theta_i)\right]^2 \\ + \frac{m^3}{16}\left[\sum_{i=1}^{N}\cos(\omega_i t + \theta_i)\right]^3 + \ldots \end{array} \right\} \end{array} \right\}. \quad (9.100)$$

From this equation, we can obtain the following results:

The additional loss due to SBS:

$$L = -10\log(1 - \eta_s)^2 = -10\log\sigma^2 . \quad (9.101)$$

The composite second-order (CSO):

$$\text{CSO} = 10\log\left\{ N_{CSO}\left[\frac{(1-\sigma)m}{4\sigma}\right]^2 \right\} . \quad (9.102)$$

The composite triple beat:

$$\text{CSO} = 10\log\left\{ N_{CTB}\left[\frac{3(1-\sigma)m^2}{16\sigma}\right]^2 \right\}, \quad (9.103)$$

where we have defined $\sigma^2 = (1 - \eta_s)^2$ as the fractional transmission coefficient, and N_{CSO} and N_{CTB} are the CSO and CTB product counts, respectively. From the measured additional loss L due to SBS, one can then estimate the corresponding CSO and CTB. Measured and calculated CSOs caused by SBS in 25 and 50 km of dispersion-shifted (DS) and conventional single-mode fiber are shown in Fig. 9.37 [75]. It can be seen that to maintain a CSO level below about −62 dBc, the backscattered power ratio must be below −20 dB, which corresponds to a launched power of ~14 dBm in that experiment.

9.6.1.3 Techniques to Increase SBS Threshold Power Two techniques have been practically used to increase the SBS threshold. The first is single-tone (whose phase

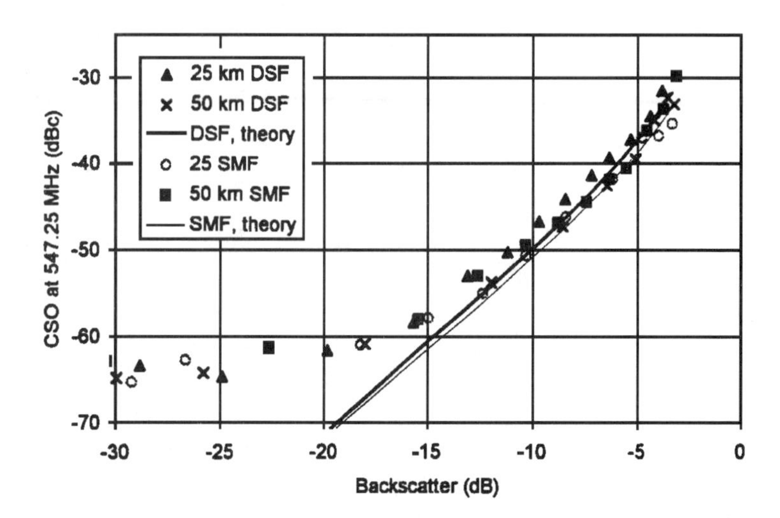

Figure 9.37: Measured and predicted CSO caused by SBS in 25 and 50 km of DS and SM fiber. After Ref. [75].

frequency is three times higher than the maximum signal frequency) modulation, and the second is dithering the optical frequency of the transmitter DFB laser [73].

When using the first approach, the SBS threshold increase in the case of single-tone phase modulation is given by [83] (see Problem 9.5)

$$\Delta P_{th,SBS} \approx -10\log\left[\max_{k\in\{0,1,2,\ldots\}}\left\{J_k^2(\beta)\right\}\right]. \tag{9.104}$$

Figure 9.38a shows the calculated and experimental data. It can be seen that, as long as the driving power of the phase modulator is high enough to reach a large β, the SBS threshold power can be increased by up to about 10 dB, that is, a high fiber injected power of 6 + 10 = 16 dBm can typically be obtained from 1550-nm commercial transmitters.

When using the second approach, the SBS threshold power increase can be obtained from Eq. (7.56) (with the assumption of Lorentzian linewidth):

$$\Delta P_{th,SBS} \approx 10\log\left(1+\frac{\Delta\nu_L}{\Delta\nu_B}\right), \tag{9.105}$$

where $\Delta\nu_L$ is the broadened effective laser linewidth, and $\Delta\nu_B$ is the SBS gain linewidth. Figure 9.38b shows the experimental data and theory for SBS threshold as a function of the effective laser linewidth $\Delta\nu_L$. Although a large SBS threshold

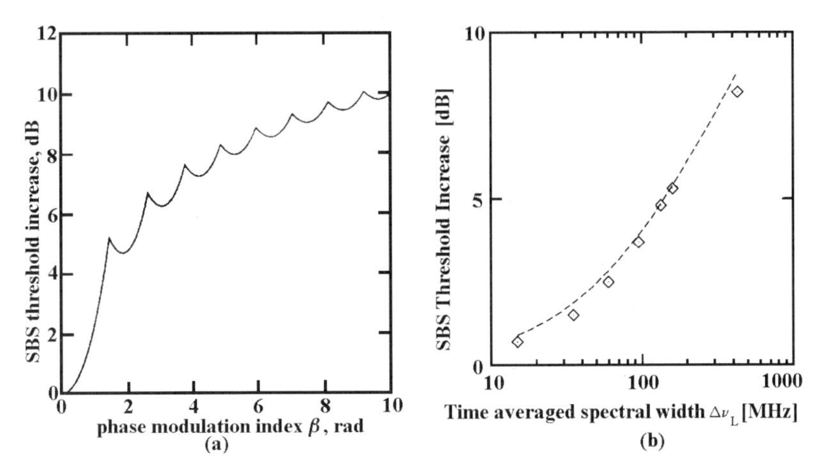

Figure 9.38: (a) SBS threshold increase versus phase modulation index β. The theoretical line is based on Eq. (9.104) (after Ref. [83]). (b) SBS threshold increase versus time averaged optical laser spectral width $\Delta\nu_L$ which was measured by a self-homodyne setup. The measured values are denoted by small diamonds. The theoretical line is plotted using a *fitted* Brillouin linewidth $\Delta\nu_B$ of 65 MHz. After Ref. [73] (© 1994 IEEE).

power increase can be obtained using this approach, there are two practical limitations.

The first is that the package parasitics of currently available commercial high-power DFB lasers limit the modulating signals to below about 3 MHZ (for laser with DIP package) or ~1.2 GHz (for laser with butterfly package). The second is that the FM components must not affect the normal CATV signals between 54 and 750 MHZ. Therefore, a broadened laser linewidth beyond 50 MHZ may affect the CNR performance of normal CATV channels.

9.6.2 Effects of Stimulated Raman Scattering on WDM-SCM Channels

The crosstalk between two high-power WDM channels which we mentioned in Fig. 9.35 is mainly due to SRS and will be briefly explained in this section.

Following the derivations by Wang *et al.* [76], we consider two WDM wavelengths λ, where each λ contains the same *M* subcarrier channels. The optical power of each λ at the fiber input is given by

$$P_p(t)\,|_{z=0} = P_o\left(1 + \sum_{i=1}^{M} m\cos(\omega_i t + \theta_{pi})\right) \tag{9.106}$$

and

$$P_s(t)\,|_{z=0} = P_o\left(1 + \sum_{i=1}^{M} m\cos(\omega_i t + \theta_{si})\right) \tag{9.107}$$

respectively. In Eqs. (9.106) and (9.107), the subscripts p and s represent the pump (shorter λ) and signal (longer λ) wavelengths, respectively. The same average optical power P_o and subcarrier frequencies ω_i are assumed for both λs. It was further assumed that while all subcarriers have the same modulation index m, the phase θ_{pi} or θ_{si} of any given subcarrier i is totally random.

For a given Raman efficiency g_{ps} between pump λ and signal λ, the SRS interaction is proportional to $g_{ps}P_pP_s$ [84]. By substituting for P_p and P_s from Eqs. (9.106) and (9.107), we see that the total crosstalk for subcarrier i in the *pump* channel is composed of three terms:

1. $mg_{ps}P_o^2\cos(\omega_i t + \theta_{pi})$: This term is due to SRS interaction between subcarrier i in the pump channel and the optical carrier in the signal channel. It is just the optical power loss, $mg_{ps}P_o^2$, of subcarrier i.
2. $mg_{ps}P_o^2\cos(\omega_i t + \theta_{si})$: This term is attributed to SRS interaction between the optical carrier in the pump channel and the subcarrier channel i in the signal channel.
3. $N_{CSO}m^2g_{ps}P_o^2$: This term is due to SRS interaction between subcarriers in the pump and signal channels.

Note that the SRS interactions between optical carriers only cause DC optical power loss (or gain) and do not contribute to crosstalk at subcarrier frequencies.

For subcarrier i in the *signal* channel, the crosstalk is the same except that the first term contributes optical power gain, $mg_{ps}P_o^2$, instead of loss. Thus, the shorter wavelength channel will suffer more in terms of power loss from SRS crosstalk than the longer wavelength channel.

The analytical results of crosstalk can be determined by solving the following equations which govern the SRS interaction in optical fiber [76]:

$$\frac{\partial P_s}{\partial z} + \frac{1}{v_{gs}}\frac{\partial P_s}{\partial t} = (g_{ps}P_p - \alpha)P_s \tag{9.108a}$$

$$\frac{\partial P_p}{\partial z} + \frac{1}{v_{gp}}\frac{\partial P_p}{\partial t} = (-g_{ps}P_s - \alpha)P_p, \tag{9.108b}$$

where v_{gs} and v_{gp} are the group velocities of the transmitted light at signal and pump wavelengths, respectively, and we have assumed that both channels have the same fiber loss coefficient α. A closed-form solution can be obtained from Eq. (9.108) by further assuming that the SRS interaction is in the linear regime [76], and the resultant electrical crosstalk level suffered by subcarrier i in the signal or pump λ due to SRS is then found to be

$$XT_i^{sp} = \frac{(2g_{ps}P_o)^2\left(\left(\frac{\alpha L_{eff}}{2}\right)^2 + e^{-\alpha L}\sin^2\frac{d_{ps}\omega_i L}{2}\right)}{\left(\alpha^2 + (d_{ps}\omega_i)^2\right)\left(1 \pm 2g_{ps}P_o L_{eff}\right)^2}, \tag{9.109}$$

where $d_{ps} = \left|\frac{1}{v_{gp}} - \frac{1}{v_{gs}}\right|$ is the group velocity mismatch between the pump and signal λ's, L is the fiber length, and $L_{eff} = (1 - e^{-\alpha L})/\alpha$ is the effective fiber length. When the high-power λs are in the 1.3-μm window, d_{ps} approaches zero, and Fig. 9.39 shows the calculated crosstalk results for two WDM channels with 20 km of standard single-mode fiber. An effective fiber core area of 50 μm^2 and an optical fiber loss of 0.4 dB/km were assumed. It can be seen that for two WDM λs (14.4 dBm per λ) with a channel spacing of 15 nm, as was used in Fig. 9.35, the calculated crosstalk level is about −30 dBc, which is still about 11 dB lower than what was measured. But the analytical results were later confirmed in a 1.55-μm two-wavelength AM-SCM system experiment with input optical power per channel < 11 dBm [85].

Note that for 1.3-μm window operation, the SRS crosstalk is independent of subcarrier frequency [as can be observed from Eq. (9.109) with $d_{ps} = 0$], but for 1.55-μm window operation with high dispersion, the SRS crosstalk level will be lower than in the zero-dispersion case, and will be lower for high subcarrier frequencies because fiber dispersion is significant and can cause walk-off between subcarriers at different λ's [85].

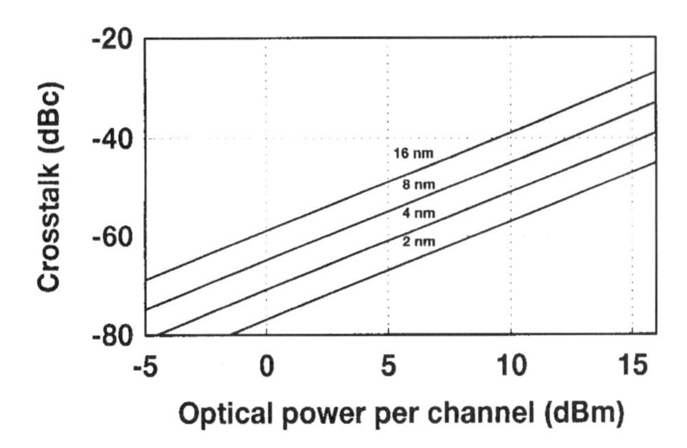

Figure 9.39: SRS crosstalk level (electrical) versus input optical power per WDM channel for 1.3-μm WDM SCM systems with two WDM channels and 20 km of standard single-mode fiber. The crosstalk dependencies for channel spacings of 2, 4, 8, and 16 nm, respectively, are shown. After Ref. [76] (© 1995 IEEE).

9.6.3 Effects of Self-Phase Modulation

Power-dependent self-phase modulation in intensity-modulated systems causes frequency chirp, which in combination with fiber dispersion generates nonlinear distortions. It has been theoretically [58] and experimentally [77] shown that in external modulation systems, this effect can cause intolerable amounts of composite second-order distortion even when the transmission distance is only a few tens of kilometers. A closed-form solution of the SPM-induced CSO and CTB can be obtained from the following wave-envelope equation in a lossy, dispersive, and nonlinear medium:

$$\frac{\partial \boldsymbol{A}}{\partial z}+\beta_1\frac{\partial \boldsymbol{A}}{\partial t}-\frac{\boldsymbol{i}}{2}\beta_2\frac{\partial^2 \boldsymbol{A}}{\partial t^2}+\alpha \boldsymbol{A}=-\boldsymbol{i}\boldsymbol{k}\boldsymbol{n}_2|\boldsymbol{A}|^2\boldsymbol{A}\,. \tag{9.110}$$

Here, $k = 2\pi/\lambda$, α is the intensity loss, $\beta_1 = d\beta/d\omega|\omega_0$ and $\beta_2 = d^2\beta/d\omega^2|\omega_0$ (β is the propagation constant); n_2 ($\approx 3 \times 10^{-20}$ m^2/W) is the power-dependent nonlinear refractive index $n = n_0 + n_2|A|^2$, where n_0 is the refractive index at low optical power and $A(z,t)$ represents the electric field envelope

$$A=\sqrt{x(z,t)}\exp\left(iy(z,t)\right), \tag{9.111}$$

where t and z are time and distance, respectively, and $x(z,t)$ and $y(z,t)$ are the intensity and phase modulations, respectively. At the input end, they are given by

$$\begin{aligned}
x(0,t) &= 1+m\sum_{i=1}^{N}\cos(\omega_i t+\phi_i)\\
y(0,t) &= \gamma m\int\sum_{i=1}^{N}\cos(\omega_i t'+\phi_i)dt' \quad \text{(direct modulation)}\\
y(0,t) &= \beta\sum_{j=1}^{M}\cos(\omega_j t+\theta_j) \quad \text{(external modulation)},
\end{aligned} \tag{9.112}$$

where $\gamma = \eta_{FM}(I_b - I_{th})$ is the laser peak chirped frequency and β is the phase modulation index, respectively. Note that in the external modulation case, the phase modulation may be due to some dithering signals that are different from the original modulating signals [i.e., $\omega_j \neq \omega_i$ in Eq. (9.112)]. By substituting Eqs. (9.111) and (9.112) into the wave equation (9.110) and requiring that $x(z,t)$ and $y(z,t)$ be real, and through a tedious manipulation [58], we can obtain the CSO (CTB is much smaller than CSO in the distance that we calculated) limited by SPM only:

$$\boldsymbol{CSO}=\boldsymbol{N}_{\boldsymbol{CSO}}\cdot\left(\frac{1}{2}\frac{\boldsymbol{P}_{\boldsymbol{in}}}{\boldsymbol{A}_{\boldsymbol{eff}}}\frac{\alpha \boldsymbol{l}-1+\boldsymbol{e}^{-\alpha l}}{\alpha^2}\beta_2\boldsymbol{k}\boldsymbol{n}_2\boldsymbol{m}\boldsymbol{f}_d^2\right)^2, \tag{9.113}$$

where N_{CSO} is the CSO product count, P_{in} is the optical power in watts entering the fiber section, A_{eff} is the effective fiber core area, α is the fiber attenuation, l is the transmission distance, and f_d is the distortion frequency the CSO falls into. Calculated results assuming CSO $= -60$ dBc using this equation are shown in Fig. 9.40 with the parameter values listed in the figure caption. It is clear that if we want to avoid the CSO due to the strong input power-induced SPM, while maximizing the repeaterless transmission distance (~66 km or ~77km, depending on n_2 and m), the optimum power launched into optical fiber is less than 13.5 ~ 15.5dBm.

Problems

9.1 Assume a downstream optically-linearized external modulator-based transmitter has an RIN $= -160$ dB/Hz and is modulated by multiple M-QAM signals. Plot a diagram similar to Fig. 9.22. Use the transfer function given in Eq. (7.55) with $\gamma = 63°$, $\phi_1 = 0$, $\phi_2 = 0$, and c $= 0.5$.

9.2 Based on Eq. (9.95), check the solid curves of CSO versus conventional single-mode fiber transmission distance in Fig. 9.33. (Note that the calculated results in Fig. 9.33 were obtained from a perturbation analysis [57]).

9.3 Use Eqs. (9.101) and (9.102) to confirm the results given in Fig. 9.37.

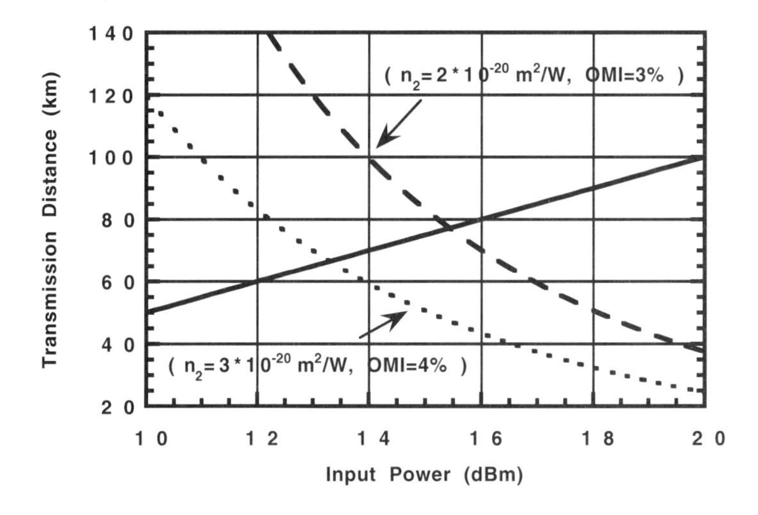

Figure 9.40: Transmission distance limited by SPM (dotted and dashed lines) and optical attenuation loss (solid line), as a function of the input optical power into the optical fiber link. Parameters used are $m = 0.03$ or 0.04/ch., AM-channel number = 80, $A_{eff} = 80\ \mu m^2$, $n_2 = 2$ or 3×10^{-20} m²/W, $\lambda_0 = 1.55\ \mu m$, D =17 ps/nm/km, α =0.2 dB/km. The received optical power is assumed to be 1 mW.

9.4 Develop a computer program using the split-step Fourier method to solve Eq. (9.110) and see if you can obtain a similar result given by the analytical results shown by the solid curve in Fig. 9.40.

9.5 Prove that Eq. (9.104) is valid.

References

[1a] J. Holobinko, "Scalability: the newest HFC network weapon," Commun. Engineering Design, pp.30–35. January 1996.

[1b] D. A. Atlas and J. J. Kenny, "Multiwavelength analog video transport network," Tech. Papers, pp.9–13, 47th Annual NCTA Convention and International Exposition, May 1998.

[2] A. A. M. Saleh, "Fundamental limit on the number of channels in subcarrier-multiplexed lightwave CATV system," Electron. Lett., vol.25, pp.776–777, 1989.

[3] M. Maeda and M. Yamamoto, "FM-FDM optical CATV transmission experiment and system design for MUSE HDTV signals," IEEE J. Selected Areas in Commun., vol.8, pp.1257–1267, September 1990.

[4] P. Y. Chiang and W. I. Way, "Ultimate capacity of a laser diode in transporting multichannel M-QAM signals," J. Lightwave Technol., vol.15, pp.1914–1924, October 1997.

[5] J. Mazo, "Asymptotic distortion spectrum of clipped, dc- biased, Gaussian noise," IEEE Trans. Commun., vol.40, pp.1339–1344, 1992.

[6] N. J. Frigo, M. R. Phillips, and G. E. Bodeep, "Clipping distortion in lightwave CATV systems: Models, simulations, and measurements," J. Lightwave Technol., vol.11, pp.138–145, January 1993.

[7] K. P. Ho and J. M. Kahn, "On models of clipping distortion for lightwave CATV systems," IEEE Photon. Technol. Lett., vol.8, pp.125–126, January 1996.

[8] J. B. Thomas, *An Introduction to Statistical Communication Theory*, Wiley, New York, 1969.

[9] G. C. Wilson, "Optimized predistortion of overmodulated Mach-Zehnder modulators with multicarrier input," IEEE Photon. Technol. Lett., vol.9, pp.1535–1537, 1997.

[10] D. A. Atlas, "On the overmodulation limit in externally modulated lightwave AM-VSB CATV systems," IEEE Photon. Technol. Lett., vol.8, pp.697–699, 1996.

[11] K. Maeda, H. Najata, and K. Fujito, "Analysis of BER of 16 QAM signal in AM/16QAM hybrid optical transmission system," Electron. Lett., vol.29, pp.640–642, April 1993.

[12] S. Lai and J. Conradi, "Theoretical and experimental analysis of clipping-induced impulsive noise in AM-VSB subcarrier multiplexed lightwave systems," J. Lightwave Technol., vol.15, pp.20–30, January 1997.

[13] Q. Pan and R. J. Green, "Amplitude density of infrequent clipping impulse noise and bit- error rate impairment in AM-VSB/M-QAM hybrid lightwave systems," IEEE Trans. Commun., vol.44, pp.1329–1334, October 1996.

[14] X. Lu, G. E. Bodeep, and T. E. Darcie, "Clipping-induced impulse noise and its effect on bit-error-rate performance in AM-VSB/64QAM hybrid lightwave systems," IEEE Photon. Technol. Lett., vol.6, pp.866–868, 1994.

[15] Q. Shi, "Asymptotic clipping noise distribution and its impact on M-ary QAM transmission over optical fiber," IEEE Trans. Commun., vol.43, pp.2077–2984, 1995.

[16] K. Pham, J. Conradi, G. Cormack, B. Thomas, and C. W. Anderson, "Impact of noise and nonlinear distortion due to clipping on the BER performance of a 64-QAM signal in hybrid AM-VSB/QAM optical fiber transmission system," J. Lightwave Technol., vol.13, pp.2197–2201, November 1995.

[17] A. Papoulis, *Probability, Random Variables, and Stochastic Processes*, McGraw-Hill, New York, 1984.

[18a] S. Keller, H. Krimmel, G. Warme, and R. Heidemann, "Hybrid transmission of analog PAL and digital 64-QAM CATV channels," Electron. Lett., vol. 31, pp.904–905, 1995.

[18b] S. Ovadia and C. Lin, "Performance characteristics and applications of hybrid multichannel AM-VSB/M-QAM video lightwave transmission systems," J. Lightwave Technology, vol. 16, pp.1171–1186, July 1998.

[19a] C. Y. Kuo and S. Mukherjee, "Clipping reduction for improvement of analog and digital performance beyond clipping limit in lightwave CATV systems," Conf. Optical Fiber Commun., Postdeadline papers, PD18, February 1996.

[19b] P. Y. Chiang, C. C. Hsiao, and W. I. Way, "Using precoding technique to reduce the BER penalty of an M-QAM channel in hybrid AM-VSB/M-QAM subcarrier multiplexed lightwave systems," IEEE Photon. Technol. Lett., vol.10, pp. 1177–1180, August 1998.

[20] M. R. Phillips and T. E. Darcie, "Numerical simulation of clipping-induced distortion in analog lightwave systems," IEEE Photon. Technol. Lett., vol.3, pp.1153–1156, 1991.

[21] K. B. Chia, M. K. Haldar, H. K. Garg, and F. V. C. Mendis, "Influence of laser parameters on statically calculated clipping distortion in subcarrier multiplexed systems," Electron. Lett., vol.33, pp.142–143, January 1997.

[22] M. K. Haldar, K. B. Chia, and F. V. C. Mendis, "Dynamic considerations in overmodulation of semiconductor lasers," Electron. Lett., vol.32, pp.659–661, March 1996.

[23] T. Anderson and D. Crosby, "The frequency dependence of clipping induced bit-error- rates in subcarrier multiplexed systems," IEEE Photon. Technol. Lett., vol.8, pp.1076–1078, August 1996.

[24] S. Betti, E. Bravi, and M. Giaconi, "Effect of the turn-on delay of a semiconductor laser on clipping impulsive noise," IEEE Photon. Technol. Lett., vol.9, pp.103–105, January 1997.

[25] J. L. Bihan, "Approximate dynamic model for evaluating distortion in a semiconductor laser under overmodulation," IEEE Photon. Technol. Lett., vol.9, pp.303–305, March 1997.

[26] B. Pucel, "Comparison between static and dynamic clipping distortion in semiconductor lasers," IEEE Photon. Technol. Lett., vol.9, pp.1532–1534, November 1997.

[27] B. H. Wang, P. Y. Chiang, M. S. Kao, and W. I. Way, "Large-signal spurious-free dynamic range due to static and dynamic clipping in direct and external modulation systems," to be published in J. Lightwave Technol., October, 1998.

[28] K. P. Ho and J. M. Kahn, "Equalization technique to reduce clipping-induced nonlinear distortion in subcarrier-multiplexed lightwave systems," IEEE Photon. Technol. Lett., vol.5, pp.1100–1103, 1993.

[29] A. Kanazawa, M. Shibutani, and K. Emura, "Pre-clipping method to reduce clipping- induced degradation in hybrid analog/digital subcarrier-multiplexed optical transmission systems," IEEE Photon. Technol. Lett., vol.7, pp.1069–1971, September 1995.

[30] Q. Pan and R. J. Green, "Performance analysis of preclipping AM/QAM hybrid lightwave systems," J. Lightwave Technol., vol.15, pp.1–5, January 1997.

[31] Q. Shi, R. S. Burroughs, and S. Lewis, "An alternative model for laser-clipping induced nonlinear distortion for analog lightwave CATV system," IEEE Photon. Technol. Lett., vol.4, pp.784–787, July 1992.

[32] J. B. Thomas, *An Introduction to Statistical Communication Theory*, Wiley, New York, 1979.

[33] K. Alameh and R. A. Minasian, "Ultimate limits of subcarrier multiplexed lightwave transmission," Electron. Lett., vol.27, pp.1260–1262, July 1991.

[34] A. J. Rainal, "Distortion spectrum of laser intensity modulation," IEEE Trans. Commun., vol.43, pp.1644–1652, February 1995.

[35] O. Shimbo, *Transmission Analysis in Communication System*, vol. I. Computer Science, Rockville, Maryland, 1988.

[36] J. L. Gimlett and N. K. Cheung, "Effects of phase-to-intensity noise conversion by multiple reflections in gigabit-per-second DFB laser transmission system," J. Lightwave Technol., vol.7, pp.888–895, June 1989.

[37] W. I. Way, C. Lin, C. E. Zah, L. Curtis, R. Spicer, and W. C. Young, "Multiple-reflection-induced intensity noise studies in a lightwave system for multichannel AM-VSB television signal distribution," IEEE Photon. Technol. Lett., vol.2, pp.360–362, May 1990.

[38] H. Yoshinaga, K. Kikushima, and E. Yoneda, "Influence of reflected light on erbium- doped fiber amplifiers for optical AM video signal transmission systems," J. Lightwave Technol., vol.10, pp.1132–1136, August 1992.

[39] A. F. Judy, "Intensity noise from fiber Rayleigh backscatter and mechanical splices," Proc. ECOC'89, Gothenburg, Sweden, paper TuP-11, 1989.

[40] S. Wu, A. Yariv, H. Blauvelt, and N. Kwong, "Theoretical and experimental investigation of conversion of phase noise to intensity noise by Rayleigh scattering in optical fibers," Appl. Phys. Lett., vol.59, pp.1156–1158, September 1991.

[41] M. O. van Deventer, "Polarization properties of Rayleigh backscattering in single-mode fibers," J. Lightwave Technol., vol.11, pp.1895–1899, December 1993.

[42] P. Wan and J. Conradi, "Impact of double Rayleigh backscatter noise on digital and analog fiber systems," J. Lightwave Technol., vol.14, pp.288–297, March 1996.

[43] A. Yariv, H. Blauvelt, and S. Wu, "A reduction of interferometric phase-to-intensity conversion noise in fiber links by large index phase modulation of the optical beam," J. Lightwave Technol., vol.10, pp.978–981, July 1992.

[44] J. H. Angenent, I. P. D. Ubbens, and P. J. de Waard, "Distortion of a multicarrier signal due to optical reflections," Tech. Digest, ECOC'91, paper WeC8-4, Paris, France, September 1991.

[45] T. Uno, M. Tanabe, and Y. Matsui, "High quality 64 channel 15 km AM-FDM transmission using low distortion characteristics DFB laser module," Conf. Digest, paper BAM9, LEOS Summer Topical Meeting Broadband Analog Optoelectronics—Devices Systems, July 1990.

[46] K. Kikushima, K. Suot, H. Yoshinaga, and E. Yoneda, "Polarization dependent distortion in AM-SCM video transmission systems," J. Lightwave Technol., vol.12, pp.650–657, April 1994.

[47] R. H. Stolen, "Polarization effects in fiber Raman and Brillouin lasers," IEEE J. Quantum Electron., vol.15, pp.1157–1160, October 1979.

[48] J. Ohya, H. Sato, M. Mitsuda, T. Uno, and T. Fujita, "Second-order distortion of amplified intensity-modulated signals with chirping in erbium-doped fiber," J. Lightwave Technol., vol.13, pp.2129–2135, November 1995.

[49] B. Clesca, P. Bousselet, and L. Hamon, "Second-order distortion improvement or degradations brought by erbium-doped fiber amplifiers in analog links using directly modulated lasers," IEEE Photon. Technol. Lett., vol.5, pp.1029–1031, September 1993.

[50] C. Y. Kuo and E. E. Bergmann, "Erbium-doped fiber amplifier second-order distortion in analog links and electronic compensation," IEEE Photon. Technol. Lett., vol.3, pp.829–831, 1991.

[51] K. Kikushima, "AC and DC gain tilt of erbium-doped fiber amplifiers," J. Lightwave Technol., vol.12, pp.463–470, March 1994.

[52] N. Otsuka, M. Kito, M. Ishino, J. Ohya, Y. Kudo, and Y. Matsui, "1.5 μm strained-layer MQW-DFB lasers with low chirp and low distortion characteristics," Proc. Integrated Optics Optical Commun., paper FB2, 1995.

[53] G. J. Meslener, "Chromatic dispersion induced distortion of modulated monochromatic light employing direct detection," IEEE J. Quantum Electron., vol.20, pp.1208–1216, October 1984.

[54] C. S. Ih and W. Gu, "Fiber induced distortions in a subcarrier multiplexed lightwave system," IEEE J. Selected Areas Commun., vol.8, pp.1296–1303, September 1990.

[55] H. Yonetani, I. Ushijima, T. Takada, and K. Shima, "Transmission characteristics of DFB laser modules for analog applications," J. Lightwave Technol., vol.11, pp.147–153, January 1991.

[56] D. B. Crosby and G. J. Lampard, "Dispersion-induced limit on the range of octave confined optical SCM transmission system," IEEE Photon. Technol. Lett., vol.6, pp.1043–1045, August 1994.

[57] E. E. Bergmann, C. Y. Kuo, and S. Y. Huang, "Dispersion-induced composite second- order distortion at 1.5 μm," IEEE Photon. Technol. Lett., vol.3, pp.59–61, January 1991.

[58] M. R. Phillips, T. E. Darcie, D. Marcuse, G. E. Bodeep, and N. J. Frigo, "Nonlinear distortion generated by dispersive transmission of chirped intensity-modulated signals," IEEE Photon. Technol. Lett., vol.3, pp.481–483, May 1991.

[59] C. Y. Kuo, "High-performance optically amplified 1550 nm lightwave AM-VSB CATV transport system," Tech. Digest, Conf. Optical Fiber Commun., paper WN2, February 1996.

[60] H. Dai, S. Ovadia, and C. Lin, "Hybrid AM-VSB/M-QAM multichannel video transmission over 120 km of standard single-mode fiber with cascaded erbium-doped fiber amplifiers," IEEE Photon. Technol. Lett., vol.8, pp.1713–1715, December 1996.

[61] J. Marti, D. Pastor, M. Tortola, J. Capmany, and A. Montero, "On the use of tapered linearly chirped gratings as dispersion-induced distortion equalizers in SCM systems," J. Lightwave Technol., vol.15, pp.179–187, February 1997.

[62] D. Pastor, J. Capmany, and J. Marti, "Reduction of dispersion induced composite triple beat and second-order intermodulation in subcarrier multiplexed systems using fiber grating equalizers," IEEE Photon. Technol. Lett., vol.9, pp.1280–1282, September 1997.

[63] W. Muys, J. C. van der Plaats, F. W. Willems, A. M. Vengsarkar, C. E. Soccolich, M. J. Andrejco, D. J. DiGiovanni, D. W. Peckham, S. G. Kosinski, and P. F. Wysocki, "Directly modulated AM-VSB lightwave video transmission system using dispersion-compensating fiber and three cascaded EDFAs, providing 50-dB power budget over 38 km of standard single-mode fiber," Tech. Digest, Optical Fiber Commun. Conf., paper WN4, 1996.

[64] M. R. Phillips, A. G. Gnauck, T. E. Darcie, N. J. Frigo, G. E. Bodeep, and E. A. Pitman, "112 channel split-band WDM lightwave CATV system," IEEE Photon. Technol. Lett., vol.4, pp.790–792, July 1992.

[65] W. Muys, J. C. Van der Platts, F. W. Willems, and P. H. van Heijningen, "Mutual deterioration of WDM-coupled AM-CATV and digital B-ISDN services in single fiber access networks," IEEE Photon. Technol. Lett., vol.5, pp.832–834, 1993.

[66] R. Heidemann, B. Junginger, H. Krimmel, J. Otterbach, D. Schlump, and B. Wedding, "Simultaneous distribution of analog AM-TV and multigigabit HDTV with optical amplifier," Tech. Digest, Topical Meeting of Optical Amplifiers and Their Applications, paper FB2, 1991.

[67] W. I. Way, S. S. Wagner, M. M. Choy, C. Lin, R. C. Menendez, H. Tohme, A. Yi-Yan, A. C. Von Lehman, R. E. Spicer, M. Andrejco, M. A. Saifi, and H. L. Lemberg, "Simultaneous distribution of multichannel analog and digital video channels to multiple terminals using high-density WDM and a broad-band in-line erbium-doped fiber amplifier," IEEE Photon. Technol. Lett., vol.2, pp.665–668, September 1990.

[68] K. P. Ho, C. Lin, H. Dai, S. K. Kiaw, H. Gysel, and R. Ramachandran, "Hybrid wavelength-division-multiplexing systems for high-capacity digital and analog video trunking applications," IEEE Photon. Technol. Lett., vol.10, pp.297–299, February 1998.

[69] K. Inoue, H. Toba, and K. Nosu, "Multichannel amplification utilizing an Er^{3+}-doped fiber amplifier," J. Lightwave Technol., vol.9, pp.368–374, March 1991.

[70] K. Kikushima and H. Yoshinaga, "Signal crosstalk due to fiber nonlinearity in wavelength multiplexed SCM-AM-TV transmission systems," Optical Fiber Commun. Conf., Postdeadline paper PD24, February 1995.

[71] X. P. Mao, G. E. Bodeep, R. W. Tkach, A. R. Chraplyvy, T. E. Darcie, and R. M. Derosier, "Brillouin scattering in externally modulated lightwave AM-VSB CATV transmission systems," IEEE Photon. Technol. Lett., vol.4, pp.287–289, March 1992.

[72] H. Yoshinaga, "Influence of stimulated Brillouin scattering on nonlinear distortion in SCM video transmission," Electron. Lett., vol.29, pp.1707–1708, September 1993.

[73] F. W. Willems, M. Muys, and J. S. Leong, "Simultaneous suppression of stimulated Brillouin scattering and interferometric noise in externally modulated lightwave AM-SCM systems," IEEE Photon. Technol. Lett., vol.12, pp.1476–1478, December 1994.

[74] D. Piehler, C. Y. Kuo, J. Kleefeld, C. Gall, A. Nilsson, and X. Zou, "Influence of SBS suppression on distortion in an optically amplified analog video transport system," OSA TOPS on Optical Amplifiers and Their Applications, vol.5, pp.377–380, 1996.

[75] M. Phillips and K. L. Sweeney, "Distortion by stimulated Brillouin scattering effect in analog video lightwave systems," Tech. Digest, Conf. Optical Fiber Commun., postdeadline paper PD23-1, February 1997.

[76] Z. Wang, A. Li, C. J. Mahon, G. Jacobsen, and E. Bødtker, "Performance limitations imposed by stimulated Raman scattering in optical WDM SCM video distribution systems," IEEE Photon. Technol. Lett., vol.7, pp.1492–1494, December 1995.

[77] F. W. Willems, W. Muys, and J. C. van der Platts, "Experimental verification of self- phase-modulation-induced nonlinear distortion in external modulated AM-VSB lightwave systems," Tech. Digest, Conf. Optical Fiber Commun., paper ThR4, February 1996.

[78] C. Densem, "Composite second order distortion due to self-phase modulation in externally modulated optical AM-SCM systems operating at 1550 nm," Electron. Lett., vol.30, pp.2055–2056, 1994.

[79] A. D. Kersey and M. A. Davis, "Brillouin bandwidth determination from excess noise characteristics of SBS signals in single mode fibers," Tech. Digest, Conf. Optical Fiber Commun., paper ThJ5, February 1993.

[80] R. G. Smith, "Optical power handling capacity of low loss optical fibers as determined by stimulated Raman and Brillouin scattering," Appl. Opt., vol.11, pp.2489–2494, 1972.

[81] Y. Aoki, K. Tajima, and I. Mito, "Input power limits of single-mode optical fibers due to stimulated Brillouin scattering in optical communication systems" J. Lightwave Technol., vol.6, pp.710–719, May 1988.

[82] H. Yoshinaga, "Influence of stimulated Brillouin scattering on nonlinear distortion in SCM video transmission," Electron. Lett., vol.29, pp.1707–1708, September 1993.

[83] F. W. Willems, J. C. van der Plaats, and W. Muys, "Harmonic distortion caused by stimulated Brillouin scattering suppression in externally modulated lightwave AM-CATV systems," Electron. Lett., vol.30, pp.343–345, February 1994.

[84] A. R. Chrplyvy, "Optical power limits in multi-channel wavelength-division-multiplexed systems due to stimulated Raman scattering," Electron. Lett., vol.20, pp.58–59, 1984.

[85] A. Li, C. J. Mahon, Z. Wang, G. Jacobsen, and E. Bodtker, "Experimental confirmation of crosstalk due to stimulated Raman scattering in WDM AM-VSB CATV transmission systems," Electron. Lett., vol.31, pp.1538–1539, August 1995.

Chapter 10

Wireless Access in HFC Systems

Wireless access in HFC systems essentially integrates two major telecommunication infrastructures, i.e., wireless and HFC networks. In this chapter, we will discuss the technology considerations and progress in this field. The advantages of using optical fiber links to transport wireless signals are that optical transceivers are (1) transparent to any modulation formats in wireless systems, (2) compact and reliable, and (3) can repeat wireless signals over a long distance with high fidelity. On the other hand, the disadvantages in using optical fiber links are (1) there may not be existing optical fiber infrastructure which is ready to be used, and the labor cost in having custom-designed optical links is very high; and (2) the cost on transmitters for uplinks (from portable handsets to a base station) in cellular or personal communication service (PCS) networks are still too high for large scale deployment.

We will first start with a general historical background introduction in Sec.10.1 and a review of current cellular and PCS systems in Sec.10.2. Radio propagation and wireless operational backgrounds are given in Sec.10.3. After these background introductions, we will follow up with reviews on two separate system considerations: Sec.10.4 reviews wireless access through optical fibers, and Sec.10.5 reviews wireless access through a combination of optical fibers and coaxial cables.

10.1 General Historical Background Introduction

Besides analog and digital CATV signals, subcarrier multiplexed lightwave systems can be used to transport downlink satellite signals [1], point-to-point digital microwave signals [2], and wireless cellular/PCS signals [3]. The main topic in this chapter is to study the feasibility of using HFC infrastructure and SCM technology to transport wireless cellular/PCS signals. The first commercial application in using optical fiber links to transport wireless cellular-phone signals was to resolve "blind-zone" problems which often occur in a cellular system. As shown in

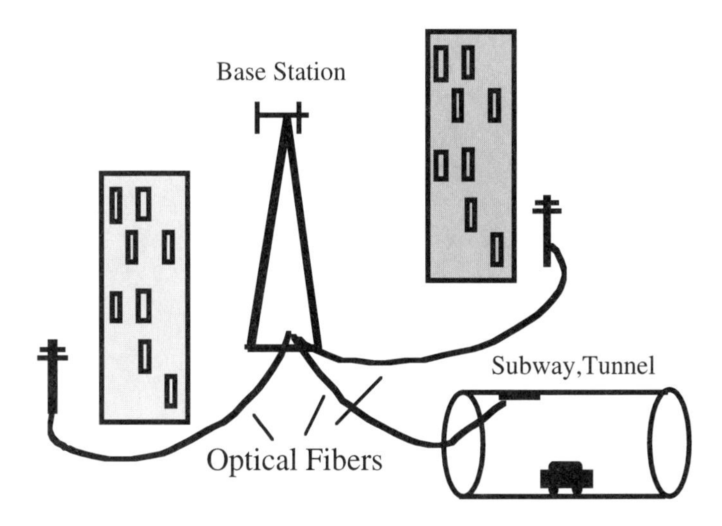

Figure 10.1: Optical fiber links used to solve "blind zone" problems in a cellular network. After Ref. [4].

Fig. 10.1, blind zones were found to be behind high-rise buildings, inside tunnels and subways, etc., and they can be eliminated simply by deploying optical fiber links between the cellular base-station and a proper spot in the blind-zone. Initially, the cost of optical links for this kind of application was not of concern because it is much less expensive than installing a new base station.

Optical fiber links were then gradually used in microcellular and picocellular systems. This has to do with the vastly increasing demands on cellular wireless communications (an average of 30–40% annual growth rate worldwide), and the fact that user traffic in many urban areas has already attained saturation. Cell splitting [5] and sectorization techniques [6] (e.g., a seven-cell system uses three 120° sectors per cell) have been used to increase the user capacity. If the cell diameter is reduced by K, the total system capacity can be increased by K^2, mainly because the frequency reuse efficiency can be increased. This is illustrated in Fig. 10.2, where we can see that in a seven-cell region, originally a specific wireless frequency can only be used by one cell. But after cell splitting, the same frequency can be used by seven smaller cells (marked as black cells in Fig. 10.2). In other words, the total capacity is increased by seven times because the distance between cells that can use the same frequency is reduced from $4.6R$ to $4.6r$, where $R = \sqrt{7}\, r$.

The cell size can be reduced to within about 300 meters of diameter for outdoor environments [8,9], called "microcells," or can be reduced to tens of meters in diameter for indoor offices, called "picocells" [9,10]. Although more users can be accommodated within a given spectrum allocation by deploying microcellular or picocellular systems, significantly more antenna radio ports (RAPs) are required, and consequently access to cell-site real estate becomes a serious problem. Current

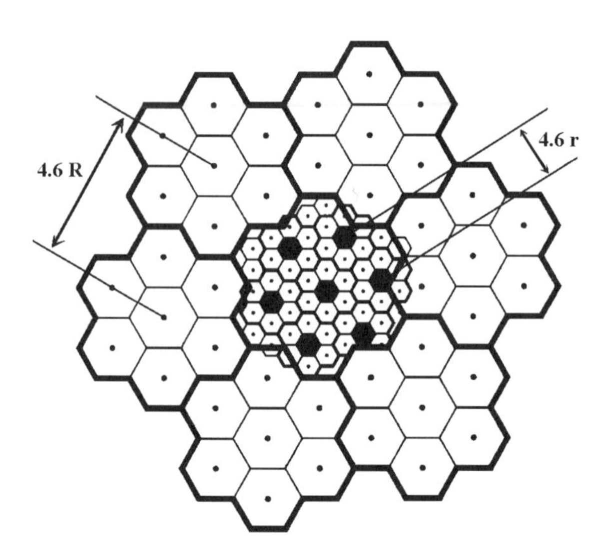

Figure 10.2: Cell splitting in a seven-cell cluster system. 4.6 R and 4.6r illustrate the distance between cells which use the same frequency before and after cell splitting. After Ref. [7].

common practice is to use T1 lines or microwaves [11] to establish links between a base station and cells. However, the first approach requires significant signal processing equipment to convert multiple voice channels at 800/1900 MHz bands to a few T1 lines, and the equipment can occupy a significant real-estate space. The second approach requires microwave frequency coordination and may suffer from multipath outage problems. To save the real-estate space, subcarrier-multiplexed (SCM) optical fiber microcellular systems and the associated technologies have been extensively studied [3,12]. SCM optical fiber technology can be used to make compact RAPs by transporting wireless signals directly at 800/1900 MHz bands or by transporting down-converted IF signals (see Section 10.4.2). Figure 10.3 illustrates the optical fiber-based microcellular/picocellular configuration, and the typical components in a compact RAP.

Fiber-to-the-microcell approach can be aligned with the optical network unit (ONU) in a fiber-in-the-loop system (FITL), or with the optical fiber distribution node in an HFC system, as illustrated in Figs. 10.4a and b, respectively. In a FITL system, each ONU serves up to 32 or 64 subscriber houses [13]; in an *advanced* HFC network, each optical node serves up to about 500 subscribers, and the coverage area should be well consistent with a "microcell" coverage range. On the other hand, in today's HFC networks, where each optical node serves up to about 2000 to 5000 subscribers, the wireless-coverage can be more complete only if the RAPs are deployed in the coaxial part of HFC networks [14,15c], as will be discussed more in Section 10.5.

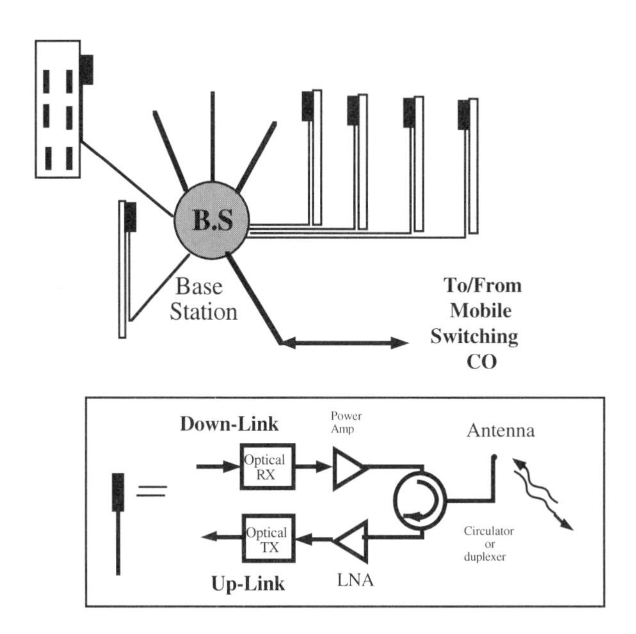

Figure 10.3: (a) Typical outdoor microcell configuration, (b) Typical components in an antenna radio ports. After Ref. [3].

Fiber-to-the-picocell can be aligned with an optical-fiber-based local-area network inside a building, as illustrated in Fig. 10.5.

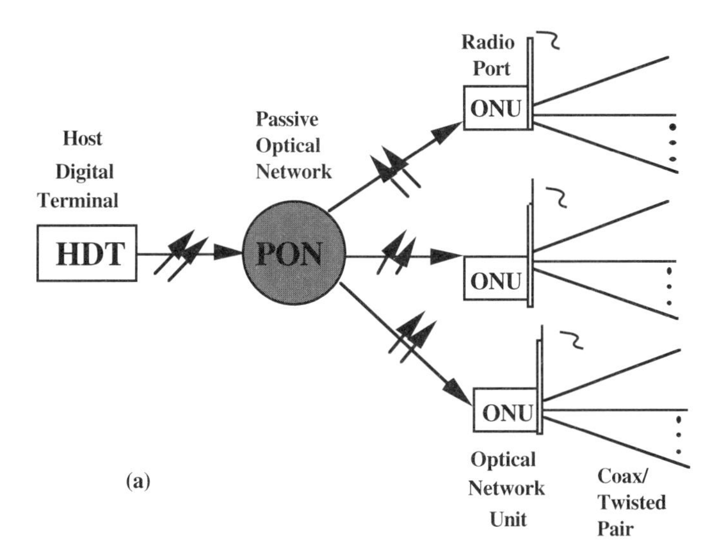

Figure 10.4: *(continued)*

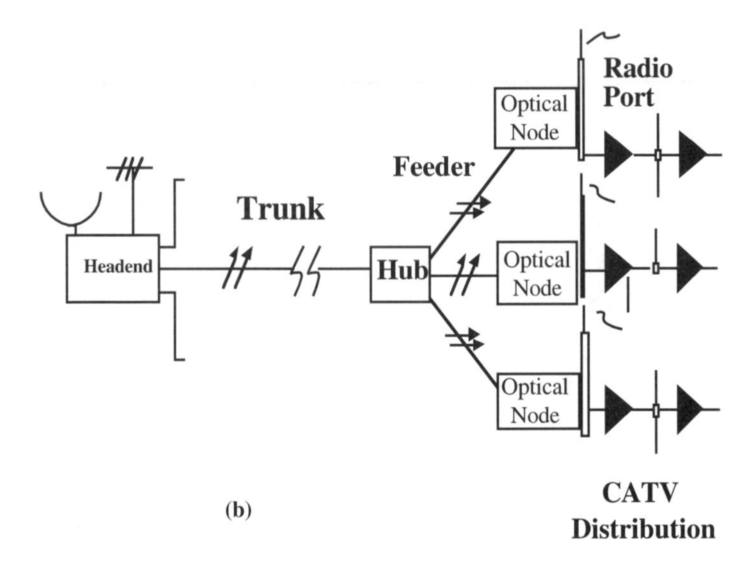

Figure 10.4: *(continued)* Integrating wireless radio antenna ports (a) in an FITL PON network and (b) in an advanced HFC network. After Ref. [3].

10.2 Review of Current Cellular/Personal Communication Service (PCS) Systems

High-power cellular systems with large cells (0.5 to 20 km in diameter) which can support mobiles traveling in excess of 100 mph are listed in Table 10.1 (sometimes termed "high-tier PCS systems"), together with their associated system parameters. Low-power systems include low-tier PCS systems such as PACS (Personal Access Communication System) and the Japanese PHS (Personal Handy System), and digital cordless systems such as CT-2 (second-generation cordless telephone) and DECT. The associated system parameters for low-power systems are shown in Table 10.2.

10.3 Radio Propagation and Operational Background

Design considerations for the optical fiber up- and downlinks are quite different. Downlink transports multiple equal-amplitude carriers from a base station to multiple RAPs. The maximum number of channels that a single laser needs to handle depends on the wireless systems. For example, the numbers of distinct-frequency carriers for IS-136, GSM (Global System for Mobile Communications), DCS-1800 (Digital Cellular System-1800), DECT (Digital European Cordless Telephone), CT-2 (Cordless Telephone 2), and PHS are 832, 125, 375, 10, 40, and 77, respectively. Depending on the CNR and nonlinear distortion requirement at the transmit-

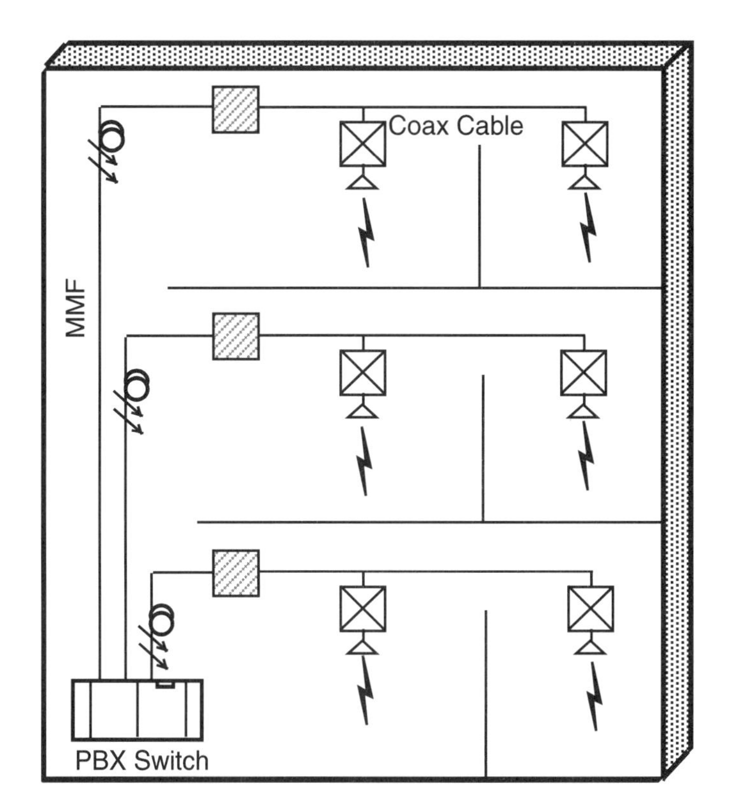

Figure 10.5: Integrating radio antenna ports in a local area voice or computer network. —(▨) optical transmitter and receiver, (☒) up- and down-converter, (MMF) multimode fibre, (PBX) private-branch exchange. After Ref. [4].

ting antenna, a single high-quality laser or multiple lasers can be used [3]. High-output-power optical transmitters may be preferred for power splitting to multiple RAPs. Note that different lasers must be used for different sets of reused frequencies. Generally speaking, the cost problem for the downlink E/O is much less severe than that for the uplink, because every RAP must have an uplink E/O while a downlink E/O can be shared by multiple RAPs. In addition, uplink E/O must be able to handle the large variations of received signals from different handsets, and consequently we have to specify the dynamic range requirement on an E/O for different radio environments and wireless systems. Before we discuss E/O dynamic range characterizations, let us first review some background about radio propagation and wireless system operation.

10.3.1 Radio Propagation Background

Uplinks transport carriers from mobile or portable telephone units to base stations. By the time these carriers arrive at the E/O input, they have experienced different

Table 10.1 Parameter overview for high-power macrocellular systems*

	AMP	**IS-136**	**GSM**	**DCS-1800**	**CDMA**	**PDC (Japan)**
Band (uplink)	824–849	824–849	890–915	1710–1785	824–849	810 826 1429 1441 1453 1465
(downlink) (MHz)	869–894	869–894	935–960	1805–1880	869–894	940 956 1429 1441 1453 1465
Multiple access	FDMA	TDMA/FDMA	TDMA/FDMA	TDMA/FDMA	CDMA/FDMA	TDMA/FDMA
Interleaved freq. spacing (kHz)	30	30	200	200	1250	25
#Voice ch/RF carrier	1	3	8	8	*	3
#RF carriers (simplex)	832	832	125	375	20	640
Channel rate (kb/s)	analog	48.6	270.833	270.833	1,228.8 (chip rate)	42
Modulation	FM	$\pi/4$ DQPSK	GMSK	GMSK	QPSK/ OQPSK	$\pi/4$ DQPSK
Duplexing	FDD	FDD	FDD	FDD	FDD	FDD
Frequency assignment	fixed	fixed	dynamic	dynamic	fixed	fixed
Typical reuse factor	7	7	dynamic	dynamic	1	4/7
Portable transmit power (max./avg.)		600 mW/ 200mW	1 W/ 125mW	1W/ 125mW	600 mW	2W/ 660mW
Handset power control (dB)	28	28	28	28	80	
		antenna diversity	frequency hopping	frequency hopping		antenna diversity
Min. *C/I* (dB)	18	Yes 12	Yes 9	Yes 9	6	Yes 13
		No 16	No 11	No 11		No 17

* For uplink CDMA, capacity $N = \frac{W/R}{E_0/I_0}(\frac{1}{d})\,gf$, where W is spreading bandwidth, R is information rate, E_0/I_0 is the required energy per bit to interference power spectral density, d is voice activity factor, g is sectorization factor, and f accounts for the outer-cell interference. See Peterson *et al.* [15a]. The typical number of voice channels per RF carrier at a BER = 10^{-3} is about 32.

Table 10.2 Parameters overview for low-power micro/picocellular applications

	PHS (Japan)	**DECT**	**CT-2**	**PACS**
Frequency band (MHz)	1895–1907	1880–1900	864–868	emerging technologies bands (1850–1910, 1930–1990)
Multiple access	TDMA/FDMA	TDMA/FDMA	FDMA	TDMA/FDMA
Interleaved freq. spacing (kHz)	300	1728	100	300
#Voice ch/RF carrier	4	12	1	8
Modulation	$\pi/4$ DQPSK	GFSK	GFSK	$\pi/4$ QPSK
#RF carriers (simplex)	77	10	40	
Channel bit rate (kb/s)	384	1152	72	384
Speech bit rate (kb/s)	32	32	32	32
Frequency assignment	dynamic	dynamic	dynamic	dynamic
Portable transmit power (max/avg.)	80/10 mW	250/10 mW	10/5 mW	200/25 mW
Handset power control	N/A	No	No	vendor defined
Min. C/I (dB)	26	12	20	17

degrees of propagation path loss, multipath fading loss, and shadowing loss. This implies that a wide range of power level variation may exist in the received carriers. We now examine the range of this power level variation, which has important implications for the dynamic range requirement on the uplink. First, the distances of portable/mobile transmitters can range from several meters (for a user near the antenna) to about 300 meters (for a user who is at the edge of a microcell). The path loss as a function of the distance d between the wireless transmitter and receiver, which may or may not include multipath fading, varies according to $d^{-1.5}$ or d^{-2} for short distances with no obstructions, and varies according to d^{-3} or d^{-4} for congested areas with obstructions [16]. In the former case, the path loss between the strongest carrier and the weakest carrier is about 30 to 35 dB, and in the latter case, about 60 to 70 dB. In the case of an indoor picocellular environment, the received power variation may be in the range of 20 to 25 dB [9]. A second loss factor is due to the shadowing by natural or manmade objects, which statistically adds up to another 10–20 dB [10,17]. To summarize, for indoor picocells or indoor/outdoor microcells with no obstructions, the input signal variation Δp_r before sending to an E/O is about 20 to 35 dB; for outdoor microcells with some non-line-of-sight coverage area, Δp_r is about 40 to 55 dB; and for outdoor microcells with congested obstructions, Δp_r is as high as 80 to 90 dB. Typical handsets for high-power TDMA/FDMA systems have an adjustable transmitter power range of 28 dB (see Table 10.1); therefore, the maximum dynamic range that an E/O needs to handle in a TDMA microcell is about 60 dB. However, the handsets of most low-power PCS systems do not have any power control capability. Therefore, the minimum Δp_r in a low-power system is at least 20 to 35 dB. Lastly, we note that repeating CDMA signals in optical fibers is a fairly easy task because of the tight power control and the large processing gain in CDMA systems [15a,15b,15c]. However, a CDMA signal can still be damaged by laser diode's strong-clipping-induced Gaussian-like intermodulation noise [15d].

10.3.2 Operational Aspects

For the same wireless radio propagation environment, the required dynamic range on uplink E/O can vary among different wireless systems. The following operational aspects must be carefully taken into design consideration to alleviate the dynamic range requirement on uplink E/O:

1. *TDMA slots:* Besides AMPS and CT-2, most wireless systems use TDMA/FDMA technology, which means that there are multiple time slots assigned to a single frequency carrier. This parameter was given in Tables 10.1 and 10.2 for high-power and low-power systems, respectively. For example, we note that low-power systems (except CT-2) in general have 4 to 12 duplex paired time slots per RF carrier, while high-power systems (except AMPS) have 3 to 8 duplex paired time slots per RF carrier. Therefore, if the number of simultaneous users in a microcell or picocell is small, which is very likely the case, only one or two

frequency carriers need to be assigned to that particular cell. Consequently, the linearity requirement on an E/O can be significantly alleviated.

2. *Dynamic channel assignment (DCA) with a lookup table for frequency plans that prevent active channels from being corrupted by intermodulation products:* DCA has been implemented in systems such as GSM, DCS-1800, DECT, and PACS at a central control unit. If a lookup table which stores the frequency plans to avoid having intermodulation products fall into any active channel can be incorporated in the central control unit, then a low dynamic range uplink E/O can be used [18].
3. *CDMA wireless systems:* A CDMA IS-95 system with its extremely tight power control (up to 80 dB power adjustment capability in a handset) and a large processing gain of 21dB (1.2288 Mbps/9.6 Kbps) can greatly alleviate the dynamic range requirement on an uplink E/O [15c,d].
4. *Macrodiversity and microdiversity:* Both macrodiversity and microdiversity are techniques that can be used to reduce the received power level variations due to multipath shadow fading, and can potentially alleviate the dynamic range requirement on an uplink E/O. Macrodiversity allows communication to a mobile/portable unit by one of multiple RAPs. By monitoring the signal transmitted from the handset via the multiple RAPs, the best link can be selected [19]. The disadvantages are that the effective capacity of a DCA scheme may be decreased because no two handsets can share the same radio channel, and because the system becomes more sensitive to co-channel interference [19]. Microdiversity uses antenna space diversity to overcome the slow multipath fading. For multipath signals arriving from all directions in the azimuth, antenna spacing of the order of 0.5–0.8λ is quite adequate [6,8]. At 2 GHz, this corresponds to a separation of 7.5 to 12 cm. A picture of the actual implementation in an HFC system is shown in Fig. 10.6 [20].

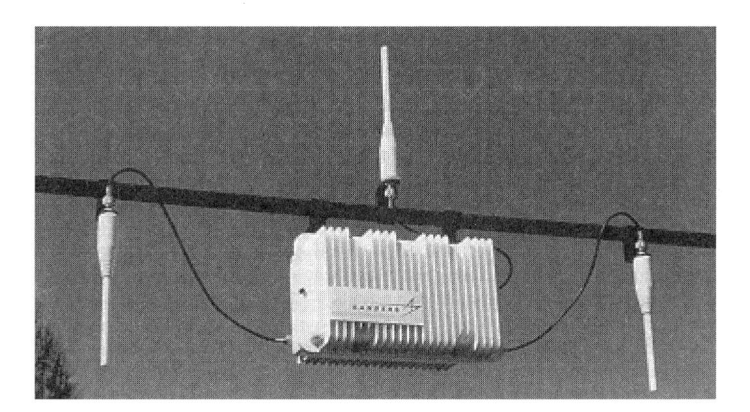

Figure 10.6: This cable microcell integrator (CMI) transceiver was built to provide the air-to-cable interface for a cable-TV-based implementation of Personal Communications Services. (Photo courtesy of Sanders), after Ref. [20].

10.4 Optical Fiber-Based Microcellular/Picocellular Systems

Having understood the approximate power variation ranges in different outdoor/indoor and urban/suburban environment, we can now determine how we can set the dynamic range requirement on an uplink E/O.

10.4.1 Spurious-Free Dynamic Range Considerations

The signal dynamic range Δp_r which can be handled by an optical fiber link is usually measured in terms of spurious-free dynamic range (SFDR, equals the sum of Δp_r and CNR_{min}, which is the minimum CNR in the system). The definition of SFDR (dB-$Hz^{2/3}$), derived in Appendix B, is essentially the power ratio of a fundamental carrier to the two-tone third-order intermodulation (IM_3) product, and IM_3 is equal to the ground noise level in the receiver bandwidth. Figure 10.7 relates SFDR to a "worst-case" wireless environment. We can see that the worst-case condition is when there are two tones corresponding to two users who are close to a RAP, and these two tones generate an IM_3 because of the nonlinearity in the transmission sys-

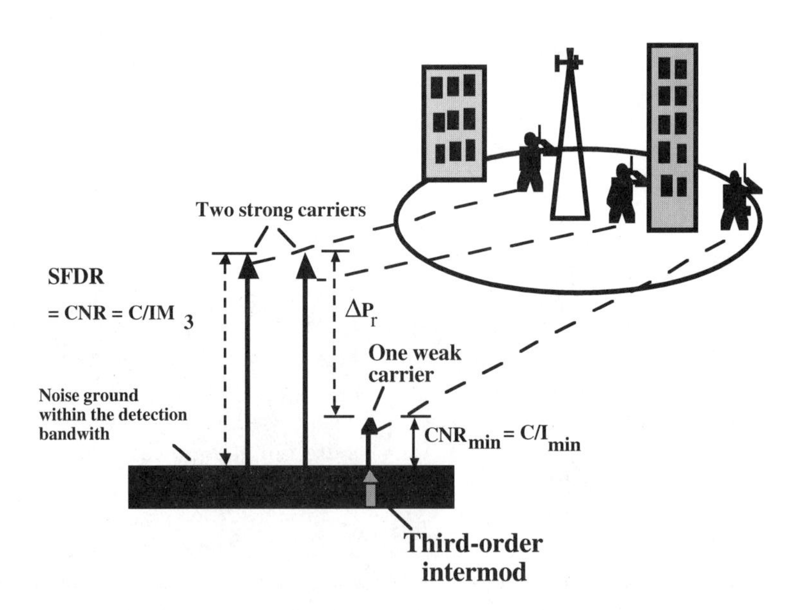

Figure 10.7: Illustration of the worst-case condition of two strong carriers and a weak neighbor carrier. Spurious-free dynamic range (SFDR) is defined as the level which equals CNR and C/IM_3 simultaneously. In addition, SFDR $= \Delta p_r + CNR_{min}$.

tem (partiularly the E/O), while in the mean-time a third active user at the edge of the cell is assigned a frequency right on top of the IM_3. The power level difference between the strong carriers and the weak carrier is the Δp_r, which we have described in Section 10.3.1. Since the weak carrier must maintain a minimum CNR in order to ensure the voice quality, we have

$$SFDR\ (dB\text{-}BW^{2/3}) = \Delta p_r + CNR_{min} - \Delta G, \tag{10.1}$$

where BW is the interleaved frequency spacing, and ΔG is the adjustable power range of a handset (typical 28 dB for high-power TDMA/FDMA systems, and 0 dB for most low-power TDMA/FDMA systems). Note that min $(\Delta P_r - \Delta G) \geq 0$. Using the CNR_{min} and ΔG for various systems in Tables 10.1 and 10.2, and using the summarized results of ΔP_r given at the end of Section 10.3.1, we obtain the $SFDR_w$ under *deterministic worst-case conditions* as shown in Table 10.3 (note that the unit in this table is dB-$Hz^{2/3}$ instead of dB-$BW^{2/3}$).

Table 10.3 Worst-case $SFDR_w$ per channel for different microcelluar/picocellular environments.

$SFDR_w$ (dB-$Hz^{2/3}$)	Indoor, no obstructions	Indoor / outdoor with some non-line-of sight obstructions	Outdoor, with congested obstructions
IS-136	—	58 to 73	98 to 108
GSM/DCS1800	—	56 to 71	96 to 106
PDC	—	59 to 74	99 to 109
PHS	83 to 98	103 to 118	—
DECT	74 to 89	94 to 109	—
CT-2	73 to 88	93 to 108	—

There are several interesting facts that can be observed from Table 10.3. The first is that the high-end SFDR requirements for high-power systems are almost comparable with those of low-power systems, thanks to the large power control range of each handset. The second is that the low-end SFDR requirements for high-power systems are even lower than those of low-power systems. Therefore, the 28-dB power control range in TDMA/FDMA handsets of high-power systems plays an important role in the overall SFDR requirements on uplink E/O.

It was later noted that the $SFDR_W$ obtained from two-tone intermodulation measurements is much tighter than necessary, considering statistical multiple access in an actual wireless environment [4, 21], because the chance of having a weak carrier just next to two strong carriers is slight. For example [4], in an outdoor cell with a diameter of 1.2 km, the $SFDR_W$ was supposed to be about 117 dB-$Hz^{2/3}$. But when we replace the two strong tones with N random tones whose amplitudes depend on where in the cell the users are located, the required two-tone SFDR (to reach a blocking probability or grade of service set at 2%)[1] can be reduced to about 84, 80, and 74 dB-$Hz^{2/3}$, for 20, 10, and 5 carriers, respectively, as shown in case A′, Fig. 10.8. This is indeed a dramatic reduction in the uplink E/O dynamic range requirement.

If we assume a microcell or a picocell environment, we can obtain even further reduction in the E/O uplink dynamic range. This phenomenon is illustrated in the following four examples (also illustrated in Fig. 10.8):

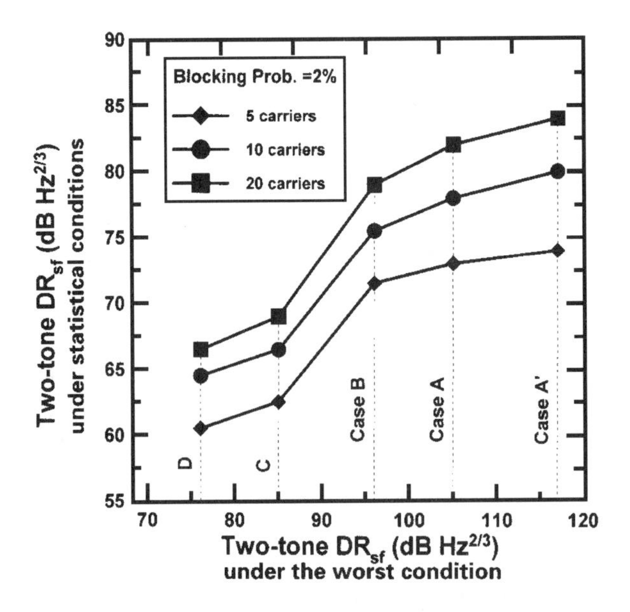

Figure 10.8: SFDR under statistical considerations versus the worst condition, for four microcellular environments (A = 5–300 m + shadowing; B = 3–100 m + shadowing; C = 5–300 m; d = 3–100 m) and for the case presented in Ref. [21] (A′ = 20–1200 m). Note that ΔG is assumed to be zero in all cases.

[1]In the computer simulation which was used to obtain results given in Fig. 10.8, we first assume that the nonlinear characteristic of an optical transmitter can be modeled as a Taylor series. We can use the third-order coefficient (a_3) of the Taylor series expansion to obtain the deterministic two-tone SFDR (see Appendix B). We then replace the two equal tones by N unequal tones whose amplitudes depend on how far each user is from the RAP. These unequal tones then generate intermodulation products, and any carrier which cannot meet a minimum C/I requirement of 16 dB (as in IS-136 system) is considered blocked. If the blocking rate is lower than 2%, we increase a_3 until the blocking rate becomes 2%. The new a_3 then gives us a new SFDR under a specific statistical wireless environment (e.g., cases A-D and A′ in Fig. 10.8).

(A)An outdoor microcell ($R < 300$ m, $H = 5$ m), 50% probability of shadowing, $P_r \propto d^{-2}$ for 5 m < d < 200 m, and $P_r \propto d^{-4}$ for 200 m < d < 300 m

(B) An indoor microcell ($R < 100$ m, $H = 3$ m), 50% probability of shadowing, $P_r \propto d^{-2}$

(C) An outdoor microcell ($R < 300$ m, $H = 5$ m), without shadowing, $P_r \propto d^{-2}$ for 5 m < d < 200 m, and $P_r \propto d^{-4}$ for 200 m < d < 300 m

(D) An indoor microcell ($R < 100$ m, $H = 3$ m), without shadowing, $P_r \propto d^{-2}$,

where R = radius of the microcell, H = antenna height, P_r = received RF power, d = the distance from the antenna to the user's handset, and the shadowing loss is uniformly distributed in the 10- to 20-dB range. We further assume that each RF carrier is assigned to a single user (as in CT-2), and that there are 5, 10, or 20 users in a given cell. If the laser RIN noise is high, an AGC amplifier is supposed to automatically boost the weakest input RF carrier level to maintain a minimum CNR of 16 dB. In the computer simulation, a user is randomly located anywhere in the cell. The product of the number of runs and the number of carriers is more than 30,000. The simulation results for the four wireless environments (A–D) are shown in Fig. 10.8. Case A had slightly lower SFDR than those given by Curter *et al.* [21] because in the paper by Curter *et al.* [21] the minimum CNR was set at 18 dB for AMPS systems.

It is clear that, for any given wireless environment, the worst-case two-tone SFDR are significantly higher than that obtained from a statistical environment (the statistical two-tone SFDR can be even lower if handset power control is included). Consequently, if a system designer sets the system performance goal as a blocking rate of 2%, as is usually the case in cellular system design, an uplink E/O with a SFDR of 90 dB-$Hz^{2/3}$ is sufficient to take care of almost all microcell and picocell environments.

From the preceding discussions, we can see that how stringent an SFDR requirement we should impose on an uplink E/O really depends on whether it is the worst-case or statistical condition that a system operator judges to be more important.

10.4.2 Uplink E/O Performances

There are two alternatives for uplink E/O: a direct modulation system and an external modulation system. The latter is not yet considered a practical candidate for optical fiber-fed wireless systems for several reasons: (1) A $LiNbO_3$ Mach–Zehnder modulator without pre-distortion circuits has poor third-order linearity performance (see Chapter 7), while a laser diode generally has a good third-order linearity. For a narrowband wireless system whose transmission signals are within an octave, an E/O with superior third-order linearity is preferred. (2) A $LiNbO_3$ Mach–Zehnder modulator requires a high driving voltage, while a linear high-power driving amplifier at the desired wireless band may consume high power and is very costly. (3) A $LiNbO_3$

Mach–Zehnder modulator could exhibit a DC drift problem which is difficult for a feedback-controlled bias to operate within its maximum setting (see Section 7.4).

Using the analysis given in Appendix B, theoretical SFDRs for laser diodes (considering only dynamic nonlinearity) can be expressed by

$$SFDR = \frac{2}{3} \bullet 10 \log \left[\frac{I_p^2 / \left| \left(\frac{f_1}{f_o}\right)^4 - \frac{f_1^2}{2f_o^2} \right|}{RIN \bullet I_p^2 + 2qI_p + \langle i_{th}^2 \rangle} \right]. \quad (10.2)$$

We assume two types of laser diodes: (a) a high-speed laser diode whose resonance frequency is 10 times higher than the operating frequency range, that is, $f_r/f \approx 10$, and a low RIN of -160 dB/Hz; and (b) an inexpensive laser diode with $f_r/f \approx 5$ and a relatively high RIN of -130 or -140 dB/Hz. The former represents a high-end laser which is of the same superior quality as a DFB laser used for >80-channel AM-VSB signal transmission. The latter represents a low-grade laser with low-cost packages and without optical isolators. Using Eq. (10.2) and the above assumptions (with receiver thermal noise given by $\langle i_{th}^2 \rangle^{1/2} = 10$ pA / $\sqrt{\text{Hz}}$), we can obtain the calculated results as shown in Fig. 10.9. If we consider statistical multiple access

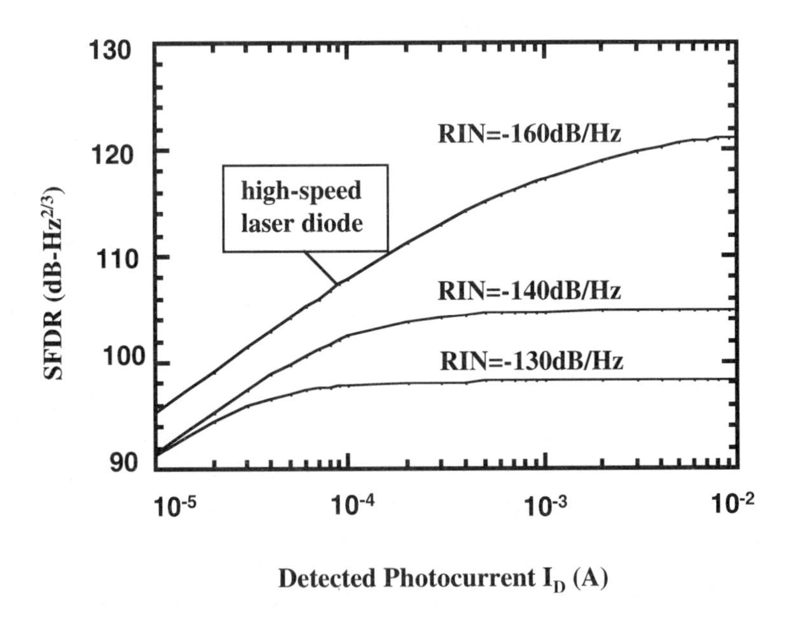

Figure 10.9: Spurious-free dynamic range of a CATV-grade laser diode with $f_r/f=10$ (top curve) and low-grade laser diodes with $f_r/f=5$ (lower two curves). f_r and f represent the laser resonance frequency and operating signal frequency, respectively.

conditions and use 90 dB-Hz$^{2/3}$ (according to the results obtained in Fig. 10.8) as the required SFDR, then all E/Os can meet the requirement at a received photocurrent $> 10\ \mu$A. If we consider the worst-case condition requirements set as in Table 10.3, we see that the high-end SFDRs for both high-power and low-power systems are hard to meet by using low-grade lasers. In addition, when using low-grade lasers, we must remember that Eq. (10.2) does not include static nonlinearity of a laser diode, and therefore it is even more difficult for this kind of laser to meet >100 dB-Hz$^{2/3}$ of high-end SFDRs when statis *L–I* nonlinearity is included.

As a last note before leaving this section, uplink E/Os must be able to function normally under temperature extremes for outdoor microcell applications.

10.4.3 Novel Circuit Techniques to Increase Link Distance

When an uplink E/O is directly modulated by RF wireless signals, the achievable receiver sensitivity is generally poor. For example, to reach a SFDR = 90 dB-Hz$^{2/3}$, a low-grade E/O with a high RIN requires about 10 μA of received photocurrent or −20 dBm of received optical power. Even a top-grade DFB laser with a low RIN requires about −30 dBm of received optical power. If there are applications which require very large power budgets to overcome splitting loss or to achieve long-distance transmission, we can use a circuit technique to convert the entire band of wireless signals into digital signals, and consequently the digital optical fiber system with a receiver sensitivity of −40 to −50 dBm can easily be achieved. This kind of circuit technique has another important advantage that it is compatible with many in-building multimode fiber (MMF) infrastructures (Fig. 10.5). The MMF modal noise and fiber dispersions can severely limit the performance of an RF optical fiber link at 900 MHz or 2 GHz, but are not a problem for a digital link at several hundred megabits.

The circuit technique uses a high-sampling-speed analog-to-digital converter to convert the entire band of wireless signals into digital bit streams [4,22]. This technique is essentially an implementation of an "FDM to TDM converter," and consequently the linearity of a laser is no longer of concern. Wala [22] demonstrated the feasibility of using digital optical transmitters and receivers by digitizing a 12.5-MHz AMPS signal via an ADC with a sampling rate of 30.72 MHz. After adding overheads and signaling information, the total transmission rate was 552.96 Mb/s. Way *et al.* [4] demonstrated a CT-2 (4-MHz bandwidth) optical fiber repeater link by using a 10-bit ADC of 10-MHz sampling rate. The total transmission rate was 125 Mb/s, which includes error-coding overheads. The experimental setup of the latter is shown in Fig. 10.10. A down-converter first converts the received wireless signals to an IF band within the 20–25 MHz passband of the bandpass filter (BPF). An AGC amplifier with a 40-dB dynamic range before the ADC was used. The theoretical maximum SNR that can be achieved by an ADC is given by [23]

$$SNR = 6.02\,B_r + 1.76 + 10\log[f_s/(2f_{max})] - 20\log N, \qquad (10.3)$$

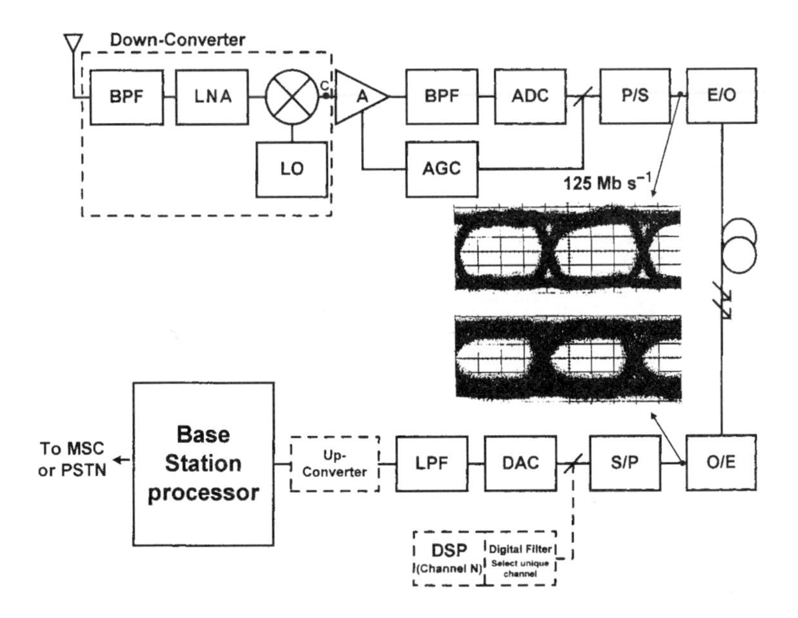

Figure 10.10: Experimental setup of an optical fiber-fed wireless system using analog-to-digital-conversion approach: BPF, bandpass filter; LNA, low-noise amplifier; LO, local oscillator; A, amplifier; PA, power amplifier; AGC, automatic gain-control amplifier; ADC, analog-to-digital converter; P/S, parallel-to-serial converter; E/O, optical transmitter and O/E, optical receiver; S/P, serial-to-parallel converter; DAC, digital-to-analog converter; LPF, low-pass filter; MSC, mobile switching center; PSTN, public switched telephone network; DSP, digital signal processing circuit. After Ref. [4].

where B_r is the number of bits of resolution of the ADC, f_s is the sampling frequency, f_{max} is the maximum frequency of the input analog signal, and N is the number of RF carriers. In the case of CT-2 signal, f_{max} = 100 kHz. Therefore, for two tones, the theoretical SNR per channel (for a 10-bit ADC) is 73 dB for a CT-2 system. The actual SNR, being limited by differential nonlinearity (DNL) [24] and the effective number of bits of the ADC, was measured to be about 62 dB (for two carriers).

The linearity performance of the ADC is determined by the integral nonlinear (INL) distortion [24] and is two least-significant-bits (LSBs) in the ADC used. The two-tone third-order intermodulation products (IM_3) are measured by sending two tones centered at 22 MHz and spaced by 100 kHz into the ADC and a 10-bit DAC. The result is shown in Fig. 10.11. It is clear that the variation of IM_3 remained within a 13-dB range for RF input driving power (measured at point C in Fig. 10.10) ranging from -70 to -24 dBm. This result is very different from the performance of a conventional nonlinear device such as an electrical amplifier or a laser diode whose IM_3 increases much faster than the fundamental carrier power. In other

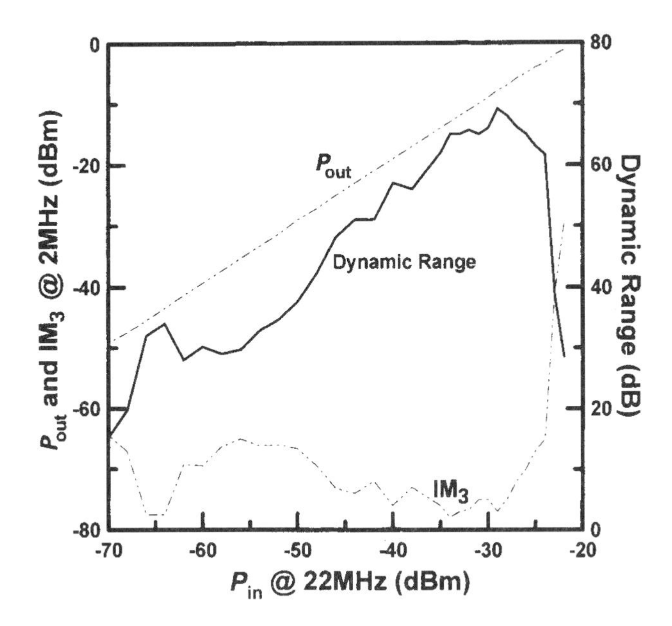

Figure 10.11: Linearity performance of a 12-bit ADC (10-MHz sampling rate, with only 10 bits used) in combination with a 10-bit DAC and a 4-MHz low-pass filter. After Ref. [3].

words, the nonlinear characteristics of an ADC *cannot* be modeled as a Taylor series expansion. The optimum dynamic range (equal to $P_{out} - IM_3$) was about 69 dB at an input power of $P_{in} = -28$ dBm.

The 125 Mb/s data rate from the ADC can be easily handled by many types of optical LAN transceivers for multimode optical fiber links or telecommunication optical transceivers for single-mode fiber links. The eye diagrams before the optical transmitter (E/O) and after the optical receiver (O/E) are shown in the insets of Fig. 10.10. The received 125 Mb/s data stream is then sent to a serial-to-parallel (S/P) converter, a DAC, and a low-pass filter (LPF). The entire band of CT-2 signals is thus recovered. Alternatively, the analog audio signal can be recovered by a digital filter in combination with a digital signal processing (DSP) circuit, as shown by the dashed lines after S/P in Fig.10.10.

In order to examine the dynamic range performance of the transmission system that used ADC, P/S, E/O, O/E, S/P, and DAC, we sent two tones into the ADC input and examined the output after the LPF in Fig. 10.10. The result is shown in Fig.10.12. We can see that the carrier-to-spur power ratio is about 62 dB. Note from Fig. 10.8 that 90 dB-Hz$^{2/3}$, which is equivalent to 57 dB per 100 kHz bandwidth, can meet almost all statistical wireless operating environment requirements. Therefore, the ADC demonstrated in Fig. 10.10 can be comfortably used in all statistical conditions.

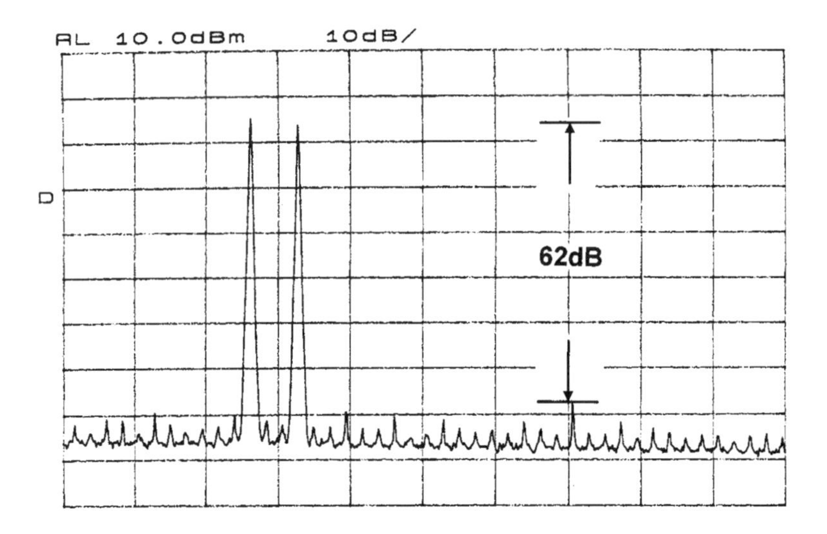

Figure 10.12: Linearity performance of the entire link in Fig. 10.10 (measured after the LPF). After Ref. [4].

10.5 Hybrid-Fiber-Coax-Based Microcellular/Picocellular Systems

In previous sections, we have focused on using only the optical fiber portion of an HFC system to transport downlink and uplink wireless signals. However, in an existing HFC system, each optical node may be serving a large number of subscribers (e.g., 2000 to 5000), which is far larger than the targeted 200 to 500 subscribers per optical node. Consequently, the service area is so large (and congested) that the architecture proposed in Fig. 10.4b is not feasible. Under such a condition, a radio port antenna must be installed on top of a high building to increase the coverage. The problem is that the antenna may be some distance away from the optical node that is usually located between utility poles or inside a hub. Another problem with such a large service area per optical node is that the dynamic range requirement on the return-path laser becomes very high.

To cope with this transitional stage problem, an alternative location for RAP installation is along the coaxial trunk/feeder lines of an HFC system. This system concept can be grasped by comparing the two networks illustrated in Figs. 10.13a and b. Figure 10.13a is a ring-trunk-line-based HFC network with wireless RAPs installed at the same location as an optical node, while in Fig. 10.13b two more RAPs in the coaxial portion of the system were added to (1) extend cell coverage or resolve a "blind-zone" coverage problem, (2) increase resource sharing in low-traffic service areas (since all RAPs use the same radio spectrum and call processing equipment), and (3) reduce the dynamic range requirement on a return-path laser diode. This kind of RAP has been practically used and was shown in Fig.10.6 previously. A very important difference between the two arrangements in Figs. 10.13a and b for wireless access is that the former is a (logically) star-based architecture (physically ring-based), and the latter is a (logically) tree-and-branch-based architecture.

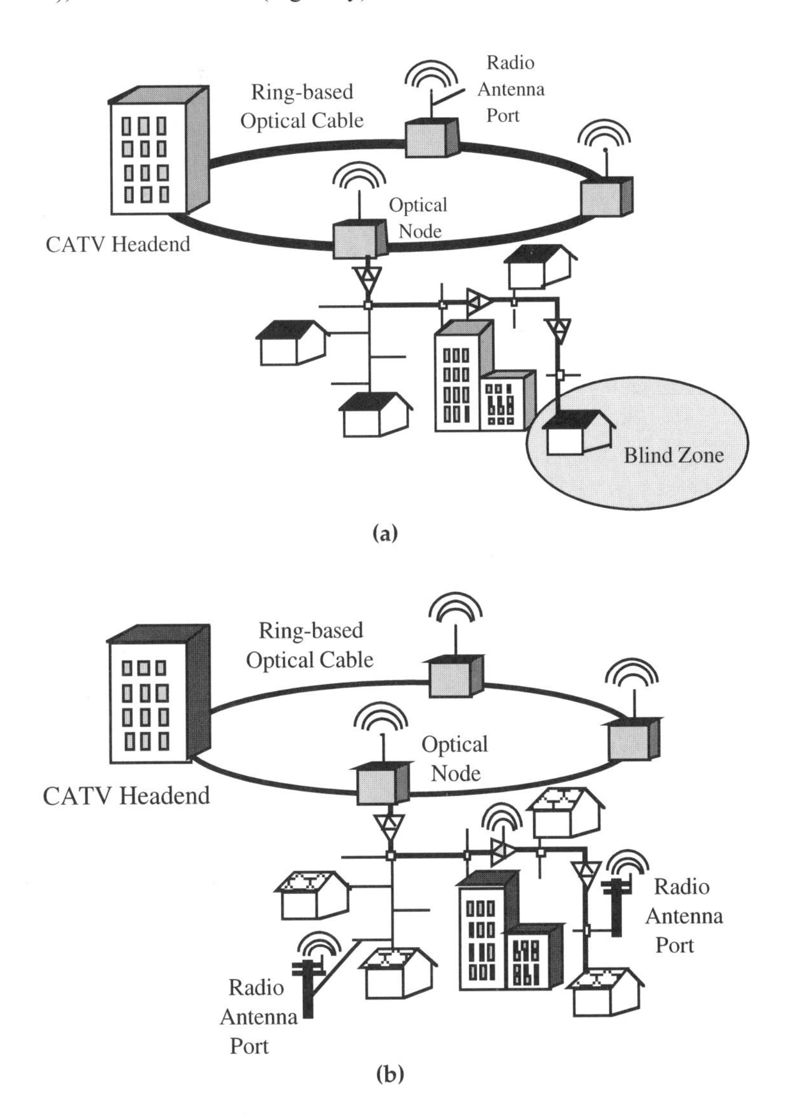

Figure 10.13: A ring-trunk-line-based HFC network. (a) Each radio antenna port is installed at the same location of an optical node. (b) Some radio antenna ports are installed in the coaxial cable portion of the network to resolve the blind-zone coverage problem.

10.5.1 Coaxial Cable Transmission Characteristics Affecting Wireless Access

Owing to the tree-and-branch architecture of the coaxial cables, the arrangement in Fig.10.13b may have one of the following problems:

1. *Simulcasting [25–27]:* The RAPs which use the same downstream and upstream transmission paths are broadcasting and receiving wireless signals on the same radio channels. As a result, the multiple antenna noise, mutual RAP interference, and a large transmission delay between any two RAPs may affect the system performance.
2. *Available cable spectra:* Currently available cable spectra in North America are 54 to 750 MHz and 5 to 42 MHz, for down- and upstream, respectively, and both do not include the cellular (around 800–900 MHz) or PCS (around 2 GHz) frequency bands. In addition, high cable loss will be incurred if the wireless signals are directly transmitted in their original frequency bands. Therefore, up- and down-converters are required at RAPs and the CATV headend, so that the wireless signals can be fit into the currently available cable spectra. These converters will increase the RAP cost.
3. *Nonlinearity and AGC range of a broadband RAP amplifier:* In a TDMA wireless system, a *broadband* RAP amplifier is required to reduce cost or to handle dynamically assigned frequency channels. However, the maximum number of simultaneous active channels is limited by the nonlinearity of the broadband RAP amplifier. This number is typically 16 or lower, depending on the local geographical obstructions and amplifier cost. The AGC range of a RAP amplifier is important to reduce the dynamic range requirement on an upstream laser diode, but an AGC with a large dynamic range may add cost.
4. *Ingress and impulse noise in upstream coaxial cables:* In addition to the multiple antenna noise mentioned in (1), there is funneling ingress and impulse noise in the 5- to 42-MHz upstream spectrum which is generated from the coaxial cable distribution plant and many subscribers' houses (see Section 2.4). The funneling noise levels may be so high that the upstream signal link performance is severely degraded, especially for FDM/TDMA systems.
5. *Multiple microreflections in the coaxial cable plant:* Microreflections from coaxial cable connectors, taps, splitters, and amplifiers due to impedance mismatch can cause multipaths in the transmission link. Relative delays among multipaths are in the range of several hundred nanoseconds [28]. If a RAP is installed along the feeder coaxial cable, we can assume the amplitude ripples and linear phase distortions are negligibly small. This assumption is not valid if a RAP is installed along a drop cable, as the peak-to-peak amplitude ripples could be as high as 10 dB [28].

Some of the problems mentioned in (1) have recently been investigated [25–27]. Their results show that, without considering problems (4) and (5), wireless signal-to-interference ratio (SIR) and signal-to-noise ratio (SNR) can both be improved by the distributed RAPs. The constraint is that, for two or more RAPs that may transmit/receive signals to/from a mobile handset with comparable power, the delays between any two RAPs must be controlled to be within $T/4$, where T is the channel bit rate. According to Table 10.2, the delay must be less than 0.65 μs (PACS), 5 μs (IS-136), and 0.2 μs (CDMA), respectively, to avoid intersymbol interference. Note that 1 μs corresponds to a distance of about 300 m. Therefore, to relieve a system from the stringent requirements on the propagation time difference between any two RAPs in CDMA or PACS, the RAPs must be separated by a sufficient distance or by significant obstructions. Problem (2) associated with limited and noisy upstream bands in 5 to 42 MHz cannot be economically resolved in the near future because of the availability of return-channel amplifiers (note that long-term solutions include deeper fiber penetration and/or using frequencies beyond 550 MHz for upstream transmission). Problems (4) and (5), however, can potentially be resolved by using CDMA signals such as the Telecommunications Industry Association (TIA) IS-95 standard [29], mainly because of its inherent anti-jamming capability and the strong viterbi decoder [15a]. A computer simulation result can be seen in a recent report [15d].

Problems

1. In an optical link used for transporting wireless signals, in addition to considering SFDR, the noise figure of the overall link is also an important system design parameter. Assume both the directly modulated laser transmitter and the p-i-n-diode-based receiver both have a matching circuit as shown in Fig. 10P.1. Refer to reference [30] and calculate the equivalent noise figure of a 2-km optical fiber link using a 1.3 μm FP-laser with an output power of 4 mW and RIN= −135 dB/Hz. Both the transmitter and the receiver are assumed to be well-matched in terms of impedance. The slope efficiencies of the laser and the p-i-n diode are 0.5 W/A (η_L) and 0.9 A/W (η_D), respectively; and the laser series resistance R_L is 3 ohms. Also, $N_D = 4$, $R_{IN} = R_{OUT} = 50\ \Omega$.

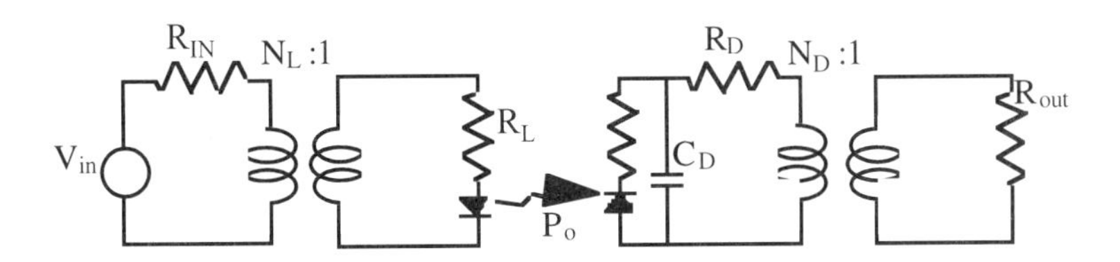

Figure 10.P.1: Small-signal circuit models of a laser transmitter and a p-i-n-diode-based receiver, both with impendance matching transformers.

2. Run a computer simulation to verify the results given in Ref. [21].

3. Explain why wireless CDMA signals (see [15a]) can robustly combat ingress noise or optical beat interference [15c, 31], but they cannot cope with the laser strong-clipping-induced nonlinear distortions (assuming single or multiple channels of CDMA signals modulate a laser).

References

[1] W. I. Way, R. S. Wolff, and M. Krain, "A 1.3 micron 35 km fiber-optic microwave multicarrier transmission system for satellite earth stations," IEEE J. Lightwave Technol., vol.LT-5, pp.1325–1332, September 1987.

[2] W. I. Way, "Fiber-optic transmission of microwave 8 phase-PSK and 16-ary quadrature-amplitude-modulated signals at 1.3 micron wavelength," IEEE J. Lightwave Technol., vol.LT-6, pp.273–280, February 1988.

[3] W. I. Way, "Optical fiber-based microcellular systems: An overview," IEICE Trans. Commun., vol.E76-B, pp.1091–1102, September 1993.

[4] W. I. Way, K. Y. Yen, and W. W Deng, "Super-FM and analog-to-digital-conversion technologies for next-generation wireless access in optical fibers," IEE J. Optical Quantum Electron., vol.28, pp.1521–1534, 1996.

[5] W. C. Y. Lee, "Smaller cells for greater performance," IEEE Commun. Magazine, pp.19–23, November 1991.

[6] W. C. Y. Lee, *Mobile Cellular Telecommunications*, McGraw-Hill, New York, 1989.

[7] M. Daoud Yacoub, *Foundations of Mobile Radio Engineering*, CRC Press, Boca Raton, Florida, 1993.

[8] D. C. Cox, "Universal portable radio communications," IEEE Trans. Vehicular Technol., pp.117–121, August 1985.

[9] L. J. Greenstein, et al., "Microcells in personal communication systems," IEEE Commun. Magazine, vol.30, pp.76–88, December 1992.

[10] S. Chia, "The universal mobile telecommunication system," IEEE Commun. Magazine, vol.30, pp.54–62, December 1992.

[11] D. Tang, "Fiber-optic antenna remoting for multisector cellular cell sites," Conf. Record, Int. Conf. Commun., paper 304.1, June 1992.

[12] T. S. Chu and J. M. Gans, "Fiber optic microcellular radio," IEEE Trans. Vehicular Technol., vol.40, pp.599–606, 1992.

[13] Bellcore Technical report, TR-NWT-000909, Generic Requirements and Objectives for Fiber in the Loop, Issue 1, December 1991.

[14] W. I. Way, "Wireless access in HFC systems," Tech. Digest, Int. Topical Meeting Microwave Photonics, Kyoto, Japan, paper TU4-1, December 1996.

[15a] R. L. Peterson, R. E. Ziemer, and D E. Borth, *Introduction to Spread Spectrum Communication*, Prentice-Hall, Englewood Cliffs, New Jersey, 1995.

[15b] V. O'Byrne, "Digital cellular in the fiber-optic access network," Tech. Digest, Conf. Optical Fiber Commun., paper TuN5, February 1992.

[15c] S. L. Woodward and S. Ariyavistakul, "Transmission of CDMA signals over an analog optical link," Conf. Record, International Conference on Communications, pp.344–348, 1997.

[15d] Y. M. Lin and W. I. Way, "Transporting wireless CDMA signals on HFC networks," Proc., SPIE Asia Pacific Symposium on Optoelectronics, paper 3420-04, July 1998.

[16] D. L. Schilling, L. B. Milstein, R. L. Pickholtz, F. Bruno, E. Kanterakis, M. Kullback, V. Erceg, W. Biederman, D. Fishman, and D. Salerno, "Broadband CDMA for personal communication systems," IEEE Commun. Magazine, pp.86–93, November 1991.

[17] V. Erceg, S. Ghassemzadeh, M. Taylor, D. Li, and D. L. Schilling, "Urban/suburban out-of-sight propagation modeling," IEEE Commun. Magazine, pp.56–61, June 1992.

[18] B. H. Wang and W. I. Way, "New frequency management algorithm minimizing intermodulation products for mobile cellular systems," International J. Wireless Information Networks, vol.2, No.1, pp.55–59, 1995.

[19] C. Harvey, I. C. Symington, and D. J. Kirsten, "Cordless communications utilizing radio over fiber techniques for the local loop," Conf. Record, Int. Conf. Commun., pp.1171–1175, June 1991.

[20] F. Dawson, "MSOs search for a role in PCS delivery," Commun. Eng. Design (CED) Magazine, pp.99–102, May 1996.

[21] D. M. Curter, J. B. Georges, T. H. Le, and K. Y. Lau, "Dynamic range requirements for optical transmitters in fiber-fed microcellular networks," IEEE Photon. Technol. Lett., vol.7, pp.564–566, May 1995.

[22] P. M. Wala, "A new microcell architecture using digital optical transport," Record, 43rd IEEE Vehicular Technol. Conf., pp.585–588, Secaucus, New Jersey, May 1993.

[23] J. A. Wepman, "Analog-to-digital converters and their applications in radio receivers," IEEE Commun. Magazine, pp.39–45, May 1995.

[24] D. F. Hoeschele, *Analog-to-Digital and Digital-to-Analog Conversion Techniques*, 2nd Edition, Wiley, New York, 1994.

[25] R. W. Donaldson and A. S. Beasley, "Wireless CATV network access for personal communications using simulcasting," IEEE Trans. Vehicular Technol., vol.43, pp.666–671, August 1994.

[26] K. J. Kerpez, "A radio access system with distributed antennas," IEEE Trans. Vehicular Technol., vol.45, pp.265–275, May 1996.

[27] S. Ariyavisitakul, T. E. Darcie, L. J. Greenstein, M. R. Phillips, and N. K. Shankaranarayaan, "Performance of simulcast wireless techniques for personal communication systems," IEEE J. Selected Areas Commun., vol.14, pp.632–643, May 1996.

[28] K. Laudel, E. Tsui, J. Harp, A. Chun, and J. Robinson, "Performance of 256-QAM demodulation/equalizer in a cable environment," Tech. Papers, 43rd Annual NCTA Convention and Exposition, pp.283–304, 1994.

[29] TIA/EIA/IS-95 Interim Standard, Mobile Station-Base Station Compatibility Standard for Dual Mode Wideband Spread Spectrum Cellular System, Telecommunications Industry Association, Washington, D.C., July 1993.

[30] C. H. Cox III, G. E. Betts, and L. M. Johnson, "An analytic and experimental comparison of direct and external modulation in analog fiber-optic links," IEEE Trans. on Microwave Theory and Techniques, vol.38, pp.501–509, May 1990.

[31] B. H. Wang, C. C. Hsiao, and W. I. Way, "Suppression of optical beat interference using synchronized CDMA technique and in-band clipping carrier," Electron. Lett., vol.33, pp.1888–1890, July 1997.

Chapter 11

Medium Access Control Protocols in Cable Modems with an Overview of Cable Modem Functionalities

A medium access control (MAC) protocol is a mechanism allowing multiple stations to share a single transmission medium. In the case of HFC networks, this common transmission medium is the upstream channel. For data over cable applications, a modern HFC MAC protocol should consider various factors such as ranging or acquisition process, frame format, support for higher layer traffic classes (ABR, CBR, and VBR),[1] bandwidth request and allocation, and contention resolution algorithms [2].

Generally speaking, an HFC MAC protocol can use one of the following forms of transmission control to resolve contentions:

- *Random Control:* Any station can transmit, and specific permission is not required. A station may check the medium to see if it is available before and while transmitting. Carrier sense multiple access with collision detection (CSMA/CD) [3], which was used by many first-generation two-way CATV networks in the early 1980s, belongs to this category. State-of-the-art cable modems based on Multimedia Cable Network System (MCNS) Data-Over-Cable Service Interface Specifications (DOCSIS) [4] or IEEE 802.14 [5] also use random access control (see Section 11.2) to join a network and request bandwidth.

---

[1]ABR (available bit rate) is for non-real-time data applications which can tolerate unpredictable time delays, and is a time-varying available bandwidth that users can share [1]. CBR (constant bit rate) and VBR (variable bit rate) data streams require the data arrives at the receiver within a narrow window of time. Typical CBR and VBR traffic is from voice and video codecs, respectively. VBR traffic, like CBR traffic, is periodic. However, the amount of data generated from a video codec may change with the motion contained within each picture frame.

- *Centralized Control:* A central headend controller (e.g., located in the CATV headend) controls the entire network, and other stations must receive permission from the headend controller in order to transmit their data. Polling, which has been widely used in today's set-top boxes for pay-per-view, and time-division multiple-access (TDMA), which is the main form of access technique (after stations join the network) proposed by MCNS [4], IEEE 802.14 [5], and DAVIC [6], belong to this category.

Traditional MAC protocols such as slotted Aloha and CSMA/CD were tried out in two-way coaxial-cable-based CATV systems in the early 1980s [7], but their performance was generally poor and unstable, and they can only be applied to limited covering ranges. These traditional MAC protocols cannot be satisfactorily applied to today's HFC networks, not only because the HFC network architecture is quite different from that of conventional local area networks, but also because very different initial physical conditions are imposed on HFC networks: (1) the longest distance from a CATV headend controller to a customer's premises may be as long as 40 km [5] or 160 km [4]; and (2) the total number of simultaneous users served by each pair of RF channels (one for upstream and the other for downstream) can be as high as a few hundred, depending on the tolerable delay versus offered load characteristics of the MAC used.

Before we touch on the state-of-the-art design of cable modem MAC protocols in Section 11.2, we shall first review some of the conventional MAC protocols in Section 11.1.

11.1 Conventional Random Access Control

In this section, we will review three conventional random access protocols which were used in the first-generation CATV data networks and are also selectively used in today's nonstandard cable modems.

11.1.1 Aloha

The earliest of random access techniques is Aloha, which was developed for ground-based packet radio broadcasting network [8]. A station can transmit a frame whenever it wishes. The station then listens for an amount of time equal to the maximum possible roundtrip propagation time on the network (twice the time it takes to send a frame between the two most widely separated stations). If the station does not obtain an acknowledgement during that time, it retransmits the frame. After repeated failures, it gives up. The station acknowledges immediately if the frame is valid (by examining the check sum); otherwise, it just ignores the frame.

It turns out that this simple Aloha access protocol is quite wasteful of bandwidth, attaining at most about 18% of the channel capacity. This limit on throughput can be easily proven as follows [9]: Let the throughput or traffic intensity be expressed as $S = N\lambda m(1 + a)$, where N is the number of stations contending for use of the channel, m is the fixed packet transmission time, λ is the average number of packets/s transmitted by each station, and a = (the one-way propagation time)/(the packet transmission time). For example, if the upstream speed of a cable modem is 108 kb/s and each upstream packet consists of 16 bytes, then the corresponding packet length m is 1 ms; if $a = 0.01$, that is, it takes an additional 0.01 ms to propagate to the receiving end, then the effective packet length is $m(1 + a) = 1.01$ ms. As the traffic intensity increases, collisions occur and many packets must be retransmitted, and the net channel traffic becomes $G = N\lambda' m$, where λ' ($>\lambda$) is the total rate of newly generated plus retransmitted packets. Assume that the arrival of newly generated and retransmitted packets at each station still obey a Poisson process with a probability given by $(\lambda t)^k e^{-\lambda t}/k!$, where t is the time interval under consideration, and k is the number of occurrences. Rigorously speaking, this assumption for the retransmitted packets is valid only if the random retransmission delay time is relatively long, but it turns out to be a good general approximation [10].

As shown in Fig. 11.1, a packet transmitted by station A will suffer a collision if B begins transmission prior to A but within a time $(1 + a)m$ of the beginning of A's transmission, or if B begins transmission after A within a time period $(1 + a)m$ of the beginning of A's transmission. Thus, the vulnerable period for a collision to occur is $2(1 + a)m$. Consequently, the probability that there is no collision in an

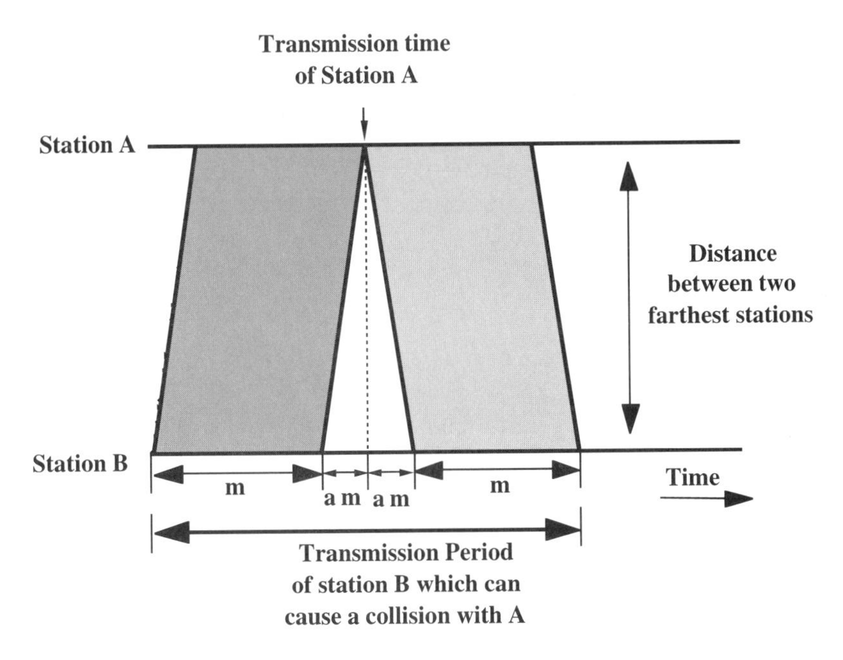

Figure 11.1: In an Aloha system, Station B cannot transmit a message in a time duration $(1 + a)m$ before and after the transmission time of Station A. Otherwise, a collision occurs.

interval of $2(1 + a)m$ is given by the Poisson probability of $(N\lambda'(2(1 + a)m))^0 \cdot \exp(-N\lambda'(2(1 + a)m))/0!$ or simply $\exp(-2(1 + a)G)$. Since the ratio S/G represents the fraction of packets transmitted over the channel that get through without collisions, it is equal to $\exp(-2(1 + a)G)$. Therefore, we have the ALOHA throughput equation as

$$S = G \cdot \exp\left[-2(1+a)G\right] \tag{11.1}$$

The maximum throughput S can be found by setting $dS/dG = 0$ to get $S_{max} = 0.5/[(1 + a)e] \approx 0.18/(1 + a)$ at $G = 0.5/(1 + a)$. It is clear that the parameter a must be kept small, that is, the propagation time << packet transmission time, in order to keep higher throughput. Figure 11.2 shows this throughput characteristic. We can see that $G \approx S$ when the traffic intensity is low, because there are few collisions. As S begins to increase and exceeds its maximum value of 0.18, the number of collisions and the corresponding retransmissions increase rapidly. The system becomes unstable, S drops, and G increases to a large value.

11.1.2 Slotted Aloha

To improve the channel utilization efficiency of 18%, slotted Aloha was developed, in which time on the channel is organized into uniform slots whose size equals m. A central clock is needed to synchronize all stations. Transmission is allowed to begin only at a slot boundary; consequently, collisions only occur when two or more users attempt transmission in the same time slot. We can follow the same derivation as that for Aloha, except that the interval for zero collision that should be considered is $(1 + a)m$ instead of $2(1 + a)m$. We can therefore obtain the slotted Aloha throughput equation as

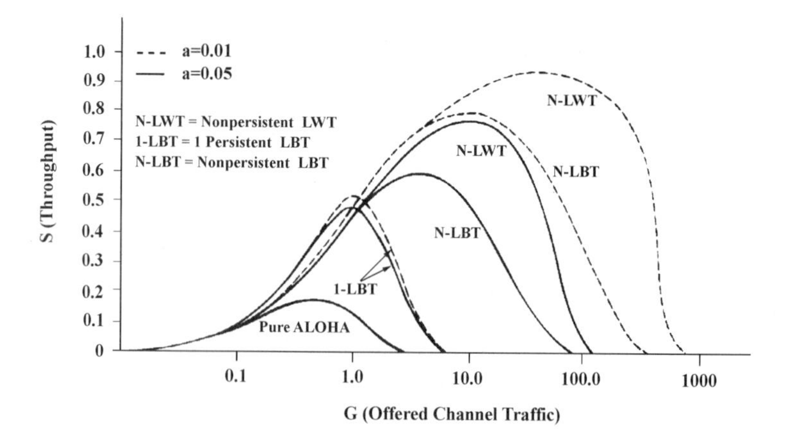

Figure 11.2: Throughput (S) versus offered channel traffic (G) for several conventional random access control protocols (a = 0.01 and 0.05). After Ref. [9].

$$S = G \cdot \exp(-(1+a)G) \ . \tag{11.2}$$

S_{max} is thus increased to about 36.8%. Slotted Aloha is proposed in DAVIC 1.1 [6] to manage contention of transmission over an HFC upstream signaling link.

The DAVIC upstream 64-byte slot consists of a unique word (4 bytes), an ATM-cell payload (53 bytes), a Reed–Solomon parity field (6 bytes), and a guard band (1 byte). Therefore, for an upstream data rate of 256 kb/s, there are 500 slots per second; for an upstream data rate of 1.544 Mb/s, there are ~3000 slots per second.

11.1.3 CSMA

The poor channel utilization efficiency of both Aloha and slotted Aloha stems from the same problem: The first portion of a packet cannot make a reservation for the rest of the packet. In other words, after a station launches a frame, it may take a long time before other stations know about it; during that time, one or more other stations may also transmit a frame, which results in collision with the earlier packet, and neither gets through. Therefore, if we can let propagation time << frame transmission time, when a station launches a frame and all other stations are listening before talk (LBT), all the other stations can know it almost immediately. Collision can occur only when more than one user begins transmitting within the period of propagation delay. This leads to the carrier sense multiple access (CSMA) or LBT technique [9].

Another question that needs to be answered when using CSMA is what a station should do if the medium is found to be busy. The three algorithms that exist are as follows:

(a) *Nonpersistent:* If the medium is busy, wait a random amount of time and try again.

(b) *1-Persistent:* If the medium is busy, continue to listen and transmit immediately as soon as the medium goes idle. If there is a collision due to absence of acknowledgment, wait a random amount of time and try again.

(c) *p-Persistent:* If the medium is idle, transmit with a probability p; otherwise, delay one slot (with probability $1 - p$) and repeat the process.

The drawback of (a) is that the medium will generally remain idle following the end of a transmission, and therefore the channel capacity is wasted. The drawback of (b) is that if two or more stations are waiting to transmit, a collision is guaranteed. (c) is a compromise between (a) and (b), but the value of p must be set low enough so that $np < 1$, where n is the number of stations contending the medium (if $np > 1$ at the end of a previous transmission, it implies that more than one station will attempt to transmit and a collision is guaranteed to occur), and this could result in unnecessary delays under light load [9].

The throughput performances of 1-persistent and nonpersistent CSMA (LBT) protocols are shown in Fig. 11.2, for $a = 0.01$ and 0.05. It is obvious that their performances are both superior to that of pure Aloha, even at a large value of $a = 0.05$.

11.1.4 CSMA/CD

Although CSMA is more efficient than Aloha or slotted Aloha, it still has the inefficiency that when two frames collide, they continue to transmit till the end of their damaged frames, and the medium remains unusable during that time. Therefore, the amount of wasted bandwidth may be significant if the frames involved are long. This waste can be reduced if a station continues to listen to the medium while it is transmitting, that is, "listen while talk" (LWT), or collision detection (CD). Therefore, two more transmission rules can be added to the CSMA rules: (a) as soon as a collision is detected, the station aborts the transmission, sends a short jamming signal, and reschedules the transmission after a randomized delay, and (b) after sending the jamming signal, the station waits a random amount of time and then tries again using CSMA. This leads to the CSMA/CD algorithm, which is commonly used in today's popular Ethernet local network. A baseband version inspired by Ethernet was issued as the IEEE 802.3 Standard [11], and a "broadband" version specified particularly for single- and dual-cable broadband media was provided as supplements to 802.3 [12]. CSMA/CD is a random access technique which was applied to many two-way CATV systems in the early 1980s [7] and in some earlier versions of commercial cable modems.

With CSMA/CD, the amount of wasted bandwidth is reduced to the time it takes to detect a collision. The throughput performance of the nonpersistent CSMA/CD (LWT) protocol is also shown in Fig. 11.2, for $a = 0.01$ and 0.05. We can see that its performance is superior to that of pure Aloha and CSMA (LBT). The maximum throughput of nonpersistent CSMA/CD can be close to 100% when a is as small as 0.01.

The maximum amount of time for collision detection in a single-cable system (see Fig. 11.3) using broadband CSMA/CD or 10Broad36 can be estimated as follows. Stations A and B are the two farthest stations and both have a distance d from the CATV headend.

Assume station A starts transmitting a packet at time t. Shortly before the packet transmitted by A reaches the headend, a packet sent by station B also arrives (which happens at $t + (d/\mathrm{v})$, where v is the propagation speed). The headend thus received garbled data or higher-than-expected data. The damaged data, after the carrier-frequency translator at the headend, reroutes back to station A. Station A performs a bit-by-bit comparison between transmitted and received data, and finds that the collision happened at the headend. This happens at $t + (2d/\mathrm{v})$. We can see that if a collision at headend is to be "sensed" while the packets involved are still being transmitted, the transmission of the shortest packet must take at least a time length of $2d/\mathrm{v}$, where $2d/\mathrm{v}$ is the roundtrip propagation time between the headend and a farthest station.

Now if we momentarily switch our attention to baseband Ethernet with a bus architecture, $2d/\mathrm{v}$ is the roundtrip propagation time between the two farthest stations. In a commercial (baseband) Ethernet, the roundtrip propagation delay is set at 510 bits and is not determined based on the actual propagation diameter of the net-

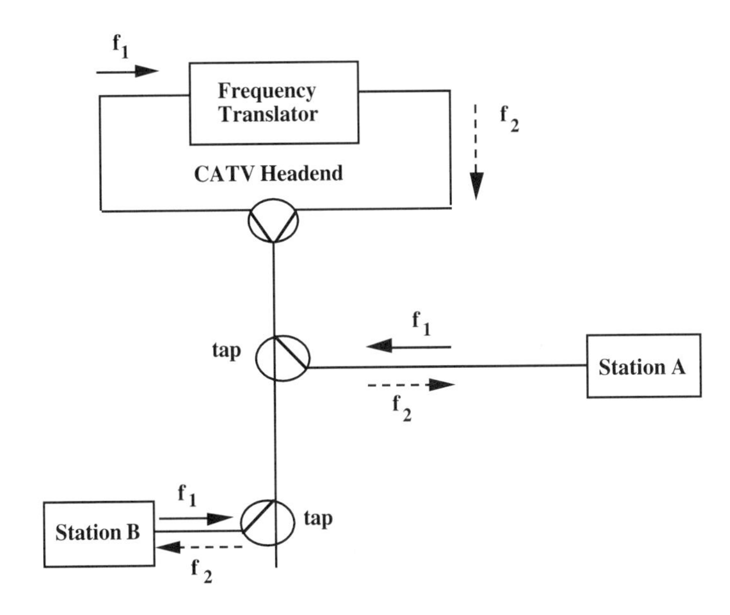

Figure 11.3: Cable system using CSMA/CD and a frequency translator in the headend.

work. According to the frame format in Ethernet (see Fig. 11.4), we can see that there is a *minimum* payload packet of 46 octets (368 bits) to assure proper collision-detection operation ($510 - 48 - 48 - 16 - 32 = 368$ bits; note that the preamble is not counted as a part of the packet header).

For broadband Ethernet, the total delay budget is divided into many segments, including delays in the roundtrip trunk cables, drop cables, and bit comparison, and only 140 bits of roundtrip delay are allowed in the trunk cable [12]. For a 10 Mb/s Ethernet and a cable propagation delay of 3.83 ns/m, this corresponds to a maximum distance from a station to the headend of $(140 \cdot 100\text{ns})/(2 \cdot 3.83 \text{ ns/m}) \approx 1800$ meters.

PR	DA	SA	ET	INFO	FCS

PR:	Preamble	8 octets
DA:	Destination MAC address	6 octets
SA:	Source MAC address	6 octets
ET:	Ethernet type	2 octets
INFO:	Information field	46~1500 octets
FCS:	Frame check sequence	4 octets

Figure 11.4: Ethernet frame format.

This obviously is a severe limitation for HFC networks, and therefore broadband Ethernet is rarely used today.

11.2 Random Access Control for Data-over-Cable

Two MAC protocols, *p*-persistent and ternary tree, have been proposed in MCNS [4] and IEEE802.14 [5], respectively, for a station to join the HFC network. Note that either MCNS or IEEE 802.14 has chosen a random access algorithm as the basis for medium access control, because it is more efficient to handle a large number of mostly quiet stations than contentionless algorithms such as pipeline polling [13]. In addition, both schemes were proposed as means to stabilize slotted Aloha performance [14].

11.2.1 *p*-Persistent with Binary Exponential Backoff

The contention resolution in MCNS is based on truncated binary exponential backoff, with the *initial backoff window* (W_i) and the *maximum backoff window* (W_m) controlled by the headend controller. When a station first enters the contention resolution process, it sets its internal backoff window equal to W_i currently in effect. The station must randomly select a number X within its backoff window, which is between 0 and R, with $R = (2^{\max(W_i + K-1, W_m)} - 1)$ and K the number of collisions. For example, if the backoff window is between 0 and 15 (with $W_m = 4$) and the station selects x = 11, then it must defer a total of 11 contention opportunities.

After a contention transmission, the station waits for the feedback information from the headend, which could be Data Grant, Data Grant Pending, or Acknowledgment. Once the feedback is received, contention resolution is complete. However, when the station finds no feedback it determines that the contention transmission was lost (e.g., because of collision) and it must now increase its backoff window by a factor of 2 (i.e., increase K by 1), so long as it is less than the maximum backoff window. The cable modem must now select a random number within its new backoff window and repeat the contention process just described. This retry process continues until 16 retries have been reached.

11.2.2 Ternary Tree Algorithm

In IEEE 802.14, stations use a ternary tree contention algorithm with variable entry persistence. When a station enters the contention process, it waits for a request minislot allocation information element (IE) (sent downstream from the headend controller) that contains a minislot group with an RQ value[2] of zero. Within this

[2]RQ stands for "resolution queue." RQ is a number provided by the cable headend and distributed to the stations indicating which collision groups are being resolved.

allocation IE will be the entry range R. The station calculates a random value R_S between 0 and R. If R_S is less than the number of minislots (minislot is the basic unit for upstream transmission) belonging to this group (RQ = 0), then the station transmits a request PDU (packet data unit) in the minislot at position R_S within that group. Otherwise, the station waits for the next opportunity to enter the contention process.

Once the process is entered, the station waits for a minislot feedback IE that describes the status of the minislot in which the station transmitted. If the feedback indicates that the minislot was empty, then the station should simply restart the contention process. If the feedback indicates the minislot contained a valid request, the station may assume that the request was the one sent by the station and the station should wait for a data grant from the headend controller. If no data grant arrives within a network specified timeout, then the station should restart the contention process.

In the case where the station collides with another station attempting to make a request in the same minislot, the feedback will indicate that a collision occurred and give a new RQ value to the station. The station then waits for a minislot group with an RQ value less than or equal to the RQ value of the station and contends within that group by picking one of the three minislots at random. Note that the range value R is only used when RQ = 0. When RQ≠0, the random value is always one of the three and there will be three minislots with a given RQ value. This process is repeated until the station receives an indication that it made a successful request.

Figure 11.5 illustrates how the contention mechanism works [15]. To simplify the illustration, contention minislots will always be grouped by three and are located next to each other. In practice, the contention minislots are not required to be adjacent to each other, but logically contention minislots are required to be done in only groups of three. To begin with, we can see that the cable headend sends MAC management PDU to indicate that there are request minislots available on the upstream link at position 481 and 495, and that any station may compete for their usage. Let us assume all stations (1–6) have information to send. Stations 1 and 6 successfully put their IDs into the first and the third minislots, respectively, in slot 481 (RQ = 0). However, stations 2 and 3 collide on minislot 1 and stations 4 and 5 collide on minislot 2 in slot 495 (RQ = 0), respectively. The headend examines the received minislots and reports back downstream what has been received. Since no further information is required for stations 1 and 6, the headend will send a DATA_GRANT message. As for the collided minislots, another two MAC messages will be sent with slot number 577 and 591 (both with RQ = 1). Stations 2 to 5 then randomly select which minislots to contend for. As shown in Fig. 11.5, station 2 chooses minislot 1 and station 3 chooses minislot 3 in slot 577, and station 4 chooses minislot 2 and station 5 chooses minislot 3 in slot 591. This time, all requests are received correctly and the data can be sent.

The 802.14 resolution protocol precludes new stations wishing to contend for minislots from entering into an ongoing resolution process. This is known as a blocking algorithm, in contrast with a free-access algorithm that allows new stations

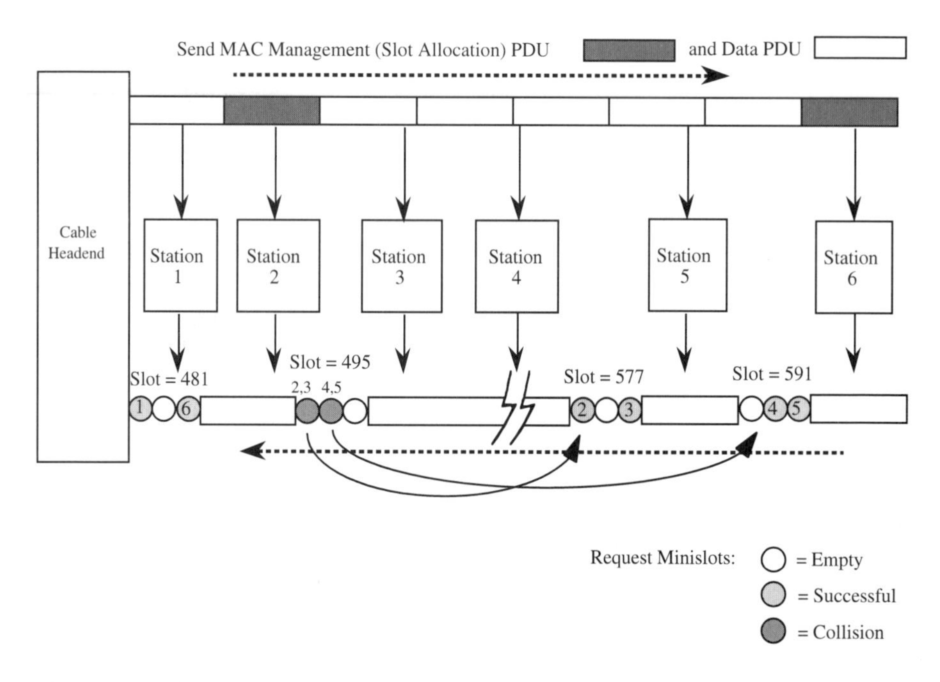

Figure 11.5: Illustration of how ternary-tree contention resolution algorithm works. PDU = packet data unit.

to contend at any time [16]. The free-access tree algorithm and p-persistence cannot guarantee how long it will take to resolve a contention, since new arrivals can join the current resolution. This means that the longest time a particular arrival may take to be successfully transmitted is not bounded. Despite this fact, it was found [17] that the difference of worst-case (when all the users contend in the same minislot) delay among p-persistence, free-access tree, and blocked-access tree is small. One of the reasons is that p-persistence and free-access have small values of variance. Another reason is that the guaranteed values of the blocked-access tree are not small.

11.3 Overview of Cable Modem Functionalities and IEEE 802.14/MCNS Standards

A general network configuration of data-over-HFC was shown in Fig. 1.5 and is repeated here in Fig. 11.6 for reference convenience. In a subscriber's premises, a cable modem transmits PDUs onto an upstream RF channel and receives PDUs from a downstream RF channel. A cable modem will likely exist as a stand-alone box in the early stage, because of issues such as potential liability and obsolescence, and RF-radiation-induced interference problems. The PC will connect to the cable modem

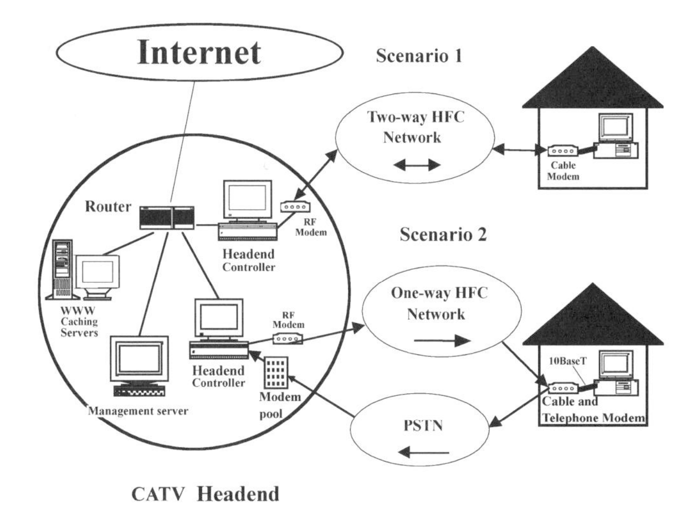

Figure 11.6: A general network configuration for data-over-HFC. Note that scenario 2 was developed to let the >80% one-way CATV systems provide the data-over-cable service as early as possible. PSTN = public switched telephone network.

using a LAN card such as the Ethernet 10BaseT interface. In the CATV headend, a headend controller transmits PDUs onto one or more downstream RF channel, and receives PDUs from one or more upstream RF channels. The headend controller serves as a bridge, as an internetworking layer router, or as an ATM switch, for any two stations in the network that wish to communicate. The headend controller transfers data between HFC systems and standard IP networks; manages bandwidth and spectrum usage; implements encryption for secure data transport over the shared HFC systems; and downloads software to subscriber's cable modems to provide feature upgrades without dispatching technicians. The RF modem in the headend can transmit data at a speed up to 30 (64-QAM) or 40 Mb/s (256-QAM), and the RF modem in the subscriber's cable modem can transmit up to 2.56 Mb/s.

A router in the headend serves as an interface between multiple headend controllers and networking devices such as WWW and billing management servers using standard LAN/WAN interfaces, and interfaces the local data with outside Internet networks. Routers function at the network layer of the OSI model and thus are protocol-specific. One major advantage of a router is that it builds a "firewall" that protects one network from packets generated by another network. Because a router is protocol-specific, it can be programmed to permit only those packets that match certain profiles. Routers can perform much more sophisticated functions than bridges. Before transmitting a packet to its destination, a router can analyze current traffic conditions and determine the best route for the packet to take. If traffic conditions change, the router can change its proposed route and redirect its packets over a revised path.

The router in the headend can be replaced by a bridge in certain cases. Bridges are usually very fast because they do not need to do any reformatting. Bridges are devices that function at the Data Link Layer of the OSI model. At this level, there is a concern with packets' sources and destination addresses, but not with any higher-level protocols. A bridge keeps a table that lists the addresses of the stations on its LAN. It simply reads a destination address and make the decision to filter or forward the packet.

A WWW caching server in the headend stores frequently accessed Internet information to reduce the possibility of having outside Internet access bottlenecks and improve subscriber response time.

A management server is used to control and monitor individual subscriber access to data services. The server also interfaces with an external billing provider, if one is used by the cable operator.

Note that data-over-cable or data-over-HFC is based on the basic routing protocol TCP/IP (Transmission Control Protocol/Internet Protocol). The IP is designed to provide a *datagram* service whereby packets can be sent and received across the Internet, although the data transfer cannot be considered reliable. The TCP provides reliable *connection-oriented* data transfer using the underlying IP. The connection-oriented full-duplex communication provided by the TCP is based on the establishment of a connection between two ports. The TCP must pass the source and destination addresses to the IP along with the TCP packet. To ensure reliable data transfer, sequence numbers are used by a receiving port to identify missing, out-of-sequence, or duplicate packets. Checksum is also used to check if there are damaged packets. Acknowledgments of transmitted TCP packets are returned, and if not, they must be retransmitted. Window size, which specifies the number of bytes that a sender is allowed to transmit before receiving an acknowledgment, provides an important feature of the TCP—*flow control.* Since each acknowledgment that is returned also contains a window size, the window size can be adjusted smaller or larger depending on whether the network is congested or not.

The two primary functions performed by the IP are (1) to route packets across an interconnected system of networks, and (2) to segment and reassemble (SAR) a packet to accommodate a network that has a smaller maximum packet size. IP function in the sending station must also correctly determine if the destination address is part of the local network.

In the following, we will present a brief overview of cable modem functionalities based on the drafts of the two important standard bodies, MCNS DOCSIS and IEEE 802.14. We will sort out the similarities and differences between the two standards and will add our own interpretations, whenever appropriate, to some plain statements in those drafts. However, the summaries below are by no means exhaustive, and the security-related issues are not covered.

11.3.1 Common Design Goals

In both standards, the common goals of designing cable modems and their corresponding central headend controllers are as follows:

1. The system must support a few hundred (depending on the tolerable delays) simultaneous users per up- and downstream RF channel pair, and cover a service area with a radius up to 40 km [5] or 160 km [4].
2. Bandwidth allocation should be based on headend-controlled time-division multiple access (TDMA). The headend controller must periodically provide a time-base to all stations so that all stations are synchronized in time. If any two stations on the network decide to transmit at a given network time, both transmissions will arrive at the headend controller at the same instant.
3. New members use a random access-based or contention-based MAC algorithm to join the network.
4. Active network members use dynamically mixed contention- and reservation-based upstream *minislots* to send request and data. The minislot is the basic unit for upstream transmission.
5. The cable modem must be able to support all TCP/IP-based applications, such as WWW, I-Phone, FTP, and Telnet.
6. Support transmissions of ATM-cells [5] , MPEG-2 packets [4], and variable-length packets (such as Ethernet packets) [4,5].
7. Transmissions of ABR, CBR, and VBR[3] must be supported.
8. Network security must be ensured.

With these common goals in mind, we can now take a look at other approaches that have been taken by both IEEE 802.14 and MCNS standards.

11.3.2 Similar Approaches Taken by Both IEEE 802.14 and MCNS

11.3.2.1 Protocol Stack The principal function of the cable modem system is to transmit IP packets transparently between the headend and the subscriber location. Cable modems and the headend controller will operate as IP and logical-link-control (LLC) hosts in terms of IEEE Standard 802 for communication over the network. As shown in Fig. 11.7, data forwarding through the headend controller may be transparent bridging, or may employ network-layer forwarding (routing); data forwarding through the cable modem is link-layer transparent bridging.

Forwarding between the upstream and downstream channels within a MAC layer differs from traditional LAN forwarding in that (1) upstream channels are essentially point-to-point shared-media, whereas downstream channels are point-to-multipoint shared-media, and (2) a single RF channel is simplex.

In Fig. 11.7, the Network Layer protocol is IP version 4, and migrating to IP version 6 [18]. The Data Link Layer is divided into sublayers in accordance with IEEE802 (including the LLC sublayer and MAC sublayer), with an additional Link-Layer security sublayer. Note that IEEE 802.14 has added ATM AAL5 to the LLC

[3]Since MPEG-2 transmission is not yet supported by IEEE802.14, VBR is currently not supported by that standard.

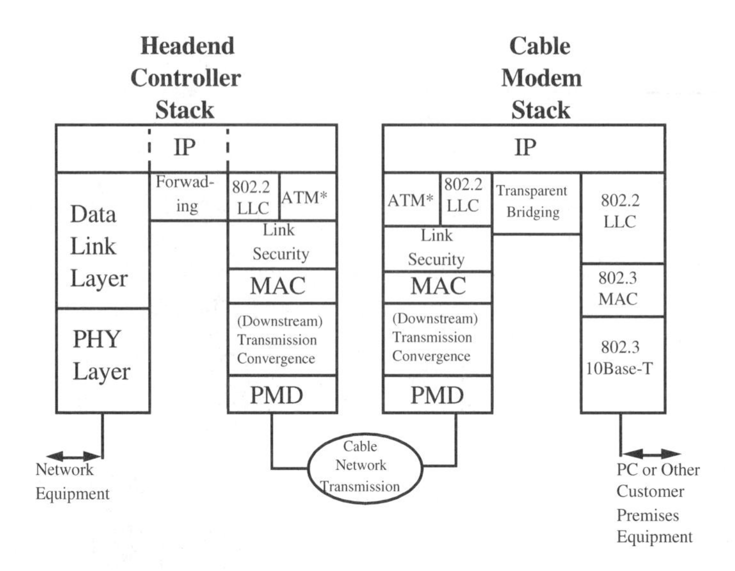

Figure 11.7: Data forwarding through a cable modem and a headend controller (*defined in IEEE 802.14 only). After Ref [4.5].

sublayer. The Physical Layer comprises two sublayers: the Transmission Convergence (TC) sublayer (downstream only) and the Physical Media Dependent (PMD) sublayer. The TC sublayer includes responsibility for converting MAC PDU and MAC logical time to and from PHY PMD data and PHY physical time, respectively. It improves demodulation robustness and facilitates common receiving hardware for both video and data, as well as providing an opportunity for the future multiplexing of video and data over the PMD sublayer.

11.3.2.2 Station Addressing Each station has a unique IEEE 802 administered 48-bit MAC address, which is typically programmed by the manufacturer of the equipment, and one or more headend controller-assigned 14-bit local identifiers (LIs). The latter is used in a *local* (IEEE 802.14 or MCNS) network as the *main* ID for MAC, security, data transfer, and physical layer control. There are two reasons why 14 bits of LI is used: One is that the 14 bits contain the conventional 12 bits VPI (virtual path identifier) in an ATM cell plus an additional 2 bits, and the other is that 14-bit LI can save overhead when compared to a 48-bit MAC address. Each station shall maintain and recognize at least three LIs, that is, (1) a dynamically assigned primary LI (bound to the station 48-bit MAC address), (2) a dynamically assigned acquisition multicast LI or a temporary service ID (for stations which have not yet joined the network during a registration process), and (3) the broadcast LI. A station may have several additional LIs corresponding to different service classes.

A station uses its 48-bit MAC address to join a multicast group in an 802.14 network. The headend controller first assigns a primary LI to this new station according to the unique MAC address. Then the headend controller sends an acquisition multicast LI to the primary LI of this new station. If the multicast group is encrypted, the headend controller must also inform the new station of the keys that are pertinent to the multicast group.

A station also uses its 48-bit MAC address to request removing itself from a multicast group. The headend controller must send the station (to its primary LI) a message instructing it to no longer recognize the LI of the multicast group. A new encryption key should be assigned to the group when stations have been removed.

11.3.2.3 Frame Formats The channel time is partitioned into fixed-size time slots, as is necessary in a TDMA network. The time slots for the downstream are in increments of PDUs and, in the upstream, in increments of *minislots* (see Fig. 11.5). Multiple minislots can be combined together to form a single PDU. If the upstream data has a size beyond a PDU, then additional minislots can be concatenated. A PDU that only occupies a single minislot is termed a mini-PDU. Mini-PDUs or minislots are primarily used for contention opportunities to request bandwidth.

To give an example of the relative sizes of PDU and minislot, a PDU can have 64 bytes, and a minislot can have 16 bytes, so that 4 minislots can fit in a PDU. If all large packets must be segmented into small packets such as 64-byte PDUs, significant overheads may be wasted. Therefore, a relatively large data slot concept called "frame," in the range of 3 to 5 ms, may be very useful to minimize fragmentation overheads, mix isochronous voice and delay-tolerant data streams on the same channel, and minimize access delay due to very large packets. This data frame concept has not yet been defined in either MCNS/DOCSIS or IEEE 802.14, but is under intensive discussion and may be the key to meet long-term goals for cable modems to support a variety of traffic types.

As mentioned, a typical frame duration is about a few milliseconds. For example, if the upstream data rate is 2.56 Mb/s, the frame size can be adjusted to 3.594 ms in order to fit up to a maximum of 18 64-byte data slots or 72 16-byte minislots, because $(1/2.56 \text{ Mb/s}) \times 18 \times 64 \times 8 = 3.594$ ms. The frame size should also be an integral number of the data slots and minislots.

The upstream and downstream frames have the *same* duration, which in most cases is longer than the maximum roundtrip delay in the network. The reasons why the frame duration must be longer than the maximum roundtrip delay are twofold: From the downstream point of view, the headend controller must wait for responses from all stations for every frame it sends downstream (generally by locating the MAC messages at the starting portion of a frame); otherwise it may be constantly interrupted by responses from various stations. From the upstream transmission point of view, a transmission time frame duration greater than the roundtrip time enables a station to receive feedback before the end of the frame provided the number of contention slots is small and if they are located at the beginning of the frame. In this case, the stations can attempt a retransmission in the next frame if its request collided. However, if the number of contention slots is large or variable, stations may have to skip a frame before retransmitting a collided request.

The contents of downstream and upstream data can be summarized as follows:

Downstream data:

- Data slots
- MAC management slots (including allocation grants, request minislot allocations, and request feedback, see Fig. 11.12)

Upstream data:

- Reserved data slots (composed of multiple minislots)
- Contention mini-PDUs (a single minislot)

MAC management slots contain synchronization information and feedback information about collision status and bandwidth allocation.

11.3.2.4 Upstream Bandwidth Allocation The upstream minislots are used for signaling requests and transmitting data from stations to the headend. The headend controller must generate the master time reference for identifying these slots. The time reference information is distributed to the stations by means of SYNC packets.

The upstream bandwidth allocation is entirely controlled by the headend. Upstream frames carry data slots and contention slots, and the headend controller must decide the proportion of contention slots and data slots per frame. The contention slots can be used for bandwidth request by a registered station, or can be used as an opportunity for new stations to join the network. Data slots contain data packets sent by stations. Data slots may be assigned to stations by reservation (no contention). When a station wants to reserve one or several data slots, it must transmit a request in a contention slot.[4] If more than one stations send their request in the same contention slot, they collide. The downstream frame contains feedback information about collision status and allocation.

The ratio of the contention and data slots is adaptive in both 802.14 and MCNS. However, it takes some thought to determine the ratio of the contention and data slots. Too few contention slots prevent requests from reaching the headend controller, and consequently reduce the amount of data that can be transmitted. However, too many contention slots can also reduce the bandwidth available for the data slots.

At a station, if a new packet is generated while the station is resolving a collision, the request size may be increased at the next transmission interval. The request size is, however, limited by the request field size.

11.3.2.5 Ranging Stations initialize with a ranging process to learn their round-trip distance from the headend. Having acquired the ranging information, stations

[4]A station may use *piggybacked* requests in data slots as an additional means for (CBR and VBR) bandwidth request. In other words, the station can request a number of minislots at the time of the request, instead of going through the contention process for every minislot requested.

then delay their subsequent transmissions in such a way that transmissions from various stations arrive at the headend in synchronization, that is, all transmissions are aligned to the correct minislot boundary such that the tail of a packet will not collide with the head of another packet. Even there are collisions, the collisions are constrained within minislots; therefore, the network throughput can be increased when compared to that of "unslotted" transmissions. However, there should still be some guard time between upstream bursts to avoid the effect of unavoidable bits slip. MCNS defines a minimum guard time of ~5 symbols, as shown in Fig. 11.8.

The ranging adjustment can also be sent to a global address, allowing the adjustment of the entire network to a longer range in the event that a station beyond the end of the current network applies for entry.

An important consideration of the ranging process is illustrated in Fig. 11.9. Assuming that we have three cable modems located at different distances from the headend controller, it can be seen in the figure that, at each cable modem, its ranging request packet (RNG-REQ) must be held back by a time duration which is equal to the maximum roundtrip delay time (RTD_{max}), in order to avoid collisions with the upstream data that was transmitted before the arrival of the management message payload (MAP PDU) which contains the ranging invitation. An interval is reserved, sometimes called "ranging window," as illustrated in Fig. 11.9, for all new stations to join the network.

11.3.2.6 Support for Higher Layer Traffic Classes If the MAC needs to provide support for ATM, it also needs to differentiate between different classes of traffic supported by ATM, such as CBR, VBR, and ABR.

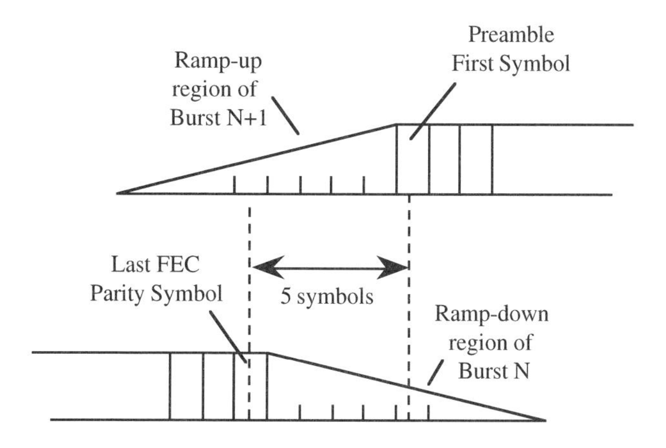

Figure 11.8: Worst-case burst timing which shows that a guard time of 5 symbols is maintained. After Ref. [4]

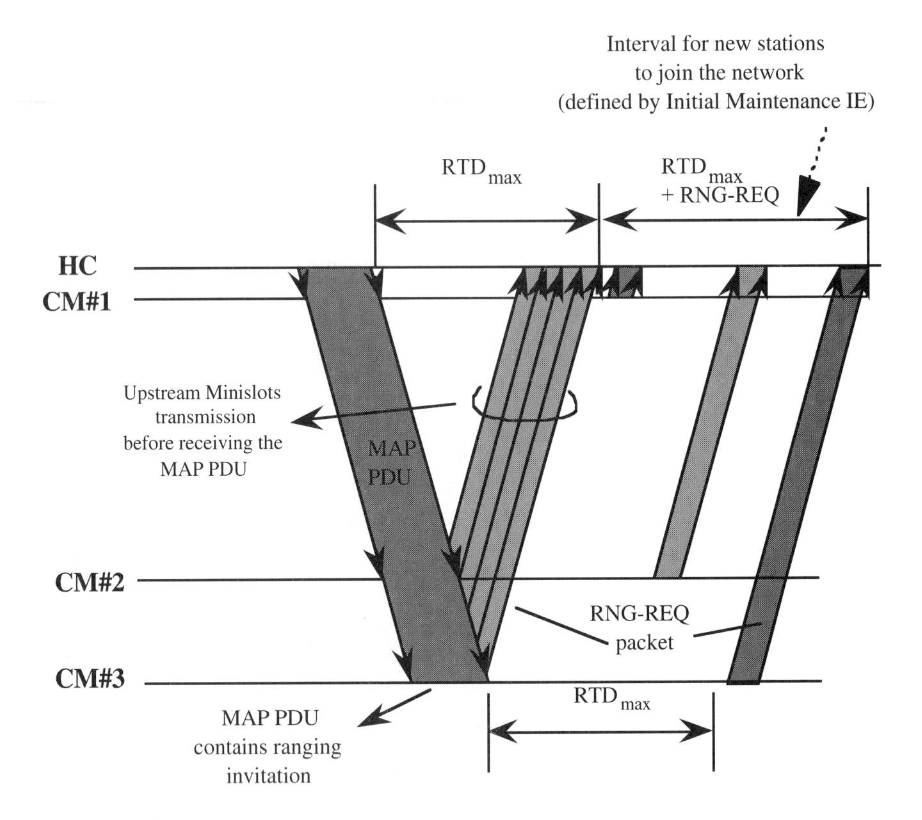

Figure 11.9: Illustration of the ranging process. A hold-back time which is equal to the maximum roundtrip delay time (RTD_{max}) must be implemented by all cable modems (CM#N, N = 1,2,3), in order to avoid collisions with those upstream minislots transmitted before the ranging invitation is received. An interval is reserved for new stations to join the network. HC and MAP stand for "headend controller" and "management message payload," respectively.

MCNS divides classes of service into four categories: (1) delay-intolerant inelastic application: The data must arrive within a perfectly reliable upper bound on delay; (2) delay-tolerant inelastic application: the data must arrive within an negotiated delay bound; (3) interactive burst elastic application; and (4) interactive bulk elastic application.

11.3.2.7 Synchronization Time Base In both IEEE 802.14 and MCNS, the headend controller maintains a master timebase. For example, in MCNS, SYNC message is measured in 6.25-μs ticks. These units were chosen as the greatest common divisor of the upstream minislot time across various modulations and symbol rates. An example was given in MCNS [4], and repeated here in Table 11.1, which relates minislots to time ticks.

Table 11.1 MCNS example relating minislots to time ticks. After Ref. [4].

Parameter	Example Value
Time tick	6.25 μs
Bytes per minislot	16 (nominal, when using QPSK)
Symbol/byte	4 (assuming QPSK)
Symbols/second	2,560,000
Minislots/second	40,000
Microseconds/minislot	25
Ticks/minislot	4

MCNS specifies that the headend controller must provide ranging offsets to cable modems with accuracy within 1/2 symbol and resolution of 1/64 ticks (6.25 μs/64 = 1/4 of the symbol duration at 2.56 Msps). The resolution is equivalent to about a distance of 20 m.

The headend controller broadcasts a time-stamped frame to all stations at regular intervals of around 100 ms.

11.3.3 Major Differences between 802.14 and MCNS

11.3.3.1 Frame Formats Besides the common characteristics of frame formats described in Sec.11.3.2.3, the design background of IEEE 802.14 and MCNS frame formats are very different. IEEE 802.14 emphasized ATM-cell transmission and Ethernet packets (although work is ongoing with respect to MPEG-2 compatibility), while MCNS includes transmissions of, in addition to ATM cells and Ethernet-type packets, digital MPEG-2 videos. The detailed descriptions of the respective frame formats are given as follows.

In IEEE 802.14, the downstream frame formats and hierarchy are shown in Fig. 11.10. It is noted that the basic unit has 6 bytes. ATM cells and variable length packet fragments are the two major types of PDU. An ATM cell has 54 bytes, including 6 bytes of header (with one byte more than conventional ATM cell) and 48 bytes of payload. The payload is composed of several information elements (IEs),[5] and each IE has a 2-byte header specifying the type and the length of the IE. The variable length packet fragment has a 6-byte header and a variable-length payload. The advantage of the variable length packet fragment is that no ATM cell header is required. The third type of PDU is the control PDU, which is an ATM cell that contains channel initialization, initial station transmission, ranging, upstream channel assignments, minislot occurrence time, and data PDU granting. All

[5]Both upstream and downstream MAC layer messages are composed of one or more IE. IEs are carried in the payload of ATM cells. Besides a payload area, each IE has an ID field which identifies the purpose of the element, and a LENGTH field which contains the length of the payload area in octets. There are 25 types of IE defined in IEEE 802.14.

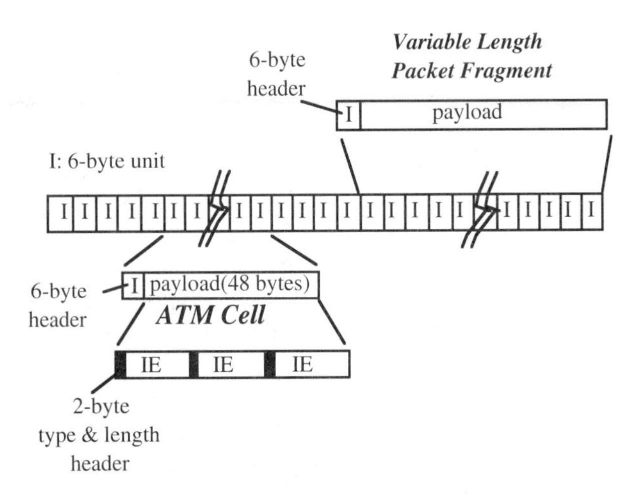

Figure 11.10: 802.14 downstream frame hierarchy and formats (including ATM cells and variable length fragments). IE = information element. After Ref. [5]. (Note: These formats are subject to changes by IEEE 802.14.)

stations are required to receive the control PDU and then determine the applicability of the message.

The upstream frame format in IEEE 802.14 is similar to that of the downstream, except that generally a 4-bit extended bandwidth request (EBR) field can be added (the EBR field allows each packet sent to request additional bandwidth without the necessity of contending for a request minislot, and is commonly known as "piggybacking"). In addition to the request minislot, an upstream message may simply be a *miniPDU*, which is just a single minislot (e.g., immediate minislots (IMS), which are used to identify the sending for immediate mode data transmission—this mode is particularly useful for station initialization, and it occurs when a station powers on). There can also be user-defined minislots.

As for the MCNS standard, the downstream bitstream is defined as a continuous series of 188-byte MPEG packets. These packets consist of a 4-byte header followed by 184 bytes of payload. The payload may be data or digital video. The MPEG payload portion of the MPEG packet will carry the MCNS MAC frames.

A generic MAC frame format is shown in Fig. 11.11. The same basic structure is used in both the upstream and downstream directions. Preceding the MAC frame is either a PMD sublayer overhead (upstream) or an MPEG transmission convergence header (downstream). The MAC header uniquely identifies the contents of the MAC frame. The MAC header may be followed by a Packet Data PDU, an ATM Data PDU, or a MAC-specific Header frame (with no PDU or with MAC management messages).

In addition to Fig. 11.11, MCNS has a very detailed description of its MAC-specific frame, as summarized in Fig. 11.12. We can see that a typical MCNS MAC-specific frame consists of three main parts: a MAC header, a MAC management

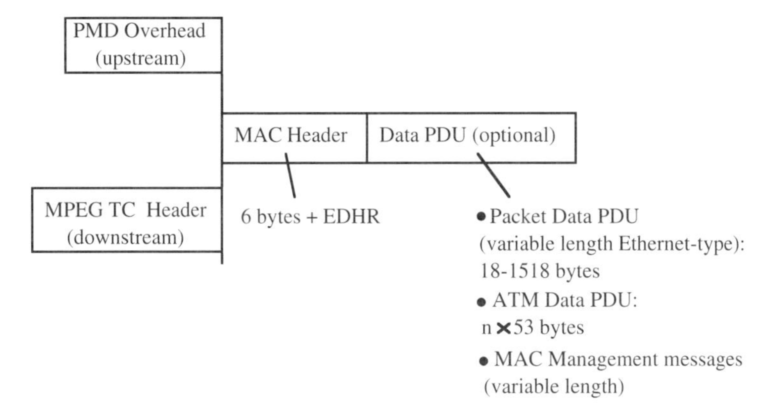

Figure 11.11: MCNS MAC frame format. EDHR = Extended Header, for future data link security use.

header, and a management message payload (MAP). Furthermore, the MAP is composed of a sync time-stamp (SYNC), an upstream channel descriptor (UCD), an upstream bandwidth allocation MAP, a ranging request (RNG-REQ) or a ranging response (RNG-RSP), a register request (REG-REQ) or a register response (REG-RSP), an upstream channel change request (UCC-REQ), and an upstream channel change response (UCC-RSP). The functions of these payloads will become clear when we discuss the MCNS operation principles in Section 11.5.

11.3.3.2 LI versus Service ID Out of the 14-bit local identifier (LI) or service ID address, MCNS uses 802.3 MAC addresses (6-bit DA and 6-bit SA), while 802.14 uses the ATM VPI (12 bits).

11.3.3.3 Downstream PHY-PMD Formats IEEE 802.14 supports two types of downstream PHY-PMD formats: type A ITU J.83 Annex A, and type B ITU J.83 Annex B (see Table 8.2, Chapter 8). However, MCNS/DOCSIS only supports type B.

Having introduced the main similarities and differences between the two data-over-cable standards, we can now go over the operation principles of each standard in the following two sections. We will describe both the IEEE 802.14 and MCNS MAC operation in two sequences. The first sequence of operation is when a station intends to enter the network, and the second sequence of operation is for steady-state conditions in which a contention resolution algorithm must be implemented.

11.4 Overview of Operation Principles in IEEE 802.14 Standard

The IEEE 802.14 working group has been formed with the responsibility of standardizing physical layer and MAC layer protocols for HFC networks. We will summarize the basic framework in the IEEE 802.14 for station entry processes and steady-state request/grant sequence descriptions, respectively. Note the following descriptions are a brief summary of the original IEEE 802.14 draft, so that the reader can grasp the main concepts of operation in IEEE 802.14.

11.4.1 Station Entry Processes

A station must go through four main entry stages before it becomes a member of the network: (1) acquire synchronization from a downstream PDU header, (2) search for signature cells and synchronize the station's and headend controller's time bases, (3) wait for an invitation to enter the network, (4) accept the headend controller-assigned LIs and upstream physical layer parameters, and implement initial ranging and power leveling. Each of these stages is described in more detail as follows:

1. *Scan downstream channels and acquire synchronization from a downstream PDU header:* 802.14 suggests that a new station begin to detect the HEC (header error check) byte in each ATM cell or variable length fragment, and when three consistent HECs can be detected, synchronization is acquired.
2. *Search for signature cells and synchronize the station's and headend controller's time bases:* To provide a time base to all stations that is synchronized correctly, the headend controller broadcasts a signature cell to all stations at periodic intervals (every few milliseconds). A signature cell is addressed to the broadcast LI and provides a time stamp (TS) which specifies the time the cell began transmission in head-end time. The TS is used to synchronize the station's local clock with the clock in the headend controller (the headend controller will adjust the station time base in the invitation response message; see step 3, so that if two stations attempt to transmit at the same network time, their transmissions will arrive at the headend controller at the same instant).
3. *Wait for an invitation to enter the network and respond with an unranged packet:* An invitation cell is addressed to the acquisition multicast address and is used to invite unregistered stations to attempt ranging and registration at a specific time. This cell contains, among other information, several important fields as follows:
   a. 2 octets of correlator number, which is subsequently used in the invitation response to identify the corresponding invitation message
   b. 1 octet of "upstream number" identifying the invitation channel
   c. 2 octets of p-persistence value: This value is used by the initialization algorithm to back off and retry when an invitation response does not succeed.
   d. 4 octets of range window start (RWS) and 4 octets of range window end (RWE): The RWS and RWE define the boundaries of the ranging window in tics of the master timebase. The range set by RWS and RWE is equal to "the interval for new stations to join the network" as was illustrated in Fig. 11.9.

   After receiving the invitation cell, a station responds with an invitation cell which contains 48 bits of the station, and a correlator number to identify the corresponding invitation message. The headend controller can then measure the roundtrip delay between the transmitted invitation cell and the received invitation response cell, and use this delay to adjust the time base between the station and the headend controller (see the following step).

4. *Accept the headend controller-assigned LIs and upstream physical layer parameters, and implement initial ranging and power leveling:* After the headend controller receives the invitation response cell from a station, it sends an "invitation response ACK" and "assign parameters information element" containing the information necessary to introduce a new station to the network. The assign parameters information element should be carried in cells addressed with the acquisition multicast LI. A station waiting for assignment must first match its 48-bit IEEE address with the MA field (48-bit station MAC address) of the information element. The assigned parameters include the 14-bit LI of the station, upstream physical layer parameter block, and initial range (32 bits) and power adjustments (16 bits).

   Note that the headend controller can also send a "PHY range adjust" element which contains an *incremental* delay (32 bits) to adjust the local station clock so that all stations effectively have the same time/distance from the headend controller. If this element is sent to a global address, the entire network is adjusted to a longer range in the event that a station beyond the end of the current network applies for entry. The headend controller also sends a power adjustment signal to adjust the power levels of a station. IEEE 802.14 allocates 16 bits, which corresponds to 48 dB of power adjustment, in the power adjust information element.

   After receiving an assign parameters message and implementing the specified adjustments, a station can join the network.

11.4.2 Request/Grant Bandwidth Processes

The steps involved in requesting permission to send data are as follows:

1. The station receives information from the headend on the downstream link informing it of when request minislots (RMSs) are available to send request messages.
2. A station selects one of the three RMSs randomly and places its identifier and the amount of information it has to send into a single RMS.
3. The station waits a fixed amount of time until the headend sends the feedback information for that RMS.
4. If the station does not receive a successful acknowledgment to its request, it enters a ternary-tree-based collision algorithm along with the other stations that contend on that particular RMS (see Section 11.2.2 for details).
5. When the headend indicates that the request was received successfully, it will provide a grant message to inform the station of when to transmit its data and the amount of data to transmit.

Note that there are three types of upstream data transfer: (a) a station uses the request/grant sequence described above; (b) a station uses the immediate mode of data transmission (as described in Section 11.3.3.2, the IMS is required for station initialization because when the station has just powered on, it has no reserved data slot to use, and the headend may provide an unallocated upstream data PDU with IMS); and (c) a station has established a CBR/VBR connection and does not need to contend each time for permission to transmit. The headend assigns a position in the upstream channel periodically and then informs the station of that time slot and the amount of data to transmit.

11.5 Overview of Operation Principles in MCNS/DOCSIS

We will first review the MAC frame formats (Fig. 11.12) and the associated functions in MCNS/DOCSIS. Next we will cover the station entry process and the request/grant bandwidth process.

11.5.1 MAC Frame Formats and Functions

As shown in Fig. 11.12, the first byte of the MAC header, that is, the Frame Control (FC) field, defines the type and format of the data PDU. The data PDU can be (1) a variable length (18–1518 bytes) packet, (2) one or more ATM cells ($n \times 53$ bytes), (3) a reserved PDU format for future use, or (4) a MAC-specific header (no PDU).

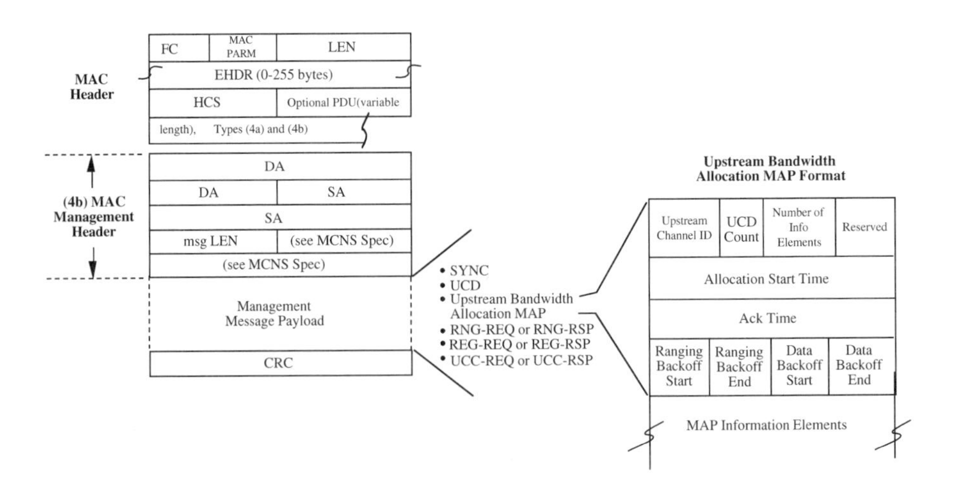

Figure 11.12: MCNS MAC header, MAC management header, MAC management payload, and one of the payloads, that is, MAC upstream bandwidth allocation MAP.

We note that (4) can be further divided into four categories (also defined by the FC field): (4a) timing header, (4b) MAC management header, (4c) request MAC header, and (4d) concatenation header.

The readers can check Ref. [4] for the functions of the rest of MAC headers, that is, parameter (PARM), extended header (EHDR), length (LEN), and header check sequence (HCS) fields.

The functions of the four types of MAC-specific headers (4a)–(4d), which include support for downstream timing and upstream ranging/power adjust, requesting bandwidth, and concatenating multiple MAC frames, are of particular importance to the understanding of the MCNS MAC. Therefore, in the following, we selectively discuss (4a)–(4c) in more detail.

11.5.1.1 Timing Header The timing header has a packet data PDU following the HCS field (see Fig. 11.12) to include downstream synchronization (SYNC) or upstream ranging-request (RNG-REQ) messages. In the downstream, the headend controller must *periodically* (on the order of tens of milliseconds, called the MAC SYNC interval) transmit the SYNC management message at precise times to provide the *global timing reference* to which all cable modems synchronize. A cable modem is considered to achieve MAC synchronization once it has received at least two SYNC messages within the maximum SYNC interval (200 ms). The message contains a time stamp that exactly marks when the headend controller transmitted the message. Cable modems must then compare the actual time the message was received with the time stamp and adjust their local clock references accordingly, so that the cable modems can time their transmissions precisely to arrive at the headend controller at the start of the assigned minislots.

In the upstream, the timing header must be a part of the ranging message needed for a cable modem's timing and power adjustments (see Section 11.5.2).

11.5.1.2 MAC Management Header The MAC management header transports all MAC management messages after the HCS field of the MAC Header (see Fig. 11.12). The DA (destination address) in Fig. 11.12 can be a unicast or multicast address, and the SA (source address) can be the MAC address of the source cable modem or headend controller system. "Msg LEN" stands for the total length of the MAC message from DA to CRC inclusive.

11.5.1.3 MAC Management Payload There are many types of MAC management messages, as shown in the middle part of Fig. 11.12:

1. *SYNC:* An incremental 32-bit time stamp based on a time-base reference clock at the cable modem with units of 1/64 of a time-based tick (i.e., 6.25/64 s) must be transmitted by cable modem at a periodic interval to establish *MAC sublayer timing.*
2. *UCD (Upstream Channel Descriptor):* Transmitted upstream by the cable modem at a periodic interval (maximum 2 s) to define the characteristics of an upstream channel, including IDs and physical layer parameters.

3. *Upstream Bandwidth Allocation MAP:* The format of the upstream allocation MAP is given in the right half of Fig. 11.12. An upstream bandwidth allocation MAP has a fixed-length header. It includes several housekeeping messages such as an 8-bit upstream channel ID and an 8-bit value indicating the number of information elements in the MAP. It also includes several important time references such as (a) a 32-bit minislot counter which shows the effective start time (from headend controller initialization) of the first entry in this MAP (Alloc Start Time, t_4 in Fig. 11.15); (b) a 32-bit Ack Time which indicates the latest time (from headend controller initialization) processed in upstream that generated a Grant, Grant Pending, or Data Ack; (c) initial and final backoff windows for initial ranging contention (Ranging Backoff Start and Ranging Backoff End); (d) initial and final backoff windows for *contention* data and requests (Data Backoff Start and Data Backoff End).

   Each of the IEs which follows the MAC management header consists of a 14-bit service ID (specifying broadcast, multicast, unicast, and null addresses), a 4-bit type code, and a 14-bit starting offset. Since all stations must scan all IEs, it is critical that IEs be short. A null IE is used to terminate the list.

   The descriptions of several important IEs are given as follows: (a) the request IE provides an invitation interval for all CMs to contend for requested bandwidth. (b) The initial maintenance IE provides an interval in which new stations may join the network (the ranging opportunity). This interval was shown in Fig. 11.9 and is equal to the maximum roundtrip delay plus the transmission time of ranging request message. (c) The station maintenance IE provides an interval in which stations are expected to perform some aspects of routine network maintenance, such as ranging or power adjustment. (d) A data grant IE provides an opportunity for a cable modem to transmit one or more upstream PDUs.
4. *Ranging Request (RNG-REQ):* A ranging request must be transmitted by a cable modem at initialization and periodically on request from cable modem to determine network delay.
5. *Ranging Response (RNG-RSP):* A ranging response must be transmitted by a headend controller in response to received RNG-REQ. This response must include encoded parameters such as the service ID (corresponding to the originator of the RNG-REQ), the upstream channel ID, and timing/power/frequency adjust information.
6. *Registration Request (REG-REQ):* A registration request must be transmitted by a cable modem at initialization after receipt of a cable modem parameter file. The request contains parameters such as an initialization ID and configuration settings for this cable modem.
7. *Registration Response (REG-RSP):* A registration response must be transmitted by the headend controller in response to received REG-REQ. The response includes parameters such as service ID from the corresponding REG-REQ, and an 8-bit response to indicate the success or failure of the corresponding re-

quest. Note that the initialization Service ID must no longer be used once the REG-RSP is received.

8. *Upstream Channel Change Request (UCC-REQ):* This request may be transmitted by a headend controller to cause a cable modem to change the upstream channel on which it is transmitting.
9. *Upstream Channel Change Response (UCC-RSP):* This response must be transmitted by a cable modem in response to a received UCC-REQ to indicate that it has received and is complying with the UCC-REQ.

11.5.1.4 Request MAC Header (Upstream Only) The request MAC header is the basic mechanism that a cable modem uses to request *bandwidth.* No data PDUs follow this header. The MAC PARM field now stands for the number of minislots or ATM cells requested, and the LEN field is replaced by a service ID to identify a particular service queue within a given station.

11.5.2 Station Entry Processes

In the MCNS Specifications, there are seven stages of initialization in a cable modem: (1) scanning and synchronization to downstream, (2) obtaining upstream parameters (from upstream channel descriptor (UCD) messages), (3) performing ranging, (4) establishing IP connectivity, (5) establishing time of day, (6) establishing security association, and (7) transferring operational parameters. We will selectively cover 1–3, and interested readers are encouraged to drill through Ref. [4] for details on 4–7.

1. *Scan for downstream channel and establish synchronization with the headend controller:* MCNS suggests that a station retry the previously stored operation parameters, and if this fails, it should begin to continuously scan the 6-MHz channels of the downstream frequency band of operation until it finds a "valid" downstream signal. A downstream signal is considered to be valid when the modem has achieved the following steps: synchronization of the QAM symbol timing, the FEC framing, and the MPEG packetization; and recognition of timing synchronization (SYNC) in downstream MAC messages. Note that SYNC must be transmitted by the headend controller at a periodic interval to establish MAC sublayer timing (via timing header). This must be followed by a 32-bit timestamp packet PDU with units in 1/64 of a time-base tick (i.e., 6.25/64 s) via MAC Management Messages.
2. *Obtain upstream parameters:* After synchronization, the cable modem must wait for an upstream channel descriptor (UCD) message from the headend controller in order to retrieve transmission parameters from the data stream. When the cable modem finds an upstream channel with acceptable transmission parameters, it must extract the parameters for this upstream from the UCD. It then must wait for the next SYNC message and extract the upstream minislot time stamp from this message.

The cable modem must then wait for a bandwidth allocation MAP (which includes a ranging opportunity) for the selected channel.

Message flows during scanning and upstream parameter acquisition are summarized in Fig. 11.13.

3. *Perform ranging:* A cable modem starts performing ranging after receiving the bandwidth allocation MAP (the headend controller periodically sends a ranging opportunity MAP to cable modems with temporary ID). A cable modem transmits (unranged) ranging request packet in contention mode with service ID parameter = 0. The headend controller receives recognizable ranging packets, allocates temporary service ID, sends ranging response, and adds the temporary service ID to its poll list. The cable modem subsequently receives the ranging response, stores the service ID, and adjusts local parameters such as upstream channel ID (the ID of the upstream channel on which the headend controller received the response request) and timing/power/frequency adjust information. The ranging and automatic adjustments procedure is summarized in Fig. 11.14.

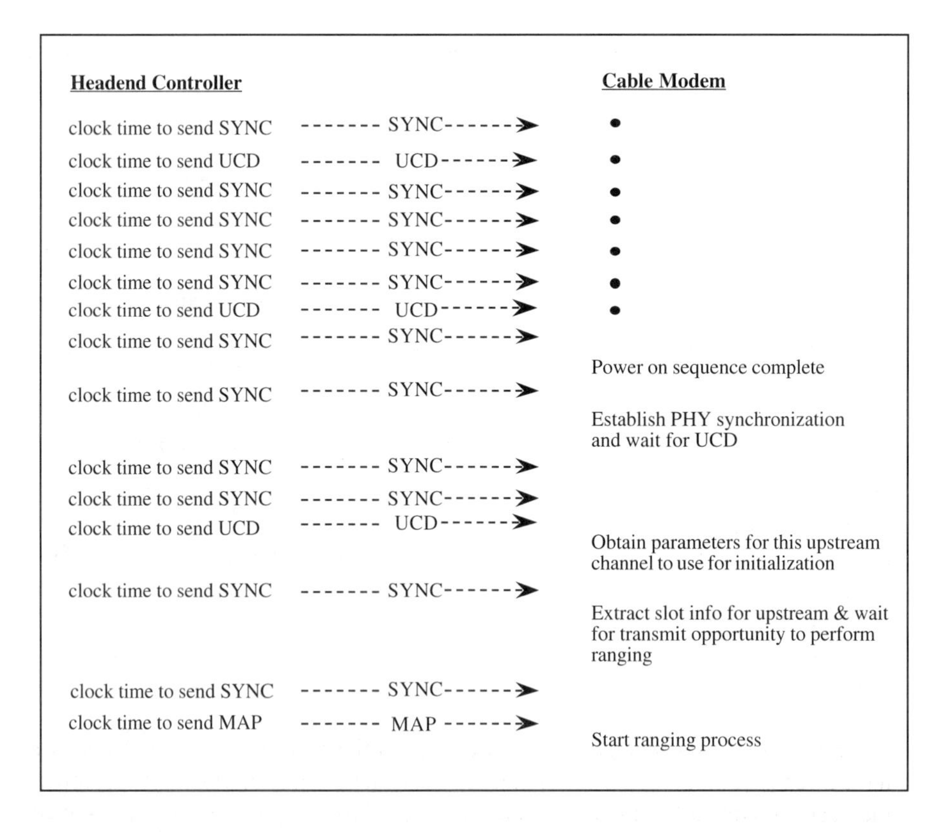

Figure 11.13: Message flows during scanning and upstream parameter acquisition. After Ref. [4].

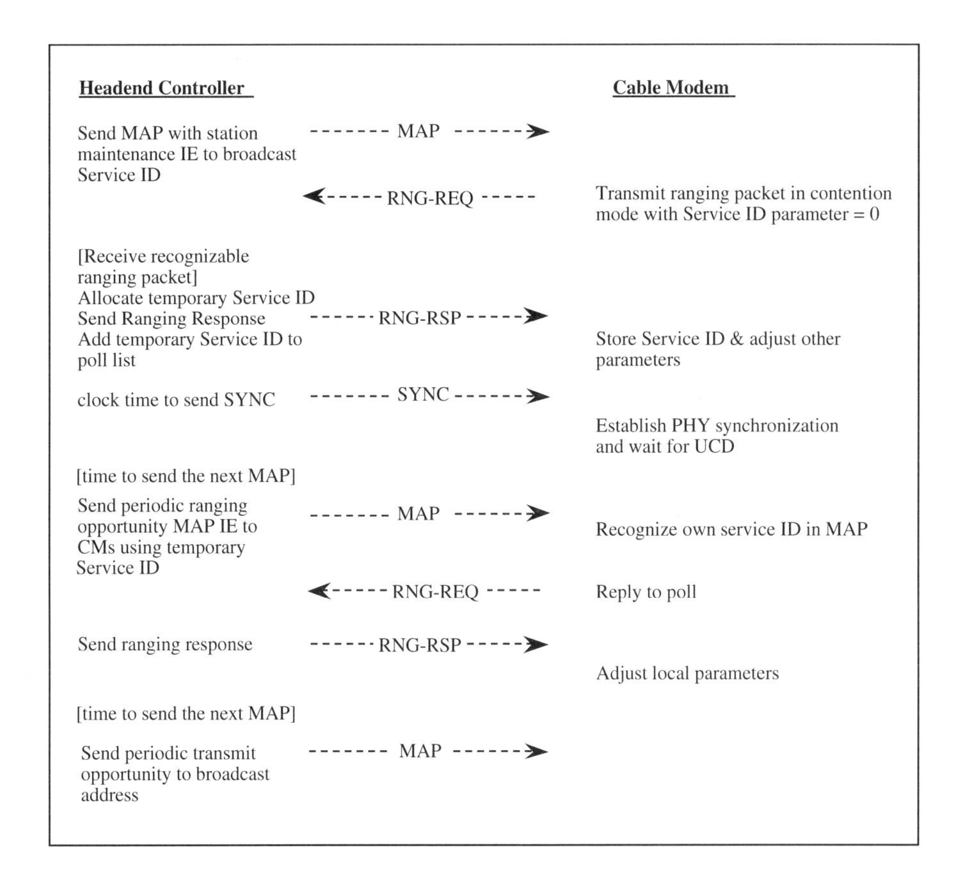

Figure 11.14: Ranging and automatic adjustments procedure. After Ref. [4].

11.5.3 Request/Grant Bandwidth Process (Using Upstream Bandwidth Allocation MAC)

As noted in Section 11.5.1.3(3), the basic mechanism for bandwidth allocation is the upstream bandwidth allocation MAP. The allocation MAP is a MAC management message transmitted by the headend controller on the downstream channel which describes the uses to which the upstream minislots must be put. A given MAP may describe some slots as grants for particular stations to transmit data in, other slots as available for contention transmission, and still other slots as an opportunity for new stations to join the link.

Figure 11.15 shows an example of the use of the MAP for cable modems (who have already joined the network) that regularly send requests and data. At time t_1, the headend controller transmits an MAP PDU. Two stations at different distances receive the MAP PDU at t_2 and t_3, respectively. Both stations implement ranging offsets so that their transmissions are aligned to the correct minislot boundary.

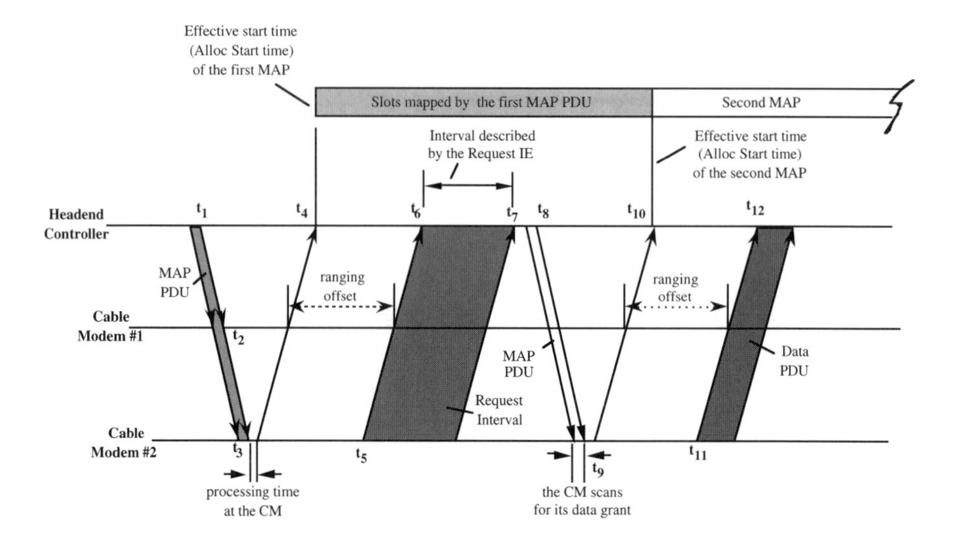

Figure 11.15: MAC protocol example for active stations (who have already joined the network).

Furthermore, to minimize request collisions, both stations select an additional random delay within the light-shaded region, which is also an interval described by the Request IE in the MAP PDU.

After the headend controller receives the requests, it schedules the requests for service in the next MAP. The cable modems scan the next MAP for their data grants, and if the data grants exist, the cable modems can transmit their data PDUs in assigned time slots, after a fixed ranging offset.

References

[1] T. M. Chen, S. S. Liu, and V. K. Samalam, "The available bit rate service for data in ATM networks," IEEE Commun. Magazine, pp.56–71, May 1996.

[2] N. Golmie, S. Masson, G. Pieris, and D. Su, "Performance evaluation of MAC protocol components for HFC networks," Proc., SPIE Broadband Access Systems, Boston, Massachusetts, pp.100–130, November 1996.

[3] Institute of Electrical and Electronics Engineers. Carrier Sense Multiple Access with Collision Detection (CSMA/CD) Access Method and Physical Layer Specifications. ANSI/IEEE Std 802.3, 1990.

[4] MCNS Data-Over-Cable Service Interface Specifications (Radio Frequency Interface Specification), SP-RFI-I02-971008, 1997.

[5] IEEE Standard 802.14 Draft1 R3, Sec.M, Media Access and Control, May 1997.

[6] Digital Audio-Visual Council, DAVIC 1.1 Specifications, 1996.

[7] M. B. Akgun and P. Parkinson, "The development of cable data communications standards," IEEE J. Selected Areas Commun., vol.3, pp.273–285, March 1985.

[8] N. Abramson, "The ALOHA system—another alternative for computer communications," Proceedings, Fall Joint Computer Conference, 1970.

[9] W. Stallings, *Local and Metropolitan Area Networks*, Macmillan, New York, 1997.

[10] M. Schwartz, *Computer Communication Network Design and Analysis*, Prentice-Hall, Englewood Cliffs, New Jersey, 1977.

[11] Institute of Electrical and Electronics Engineers. Carrier Sense Multiple Access with Collision Detection (CSMA/CD) Access Method and Physical Layer Specifications. American National Standard ANSI/IEEE Std. 802.3, 1985.

[12] Institute of Electrical and Electronics Engineers. Broadband Medium Attachment Unit and Broadband Medium Specifications, Type 10Broad36 (Section 11). Supplements to ANSI/IEEE Std 802.3, 1988.

[13] M. Momona and S. Yamazaki, "Framed pipeline polling for cable TV networks," IEEE 802.14 contribution, March 1995.

[14] C. C. Bisdikian, "A review of random access algorithms," IEEE 802.14-96/019a, January 1996.

[15] F. Koperda, "An example of a MAC based on the convergence agreements (CMAC)," IEEE 802.14–96/217.

[16] R. Citta and J. Xia, "Adaptive p-persistence algorithm with soft blocking for contention resolution," IEEE 802.14/97-037, March 1997.

[17] D. Sala, J. O. Limb, and S. Khaunte, "Performance of contention resolution algorithm using continuous-mode operation," IEEE 802.14/97-048, March 1997.

[18] W. Stallings, "IPv6: The new Internet protocol," IEEE Commun. Magazine, pp.96–108, July 1996.

Appendix A

Coding Basics

The most important code which is used in ITU J.83 is the Reed–Solomon (RS) code. Since RS code is a subclass of cyclic code, and cyclic code in turn is a subclass of linear block codes, we will introduce the coding basics from binary linear block codes.

Binary codes are constructed from a number system called a Galois field. For each positive integer m, there is a Galois field called GF(2^m) that has 2^m elements in it. The set of elements can be represented as the set of m-bit binary numbers. Thus, GF(2) consists of 2 1-bit elements, that is, {0,1}; GF(256) consists of 256 distinct 8-bit bytes; and so forth. Addition of two elements in GF(2^m) is defined as bit-by-bit modulo-2 addition (exclusive OR), while multiplication of two elements in GF(2^m) can be described as a polynomial multiplication modulo an *irreducible polynomial* $p(x)$ of degree m *over* GF(2). This means that $p(x)$ can only have coefficients from GF(2), that is, zero or one, and is not divisible by any polynomial over GF(2) of degree less than m but greater than zero. For example, the elements in GF(2^4) are {0000, 0001, 0010, ..., 1111} and we can use an irreducible polynomial $p(x) = x^4 + x + 1$ to define multiplication in the following way: To multiply, say, 0110 by 0101, the result is $c(x) = (x^2 + x)(x^2 + 1) \bmod p(x) = (x^4 + x^3 + x^2 + x) \bmod (x^4 + x + 1) = x^3 + x^2 + 1 = (1101)$.

An irreducible polynomial $p(x)$ of degree m is said to be *primitive* if the smallest positive integer n for which $p(x)$ divides $x^n + 1$ is $n = 2^m - 1$. For example, over GF(2^4), $x^8 + x^4 + x^3 + x^2 + 1$ divides $x^{15} + 1$ but does not divide any $x^n + 1$ for $1 \leq n \leq 15$. Therefore, $x^8 + x^4 + x^3 + x^2 + 1$ is a primitive polynomial. The polynomial $x^4 + x^3 + x^2 + x + 1$ is irreducible but it is not primitive, since it divides $x^5 + 1$.

A.1 Linear Binary Block Codes

The encoder of a linear block code transforms a block of k message bits into a longer block of n codeword digits, and is usually characterized by the (n,k) notation. If the message and code bits are constructed only from 0 and 1, the code is a binary code. The mapping transformation between the 2^k distinct k-bit message vectors and the 2^n distinct n-bit code vectors is unique and linear. The code is said to be linear if and only if the sum of any two code vectors is also a legitimate code vector. An example of a (6,3) linear block code is shown in Table A.1 (Ref. 6, Chapter 8).

It can be checked that the sum of any two code vectors yields another legitimate code vector. In the tabulated example, the mapping between the message and code vectors is accomplished through a lookup table. However, when k is large, as in a (204, 188) code, an extraordinary amount of memory may be required to store the 2^{204} code vectors, and the table lookup implementation of the encoder becomes prohibitive. Fortunately, it is possible to find a set of k linearly independent n-bit vectors $\mathbf{V}_1$, $\mathbf{V}_2$, ..., $\mathbf{V}_k$ to generate all the 2^k members of the code vectors. For the example given in Table A.1, a possible generator matrix is

$$G = \begin{bmatrix} V_1 \\ V_2 \\ V_3 \end{bmatrix} = \begin{bmatrix} 110100 \\ 011010 \\ 101001 \end{bmatrix}. \tag{A.1}$$

We can then see that

$$\mathbf{C}_i = \mathbf{m}_i \, \mathbf{G}, \tag{A.2}$$

that is, any code vector $\mathbf{C}_i$ ($i = 1,2,...,8$) in Table A.1 can be generated from the matrix multiplication between $\mathbf{m}_i$ and $\mathbf{G}$. Note that the generator matrix can be separated into a parity array portion $\mathbf{P}$ and the identity matrix $\mathbf{I}_3$ as follows:

$$G = \left[\begin{array}{c|c} 110 & 100 \\ 011 & 010 \\ 101 & 001 \end{array}\right] = \left[P \,\middle|\, I_3\right]. \tag{A.3}$$

Table A.1: A (6,3) linear binary block code.

Message vector	Code vector
$\mathbf{m}_1 = 0\,0\,0$	$\mathbf{C}_1 = 0\,0\,0\,0\,0\,0$
$\mathbf{m}_2 = 1\,0\,0$	$\mathbf{C}_2 = 1\,1\,0\,1\,0\,0$
$\mathbf{m}_3 = 0\,1\,0$	$\mathbf{C}_3 = 0\,1\,1\,0\,1\,0$
$\mathbf{m}_4 = 1\,1\,0$	$\mathbf{C}_4 = 1\,0\,1\,1\,1\,0$
$\mathbf{m}_5 = 0\,0\,1$	$\mathbf{C}_5 = 1\,0\,1\,0\,0\,1$
$\mathbf{m}_6 = 1\,0\,1$	$\mathbf{C}_6 = 0\,1\,1\,1\,0\,1$
$\mathbf{m}_7 = 0\,1\,1$	$\mathbf{C}_7 = 1\,1\,0\,0\,1\,1$
$\mathbf{m}_8 = 1\,1\,1$	$\mathbf{C}_8 = 0\,0\,0\,1\,1\,1$

From this we can define a general *parity check matrix* $\mathbf{H}$ as

$$\mathbf{H} = [\, \mathbf{I}_{n-k} \,|\, \mathbf{P}^{\mathrm{T}}], \tag{A.4}$$

and it can be verified that $\mathbf{C}_i\, \mathbf{H}^{\mathrm{T}} = 0$ from Table A.1 and Eqs. (A.3) and (A.4). Therefore, from a received and corrupted code vector $\mathbf{R}_i$ which is the sum of the correct code vector $\mathbf{C}_i$ and an error vector $\mathbf{e}_i$ introduced by the transmission channel, we can define its *syndrome* as $\mathbf{S}_i = \mathbf{R}_i\, \mathbf{H}^T = (\mathbf{C}_i + \mathbf{e}_i)\, \mathbf{H}^{\mathrm{T}} = \mathbf{e}_i\, \mathbf{H}^{\mathrm{T}}$. An important property of linear block codes is that the mapping between correctable error patterns and syndromes is one to one, that is, once the syndrome is calculated from $\mathbf{R}_i\, \mathbf{H}^{\mathrm{T}}$, the corresponding error vector $\mathbf{e}_i$ can be found and the correct vector can be retrieved from $\mathbf{C}_i = \mathbf{R}_i + \mathbf{e}_i$. A typical decoding step of a linear block decoder is shown in Fig. A.1.

A.2 Cyclic Codes

Cyclic codes are a subset of the class of linear codes which satisfy the following cyclic shift property: If $C = [c_0, c_1, c_2, \ldots, c_{n-1}]$ is a codeword of a cyclic code, then $[c_{n-\mathrm{i}}, c_{n-\mathrm{i}+1}, \ldots, c_{n-1}, c_0, c_1, \ldots, c_{n-i-1}]$, obtained by i cyclic shifts, is also a codeword. We can generate a cyclic code using a generator polynomial in much the same way that we generated a block code using a generator matrix, that is, a cyclic codeword is given by $C(X) = m(X)g(X)$, where $m(X)$ is the message polynomial of the form

$$m(X) = m_o + m_1 X + m_2 X^2 + \ldots + m_{k-1} X^{k-1} \tag{A.5}$$

and $g(X)$ is the generator polynomial of the form

$$g(X) = g_o + g_1 X + g_2 X^2 + \ldots + g_{n-k} X^{n-k}. \tag{A.6}$$

Note that $g(X)$ of an (n,k) cyclic code must be a factor of $X^n + 1$.

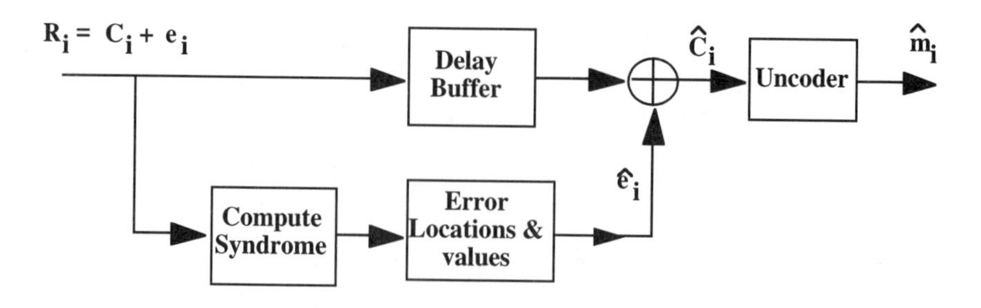

Figure A.1: Decoding steps for a linear block encoder.

A.3 Nonbinary Reed–Solomon Codes

RS codes are a special subclass of nonbinary cyclic codes. In a finite field GF(2^m), a nonzero element is said to be *primitive* if the powers of α generate all the nonzero elements of GF(2^m), and in addition, $\alpha^{2^m-1} = 1$ and no other $\alpha^j = 1$ for $j < 2^m - 1$. For all these elements to form a closed set under the multiplication operation, we assume $p(\alpha) = 0$, where $p(X)$ is a primitive polynomial of degree m over GF(2). The code generator polynomial $g(X)$ of a t-error-correcting RS code of length $2^m - 1$ is the lowest-degree polynomial over GF(2) which has $\alpha, \alpha^2, \alpha^3, \ldots, \alpha^{2t}$ as its roots, and is equal to

$$\boldsymbol{g}(\boldsymbol{X}) = (\boldsymbol{X} + \alpha)(\boldsymbol{X} + \alpha^2)\ldots(\boldsymbol{X} + \alpha^{2t}). \tag{A.7}$$

Appendix B

Spurious-Free Dynamic Range (SFDR) Quantified Using Two-Tone Intermodulation Products and Total System Noise

For a typical memoryless (static) nonlinear system, the system response to an input signal x can be expressed in terms of a Taylor series expansion $F(x) = a_0 + a_1 x + a_2 x^2 + a_3 x^3 + \ldots$. Therefore, for a two-tone testing signal which is composed of $\cos(\omega_1 t)$ and $\cos(\omega_2 t)$, we can easily see that when $\cos(\omega_1 t)$ or $\cos(\omega_2 t)$ is increased by 1 dB, the resultant third-order intermodulation products (P_{IM3}) at $2\omega_1 \pm \omega_2$ or $2\omega_2 \pm \omega_1$ are increased by 3 dB (this is illustrated in Fig. B.1). If we plot P_{IM3} and the output RF power P_{out} (at ω_1 or ω_2) as a function of the input RF power P_{in} (at ω_1 or ω_2), we can obtain the curves shown in the figure.

The curves may bend at high input RF power levels, but we can use the linearly extrapolated cross points, $IP_{3,out}$, as a useful reference point. Once this parameter is specified, and if we know the input or the output RF signal power level, we can deduce the intermodulation product levels at any other input signal power (except the curve-bending region).

The SFDR, indicated in Fig. B.1, is defined as the input signal range in which only the carrier-to-noise ratio needs to be considered and the third-order nonlinearities can be neglected. The relationship between SFDR and $IP_{3,out}$ can be derived from Fig. B.1 and is given by

$$SFDR = \frac{2}{3} 10 \cdot \log\left(\frac{IP_{3,out}}{N_{tot}}\right) \qquad (dB - Hz^{2/3}), \qquad \text{(B.1)}$$

where $IP_{3,out}$ is given by

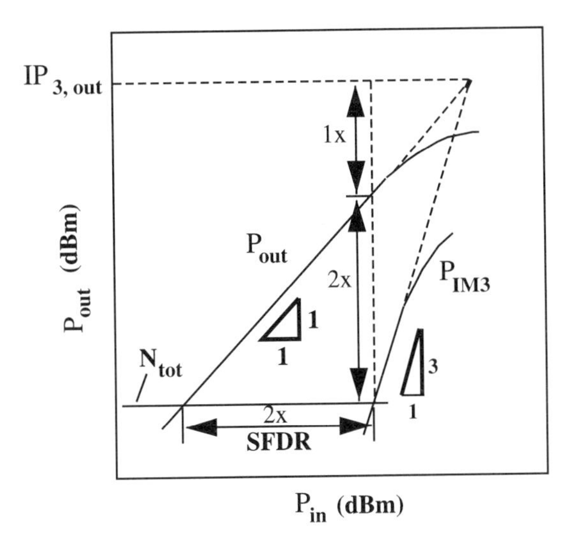

Figure B.1: Electrical P_{out} (output signal power) and P_{IM3} (output third-order intermodulation products) as functions of P_{in} (input signal power).

$$IP_{3,out} = P_{out} \cdot \left(\frac{P_{out}}{P_{IM3}}\right)^{1/2}. \tag{B.2}$$

N_{tot} is the total output noise per unit bandwidth, given by (see Chapter 5)

$$\boldsymbol{N_{tot}} = [\boldsymbol{RIN} \bullet \boldsymbol{I_p^2} + 2q\boldsymbol{I_p} + < \boldsymbol{i_{th}^2} >] \bullet \boldsymbol{R_D}, \tag{B.3}$$

where R_D is the load resistance of the photodetector, and I_p is the received photocurrent. Note that Eq. (B.3) does not include the beating noise terms due to optical amplifiers.

B.1 Direct Modulation System

The RF signal output power P_{out} in a direct modulation system can be written as

$$\boldsymbol{P_{out}} = \frac{1}{2}\boldsymbol{m}^2\boldsymbol{I_p^2}\boldsymbol{R_D}, \tag{B.4}$$

where m is the optical modulation index. We assume P_{out}/P_{IM3} in Eq. (B.2) is dominated by the laser dynamic nonlinearity such that (see Eq. (4.51))

$$(\boldsymbol{P_{out}} / \boldsymbol{P_{IM3}})^{-1} \cong \frac{\boldsymbol{m}^4}{4} \cdot \left[\left\{ \left(\frac{f_1}{f_o}\right)^4 - \frac{f_1^2}{2f_o^2} \right\}^2 + \left(\frac{f_1}{f_o}\right)^2 \left\{ \frac{1}{4\pi f_o \tau_n} \right\}^2 \right], \tag{B.5}$$

where f is the modulating signal frequency well within the 3-dB bandwidth of a laser diode, τ_n is the carrier lifetime, and f_0 is the relaxation oscillation resonance frequency of the laser diode. Equation (B.5) is a simplified form of Eq. (4.51). From Eqs. (B.2), (B.4), and (B.5), we can obtain

$$IP_{3,out} = R_D \cdot I_p^2 \Bigg/ \left[\left\{ \left(\frac{f_1}{f_o}\right)^4 - \frac{f_1^2}{2f_o^2} \right\}^2 + \left(\frac{f_1}{f_o}\right)^2 \left\{ \frac{1}{4\pi f_o \tau_n} \right\}^2 \right]^{1/2} . \quad \text{(B.6)}$$

According to Eqs. (B.1), (B.3), and (B.6), we have the SFDR for a directly modulated system whose dominant nonlinearity is caused by laser relaxation oscillation frequency:

$$SFDR = \frac{2}{3} \cdot 10 \log \left[\frac{ I_p^2 \Bigg/ \left[\left\{ \left(\frac{f_1}{f_o}\right)^4 - \frac{f_1^2}{2f_o^2} \right\}^2 + \left(\frac{f_1}{f_o}\right)^2 \left\{ \frac{1}{4\pi f_o \tau_n} \right\}^2 \right]^{1/2} }{ RIN \cdot I_p^2 + 2qI_p + \langle i_{th}^2 \rangle } \right] . \quad \text{(B.7)}$$

Note that Eq. (B.7) can be revised if the dominant third-order intermodulation products are due to static nonlinearity or clipping (with or without memory) nonlinearity. However, there may not exist a closed form for P_{out}/P_{IM3}.

B.2 External Modulation System

Based on the derivations in Section 7.3, we have for an externally modulated system the following results:

$$\text{Normalized fundamental signal mean-squared current} = 2J_0^2(\chi)J_1^2(\chi). \quad \text{(B.8)}$$

$$\text{Normalized two-tone intermod mean-squared current} = 2J_1^2(\chi)J_2^2(\chi). \quad \text{(B.9)}$$

Therefore,

$$\frac{P_{IM3}}{P_{out}} = \frac{J_2^2(\chi)}{J_0^2(\chi)} \approx \frac{\chi^4}{64} \quad \text{(B.10)}$$

$$P_{out} = \frac{1}{2} m^2 I_p^2 R_D = \frac{1}{2}(2J_0(\chi)J_1(\chi))^2 I_p^2 R_D \approx \frac{\chi^2}{2} I_p^2 R_D \quad \text{(B.11)}$$

where we have used the following approximations: $J_0(\chi) = 1$, $J_1(\chi) = \chi/2$, and $J_2(\chi) = \chi^2/8$ for $\chi \ll 1$.

From Eqs. (B.2), (B.10), and (B.11), we have

$$IP_{3,out} = 4I_p^2 R_D. \tag{B.12}$$

Substituting Eq. (B.12) and (B.3) into Eq. (B.1), we finally obtain the SFDR for an unlinearized MZI modulator:

$$SFDR = \frac{2}{3} \bullet 10 \bullet \log\left[\frac{4I_p^2}{RIN \bullet I_p^2 + 2qI_p + \langle i_{th}^2 \rangle}\right]. \tag{B.13}$$

As a final note, the above analyses are mainly for small-signals. For large-signal SFDR due to static and dynamic clipping in direct and external modulation systems, interested readers are referred to Ref.[B.1].

Reference

[B.1] B. H. Wang, P. Y. Chiang, M. S. Kao, and W. I. Way, "Large-signal spurious free dynamic range due to static and dynamic clipping in direct and external modulation systems," J. Lightwave Technology, vol. 16, October 1998.

Index

D

E

F

L

M

N

O

P

Q

T

U

V

W

Y